Cultural Geography

People, Places and Environment

Cultural Geography
People, Places and Environment

Richard H. Jackson
Department of Geography
Brigham Young University

Lloyd E. Hudman
Department of Geography
Brigham Young University

West Publishing Company
St Paul ◆ New York ◆ Los Angeles ◆ San Francisco

Cartographic Design

and Production   Cartography Lab, Brigham Young University, Jeff Bird, Supervisor and DIGIT Lab, University of Utah

Composition   Carlisle Communications

Copyediting   Pat Lewis

Cover Design   Diane Beasley Design

Cover Image   Stonehenge, England Weinberg/Clark—Image Bank

Text Design   Diane Beasley Design

COPYRIGHT © 1990 By WEST PUBLISHING COMPANY
50 W. Kellogg Boulevard
P.O. Box 64526
St. Paul, MN 55164–1003

LIBRARY OF CONGRESS

CATALOGING–IN–PUBLICATION DATA

Jackson, Richard H., 1941–
    Cultural Geography: people, places, and environment / Richard Jackson, Lloyd Hudman.
        p.   cm.
    Includes bibliographical references.
    ISBN 0-314-66316-9
    1. Anthropo-geography. I. Hudman, Lloyd E.   II. Title.
GF43.J33   1990
304.2—dc20

89-38343
CIP

Photo Credits
(All text photos not included in this listing were photographed by the authors.)
1 Sovfoto; **3** Cary Wolinsky/Stock, Boston; **21** Cary Wolinsky/Stock, Boston; **28** Alan Oddie/PhotoEdit; **43** Louis Goldman/Photo Researchers, Inc.; **45** James R. Holland/Stock, Boston; **58** Fred Kong/PhotoEdit; **59** Lila Abu-Lughod/Anthro-Photo File; **71** James R. Holland/Stock, Boston; **78** Eddie Adams/Sygma; **82** Anna Zuckerman/PhotoEdit; **90** William Gladstone/Anthro-Photo File; **97** Paul Conklin/PhotoEdit; **98** (Lower) Tim Asch/Anthro-Photo File **99** Bill Thomas/PhotoEdit; **123** Mayberry-Lewis/Anthro-Photo File; **135** Mayberry-Lewis/Anthro-Photo File; **137** (Left) Inge Morath/Magnum; **137** (Right) Peter Menzel/Stock, Boston; **148** Baldev/Sygma; **149** J. P. Laffont/Sygma; **150** Andanson/Sygma; **158** D. Otfinowski/United Nations Photo Library; **159** A. Tannenbaum/Sygma; **163** Peter Menzel/Stock, Boston; **165** Peter Menzel/Stock, Boston; **167** J. P. Laffont/Sygma; **171** Owen Franken/Stock, Boston; **172** John Isaac/United Nations Photo Library; **173** Owen Franken/German Information Center; **182** Tony Freeman/PhotoEdit; **184** Peter Menzel/Stock, Boston; **188** Ian Berry/Magnum; **195** Owen Franken/Stock, Boston; **196** Sebastio Salgado/Magnum; **202** Carrion/Sygma; **207** George Bellerose/Stock, Boston; **208** K. Cannon-Bonventre/Anthro-Photo File; **209** Peter Menzel/Stock, Boston; **212** Arnaud de Wildenberg/Sygma; **217** (Upper) Tony Freeman/PhotoEdit; **217** (Lower) Irven DeVore/Anthro-Photo File; **218** (Upper) Sebastiao Salgado/Magnum; **218** (Lower) Tim Carlson/Stock, Boston; **220** Ira Kirschenbaum/Stock, Boston; **222** Jean-Pierre Laffont/Sygma; **225** George Bellerose/Stock, Boston; **227** J. P. Laffont/Sygma; **230** Debbie Rogow/PhotoEdit; **232** (Left) Lionel Delevingne/Stock, Boston; **232** (Right) Owen Franken/Stock, Boston; **237** Jack Stein Grove/PhotoEdit; **238** Peter Menzel/Stock, Boston; **245** Robert Brenner/PhotoEdit; **250** Jack Stein Grove/PhotoEdit; **252** Alan Oddie/PhotoEdit; **257** Ellis Herwig/Stock, Boston; **263** (Upper) Owen Franken/Stock, Boston; **263** (Lower) Kevin Horan/Picture Group; **270** Tony O'Brien/Picture Group; **272** Allan Tannenbaum/Sygma; **277** Harry Wilks/Stock, Boston; **279** Rick Smolan/Stock, Boston; **280** Patrick Zachmann/Magnum; **282** (Left) Ira Kirschenbaum/Stock, Boston; **282** (Right) Rick Smolan/Stock, Boston;

continued on page 546

To our wives
Mary and Claire

Contents

List of Figures

Preface

Cultural geography is deeply rooted in the human experience. From earliest times people have been curious about the combination of people, place, and environment that makes each place unique. Geography is the science of place, the discipline dedicated to understanding the combination of human and natural features and relationships that characterize places. Everyone recognizes that each location on the earth is both similar and different from others. Physical features such as climate or landforms are central to the distinctive character of some places, as a ski resort in the American Rockies or the beaches of the French Riviera. In other places cultural features like language, architecture, land use, or field patterns may be the distinguishing characteristics, as in the features that make a Chinese village distinct from rural Mexico. Whether physical or cultural factors seem to dominate the character of a specific location, each place represents the interaction of people and environment over time.

Geography studies the physical and cultural factors that form the character of the world in which we live. The broad subject matter of geography has prompted specialization within the discipline, broadly dividing it into cultural or human geography as opposed to physical geography. Although both study the earth as the home of human beings, they differ in the degree to which they focus on people and their geographical impact versus the natural environment. When geography concentrates on the earth's physical features, it may appear to overlook the cultural or human factors, just as books focusing on human or cultural geography may pay less attention to the physical geography. The focus on cultural or physical geography in a particular book should not obscure the fact that the separation between the two branches of geography is only one of emphasis, not an absolute division.

This book emphasizes cultural or human geography. It is intended as the basic text for a college-level introductory course in Cultural or Human Geography one quarter or semester in length. The book uses the interaction between the physical and cultural factors affecting the earth as its central theme, allowing students to begin to understand how cultural variations combine with physical characteristics to create the world's complex mosaic of people and place. To provide a systematic framework within which to examine the world's cultural geography, we utilize the broad themes of culture, culture origins and diffusions, cultural ecology, cultural landscapes, and cultural regions. The five themes are used to integrate the material presented in each chapter, which illustrates their interrelationship in explaining the characteristics of the world's cultural geography.

The text utilizes a topical approach in which important cultural factors such as population, political systems, or religion are first identified and then analyzed to illustrate their environmental impact and regional pattern. Part 1 introduces the subject of geography and explains how cultural geography fits into the broader discipline. Included in this section are the basic concepts of geography, the development of the discipline from earliest times to the present, and a discussion of the things that geographers do. The latter is presented in the context of the broad range of careers chosen by individuals with a background in geography.

Part 2 examines the human modification of the world, beginning with the earliest people. Changes to the cultural landscape associated with technological advances and population growth and migration are discussed to illustrate the broad division between regions dominated by either rural or urban landscapes. Problems associated with rapid population growth, regional variations in level of economic development, urban sprawl, and environmental degradation are examined to illustrate the ongoing processes of cultural geography.

Part 3 examines culture origins and diffusions, human-land relationships, and regional patterns created by the distribution of the specific cultural features of ethnicity and race, language, religion, and political systems. Since these variables are important in explaining both the patterns of cultural geography and the issues and problems facing the world, they provide concrete, exciting examples of how cultural geography helps us understand the world. The distribution of specific cultural factors, their impact on the land and people of each region, and their contribution to the uniqueness of each place are examined to demonstrate the major cultural regions of the world.

Part 4 consists of Chapter 13, which summarizes the major issues facing the world and examines their relationship to cultural geography. This chapter analyzes the challenges created by the continuing environmental modification associated with cultural ecology, the changes in population distribution and related cultural regions, and the problems of culturally related conflict. It presents students with an overview of the issues and challenges that face a global society in the last decade of the twentieth century and first decades of the twenty-first to demonstrate that today's complex cultural geography is itself changing and evolving into tomorrow's world.

This book differs from other introductory cultural geography books in several major ways. For one thing, it stresses that adequate understanding of the cultural geography of the present depends on understanding key advances made in the past. The emphasis on the visual landscape that represents the end product of the interaction between people and their environment at a place is another difference. A final major difference results from our decision to integrate cultural variables such as economic activity into chapters on landscape rather than focusing on the types of economies

themselves since they are normally discussed in courses devoted specifically to economic geography.

The book offers several learning devices to help students master the course content and encourage their continued interest in the world around them:

1. The major themes and concepts discussed in each chapter are indicated at the beginning of the chapter.
2. A glossary at the end of the book gives students concise definitions for each term or concept introduced in the book.
3. Brief case studies included in each chapter illustrate the impact of cultural geography on individual lives.
4. Boxes introduce major models in geography in appropriate chapters and demonstrate their use in interpreting the world.
5. Review questions at the end of each chapter help students evaluate their comprehension of the material and organize their study.
6. Suggested readings at the end of each chapter include both seminal works suitable to an undergraduate course and specialized books and articles that allow students to examine topics of interest in greater depth.

A comprehensive instructor's manual with test bank to accompany *Cultural Geography: People, Places and Environment* has been developed to correspond with each chapter in the text. The extensive test bank provides substantive review of major themes and concepts discussed in each chapter, and is also available on WESTEST, a computerized testing system.

Acknowledgments

The intellectual debt associated with the development of a textbook that encompasses the entire world is complex and impossible to acknowledge fully. The colleagues, students, and teachers who have contributed ideas and insight into our own understanding of cultural geography are too numerous to catalog. It is important, however, to mention the influence on our own thinking of the writings of Vidal de la Blache, George Perkins Marsh, and Carl O. Sauer. In combination with the works of numerous living geographers, their insights have helped us define our own view of the world.

Of central importance in writing this text are the hundreds of students we have taught in the past two decades. We are grateful for their enthusiasm for learning and for the patience they exhibited as we tried to share with them our enthusiasm for studying the world and its peoples. Education is of necessity a two-way street, and the comments, suggestions, and criticism of students in class and informal discussions have provided us with a wealth of ideas.

We have also drawn heavily on the published and unpublished work of numerous colleagues. Where we could ascribe a specific idea to one person we have indicated our reliance upon his or her insights. If we have misinterpreted, oversimplified, or erroneously stated anyone's theses, we apologize. Numerous geographers have reviewed portions of the manuscript as we prepared this book, and all of their suggestions were helpful. Many colleagues will recognize specific places where their ideas have been incorporated into the text. The following individuals need to be recognized for their critical review of all of the text:

Thomas D. Anderson, Bowling Green State University; Michelle Behr, New Mexico State University; Brock J. Brown, University of Colorado; David L. Clawson, University of New Orleans; Charles O. Collins, University of Northern Colorado; Christopher H. Exline, University of Nevada at Reno; William J. Gribb, University of Wyoming; Robert S. Hilt, Pittsburgh State University; James W. King, University of Utah; Keith W. Muckleston, Oregon State University; James W. Scott, Western Washington University; Joel Splansky, California State University, Long Beach; Barbara A. Weightman, California State University, Fullerton; Nancy L. Wilkinson, San Francisco State University; Perry Wood, Mankato State University.

Jeff Bird, director of the cartography laboratory at Brigham Young University, designed and produced most of the maps in the book. His skilled hand will be seen in the numerous maps that are used to show the spatial aspects of the issues discussed in the text. Laura Wadley offered timely insights concerning organization and syntax. Ruth Sessions, Shauna Tong, Denise Clark, and Gaylene Powell provided cheerful, dedicated, and invaluable assistance in typing and retyping the manuscript, and Gaylene was also instrumental in creating many of the graphs in the book. We value their assistance and their friendship.

Our thanks also to the staff of West Publishing Company, who have provided guidance and encouragement as we attempted to complete the book. Although it is impossible to mention all who were directly or indirectly involved in the production of this volume, a few played such a central role that they must be mentioned. Editor Tom LaMarre first encouraged us to write a cultural geography text, and he has continued to prod us gently to complete the project. Theresa O'Dell, our Developmental Editor, facilitated completion of the manuscript and tried valiantly to get us to meet deadlines. Stacy Lenzen and Laura Nelson acted as production editors for the book. They coordinated our work with that of the West staff and also helped us with maps, photographs, and page design. Pat Lewis completed the difficult task of copyediting with all that entails. The excellent quality of the finished book reflects her outstanding work in polishing the authors' writing. Beth Hoeppner has directed the marketing of the text and helped us focus it on its intended audience.

To all those who have encouraged, chided, or cheered us in this task, we express our deepest gratitude. Without their continued support we would not have completed this book. We acknowledge their continued cheerful tolerance of our repeated requests for assistance.

Richard H. Jackson
Lloyd E. Hudman

Introduction to Cultural Geography

Part 1 discusses geography and one of its major subdivisions, cultural geography. Geography, which is one of the world's oldest disciplines, explains how and why people use the earth as they do. Cultural geography emphasizes the role of culture in explaining the similarities and differences between individual areas or places on the earth.

Maps are primary tools used by all geographers. Part 1 introduces the major concepts of mapping, including scale, pattern, and density. Voyages of exploration and discovery centuries ago resulted in maps, whose increasing accuracy over time reflected greater understanding of the earth as the home of human beings.

Part 1 also explains how cultural geography contributes to the continued discovery of the world's geography by indicating how people transform the natural environment into the towns and countrysides that characterize the distinct regions of the world. The section concludes with a discussion of how the tools and insights of geography are used in careers related to geography.

Geography: A Global Discipline

Geography . . . is not just the backcloth on which the traffic of the ever changing cultural mosaic is drawn. Geography, or space and environment, is an integral part of the making and remaking of societies.   R. J. JOHNSTON (1988)

THEMES

The importance of maps in geography.
The expansion of geographic knowledge over time.
Diffusion and human use of the earth.
The role of cultural geography in explaining the world.
Geographic theories explaining human-environmental relationships.

CONCEPTS

Mental maps
Culture
Hearth
Diffusion
Independent invention
Relocation diffusion
Hierarchical diffusion
Contagious diffusion
Barriers
Environmental determinism
Environmental possibilism
Cultural determinism
Environmental perception
Landscape
Cultural landscape
Culture regions
Domain
Core
Sphere
Map scale
Von Thünen model

Geography is the science of place, the discipline that focuses on places on the earth's surface and their interrelationships. Geography is the organized knowledge of the earth as the home of mankind. Geography's domain extends from the minute relationships of a single home and its place in a community to the global interconnections of the complex world of the twentieth century. Geography explains how the physical contours and features of the continents and oceans are related to human cultures with their traits, economies, and associated structures and landscapes.

Geographers attempt to explain the processes that create both natural and humanmade features. They analyze the changing character of the earth's surface, human interaction with natural phenomena, and the resulting relationships between people and their environment. Geographers want to understand the character of individual places, but realize that no place can be understood in isolation. Individual sites must be analyzed in relation to other places, as well as for their own characteristics. The production of corn in the American Midwest, for instance, is related both to local conditions of physical geography and to national and world demand for corn and meat. Studying the complex geography of the world provides insights into a variety of problems, such as how and why the central districts of many American cities are decaying while the suburbs are experiencing rapid growth; it can also help us understand the bases for many world conflicts, natural disasters, and global ecological issues such as the destruction of the tropical rain forest. The diversity of geography is reflected in the variety of spatial relationships it includes: population distribution, urban growth, drought, glaciers, landforms, and settlement forms and patterns are but a few of the many topics geographers explore as they attempt to unravel the complex relationships that explain the earth as the home of mankind.

Historically, geography has been viewed as encompassing three types of global realms:

1. The physical world of landforms and natural landscapes and the human alterations of the physical environment that create the cultural landscape. Study of drought in Africa, for example, involves both the environmental characteristics of the region and human actions that have contributed to drought to create a distinctive African cultural landscape.

2. The cultural realm of human beliefs and values concerning the environment. An examination of the different environmental values held by such groups as American Indian tribes, western ranchers, and Sierra Club members is an example of this type of geographical analysis.

3. The cultural world created by the interaction among people. Examples of this type of geographical analysis include studies of the distribution of major languages and religions or conflicts between nations.

An example of a European landscape.

An example of an Asian landscape.

As they examine the diverse geography of the world, geographers attempt to answer "where" and "why." They apply scientific methods to explain or predict the human use of the earth. Individuals use many of the tools of the geographer without necessarily being aware they are doing so.

All individuals develop a personal geography of the area in which they live. Visualize your own hometown and the relationships that make it different from all other places. You began developing an understanding of your hometown almost from the time you were old enough to go outside your home. Each person does the same. He must learn the relationships within his world and how those relationships affect his goals and desires. If a child wants to swing, play, or swim, she must know the routes, paths, barriers, and location of the playgrounds in her own private geography. With the passage of time, children develop a more formalized and broadened geographic awareness of important landscape features in their community, region, nation, and world. This "personal" geography continues to expand and change as people mature.

The geographer deals with these same types of issues in a professional setting. Geography means "the study of the earth," which has usually been interpreted to mean the analysis of the surface features of the earth and the human perception and modifications of that surface. Geographers' studies include natural and human processes and resulting patterns, structures, and landscapes that are part of the world as the home of the human population. The global nature of geography means that the discipline encompasses a vast array of interests. Individual studies may focus on different specific aspects of the world, but a central theme in all geographic study is an attempt to explain and interpret the earth so it will be used more wisely and maintained for posterity.

Geography: Origins and Change

Geography is one of the oldest intellectual disciplines. Early men and women needed to know where they could be safe, where food could be obtained, which places were dangerous, and

An example of an African landscape.

An example of a Latin American landscape.

where family and friends were located. People today need a similar understanding of the geography of their home territory. We become acquainted not only with the familiar space of our home and its immediate neighborhood, but also with places near and distant, places of interest through family ties and history, and places made objectionable by unpleasant experiences. The need to comprehend the space we occupy is basic to geography. Just as early humans needed a knowledge of geography in order to survive, the combination of the physical limitations of the

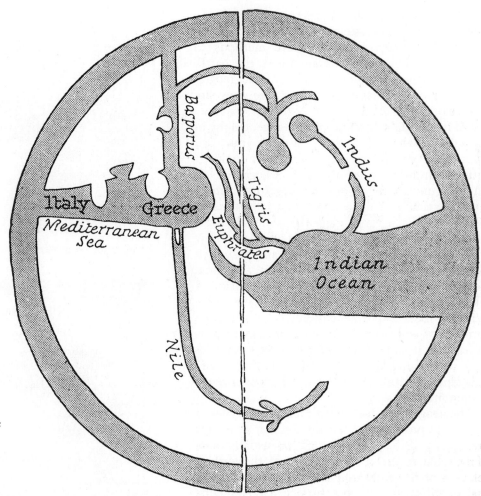

FIGURE 1–1 Early Historic Maps Early cartographers demonstrated a fairly accurate knowledge of the world of which they were aware.

Source: Both maps used by permission of John Murray (publishers) Ltd., London.

earth and the growing numbers who must rely upon it has created a modern imperative to comprehend the world to prevent its destruction as the home of mankind.

Maps: Tools of Geography

The curiosity of the individual concerning places is basic to geography. The ancient Greeks were the first to systematically and critically evaluate geographic relationships that affect places used by groups of people, but we do have a few examples of earlier geographic inquiry.

The existence of prehistoric maps not only provides proof that early humans were curious about their environment but also offers some insights into their cultures. Maps are a means of symbolically representing the three-dimensional world on a two-dimensional surface. Thus when we find crude maps representing a mountain by an inverted "V," vegetation by vertical lines, or water by wavy lines, we know that the people who produced them were capable of a high degree of abstraction. These early maps include maps on clay tablets from Babylon, a map of a village carved on a flat rock found in northern Italy, and a map engraved on a mammoth bone.

Early map fragments also reflect the importance of places to humans as well as the importance of maps in geography. Each of us could conceive maps similar to these early ones (Figure 1–1). After we have experience with a place, whether it be our home, neighborhood, community, or world, we develop a *mental map* of that

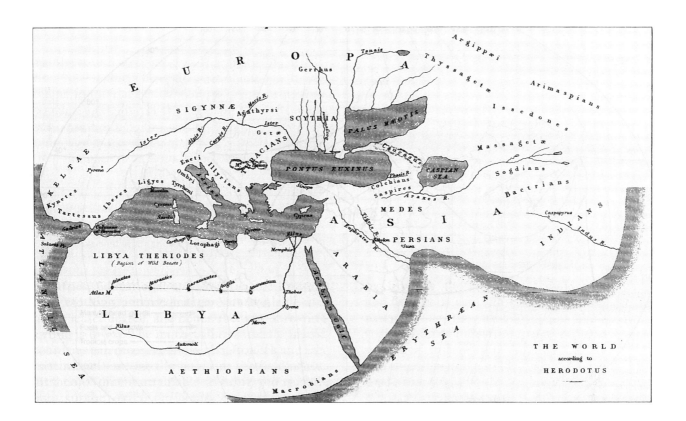

area. This map consists of landmarks and a frame of reference that allows us to visualize the spatial relationships that comprise the area. Early humans developed simple maps of the places with which they were concerned in the same way. These simple maps contained sufficient geographical information to allow the user to recongize the geographical phenomena and relationships in the world they depicted.

The oldest known map is from an ice age campsite on the banks of the Dnieper River in the Soviet Union. Details of a community are scratched on a piece of mammoth bone (Figure 1–2). The features of the community are represented by symbols, which have been interpreted by archaeologists. This map is simple, but it shows that the people who resided in the camp were concerned with spatial relationships.

The map from Bresicia in northern Italy depicts a village and the relationship of its houses, paths, and fields (Figure 1–3). Scratched onto a flat rock by someone overlooking the village, it is little different from a child's map of his or her neigh-

FIGURE 1–2 Ice Age Map This carving on a piece of mammoth bone shows that mapping skills were used by early people.

borhood. Nevertheless, it reflects an increasing level of abstraction and symbolization.

Other cultures also mapped their known worlds. Surviving archaeological relics of the Babylonian period include an early map of the

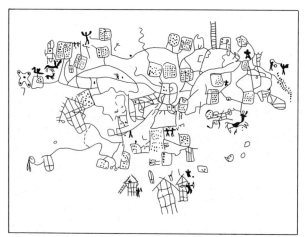

FIGURE 1–3 Map From Bresicia This map illustrates how the three dimensional world was symbolically represented to show the location and relationships of important features in the community.

world drawn on a clay tablet and maps depicting boundaries and land ownership in geometric patterns concerned with accurate location.

Central to these early efforts to symbolize the world were principles of geometry. Geometric principles were later an integral part of the geography of the Greeks and are still a part of the modern geography of mapmaking, land-use planning, industrial location, and other activities calling for the precise identification of the locations of human activity. Over time, complex maps have evolved using standardized symbolic reverential systems that achieve greater abstraction by digitally representing geographic phenomena and relationships in the form of numeric data (Figure 1–4). The increasing abstraction associated with maps should not mask the fact that each map is a symbolic representation of the real world and the geographical relationships of concern to the mapmaker.

Space and Place in Classical Geography

The Greeks formalized much early geographic knowledge; using and building upon prior efforts by earlier cultures, they created an understanding of the world unrivaled until Europeans, such as Columbus, began their great voyages of discovery.

This was the period of classical geography. One of the earliest Greek geographers, Herodotus, sometimes called the "father of geography," wrote his *Historia* in about 480 B.C.. The title does not fully reflect its content; Herodotus was as interested in geography as in history and examined the earth as the home of mankind. He included information about the oceans, rivers, transportation systems, and natural phenomena (such as earthquakes) as well as peoples and cultures. For the Greeks, the holistic nature of the universe was central to intellectual inquiry. *Historia* deals with peoples and place through time, anticipating the *holistic* nature of geography as practiced today.

Building upon the works of Herodotus and other scholars of the classical period, the later Greek astronomer Aristarchus (c. 270 B.C.) hypothesized that the earth revolves around the sun. This correct understanding of the relationship of the earth and sun allowed the Greeks to measure space and place accurately. Eratosthenes (c. 276–c. 194 B.C.) used this idea to measure the circumference of the earth more than two thousand years ago. On the longest day of the year, he measured the angle of the sun's ray at noon in the Egyptian cities of Syrene (modern Aswan) and Alexandria. Measuring the distance between the two cities, he calculated that the earth was approximately 25,000 miles in circumference, a calculation that has been substantiated by mod-

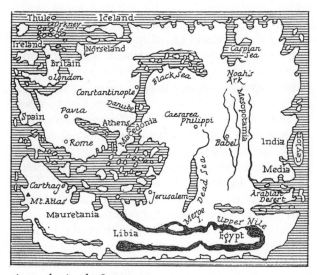

An early Anglo-Saxon map.

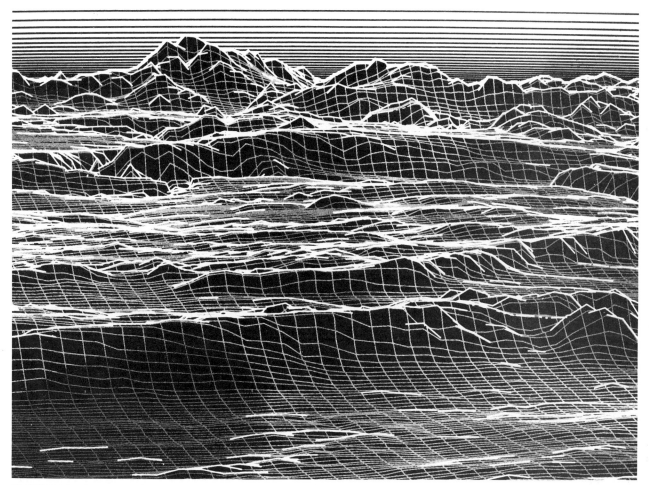

FIGURE 1–4 Digitized Computer Map Computer generated maps utilize digital data to produce maps, as in this representation of the Sierra Nevada Mountains.

ern measurements. He wrote the first book called *Geography*, which explained his methods and the mathematical systems he used to accomplish his measurement. Eratosthenes developed trigonometry in order to calculate the circumference of the earth, an illustration of how numerous fields of inquiry were closely related under the holistic approach of the Greeks (Figure 1–5).

Building upon the work of Eratosthenes, Claudius Ptolemy, who lived in Roman Alexandria in the second century A.D., prepared his great work *Geography*. In it he used a coordinate system of latitude and longitude to map the known world. His map included the location of eight thousand places and is an indication of the growing knowl-

edge of the earth's surface available to early scholars (Figure 1–6). The system of latitude and longitude is still used today.

The Decline of Classical Geography

Unfortunately, much of the geographic knowledge acquired by the Greeks was destroyed during the Roman Empire. Julius Caesar burned the great library at Alexandria in 47 B.C., destroying most of the 400,000 manuscripts on file there. The library at Alexandria represented the greatest concentration of human knowledge the earth had known to that time. Later the temple of Serapis, which held 300,000 more volumes, was destroyed by Chris-

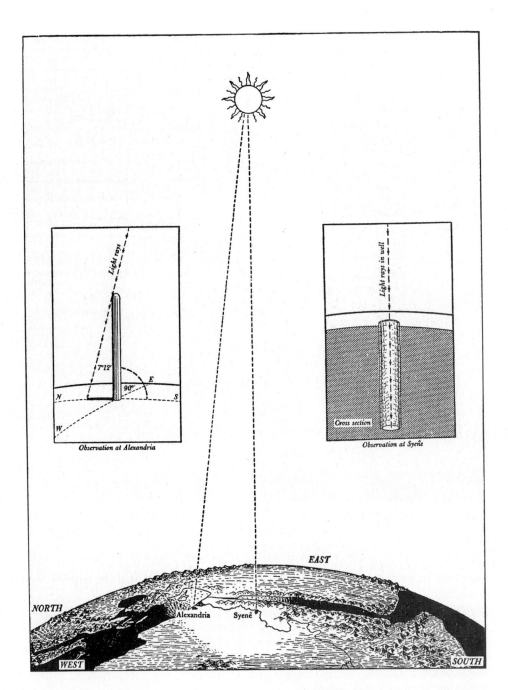

FIGURE 1–5
Eratosthenes' Technique for Measuring Circumference of the Earth Eratosthenes technique for measuring the circumference of the earth relied on his knowledge of earth-sun relationships.

Source: Used by permission of Crown Publishers, New York.

tians in an attempt to eliminate pagan religions. Fortunately, other people continued the inquiry into geographic relationships (Figure 1–7).

Other early civilizations studied and described the geography of their known worlds. After A.D. 600, the Islamic civilization of the Middle East and North Africa developed centers of learning that emphasized the understanding of the geography of the world. Much of the geographic knowledge acquired by the Greeks and Romans was translated and preserved by Arab scholars. The Arabs also traveled the world from the island of

FIGURE 1–6 Ptolemy's Map Ptolemy's map accurately located hundreds of known locations, demonstrating the level of geographic information available to him.

Source: Used by permission of John Murray Publishers Ltd., London.

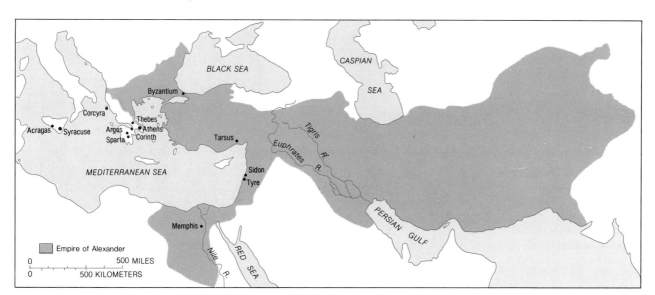

FIGURE 1–7 Map of Changing Greek Empire, Major Cities The expansion of the Greek empire to most of the Mediterranean Sea region was accompanied by development of important cities.

Indonesia to the southeast coast of Africa, gaining insights into the characteristics of lands and peoples. The Chinese were also expert sailors who made important technological advances in sailing ships, which allowed them to travel along the coasts of Asia. Some scholars believe the Chinese ventured as far as the western coast of North America. Polynesians sailed the Pacific using maps made from a lattice of sticks and shells that showed the location of individual islands of archipelagoes (Figure 1–8).

The Age of Discovery

In the Western world, the level of geographic knowledge reached by the Greeks and the Arabs did not return until the fifteenth century. From about A.D. 1450, there was a great interest in learning more about the world. This period,

known as the Age of Discovery, began largely as a result of the efforts of Henry the Navigator in Portugal. Europeans from England, Holland, France, Portugal, and Spain sailed the world in the fifty years between 1472 to 1522, learning of the existence of North America, discovering a sea route to Asia, and circumnavigating Africa (Figure 1–9). These adventurous sailors from Europe made an immense contribution to geographical knowledge. Setting sail with limited knowledge, in ships so small that today we marvel that they were able to cross the ocean, these hardy sailors traveled throughout the world and gleaned geographic information about the continents and major islands. The reliance on ships limited their knowledge chiefly to the coastal regions of the areas they visited, but they mapped the major geographical framework of the world in broad brush strokes (Figure 1–10).

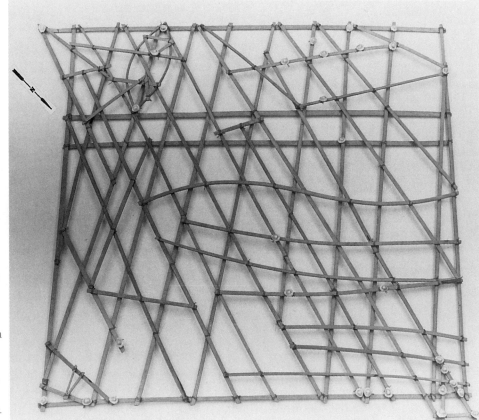

FIGURE 1–8 Polynesian Map The Polynesians used sophisticated charts made from sticks and shells to show the location of islands and the course to sail to reach them. A key to the map follows on page 13.

Source: From the collection of Dr. Thomas Hinckley, Provo, Utah.

Geographical Advances Leading to Modern Geography

The absence of geographic information about the interiors of the continents prompted a second period of unparalleled exploration in the seventeenth through nineteenth centuries. The list of explorers who contributed to geographical knowledge, filling in the empty spaces of the early maps, is long and includes such men as Alexander von Humboldt, David Livingstone, and Admiral Robert Byrd. Some geographers proclaim von Humboldt the "father of modern western geography" because of his insightful analysis and description of the physical landscapes and human societies he found.

The rising power of Europe on the world stage led to the European colonial period, during which Europeans dominated much of the Americas, Africa, and portions of Asia. Central to the colonial era was the need of mother countries to know the geography of their colonies. Consequently, they provided funding for geographic expeditions, the collection of data, and the publication of geo-

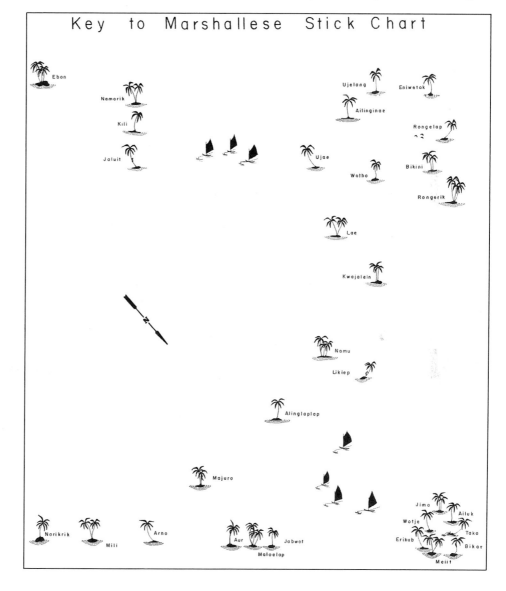

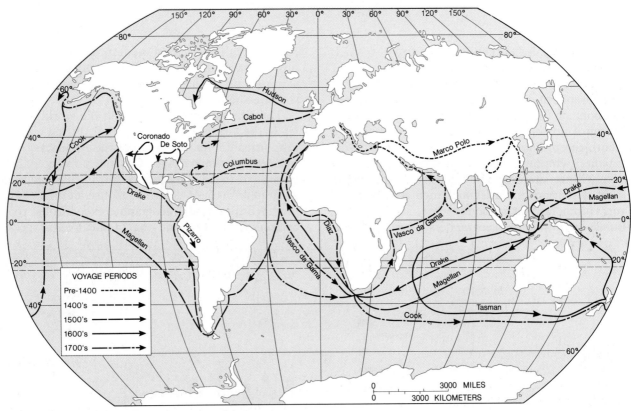

FIGURE 1–9 Voyages of Exploration During the Age of Discovery European sailors voyaged to most of the coastal areas of the rest of the world.

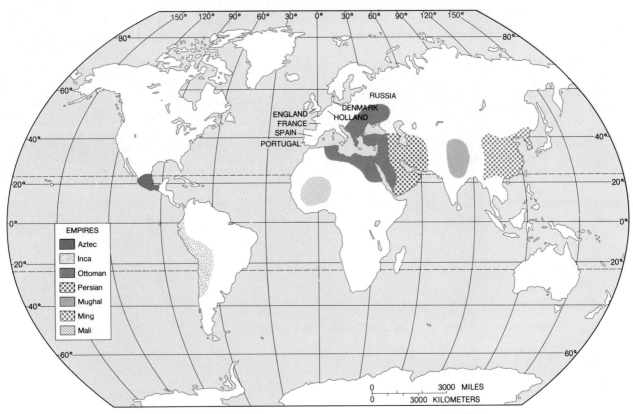

FIGURE 1–10 Map of 1550 This map of major cultural groups in A.D. 1550 also highlights the European homelands of important explorers.

graphic information about new lands, all of which played an important role in the emergence of geography as an independent discipline. As information about the entire world was combined with the ongoing careful study of the geography of European lands in the eighteenth and nineteenth centuries, geography emerged as a fundamental science in European universities. Thus the importance of geography to European states bent on commerce and conquest ensured its place as an important field of study.

The role of geography as a source of information about places was reinforced through regional analysis, resource inventory, and examination of the political and cultural characteristics of places in response to the needs of the colonial powers. Descriptions of differences in resources, standards of living, and levels of technological achievement were an important part of the data provided by geographers. Attempts to explain differences among places led to early geographical theories, which attempted to explain such differences on the basis of physical environment.

Cultural Geography: Development and Change

The recognition by some geographers that the environment did not *determine* human action led to the beginnings of analysis of how the earth has been transformed by its human occupants. Instead of asking how the physical environment forced human groups to adapt, cultural geographers asked how the earth had been modified by humans. Cultural geographers asked several basic questions:

1. How have people modified the physical geography of the earth to create today's landscapes?
2. What types of distinctive patterns have emerged from the human occupation of the earth's surface?
3. What is the significance for the earth and its inhabitants of the use and change of the earth by humankind?

The basic questions of all science are the what, where, how, and why of phenomena. The cultural

A replica of one of Colombus' ships in Barcelona, Spain.

geographer examines origins, diffusions, distributions, and patterns related to the earth as the home of mankind to answer these questions. Answers to questions raised by cultural geographers are significant both because they satisfy human intellectual curiosity and because they provide tools for solving diverse problems. Cultural geography examines the spatial variations among cultural groups and the spatial relationships associated with society. It focuses on the way human activities produce the variety of lifestyles among the earth's peoples. Cultural geography is interested in the spatial distribution and functioning patterns of all culture systems, whether they reflect religion, politics, livelihood, language, or other cultural phenomena. Cultural geographers typically examine the long-term character of human development, viewing contemporary issues as only one point on a longer continuum of development and change associated with the human occupation of the earth. Several important themes are of central interest to cultural geographers as they attempt to explain the world's cultural geography:

1. Culture
2. Cultural origins and diffusions
3. Cultural ecology and process
4. Cultural landscape
5. Culture regions

CULTURAL DIMENSIONS
A Question of Definition*

The centrality of culture in human existence makes its definition essential. Scholars from a variety of disciplines, philosophers in all ages, and thoughtful individuals from time immemorial have reflected on just what constitutes culture. The following definitions are representative of their conclusions:

Hiller: "The beliefs, systems of thought, practical arts, manner of living, customs, traditions, and all socially regularized ways of acting are called culture. So defined, culture includes all the activities which develop in the association between persons or which are learned from a social group, but excludes those specific forms of behavior which are predetermined by inherited nature."

Linton: ". . . the sum total of ideas, conditioned emotional responses, and patterns of habitual behavior which the members of that society have acquired through instruction or imitation and which they share to a greater or less degree."

Lowie: "By culture we understand the sum total of what an individual acquires from his society—those beliefs, customs, artistic norms, food-habits, and crafts which come to him not by his own creative activity but as a legacy from the past, conveyed by formal or informal education."

Panunzio: "It [culture] is the complex whole of the system of concepts and usages, organizations, skills, and instruments by means of which mankind deals with physical, biological,

and human nature in satisfaction of its needs."

Herskovits: ". . . culture is essentially a construct that describes the total body of belief, behavior, knowledge, sanctions, values, and goals that mark the way of life of any people. That is, though a culture may be treated by the student as capable of objective description, in the final analysis it comprises the things that people have, the things they do, and what they think."

Park and Burgess: "The culture of a group is the sum total and organization of the social heritages which have acquired a social meaning because of racial temperament and of the historical life of the group."

Mead: "*Culture* means the whole complex of traditional behavior which has been developed by the human race and is successively learned by each generation. A culture is less precise. It can mean the forms of traditional behavior which are characteristic of a given society, or of a group of societies, or of a certain race, or of a certain area, or of a certain period of time."

Sutherland and Woodward: "Culture includes everything that can be communicated from one generation to another. The culture of a people is their social heritage, a 'complex whole' which includes knowledge, belief, art, morals, law, techniques of tool fabrication and use, and method of communication."

Dawson: "A culture is a common way of life—a particular adjustment of man to his natural surroundings and his economic needs."

Although the specific wording of the definitions varies, each concludes that culture is learned, that it relates to human-land relationships, and that it is common to a group of people.

*Adapted from A. L. Kroeber and Clyde Kluckholm, *Culture* (New York, Vintage Books, 1952) pp. 82–84, 89–90, 105.

Culture

Culture is defined as the total assemblage of values, beliefs, symbols, and technology of a group of people; the complex set of *learned* (as opposed to instinctive) behavior associated with the group. Culture consists of the *beliefs* (reli-gious, political), *institutions* (legal, governmental), *technology* (skills, abilities), and *tools* (machines, equipment, crops) creating distinctive elements and patterns in the environment.

Culture encompasses ideas and information that must be learned as an individual becomes part of a cultural group. These include symbolic language, house styles, clothing styles, food preferences, technology, legal systems, the role of the individual in the social group, world view (human beings' relationship to the deity or deities of the universe), political systems, and racial relationships. Encompassed in this wide range of human activities is an important concept of cultural geography: the question of how human cultures shape and direct human action, resulting in the uniqueness of each individual place on the surface of the earth. The most important geographic aspect of culture is that its elements (language, religion, economy, political systems, and so forth) create a pattern that can be described, mapped, and examined to attempt to predict its impacts on humans and the environment. The manifestation of the cultural elements allows the creation of maps pertaining to language, political systems, race, religion, or other phenomena. (Figure 1–11).

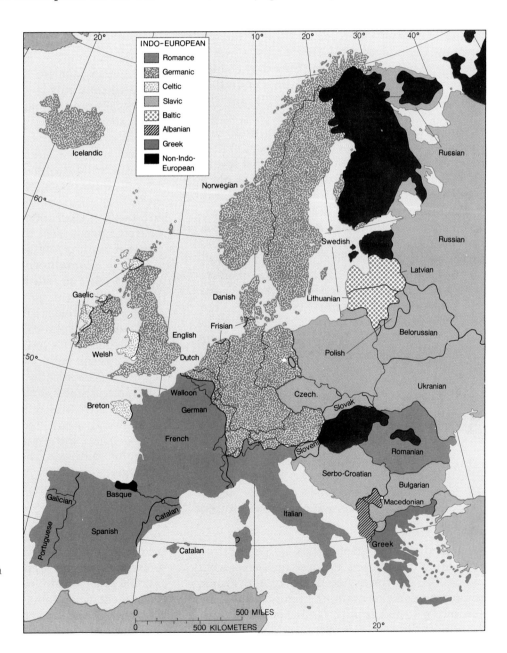

FIGURE 1–11 Map of Languages of Europe A map of the languages of Europe demonstrates both the complexity of the culture of the region and the use of language to create culture regions.

Culture Origins and Diffusions: Explaining the Pattern of Cultural Geography

A central theme in geography is the study of differences and similarities that exist between places upon the face of the earth. Cultural geography also studies the origins and diffusions of culture groups or individual cultural traits to explain how similarities and differences among places in the world occurred. Cultural traits can usually be traced to a place and time of origin.

Cultural geographers examine the origins of cultural phenomena in place and time to determine how they spread; how they intensify, wane, or disappear from a culture; and how they change the environment of both their *hearth* and other places that subsequently adopt the trait. The explanation of the origins of cultural phenomena is found in the history of cultures. Cultural geographers study the origin of traits at the individual, community, regional, national, and global level.

The cultural geographer uses documents, place names, and other evidence to determine the way in which different groups use the environment, the movement of people, and the interaction between groups with different technologies and environments. Technological advances do not always originate in a specific hearth, but instead are the products of multiple *independent invention.* Other advances may begin in one hearth, then spread and be improved in other places. Through analysis of cultural origins and *diffusions,* the cultural geographer hopes to determine four main factors:

1. The origin in time and place of specific cultural features.
2. The routes, chronology, and manner of the diffusion of those features.
3. The regional pattern of former cultures and cultural areas.
4. The nature of the landscapes that developed in association with specific cultures, past and present.

The cultural geographer has a harder time discovering the answers to these questions with the passage of time. Recent written evidence simplifies efforts to examine origins, diffusions, distributions, and former landscapes. The cultural geographer must rely heavily on the work of archaeologists, anthropologists, agronomists, and others to study the spread of human activity before the period of written records. Migrants to Canada, for example, brought hard red winter wheat from czarist Russia in the late nineteenth century. The recent date of this diffusion, coupled with written records and oral histories, allows the routes of hard winter wheat diffusion to be readily reconstructed (Figure 1–12). Examination of the original domestication of wheat, however, is made more difficult by the absence of written records at the time it occurred. In this case, cultural geographers rely on archaeological findings, which indicate the emergence of domesticated wheat forms in the Middle East.

In examining diffusion, geographers try to discover the time and place of origin, the routes of diffusion, the rates of adoption, and the environmental effects of the idea or element being diffused. When diffusion occurs, it does so in one of several ways. Geographers distinguish between *expansion diffusion* and *relocation diffusion* (Figure 1–13). Expansion diffusion occurs as an idea or trait spreads from its hearth to adjacent areas. Expansion diffusion can be seen in the diffusion of many common domesticated plants, which spread from a hearth in the Middle East to contiguous groups of people. Relocation diffusion occurs when an idea or innovation moves with an individual or group to a new location. For example, the popularity of Levis jeans in Europe in the late 1970s and early 1980s can be mapped first as a process of expansion diffusion as it spread across the United States, and then as relocation diffusion when wearing of jeans spread to Europe and the Soviet Union.

Expansion diffusion can be subdivided into stimulus diffusion, hierarchical diffusion, and contagious diffusion. *Stimulus diffusion* occurs when a specific innovation or cultural trait is rejected, but the underlying idea is accepted. The idea stimulates development of a trait or element related to the original innovation, but more suited to the local area. It is assumed, for example, that domestication of rye and oats occurred in northern Europe as substitute crops more suited to the environment were domesticated in response to the diffusion of the idea of domesticated wheat.

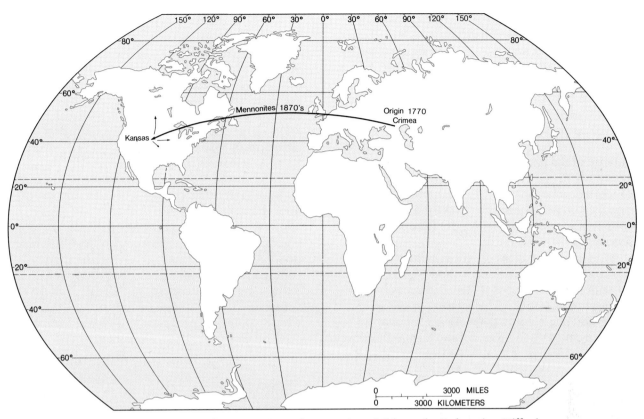

FIGURE 1–12 Diffusion of Wheat from Russia to the American Midwest by Relocation Diffusion

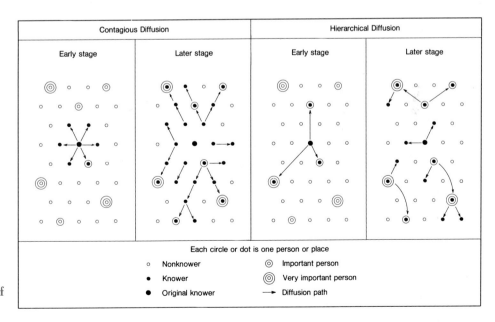

FIGURE 1–13 Models of Diffusion

Hierarchical diffusion refers to the tendency of many ideas or innovations to move from a larger city to a city of similar size, bypassing smaller centers. The idea then diffuses from the larger centers to smaller centers, creating a hierarchy of adopters. Hierarchical diffusion can also occur at the individual level, with ideas moving from one important person to another, skipping the majority of the population. The ideas diffuse to the balance of the population more slowly. Hierarchical diffusion can be seen in the spread of a variety of ideas or innovations, from fashions to technology. By contrast, *contagious diffusion* refers to the spread of ideas throughout a population without any hierarchy of adoption.

Innovations or ideas are most widely accepted in the area surrounding their hearth. The acceptance of an innovation decreases with increasing distance. The combination of *distance* and the *time* it takes for an innovation to travel weakens adoption of innovations because of *time-distance* decay. *Barriers* are also an important factor affecting all aspects of diffusion. *Permeable* barriers allow certain elements of ideas or traits to cross while preventing others. *Absorbing* or *impermeable* barriers may prevent diffusion entirely. Cultural phenomena such as religion, language, and political boundaries often serve as barriers to diffusion. The diffusion of jeans to the Soviet Union, displays that political barriers are permeable to some traits from the West, but impermeable to others. Culture traits that the Soviet leadership opposes may be excluded, but other traits may be allowed into the country without rigorous objections. Levi jeans were not officially prohibited. Anyone in the USSR with money or contacts could obtain them, but the government did not import quantities sufficient to meet the demand. This created a stimulus diffusion where copies of American jeans were imported (see the Cultural Dimensions box).

CULTURAL DIMENSIONS
Soviet People: Western Veshch (Thing)

In Russian the term *veshchism* is used to refer to the desire to have things, in this case material possessions such as a car, apartment, furniture, jewelry, and so forth. More important, the term signifies desire for Western "things," whether blue jeans, T-shirts, records, or even cars. In part, the desire for Western things reflects the desire for status that comes with having scarce, and thus expensive, items. In part, it may reflect the prevalent view that styles are set in the West. (Even the Russian word for jeans is Western, pronounced jeans-ee). But in large part, the desire for Western things represents a recognition that the centrally planned economy produces a plethora of low-quality goods that may also be expensive.

The Soviet Union produces ten brands of refrigerators: the Minsk is perceived as the best, and the Vega as the worst. The result is that Vegas are unsold, and Minsks are unavailable. Continued production of items that are not in demand reflected the past government commitment to full employment. Closing the Vega production plant would idle workers, so the Vega continued to be produced until the 1986–1990 economic plan began.

Another example of the inadequacies of government planning, Soviet style, is school uniforms. The typical uniform for eighth and ninth grade girls is a dark brown dress with a black apron (white on holidays). In 1983 a new stylish uniform with blue pleated skirt, vest, and matching jacket was introduced. Unfortunately most were manufactured in large sizes, and jackets were rarely the same size as the skirt in an individual uniform.

The desire for Western goods that results from inadequate quality makes even goods with Western-sounding names popular. Denim clothing made in Eastern European countries and sold in the Soviet Union under the brand name "Montana" is popular. Not only does the name Montana evoke images of the geography of America's Wild West, the quality is better than the Soviet-made goods.

Old Order Amish of Pennsylvania, who are very strict, will not accept electricity. Thus religion may become a completely impermeable barrier to the diffusion of electricity. The Amish may adopt other forms of technology not tied to electricity, however. Each culture and each individual accepts or rejects innovations in the same way, based on their own values, needs, and cultural constraints.

Examinations of origins and diffusions allow cultural geographers to attempt to explain why the earth's present landscapes take the forms that they do. This focus of study may include the ability of a cultural group to change an environment over time, the manner in which that ability changes over time, or the impact of the environment on change in the group. The adoption of innovations such as domesticated plants, iron and steel production, religion, or other ideas affects the ability of a group to modify its landscape. Understanding the diffusion of ideas helps to explain the relationships that exist between individual groups and their environment.

Cultural Ecology: The Process of Environmental Change

The relationship between people and the environment is an ongoing interaction. Humans modify the environment. In turn, their modifications are affected by the environment. It is more difficult to grow rice on steep slopes than on level land, but groups use both types of landform in Asia. The relationship between people and their environment where rice is grown on steep slopes represents great environmental manipulation to create terraces. On more level land, the relationship between the group and their environment reflects less drastic environmental restructuring. *Cultural ecology* is the study of the causes and effects of the interaction between a group (society) and its environment. Cultural ecology examines how the culture of a specific group impacts the entire environmental complex that they occupy and the way in which the environment affects the people who reside within it.

The ecological approach in cultural geography poses the question "How do we describe and analyze what is happening here?" Examining something as obvious as the growth of suburbs around modern American cities, a cultural geog-

Rice terraces demonstrate the extent to which the natural landscape has been modified by human action.

rapher may try to determine the process by which associated environmental and cultural change occurred. The widespread adoption of the private automobile, the expansion of highway networks facilitating the use of automobiles, changes in types of employment and location of workplaces and changing attitudes concerning life in the city all contributed to the process of suburbanization.

A cultural geographer may also ask, "How did the rice terraces of the mountain slopes in the Philippines come to be constructed?" Analysis of the cultural ecology indicates that population growth in areas dependent on wet rice production but with limited level land caused the expansion of rice production upslope through the use of terraces to enable production levels to support the population.

Cultural geographers deal with current questions such as how the loss of the tropical rain forest in less industrialized countries relates to the increased cutting of forests for fuel and agriculture, and how industrialization and urbanization processes are now changing the societies of the less industrialized world. Cultural geographers also examine the process by which migrants from one country (e.g., Mexico or Vietnam) move to others (e.g., the United States). They examine the flow of resources (e.g., oil from the Middle East to Japan and Western Europe). Cultural ecology deals with environmental issues such as the recurrent droughts of Africa and their subsequent impact on the population of sub-Saharan Africa.

When cultural geographers ask such questions to explain the cultural ecology, they are trying to understand the human-environment relationship of that place. The spatial interaction between humans and the land they occupy is so central to geography that cultural geographers have developed four general explanations of this human-land relationship: environmental determinism, environmental possibilism, cultural determinism, and environmental perception, with humans as modifiers of the earth.

Environmental determinism was a popular explanation for the human-land relationship during the first three decades of the twentieth century. Proponents of this theory argued that the cause of differences between peoples and places was to be found in the physical environment. Similar environments should produce similar cultures and levels of development because people were basically molded by the environment like clay. An old theory, environmental determinism was used by Aristotle to explain world geography. A good example of environmental determinism in the twentieth century is the work of Ellsworth Huntington. He maintained that the ideal climate for intellectual development was the marine west coast climate. He believed that temperatures of about 68° to 70° were ideal for thinking and creativity. Huntington concluded that advances in science and technology occurred in England and Western Europe because of the marine west coast climate in that region. Other environmental determinists argued that people who live along

coastlines become great navigators and relied on the ocean whereas those who lived inland tended to be sedentary and less inventive. Ellen Churchill Semple even claimed that the physical environment controlled the development of political and religious thought. According to the environmental determinist explanation of the human-environment relationship, England played such an important global role in the nineteenth and twentieth centuries because the country's island nature required seamanship and its ideal climate led to advances in all branches of human endeavor.

The idea that the environment *determines* human action overstated the role of nature in human affairs, although the environment does play a role in the cultural ecology of an area and affects a wide range of human activity. The relative importance of the role of environment versus culture in explaining the human-environment relationship of a place prompted Paul Vidal de la Blache, a French cultural geographer, to propose an alternative view, *environmental possibilism*.

Possibilists argue that although the environment plays a role in establishing the choices open to cultural groups, culture is an equally important influence on decision making. The level of technology, experiences of the individual, types of tools and crop choices available, and other elements of a group's culture all influence choices of land use by culture groups. The possibilist school maintains that the environment is restrictive, but provides a range of possible choices. Culture is the determining factor in the selection process (see the Cultural Dimensions box). Depending on the level of technology available to the culture, the environment may provide a large variety of choices or only a limited number. To the possibilist, the specific relationship of a society to its environment represents culturally based decisions made from the possibilities offered by the environment.

Recently, another school of thought has emerged based on a model of human-environment relationships. Known as *environmental perception*, it attempts to explain human actions as an outcome of both physical and cultural interactions. Environmental perception is based on the fact that we all have mental images of the world, which are largely shared within a

CULTURAL DIMENSIONS
Paul Vidal de la Blache on Geography

Paul Vidal de la Blache (1854–1922) combined the ideas of physical geography with ideas about the role of culture in geographical explanation. French geographers had long recognized the distinctive regions (*pays*) that characterized France. Each *pays* had a distinctive way of life, fiercely clung to by its inhabitants. Vidal de la Blache recognized that although each way of life reflected certain elements of the physical geography, it also reflected cultural variables. In 1888 Vidal de la Blache described the cultural regions of France one would encounter in an imaginary journey across the country:

> Between Etampes and Orléans, we cross by rail a pays called the Beauce. Without leaving the carriage door, we can distinguish certain characteristics of the landscape; a soil absolutely level, over which stretch long ribbons of cultivated fields; very few trees or streams (in 45 kilometres not one has been crossed), no isolated dwellings; all settlements grouped in towns or villages. If we cross the Loire River, we find to the south a country just as flat but with soil of a different colour, where woods and pools abound; this is the Sologne. To the west of the Beauce, between the sources of the Loire and the Eure, there is a hilly country, green, broken by small enclosures and rows of trees, with dwellings scattered far and wide. This is the Perche. We enter Normandy. If, in the Department of Seine-Inférieure, we examine the two neighbouring *arrondissements* [districts] of Yvetot and Neufchâtel, what differences we find. In the first, all is flat; fields of corn, farms enclosed within squares of great trees, wide horizons. In the second, nothing save little valleys,

quick-set hedges, and pastures. We have passed from the Pays de Caux to the Pays de Bray. The mode of living has changed with the soil.*

Vidal de la Blache's imaginary journey includes all of the elements that came to typify the French school of geography: the physical basis of soils, hydrography, and vegetation and the cultural elements of field, farm, and village. Over time, Vidal de la Blache was instrumental in creating a view of geography that endures today in the French school and is an important part of geography wherever it is practiced.

Vidal de la Blache's contribution to geography included not only the idea of culture, but the understanding that the landscapes of the *pays* were the result of several centuries of stability in the way of life in rural France. He intuitively recognized that cultural landscapes are dynamic and speculated on the changes that would affect them in the next generation: industrialization, population growth, changes in technology (especially in transportation), and growth in urban areas. The ideas of Vidal de la Blache are still current. Forces for change still affect the landscapes of the world, but at a pace and scale he did not foresee. The combination of physical and cultural geography remains the primary tool for explaining and understanding the landscapes and regions of the world.

*E. R. Crone, *Background to Geography* (London, Northumberland Press Limited, 1964), p. 71.

cultural group. Environmental perception tries to explain the human use of the earth by discovering the way individuals view their natural environment. Environmental perception is based on the premise that it is impossible for an individual to see the environment as it actually exists since an individual's perception is shaped by culture. All inputs perceived by an individual are filtered by his or her culture with its associated technology and goals (Figure 1–14).

Both the environment and culture are active elements in explaining the human-land relationship in the environmental perception model. Individuals perceive the environment through their own experiences, values, and goals and modify the environment using their technology. In return, the individuals are influenced by the environment in that their goals are either satisfied or denied. In response to this feedback, the individuals then modify their interactions with the en-

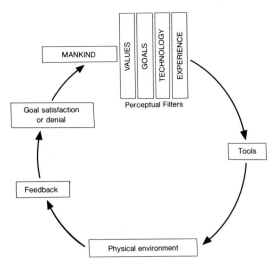

FIGURE 1–14 Environmental Perception Model
Peoples' perception of the environment reflects
the filter of culture through which they observe it.

vironment until they reach a state where the
goals are satisfied at an acceptable level.

The environmental perception model helps explain why different groups occupying similar environmental settings may create different landscapes, life-styles, and environmental impact. A comparison of the Nile Valley of Egypt with the Imperial Valley of the Colorado River of southern California and northern Mexico, reveals tremendous differences in landscape and life-styles, yet their climates are almost identical, suggesting that their residents view the environment differently (see the Cultural Dimensions boxes on pages 25–27).

Examination of these cultures reveals different histories, values, crops, and population densities. The residents of California's Imperial Valley perceive the environment as a source of profit. The resultant landscape emphasizes large farms engaged in commercial agriculture. The farmers of the Nile Valley historically perceived the environment as a site for a semisubsistence economy rather than a for-profit economy. The resultant pattern of small farms with emphasis on producing subsistence crops has persisted for over four thousand years.

The environment we live in, which affects us consciously or unconsciously, is called the *operational environment.* That part of the operational environment of which we are aware is called the

perceptual environment. We are familiar with the perceptual environment either because of direct contact (as with weather) or through learning and experience. The perceptual environment is perceived individually, but persons belonging to the same culture share basic attitudes about it. The perceptual environment represents that part of the external world that we see through the perceptual filters of our culture. The difference between the perceptual and operational environments can be illustrated by the resident of the American Midwest. She may understand correctly the local impact of a winter snowstorm, but be unaware of the global climatic relationships that caused it. The snowstorm is part of her perceptual environment, global climatic relationships are part of the operational environment (Figure 1–15).

Within the perceptual environment is the more limited *behavioral environment* to which people respond. The snowstorm discussed above is part of the behavioral environment because the Midwesterner must respond by wearing warm clothing, staying inside, or adopting other coping mechanisms. Human reactions to floods or other natural disasters, their images of places, or the ranking of different places in terms of their residential desirability are all examples of responses to the behavioral environment. The perceptual environment represents that part of the environment to which we assign positive or negative values that impact human decisions and actions (Figure 1–16).

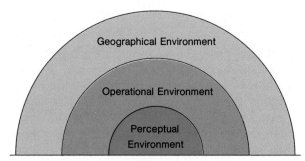

FIGURE 1–15 Levels of Environmental Perception
The environment which affects each individual
consists of both perceived phenomena and elements
of which we are unaware.

Two Rivers: The Nile and the Colorado

The Nile

The Nile River flows from south to north for 4,100 miles in northeast Africa (Figure 1–A). The Nile's principal sources, near the equator, receive a peak of heavy tropical rains from March to September. By July this peak of water flowing north has converged on the main channels of the river as they cross the Sudan and Egypt. Historically, the volume of water was much greater in July than the channels could hold, causing the river to overflow its banks. The flood stage would last until October. Until people began to modify the landscape, the Nile Valley was under a sheet of water with only ridges of higher ground rising above the flood. Since Egypt receives almost no rainfall, the valley of the lower Nile received most of its water for the entire year during the few months of the river's high water flow.

Rather than being a natural disaster, the annual flood permitted the rise of one of the richest, most advanced ancient civilizations by 3000 B.C. The yearly flood enabled the people to grow food to support a large population and freed people for building, scholarship, and art. Herodotus, the Greek geographer-historian of the fifth century B.C., wrote that Egypt was the "gift of the Nile."

The flood waters of the Nile carried soil particles from upstream and deposited them over the flood plain. For thousands of years, the silt deposited each year by the flood partly renewed the fertility of the Egyptian fields along the banks of the Nile. After the water receded in the fall, fields were planted. Crops grew during the warm dry winter and were harvested in the spring. Humans had to do little except plant after the water receded. Fields and the villages built on stretches of high ground that seldom were flooded, or on the desert margin of the flood plain, were the major changes ancient humans made in the landscape of the Nile Valley and the Nile delta.

No other major modifications were made in the Nile habitat until about 3400 B.C. when Egyptian farmers learned to use plows instead of planting seed in the natural furrows left by departing flood water. Earth embankments were built to control the movement of flood waters so that water could

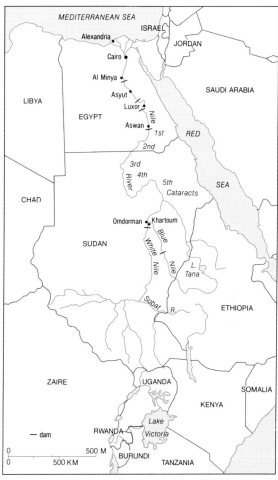

FIGURE 1–A The Nile River with Major Diversion Structures

be led to more of the shallow natural basins within the flood plain and therefore cover more land. Thus *basin irrigation* gradually evolved.

In a few places where water was available all year round, more and varied crops could be grown. Beginning about A.D. 1820, changes that greatly lengthened the growing season were made. First, irrigation canals were dug deeper so that even low water in summer before the floods

continued

CULTURAL DIMENSIONS

Two Rivers: The Nile and the Colorado *continued*

came could be diverted into many fields. The Egyptians also built small dams to raise the summer water level at intervals all along the river, forcing the water to spill over into shallow canals. By 1890 this system had lifted the general summer level of water more than twelve feet.

Early in the twentieth century, more ambitious water control started with the building of the first Aswan Dam in 1902. The dam was designed in part to store some of the Nile's flood waters instead of simply raising the level of river water so that water would flow into intake canals. The Aswan Dam was heightened in 1912 and again in 1933. Now the Aswan High Dam has been constructed to hold back an entire annual flow of the Nile. It also provides power to make cheap fertilizers, badly needed by the intensively cultivated farms that are no longer covered by silty Nile water, which had previously renewed soil fertility.

At present four-fifths of Egypt's farms are irrigated for summer crops, and the appearance of the flood plain has been greatly altered. Never does a sheet of water cover it. That once life-giving aspect of the Nile would be a major disaster, destroying the summer crops that are the most valuable of all. The flood's mud and silt, once regarded as a necessity for recharging the fertility of the land, are now considered a nuisance by many because they fill up the canals. The river itself is kept in its channel by artificial high banks pierced by hundreds of canals that take the water to the fields. Much of the land is cropped twice and even three times a year, so that Egypt in effect has more than 100 million single-crop acres under cultivation. Population density is over 3,000 per square mile, with 19 of every 20 Egyptians still living either on the flood plain or the river's delta. Most of the farms are extremely small by American standards, and most of the farmers come to the fields from nearby villages just as they have for centuries.

Salton Sea Area

The Colorado River flows 1,400 miles southwestward through mountains and desert from its

FIGURE 1–B **Colorado River Drainage of the American Southwest**

source in the Colorado Rockies before it empties into the Gulf of California. It is the largest river in the American Southwest (Figure 1–B). People living in seven states and in Mexico use its water. Like the Nile, it is an international river.

Indians living along the lower Colorado developed a form of agriculture similar to that of the early Egyptians long before white men ever saw the area. But because the Colorado flooded in late spring, the crops were planted and grown in the hot summer months. Corn, beans, and squash were staple items.

Part of the Colorado River water is used in a hot, dry depression, or basin, toward the center of which is the Salton Sea. Although the area appears to be similar to the Nile Valley, the development here has been very unlike that of Egypt. The Americans began to modify the habitat in a region of sparse population thousands of years later than did the Egyptians.

The Salton Sea basin is an elongated valley, an extension of the trough that forms the Gulf of

continued

CULTURAL DIMENSIONS

Two Rivers: The Nile and the Colorado *continued*

California. The part of the basin that is north of the Salton Sea is called the Coachella Valley, while that part south of the Salton Sea is called the Imperial Valley. Much of the basin lies below sea level; the lowest point is 277 feet below the surface of the Gulf of California.

Only a few miles to the southeast, the Colorado River flowed across its delta more than fifty feet above sea level. The first canal to bring Colorado River water to the valley was cut through the riverbank near the Mexican border in 1901. The valley was renamed the Imperial Valley to enhance its image and attract farmers and settlers. In 1905 an exceptionally high Colorado flood poured into the valley along this canal and created the Salton Sea.

The building of Hoover Dam and other large dams along the Colorado ended the danger from flooding or drought. A system of canals has been constructed to carry water to the fields. Underground drains have been installed to carry away excess water, allowing much of the area to be

farmed without the danger of waterlogging or salting from which the Nile Valley has suffered. The main intake canal has a series of settling basins in which the river silt is taken out before it gets to the fields.

The typical Imperial Valley ranch is an industrial farm of several hundred acres, which a farmer may lease from someone else. Few people live on the farms. Owners and workers live in the small towns and cities of the valley. Most settlements are shipping points and market centers located on rail lines connecting them to eastern markets.

The changes Americans have made in the Salton Sea area's landscape are even greater and took place in a far shorter time than those the Egyptians have made in the Nile Valley. The Imperial Valley was transformed from an almost uninhabited desert to a highly productive farming region. Although it is far from being as densely populated as the Nile flood plain, its 500,000 acres support thousands of people prosperously.

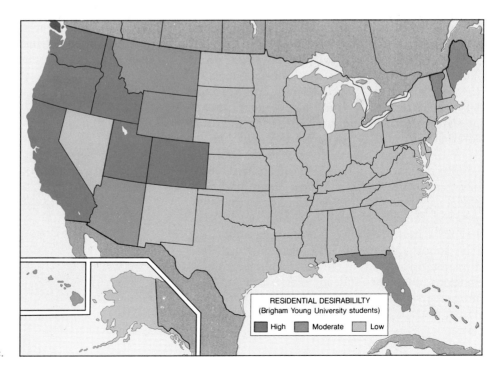

RESIDENTIAL DESIRABILILTY
(Brigham Young University students)

High   Moderate   Low

FIGURE 1–16 Map of Residential Desirability The way we perceive specific places affects our attitude about living there.

The environmental perception model helps provide answers to such questions as why people insist on living in flood plains, how they deal with natural disasters, and why they decide to migrate. Analyses of perception of natural hazards reveal that people living in areas threatened by hazards, such as floods, hurricanes, or earthquakes, deal with the uncertainty of hazards with coping strategies. They attempt to regularize hazardous occurrences by assigning a periodic cycle to their appearance, by denying or playing down their existence, or by attributing them to a divine source. Residents of the Gulf Coast of the United States deal with severe hurricanes by concluding that they occur only once every fifty or so years. People living in the Mississippi's flood plain maintain that they do not have floods, they only have high water. Those who have strong religious beliefs deal with natural hazards by simply stating that the hazard is beyond the control of humans and is in the hands of some higher power. Regardless of how people perceive natural hazards, by examining their perception, it is possible to gain some understanding of attitudes concerning natural hazards and persistence of living in hazardous places.

Examination of people's perception of place also helps to explain population migration. Historically, emigrants from Europe to the United States had an overly optimistic view of what they would find on arrival. The expectation that life would be better, jobs plentiful, and social problems of their homelands minimized motivated the great migrations of people from Europe to the United States. The same perceptions impact present migrations from Latin America and Asia. Internal movement of population in the United States in the past three decades has been from the central city to the suburbs, and from the North and East to the South and West. Examination of the perception of the migrants indicates that they perceive the suburbs and the South and West as having a more desirable environmental setting. The suburbs are regarded as less crowded, less threatening in terms of crime, and more favorable in terms of living conditions. The South and West are perceived as a "sun belt" in comparison to the "frost belt" of the North and East. Parts of the West, such as California or Colorado, are perceived as areas where life-styles are more informal and relaxed. Thus examining how people perceive the environment helps explain why people move.

The Cultural Landscape

The culture of a place is reflected in its appearance. *Cultural landscape* refers to the human-made landscape created by a cultural group occupying an area. The cultural landscape is the visual manifestation of the culture and its interaction with the environment. The German term *landschaftsbild* is the basis for the American cultural geographers' term, *landscape. Landschaftsbild* refers to the appearance of an area, not to the area itself. The cultural landscape of the Intermountain West, with its broad spaces of ranch lands and ranch estates, is visually much different from that of the Los Angeles suburbs with their tract homes with lawns, streets, and pools connected by the ubiquitous freeway.

The cultural landscape that typifies any place is a concrete, characteristic end product of the complex relationship between a given group of humans, their culture, and the natural environment. The cultural landscape reflects the net effect of a group's technology, values, beliefs, tools, and goals on the natural environment. The major changes for most of the environment in the western ranch setting of the American West have been the erection of fences, changes in the natural

Ranch lands in the American West are modified by grazing and erosion.

Suburban estates in the United States typify the built environment of American suburbia.

vegetation, and, in some cases, overgrazing, erosion, and construction of dams, windmills, or other devices to provide water for animals.

The Los Angeles suburbs, a currently visible cultural landscape, are markedly different from the original landscape. The present suburbs of Los Angeles County are the product of a complex process of cultural occupancy. This process began with Native American sporadic use of the Los Angeles basin for hunting and gathering and continued with Spanish missions and land grants. Los Angeles County has experienced a sequence of landscape change in the twentieth century involving the film industry, an oil boom, the expansion of suburban tract housing since World War II, and the growth of the aerospace industry. The term *sequent occupance* is used to refer to such a sequence of landscapes. Each sequence represents a distinct use of the environment with its associated cultural landscape.

Cultural landscapes can be described on the basis of their visible appearances and the processes by which that landscape came into being. The human impact on the landscape may not change the shape of the mountains or the general course of streams, but in replacing wild vegetation with cultivated plants or buildings, the landscape is dramatically changed. The cultural geographer attempts to determine the arrangement, style, and materials of the cultural landscape that reflect the presence of a distinctive way of life. All human constructs, whether roads, buildings, mining, farming, or waste heaps, are part of that landscape.

The cultural geographer is interested in many aspects of the cultural landscape. How old are the individual elements of the landscape? What is the relative rate of change? What things are typical and which are exceptional? Which landscape elements are accidental (such as erosion), and which are intentional (such as fences)? Which elements are temporary and which are permanent? Which are human made and which occur naturally? All of these questions, and others, help explain the cultural landscape.

The tools used by cultural geographers to examine the cultural landscape are those common to all geographers. The geographer plots the distribution and densities of given features, maps their relationships and organization, examines the technology and organization that characterize the cultural ecology, and tries to determine the dynamics of change that exist within that particular landscape (Figure 1–17).

The cultural geographer attempts not only to describe the differences that make cultural landscapes distinct, but also to explain how and when these features might continue to change as a result of the culture that occupies that particular locale. Consider your own hometown. Does it

American suburbs are often perceived as the stereotype of the American landscape.

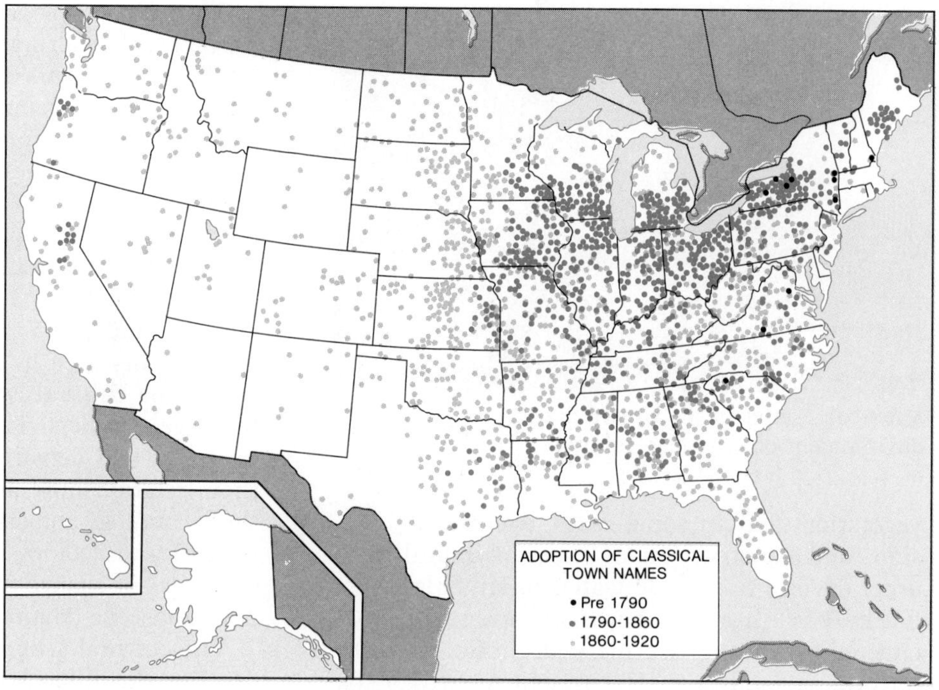

FIGURE 1–17 Map of Example of Classical Place Names The use of classical names for places in the United States illustrates the process of diffusion and the use of place name evidence to map cultural history.

Source: Adapted from Wilbur Zelinsky "Classical Town Names in the United States: The Historical Geography of An American Ideal," *Geographical Review*, 57 (1967), pp. 463–495. By permission.

have distinctive cultural landscapes that can be recognized? What changes are taking place that affect that landscape? If it is a typical American city, it has both suburbs and an older commercial core city. The processes of change are those associated with expanding suburbanization, expanding transportation networks, and changing technology that affects both urban and rural dwellers.

Culture Regions: Defining the Essence of Place

Culture regions are another theme in culture geography. A region is an area on the earth's surface marked by certain characteristics that distinguish it from surrounding regions. Culture regions are regions that have recognizable cultural characteristics that are specific to that area. The definition of a region may be as simple as determining the distribution of a characteristic, such as religion, or plotting the diffusion of country and western music to delimit the region where it is concentrated (Figure 1–18). The cultural geographer attempts to find the spatial distribution of a specific item or complex of items associated with a culture in order to discover

those areas of the earth's surface that share a common cultural characteristic or complex of characteristics.

In trying to map culture regions, the geographer is attempting to define similarities of culture. These similarities may range from absolutes, such as language or religious taboos on food, to less readily defined similar values, such as patriotism. The cultural geographer is faced with the problem of determining the boundaries of the region, the similarities that make the region distinct, the characteristics that distinguish it from surrounding regions, and the level of concentration of the trait that must exist for it to be recognized as a distinctive region. Cultural geographers recognize two major types of culture regions, formal culture regions and functional culture regions. *Formal culture regions* are inhabited by a group of people who share one or more cultural traits. Examples of formal culture regions are the language regions of Europe and the wheat and corn belts of the United States. Each region is based on the presence of some recognizable cultural trait.

A good example of a formal culture region is found in the concentration of Mormons, a reli-

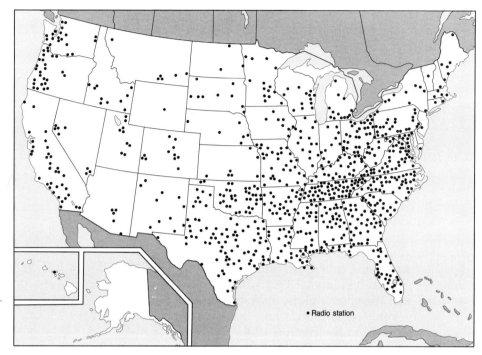

FIGURE 1–18
Distribution of Country Music Stations Country music diffused from a hearth in the South centered on Tennessee.

Source: Adapted from George O. Carney, "From Down Home to Uptown: The Diffusion of Country Music Stations in the United States," *Journal of Geography.*

FIGURE 1–19 Mormon Cultural Region The Mormon Culture Region in the United States is one of the most obvious cultural regions in the country.

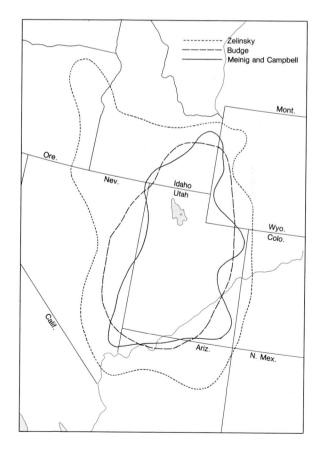

gious group in the United States (Figure 1–19). The factor that defines this region is adherence to the Mormon religion, and the map of this trait shows the area where Mormons are concentrated. Not all Mormons are found in the Mormon culture region, but it has the highest concentration of Mormons of any area of the United States.

The boundary drawn around the culture region may be misleading because of the variation in intensity of the culture traits being examined. To deal with this problem, the geographer looks at places that serve as cultural hearths or centers, routes of dissemination and diffusion, and regions where the cultural trait dominates. The hearth of the Mormon culture region is Salt Lake City. The routes of diffusion have been valleys at the base of the mountains of the Intermountain West where Mormon pioneers settled and where subsequent roads and railroads provided avenues of transportation.

The delineation of regions may mask important differences within areas in the frequency or intensity of occurrence of the traits being examined. Donald Meinig has suggested that it is possible to divide a culture region into three levels of intensity: *core, domain,* and *sphere.* Meinig applied these principles to the Mormon culture region. He defined the *core* as the area where the religion is overwhelmingly dominant. The landscape in the core has been modified by Mormons to create a specific set of features. The values and attitudes of the population reflect the religion. Beyond the core is the *domain,* where the Mormon religion is still dominant. Much of the area is unsettled, however, and consequently the "Mormonness" of the region is ephemeral. Finally, there is the *sphere,* an area where Mormonness is even less intense, but where important nodes of Mormon activity may be located, and where some communities may have a predominance of Mormons (Figure 1–20).

Geographers also look for formal culture regions of the past and their associated landscapes. It is possible to reconstruct the world as it was in 1492 when Columbus began his voyage of discovery (Figure 1–21). An example of how the world's cultural patterns have changed since then is the distribution of language. Today the English, Spanish, and Portuguese languages dominate North and South America, whereas Native American languages dominated prior to the fifteenth century. Australia and Siberia were also dominated by non-European languages until the eighteenth century. Thus analysis of past culture realms, regions, and their related hearths and routes of diffusion allows the cultural geographer to trace the process by which a culture and its associated complex of traits diffused to the rest of the world.

Functional culture regions are different from formal culture regions in that they emphasize the interconnections of an area rather than a specific cultural trait. Functional culture regions include areas that have been organized to function socially, politically, or economically. Cities, political units, school districts, trade areas, and farms are functional regions. The functional culture region is defined by the specific function involved. A farm is that area farmed by a farmer. A school district is a region administered by a particular school board. A federal reserve banking

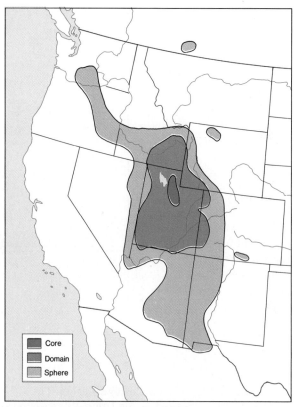

FIGURE 1–20 Examples of Core, Domain and Sphere The concepts of core, domain, and sphere relate to the intensity of occurrence of the cultural trait being mapped.

district is the region served by one federal reserve bank (Figure 1–22). The boundaries of independent countries are a global example of functional culture regions. In some cases countries are nearly formal culture regions, as in the case of Japan and the Japanese language. In others, such as India, the functional region does not correspond to any single cultural trait such as language.

The study of culture regions is a tool for grasping the essence of the characteristics that in combination make each place unique. While each place is different from all others, places sharing the same cultural traits do have common characteristics. Examination of the regional pattern of cultural traits reveals their distribution and suggests the character of a place. The culture region allows the geographer to interpret the patterns of places as they reflect human activity over time.

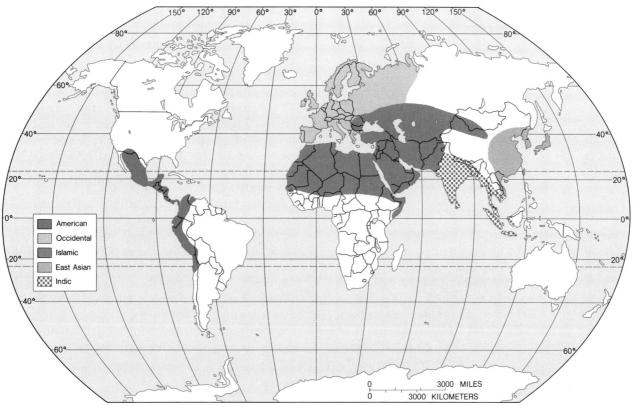

FIGURE 1–21 Cultural Regions of 1492 The cultural regions of 1492 were changed dramatically in the next 200 years as the European cultural region expanded its influence.

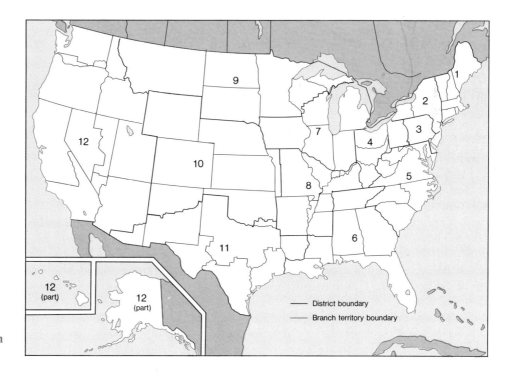

FIGURE 1–22 Federal Reserve Bank Districts in U.S.

Cultural Geography and Geographic Knowledge

The themes of cultural geography, culture, cultural origins and diffusions, cultural ecology, cultural landscape, and culture region collectively help to explain the human world—the earth as modified by mankind. Geographic knowledge obtained from systematic examination of the world's cultural geography is based on concepts that are central to all geographic analysis.

Location

One of the most important concepts in geography is location, the specific reference point that identifies the "where" of a place. Location is defined by showing the spatial relationship of a place. A specific place can only be located in reference to other places. Three concepts important to location are *mathematical* (absolute) location, *site*, and *situation* (relative location). The *mathematical* location of a place is given in terms of the mathematical coordinate grid of latitude and longitude on the globe. Absolute location establishes a place in reference to the equator (latitude) and to Greenwich, England, a suburb of London from which longitude (distance east or west of Greenwich) is measured. Longitude is measured by using lines drawn from the North Pole to the South Pole known as *meridians*. The Greenwich meridian passes through a point in the Royal Observatory in Greenwich and is known as the prime meridian. Other meridians have been used as the beginning point for measuring longitude, but Greenwich is the most commonly accepted. The mathematical location tells little about the geographic characteristics of a place, which are identified by site and situation.

Site refers to the internal characteristics and relationships of a place. The site location of New York City, for example, includes its location on the Hudson River, its climate, its landforms, its soil, and the interrelationships among these and the other geographical characteristics of the site of the city. Aspects of location related to site describe the local characteristics.

Situation or relative location, shows how a place is related to other places. Situation defines location by reference to external relationships.

New York City, for example, is approximately 2,500 miles from Los Angeles and 200 miles from Boston and Washington, D.C. Situation recognizes contact with other places that help to explain a specific place's character. Timbuktu on the southern margin of the Sahara Desert of Africa has site characteristics that include a desert climate, but it became an important city because of its situation relative to major caravan routes across the desert to North Africa. When trade shifted to sea routes, Timbuktu lost much of its importance.

Characterization of locational relationship as site or situation depends on how the area is defined. In referring to New York, the site relationships relate to the characteristics within the city, the situation to its location relative to the remainder of the state, the United States, and the world. The site characteristics of the United States include its resources, climates, landforms, economy, society, and peoples. Its situation includes its neighbors and surrounding seas.

Scale

Site and situation change with the *scale* of the area examined. *Map scale* expresses the proportional relationship between a map and earth reality. Scale is important in cultural geography because the size of the area we examine affects both the detail we can observe and the conclusions we draw. If we examine a small area, such as a city block, the resulting map is a large-scale map because the ratio that expresses the proportional relationship of the map to the world is large. On such a map, 1 inch might be equal to 100 feet of the actual block, a proportion of 1 inch on the map equaling 1,200 inches on the ground, or 1:1,200. At such a large scale, individual houses, streets, paths, and other features can be shown. If we were to examine an area covering several hundred square miles, one inch might represent one mile, a ratio of 1:63,360. This is a smaller scale map because the ratio showing the proportional relationship is smaller. Maps showing entire countries or the world are at an even smaller scale, with one inch representing 100 or more miles. The smaller the scale of the map the less detail that can be included, and the greater the level of generalization necessary (Figure 1–23).

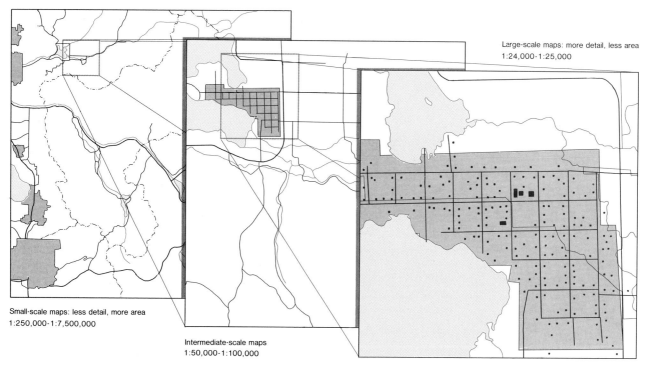

Small-scale maps: less detail, more area
1:250,000-1:7,500,000

Intermediate-scale maps
1:50,000-1:100,000

Large-scale maps: more detail, less area
1:24,000-1:25,000

FIGURE 1–23 Maps Showing Large Scale and Small Scale The concept of scale is central to geography.

The specific scale used on a map is related to the purpose of the map. A map showing the cultural landscape of a functional region such as a farm would be at a larger scale than one showing a formal region such as the distribution of the Chinese language. Understanding the scale of a map helps prevent misunderstanding of the data it presents. Failure to understand scale can result in a misperception of the world or the individual phenomena presented in maps.

Spatial Distribution

One of the important uses of maps is to show *distribution,* the way specific elements affecting the location of places are distributed. Every place has a specific location, but each place shares common elements whose distribution can be mapped. A map of the United States showing all cities with populations greater than one million would be a map of distribution. The location of towns with Spanish names is another example of distribution (Figure 1–24).

Examination of distributions reveals the way different variables are associated. The concentra-
tion of Spanish-speaking peoples in the Southwest may also be associated with dry climates. This does not necessarily mean that Hispanics settle where it is arid, but may merely reflect the fact that the border between Mexico and the United States happens to pass through a largely arid area. Examination of the distribution of cities of over one million population in the United States reveals a concentration along the coasts and in the northeast and north central portions of the country. Visual or statistical analysis of the correlation between distributions can reveal whether the relationship between them is statistically significant or accidental. Statistical analysis does not answer questions as to the merit of relationships.

Spatial distribution deals with *density, dispersion,* and *pattern.* Density refers to the degree of frequency of a phenomenon relative to the size of the area. The population density of China is much greater than that of the United States, although the two countries have almost the same area (68 per square mile in the United States, 297 in China). Dispersion is the extent of the spread of the phenomenon being studied, relative to the

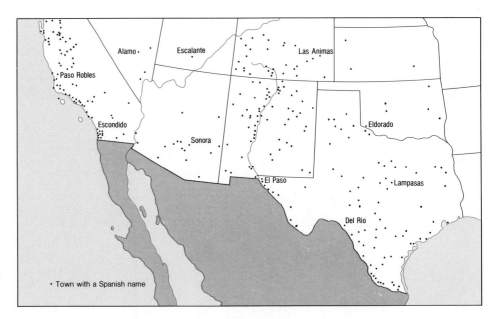

FIGURE 1–24 The Distribution of Spanish Place Names In the American Southwest, this distribution illustrates the use of linguistic evidence to map cultural regions.

size of the study area. Analysis of the dispersion of people in China and the United States shows the United States has a greater dispersion to the western part of the country. Pattern refers to the geometric relationships presented by a distribution. Examination of China and the United States shows settlement patterns that are very different. Most Chinese live in villages in rural areas, but the majority of Americans live in towns and cities (Figure 1–25).

Geographers at Work: Applying Geographic Principles

Geographers are employed in a wide range of careers that use their training in human-environment relationships. Geographic expertise is used in activities ranging from analyzing international relations to selecting sites for fast food restaurants; from planning land use for individual communities to regional planning for development projects affecting a wide area; from teaching elementary school to explaining and minimizing damage from natural hazards. The training of the geographer, emphasizing consideration of both culture and environment, provides essential ingredients to assist in making wise decisions.

The geographer involved in business-related employment may work for a large firm, helping it determine the best location for stores or factories. Geographers may also be engaged in determining the best markets for a product. Geographers delimit regions showing market areas, locate best sites for factories and retail outlets, and analyze the profit or loss potential in differing locations and regions. The formal job title of these professionals may not include geography. Rather the geographer working in business is known as a facilities planner, a market researcher, a research analyst, a locational specialist traffic manager, cartographer (mapmaker) or geo-information processor. Although the titles vary, each reflects the need for an individual trained to deal with real-world problems that involve both culture and environment.

Another large area of employment for geographers is in government activities. At the local level, geographers are engaged in land-use and city planning, transportation planning, real estate, housing, economic development, and cartography. Their work involves a wide range of activities, including research, policy formulation, program development and administration, and data collection and analysis relating to people and their use of the environment. Geographers in city planning provide guidance for daily decision making concerning approval of subdivision plats and enforcement of zoning codes to prevent inappropriate land uses. They are also engaged in long

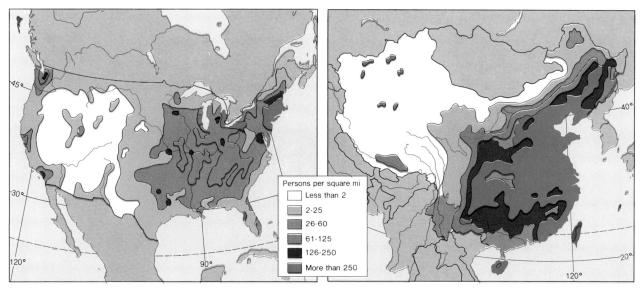

Persons per square mi

	Less than 2
	2-25
	26-60
	61-125
	126-250
	More than 250

FIGURE 1–25 Map Showing Examples of Density, Dispersion, and Frequency The difference between density, dispersion and frequency can be observed by mapping nearly any specific type of data.

range planning to ensure that the future development of the city will result in a desirable environment in which to live. Geographers help plan the routes and flows of traffic and provide guidelines for housing and business for individual communities and regions. Geographers also provide expertise to help local, state, and federal agencies minimize destructive natural hazards. Understanding people's perceptions of natural hazards has fostered federal programs to guide the location of housing, business, and industry to minimize the damage from floods, earthquakes, or hurricanes. Geographers are actively involved in such activities under a variety of titles such as hazard mitigation specialist, natural hazards planner, and cartographer.

Geographers play a major role in teaching at all levels of education from the elementary school to the university. Teachers have the opportunity to conduct research on how people use the earth and to teach and assist others in understanding the processes of the human occupation of the earth.

Central to all of a geographer's work is the theme of understanding, explaining, and helping to channel the human relationship to the physical environment. The cultural geographer is committed to minimizing the misery of mankind and ensuring that the human use of the earth is constructive rather than destructive.

Conclusion

The themes of cultural geography provide a framework for understanding the cultural elements of the world's geography. Explanation of the present, however, is possible only by understanding something of the past. The geography of the world is constantly changing, and modern cultural geography includes elements from past geographies as well as the forces for change that created today's world. Studies of culture origins and diffusion provide clues for discovering the relationships between today's geography and the past.

The cultural geography of today is as different from what it will be a century from now as it is from past centuries. The present cultural geography represents one point on a continuum of change that began with the earliest human modifications of the environment. Examination of the events and processes that helped shape the world of today indicates something of the nature, extent, and pace of change that is constantly reshaping our world. Cultural ecology, the study of the process by which cultural landscapes are created, provides important understanding of how the process of change will impact the world of the future.

The details of the earliest environmental change associated with the human occupants of

Explaining Spatial Interaction

In explaining the world, geographers use models. A model is a simplified, generalized representation of a specific portion of the world around us. Geographic models can be used to explain a specific type of geographic relationship, such as the spatial relationship of an individual town or city. One of the earliest attempts to develop a model of spatial relationships was the model of Johann Heinrich von Thünen. Von Thünen was a German estate owner who examined the spatial relationships between a market and types of farming. His book, *The Isolated State* (1826), hypothesized a circular state, occupying a level plain with uniform climate and soil characteristics throughout and surrounded by wilderness that isolated it. In his model all the people were trying to maximize their individual profits in a commercial economy. The one city at the center of the country was the only market, and prices for agricultural products were set at this market.

Von Thünen introduced one variable into his isolated state—transportation. All goods were moved by horse-drawn wagons. Under these simple circumstances, land use developed as a series of concentric rings around the city (Figure 1–C). Closest to the city a zone of perishable products (vegetables, milk) or heavy products with a low value in proportion to their weight (lumber, fuel, wood) developed. Farthest from the market, in the periphery, grazing of livestock that required little investment occurred. Between these zones there developed a progression from more intensive land uses closer to the city to less intensive

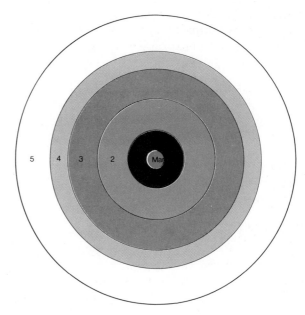

FIGURE 1–C Von Thünen's Model of Land Use

uses farther out. The specific land use represented the best return from the amount of investment. When another variable is introduced, such as a river providing cheap transportation, the pattern of land use changes. Instead of concentric zones around a central point, the zones are distorted along the cheaper transportation route.

Von Thünen's model is highly simplified and represents ideal conditions found in few areas, but it demonstrates that under his specific assumptions, sites with identical characteristics will have different uses, depending on their situation relative to the market. Although von Thünen's one variable, transportation, has changed dramatically, the underlying idea of modeling spatial interaction has not. The spread of suburbs through Orange County, California, in part reflects growth in transport accessibility that allowed urban residents to move to the suburbs, dis-

placing the orange groves, which moved farther out. Milk for the major urban areas of southern California is still perishable and bulky, and thus is produced closer to the urban markets than is grain, which is less perishable, and can be produced in the Midwest.

Geographers have developed other models to explain other spatial relationships. These are discussed in subsequent chapters, but it is important to remember that each of these models is an abstraction. Von Thünen's model ignores many other variables affecting land use, such as variations in soil fertility, technical variation, or cultural differences, but it does provide a useful means of demonstrating that there are universal factors affecting the use of land that can lead to similar land uses. Other geographical models attempt to show similar processes for different geographical relationships.

the world are only partially understood because of their great antiquity. Critical changes allowing widespread geographical impact can be recognized in the geographic record. These critical changes are analyzed in the next chapter to provide a basic understanding of the beginnings of the process that created modern cultural geography.

QUESTIONS

1. Define geography. How does it relate to you?
2. Why are maps of central importance in understanding geography?
3. What are some of the major contributions of the Greeks to geographic knowledge?
4. Identify and discuss the five themes of cultural geography.
5. Define culture and list five major elements of American culture.
6. Compare and contrast contagious and relocation diffusion.
7. What is the relationship between culture ecology and diffusion?
8. Explain the meaning of cultural landscape using examples from your local area.
9. Discuss the concept of region in geography, including an explanation of cultural landscape.
10. What are three theories that attempt to explain the interaction between environment and culture?
11. What role does environmental perception play in understanding the geography of an area?
12. How does the Von Thünen model help to explain the distribution of human activities?

SUGGESTED READINGS

Amadeo, Douglas, and Reginald G. Golledge. *An Introduction to Scientific Reasoning in Geography.* New York: John Wiley & Sons, 1975.

Bailly, Antoine S. *Les Concepts de la Geographie Humaine.* Paris: Presses Universitaires de France, 1981.

Blaut, J. M. "Diffusionism: A Uniformitarian Critique." *Annals of the Association of American Geographers* 77 (March 1987): 48–62.

Block, Robert H. "Robert Park's Human Ecology and Human Geography." *Annals of the Association of American Geographers* 70 (March 1980): 31–42.

Blouet, Brian W., ed. *The Origins of Academic Geography in the United States.* Hamden, Conn.: Shoestring Press, 1981.

Broadway, Michael J. (with Marilyn A. Brown). "The Cognitive Maps of Adolescents: Confusion about Inter-Town Distances." *The Professional Geographer* 3 (August 1981): 315–25.

Brown, E. H., ed. *Geography Yesterday and Tomorrow.* Oxford and New York: Oxford University Press, 1980.

Brown, Lawrence A. *Innovation Diffusion: A New Perspective.* New York: Methuen, 1981.

Bunting, Trudi E., and Leonard Guelke. "Behavioral and Perception Geography: A Critical Appraisal." *Annals of the Association of American Geographers* 69 (September 1979): 448–62.

Buttimer, Anne, and David Seamon, ed. *The Human Experience of Space and Place.* New York: St. Martin's Press, 1980.

Buttimer, Anne. *The Practice of Geography.* London and New York: Longman, 1983.

Clarkson, James D. "Ecology and Spatial Analysis." *Annals of the Association of American Geographers* 60 (1970): 700–716.

Cromley, Robert G. "The Von Thünen Model and Environmental Uncertainty." *Annals of the Association of American Geographers* 72 (September 1982): 404–10.

Cunningham, Susan M. "Multinational Enterprises in Brazil: Locational Patterns and Implications for Regional Development." *The Professional Geographer* 1 (February 1981): 48–62.

Curtis, James R. "McDonald's Abroad: Outposts of American Culture." *Journal of Geography* 81 (1982): 14–20.

de Blij, Harm J. *Wine: A Geographical Appreciation.* Totowa, N.J.: Barnes & Noble, 1983.

Duncan, James S. "The Superorganic in American Cultural Geography." *Annals of the Association of American Geographers* 70 (1980): 181–98.

Duncan, James, and David Ley. "Structural Marxism and Human Geography: A Critical Assessment." *Annals of the Association of American Geographers* 72 (March 1982): 30–59.

Entrikin, J. Nicholas. "Contemporary Humanism in Geography." *Annals of the Association of American Geographers* 66 (1976): 615–32.

Forman, R. T. T., and M. Gordon. *Landscape Ecology.* New York: John Wiley, 1986.

Gale, Stephen. "Ideological Man in a Nonideologician Society." *Annals of the Association of American Geographers* 67 (June 1977): 267–72.

Garreau, Joel. *The Nine Nations of North America*. Boston: Houghton Mifflin, 1981.

Gillard, Quentin. "The Effect of Environmental Amenities on House Values: The Example of a View Lot." *The Professional Geographer* 2 (May 1981): 189–96.

Glacken, Clarence J. *Traces on the Rhodian Shore*. Berkeley, Calif.: University of California Press, 1967.

Gould, Peter R. *The Geographer at Work*. Boston: Routledge and Kegan Paul, 1985.

Gregory, Derek. *Ideology, Science and Human Geography*. New York: St. Martin's Press, 1978.

Gritzner, Charles F., Jr. "The Scope of Cultural Geography." *Journal of Geography* 65 (January 1966): 4–11.

Gross, Jonathan L., and Steve Rayner. *Measuring Culture*. New York: Columbia University Press, 1985.

Grossman, Larry. "Man-Environment Relationships in Anthropology and Geography." *Annals of the Association of American Geographers* 67 (1977): 126–44.

Hagerstrand, Torsten. *Innovation Diffusion as a Spatial Process*. Chicago: University of Chicago Press, 1967.

Hanson, Susan (with Janice Monk). "On Not Excluding Half of the Human in Human Geography." *The Professional Geographer* 1 (February 1982): 11–23.

Hartshorne, Richard. *The Nature of Geography*. Lancaster, Penn.: Association of American Geographers, 1939.

_____ . *Perspective on the Nature of Geography*. Skokie, Ill.: Rand McNally, 1959.

Hawley, Amos H. *Human Ecology: A Theoretical Essay*. Chicago: University of Chicago Press, 1986.

Holt-Jensen, Arild. *Geography: Its History and Concepts*. New York: Barnes & Noble, 1980.

Huckle, Peter R. (with Geoffrey C. Smith and Denis J. B. Shaw). "Children's Perception of a Downtown Shopping Center." *The Professional Geographer* 2 (May 1979): 157–64.

Jackson, John Brinkerhoff. *Discovering the Vernacular Landscape*. New Haven, Conn.: Yale University Press, 1984.

James, Preston, and C. F. Jones, eds. *American Geography: Inventory and Prospect*. Syracuse, N.Y.: Syracuse University Press, 1954.

James, Preston, and Geoffrey J. Martin. *All Possible Worlds: A History of Geographical Ideas*. New York: John Wiley & Sons, 1981.

Johnston, R. J., ed. *The Dictionary of Human Geography*, 2d ed. Oxford: Blackwell, 1986.

_____ . *Essay: Geography and Geographers: Anglo-American Human Geography since 1945*. New York: John Wiley & Sons, 1979.

_____ , ed. *The Future of Geography*. London and New York: Methuen, 1985.

_____ . *Philosophy and Human Geography*, 2d ed. London: Edward Arnold, 1986.

The Journal of Cultural Geography. Department of Geography, Bowling Green State University, Ohio. 1980 forward.

Journal of Regional Cultures. The Popular Culture Association, Bowling Green State University, Ohio. 1983 forward.

Kates, Robert W. "The Human Environment: The Road Not Taken, The Road Still Beckoning." *Annals of the Association of American Geographers* 77 (December 1987): 525.

Kroeber, Alfred L., and Clyde Kluckhohn. "Culture, a Critical Review of Concepts and Definitions." *Papers of the Peabody Museum of American Archaeology and Ethnology* 47 (1952).

Landscape. Berkeley, California; an interdisciplinary journal devoted to the cultural landscape. 1951, forward.

Leighly, John, ed. *Land and Life: A Selection from the Writings of Carl Ortwin Sauer*. Berkeley and Los Angeles: University of California Press, 1967.

Ley, David. "Cultural/Humanistic Geography." *Progress in Human Geography* 5 (1981): 249–57; 7 (1983): 267–75.

Ley, David, and Marwyn S. Samuels. *Humanistic Geography: Prospects and Problems*. Chicago: Maaroufa Press, 1978.

Lloyd, Robert. "Cognitive Maps: Encoding and Decoding Information." *Annals of the Association of American Geographers* 79 (March 1989): 101.

Lornell, Christopher (with W. Theodore Mealor, Jr.). "Traditions and Research Opportunities in Folk Geography." *The Professional Geographer* 1 (February 1983): 51–56.

Lowenthal, David, and Hugh C. Prince. "English Landscape Tastes." *Geographical Review* 55 (April 1965): 186–222.

Lowenthal, David, and Martyn J. Bowden. *Geography of the Mind*. New York: Oxford University Press, 1976.

MacEachren, Alan M. "Travel Time as the Basis of Cognitive Distance." *The Professional Geographer* 1 (February 1980): 30–36.

Meinig, Donald W. "The Mormon Culture Region: Strategies and Patterns in the Geography of the American West, 1847–1964." *Annals of the Association of American Geographers* 55 (June 1965): 191–220.

_____ , ed. *The Interpretation of Ordinary Landscapes: Geographical Essays*. New York: Oxford University Press, 1979.

_____ . *The Shaping of America: A Geographical Perspective on 500 Years of History*. London: Routledge & Kegan Paul, 1986.

Meir, Avinoam. "A Disparity-Based Diffusion Model." *The Professional Geographer* 4 (November 1979): 382–87.

Meyer, Alfred H., and John H. Strietelmeyer. *Geography in World Society*. Philadelphia: J. B. Lippincott (1963), 31.

Meyer, Judith W. "Diffusers and Social Innovations: Increasing the Scope of Diffusion Models." *The Professional Geographer* 1 (February 1976): 17–22.

Mikesell, Marvin W. "Tradition and Innovation in Cultural Geography." *Annals of the Association of American Geographers* 68 (March 1978): 1–16.

Mitchell, Robert D. "The Formation of Early American Cultural Regions: An Interpretation," in James Gibson, ed. *European Settlement and Development in North America: Essays on Geographical Change in Honour and Memory of Andrew Hill Clark*, pp. 66–90. Toronto: University of Toronto Press, 1978.

Morgan, W. B., and R. P. Moss. "Geography and Ecology: The Concept of the Community and Its Relation to Environment." *Annals of the Association of American Geographers* 55 (1965): 339–50.

Norwine, Jim. "Geography as Human Ecology? The Man/Environment Equation Reappraised." *International Journal of Environmental Studies* 17 (1981): 179–90.

O'Sullivan, Patrick, and Bruce Ralston. "On the Equivalence of Consumer Surplus and Von Thünian Rent." *Economic Geography* 56 (1980): 73–78.

Pattison, William D. "The Four Traditions of Geography." *Journal of Geography* 63 (May 1964): 211–16.

Peet, Richard. "The Social Origins of Environmental Determinism." *Annals of the Association of American Geographers* 75 (September 1985): 309–33.

Reichman, Shalom, and Shlomo Hasson. "A Cross-cultural Diffusion of Colonization: From Posen to Palestine." *Annals of the Association of American Geographers* 74 (March 1984): 57–70.

Ristow, Walter W. *American Maps and Mapmakers: Commercial Cartography in the Nineteenth Century.* Detroit: Wayne State University Press, 1985.

Rogers, Everett M. *Diffusion of Innovations,* 3d ed. New York: Free Press, 1983.

Rooney, John F., Jr., Wilbur Zelinsky, Dean R. Louder, et al., eds. *This Remarkable Continent: An Atlas of North American Society and Culture.* College Station, Tex.: Texas A&M University Press, 1982.

Rowntree, Lester B., and Margaret W. Conkey. "Symbolism and the Cultural Landscape." *Annals of the Association of American Geographers* 70 (1980): 459–74.

Saarinen, Thomas F. *Perception of Environment,* Resource Paper No. 5. Washington, D.C.: Association of American Geographers, Commission on College Geography, 1969.

Salter, Christopher L. *The Cultural Landscape.* Belmont, Calif.: Duxbury Press, 1971.

Santos, Milton. "Geography in the Late Twentieth Century: New Roles for a Theoretical Discipline." *International Social Science Journal* 36 (1984): 657–72.

Sauer, Carl O. "Morphology of Landscape." *University of California Publications in Geography* 2 (1925): 19–54.

Shortridge, James R. "Changing Usage of Four American Regional Labels." *Annals of the Association of American Geographers* 77 (September 1987): 325.

Smith, Susan J. "Practicing Humanistic Geography." *Annals of the Association of American Geographers* 74 (September 1984): 353–74.

Solot, Michael. "Carl Sauer and Cultural Evolution." *Annals of the Association of American Geographers* 76 (December 1986): 508–20.

Sonnenfeld, Joseph. "Egocentric Perspectives on Geographic Orientation." *Annals of the Association of American Geographers* 72 (1982): 68–76.

Stilgoe, John R. *Common Landscape of America, 1580 to 1845.* New Haven, Conn.: Yale University Press, 1982.

Taylor, Peter J. "The Value of a Geographical Perspective," in R. J. Johnson, ed., *The Future of Geography,* pp. 92–110. London and New York: Methuen, 1985.

Taylor, S. Martin. "Personal Dispositions and Human Spatial Behavior." *Economic Geography* 55 (1979): 184.

Thomas, William L., Jr., ed. *Man's Role in Changing the Face of the Earth.* Chicago: University of Chicago Press, 1956.

Tivy, Joy, and Greg O'Hare. *Human Impact on the Ecosystem.* Edinburgh and New York: Oliver & Boyd, 1981.

Tuan, Yi-fu. "Humanistic Geography." *Annals of the Association of American Geographers* 66 (1976): 266–76.

———. *Landscapes of Fear.* New York: Pantheon Books, 1979.

———. *Man and Nature,* Resource Paper No. 10. Washington, D.C.: Association of American Geographers, Commission on College Geography, 1971.

———. *Topophilia, A Study of Environmental Perception, Attitudes and Values.* Englewood Cliffs, N.J.: Prentice-Hall, 1974.

Ursula, Ewald. "The von Thünen Principle and Agricultural Zonation in Colonial Mexico." *Journal of Historical Geography* 3 (1977): 123–33.

Van der Laan, Lambert, and Adries Piersma. "The Image of Man: Paradigmatic Cornerstone in Human Geography." *Annals of the Association of American Geographers* 72 (September 1982): 411–26.

Wagner, Philip L. "Cultural Landscapes and Regions: Aspects of Communication," in H. J. Walker and W. G. Haag, eds., *Man and Cultural Heritage,* pp. 133–42; vol. 5 of *Geoscience and Man.* Baton Rouge: School of Geoscience, Louisiana State University, 1974.

Wagner, Philip L. "The Themes of Cultural Geography Rethought." *Yearbook of the Association of Pacific Coast Geographers* 37 (1975): 7–14.

Wagner, Philip L., and Marvin W. Mikesell. *Readings in Cultural Geography,* pp. 1–24. Chicago: University of Chicago Press, 1962.

Ward, David. "The Victorian Slum: An Enduring Myth?" *Annals of the Association of American Geographers* 66 (June 1976): 323.

Zelinksky, Wilbur. *The Cultural Geography of the United States.* Englewood Cliffs, N.J.: Prentice-Hall, 1973.

The Human Modification of the World

Part 2 traces the way in which people have changed the surface of the earth in creating its cultural geography. Chapter 2 discusses the origins and spread of humans from an apparent initial hearth in Africa. Over time the technological advances of different groups led to the domestication of plants and animals, allowing people to transform the earth's flora and fauna. Chapter 3 examines current life-styles that reflect differences in technology and related use of the environment.

Changing technology also increased the number of people the earth could support. Chapters 4 and 5 discuss issues related to population in cultural geography, including changes in numbers and distribution. These issues are central to cultural geography since they help to explain broad global differences in both land use and life-styles.

Chapters 6 and 7 discuss how variations in life-styles are associated with distinctive cultural landscapes and regions. The broad dichotomy between rural and urban landscapes, life-styles, and environments is the most obvious contrast in cultural regions, but these broad regions contain other equally important differences as well. In combination, these differences help make the geography of each place unique.

43

The Human Occupance of the Earth

Domestication was the invention that made possible the development of pastoral and agricultural economies and therefore made populous and complex human society viable. It has thus proved to be the single most important intervention man has ever made in his environment.   E. ISAAC

THEMES

The changing human ability to modify the environment.
The impact of the Agricultural Revolution on the human use of the earth.
The impact of the Industrial Revolution on cultural geography.
Colonialism and world interrelationships

CONCEPTS

Tripod of culture
Hunting and gathering
Carrying capacity
Domestication
Sedentary
Secondary domesticates
Culture hearths
Industrial Revolution
Colonialism
Neocolonialism
Cultural convergence

The human occupance of the earth and the environmental changes associated with human activities are a central focus of cultural geography. The earliest archaeological evidence for the origin of mankind dates from three to four million years ago and indicates a hearth in East Central Africa. The alteration of the earth (cultural ecology) has been taking place ever since the first human selectively considered the environment as a resource. The growth in the human population and increasing technological ability led to ever greater environmental change, expanded the area of the earth occupied by mankind, and began a process of sequent occupance that is still occurring. The cultural geography of the earth today is the product of cultural ecology, the human use and modification of the earth.

The Human Occupance of the Earth: Origins in Place and Time

An important aspect of the human population is that mankind is adaptable to a variety of environments. Compared to many animals that have specialized to fit into specific *ecological niches,* humans seem to have very crude survival capabilities, yet they are able to survive in nearly all environments. Such survival is generally possible only because humans have adapted their actions in harsh environments or have modified the environment.

One reason humans have been able to modify the environment is their capability for reason, imagination, and creativity. People plan, they make new discoveries, they invent new tools, they discover subtle interrelationships. These skills enable people to change the environment rather than to remain at its mercy. Over time, human ability created the learned behavior we call culture and what we know as civilization (the advanced stage of human activity characterized by organization of society).

Another distinguishing feature of humans is their association with what is called the *tripod of culture.* The tripod of culture involves the ability to make and use tools, the use of symbolic language, and the ability to use fire. The ability to make and use tools allowed humans to modify the environment. Tools increased the ability of early people to use wildlife or vegetation and contributed to the processes of environmental change that have created today's cultural geography.

Symbolic language is of critical importance since it allowed people to accumulate knowledge and to share it with others through symbols. Association of certain sounds and symbols with individual objects allowed communication. Ultimately, symbolic communication included the transmittal of accumulated knowledge and culture from one generation to another.

Fire was one of the most important aspects of the cultural ecology of early humans. Through the use of fire, vegetation was modified, plants and animals could be better used for food, and human migration into harsher environments outside Africa was facilitated.

A number of other human characteristics also contributed to their expanding world role. These characteristics include the ability to walk upright on two legs (*bipedalism*), the lack of specialization (*adaptability*), and the *omnivorous* nature of humans. The ability to walk upright gave our earliest ancestors distinct advantages. Upright individuals could see farther and could use their hands for carrying, grasping, and modifying things. Lack of specialization prevented humans from being limited to a single environment and enabled them to adapt to climatic and other changes repeatedly. Because they are omnivorous, humans are able to use both vegetation and meat for food, expanding the food base upon which they rely.

The Origins of Humans: Out of Africa

There are conflicting views on the origins of mankind: the *evolutionary school,* which attempts to trace humans from ancestral human-like creatures related to the great apes and chimpanzees, and the *creationist school,* which believes humans were an instantaneous creation by God. The scientific community generally accepts the evolutionary school that shows humans evolving over the last two to three million years from early hominoids in Africa. The creationists reject evolution and maintain that human occupancy of the earth is measurable in thousands rather than millions of years.

Evidence indicates that approximately 3.75 million years ago an early hominoid (*Australopithecus afarensis*) walked upright and was clearly less specialized than any of the great apes or chimpanzees (Figure 2–1). Anthropologists have found fossilized footprints, simple stone and bone tools, and fossil evidence for early homonoids in the Olduvai Gorge area of the Serengeti Plains of East Africa. This region is hypothesized as the hearth for these early humanlike animals.

In the same location even greater evidence has been found of the emergence between 1.5 and 2 million years ago of *Homo habilis* and *Homo erectus*. *Homo habilis* seems to have evolved as a variant of *Australopithecus*, who became an evolutionary dead end. *Homo habilis* is the first of the true *homo* genus and is characterized by association with simple stone tools. *Homo erectus*, who used sophisticated stone tools and fire, is evident by 1.5 million years ago. Fire and the more complex stone tools of *Homo erectus* combined with growing numbers to increase the environmental impact of these early homonoids.

Approximately 300,000 years ago, the hominids, *Homo sapiens*, appear in the archaeological record. Changes in hominoids from the earliest *Australopithecus* to modern *Homo sapiens* may have been associated with the recurring cycles of glaciation experienced by the earth. The ability of humans to adapt in the face of such environmental change was a crucial factor contributing to the development of the tool-making skills that ultimately led to *Homo sapiens*.

The earliest evidence of *Homo sapiens* dates from 800,000 B.P. in the Massif Centrale of France, but virtually all remains of *Homo sapiens* are more recent. Some hypothesize that *Homo sapiens* represents the evolutionary progression of a lineage running back to *Homo erectus*, while others postulate an evolutionary change that originated in Africa and pushed north into Asia and ultimately into Europe some 30,000 years ago. Fossil evidence in North America and Australia dating from 20,000 to 30,000 years ago, does indicate that by then *Homo sapiens* (*modern*) had spread to all of the world except Antarctica. The conventional view is that early humans reached

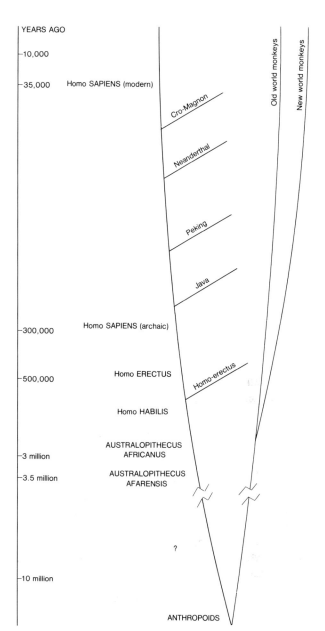

FIGURE 2–1 Evolution of Humans One widely accepted hypothesis about the origin of humans is illustrated by this diagram.

Siberia approximately 30,000 years ago and crossed much later into North America over a land bridge that was exposed as the great ice sheets lowered the sea level. It is hypothesized that similar land bridges would also have allowed *Homo sapiens* to move into Australia.

Ever-increasing intelligence and tool-making ability were associated with the spread of the human population to all parts of the world. Until the end of the Pleistocene era, all humans relied on *hunting and gathering,* or fishing for their existence. Environmental modification creating a cultural landscape was generally limited. Depending on their food preferences, human hunting habits affected certain animal species more than others and gathering practices affected the quantity of certain vegetative species, but widespread destruction of species was probably the exception.

The blade tools from the earliest hominoids to *Homo sapiens* (*modern*) indicate an increase in human efficiency in harvesting the natural resources of the earth. Increased efficiency led to greater population densities, which affected the *carrying capacity* of an individual area. Increased population pressure and better tool technology promoted more efficient harvesting, perhaps destroying some preferred plant or animal species.

The greatest human change in the environment until approximately 10,000 years ago was associated with either purposeful or accidental burning. Repeated burnings in the Pampas of Argentina, the Great Plains of North America, and the savanna lands of Africa may have changed forest or mixed forest and grassy plains into grasslands. The evolution from *Homo habilis* who lacked fire to *Homo erectus* who used fire represented an abrupt change in humans' ability to use the environment and modify the earth's surface. The cultural ecology of these people led to changes in vegetation and animal life as repeated burnings forced forest-dwelling animals into smaller ecological pockets. At the same time the expansion of grasslands led to rapid expansion in the numbers and distribution of grazing animals. Fire also allowed humans to move into colder climates where warmth was essential for survival. Fire was the primary tool of humans, a tool supplemented by crude stone tools that evolved into more sophisticated blade tools.

Despite these developments, the population of the earth remained low as long as people relied on hunting and gathering for their livelihood (see Figure 2–2 and the Cultural Dimensions box on page 50). Ranging over a specific area, the hunters and gatherers of the Pleistocene era harvested plants, animals, and fish at differing seasons of the year, creating a cultural landscape only marginally different from the natural environment. The rate and nature of environmental change were normally slow except where repeated burning caused permanent changes in plant and animal life. Environmental modifications are assumed to have been temporary in nature.

The Agricultural Revolution: Impetus for Change

The rate, nature, and permanency of human modification of the world has increased rapidly in the last 10,000 to 12,000 years, during which time the human population has figuratively exploded, transforming the entire face of the earth. The impetus for this tremendous change in numbers and impact is associated with the *domestication* of plants and animals.

Approximately 10,000 to 12,000 years ago, the population of the earth consisted of small bands of hunters and food gatherers. Within the next 2,000 years, villages and structured civilizations emerged. Within another 2,000 years (by 4,000 B.C.), towns and cities had been established, laying the foundations for city states and empires. Two thousand years before the present, the world was largely inhabited, and human economic activity based on agriculture was common. The domestication of certain plants and animals, technological advances associated with population growth, and the emergence of permanent settlements is known as the *Neolithic Revolution.*

The Neolithic Revolution: Causes and Locations

The causes of the Neolithic Revolution are unclear to scholars who have studied it. Three general theories have been developed to explain its occurrence: these theories can be called the thesis of the alert fisherman, the riverine thesis, and the archaeological record. The *thesis of the alert fisherman* was first developed by the geographer Carl Sauer at the University of California at Berkeley. Sauer hypothesized that peoples en-

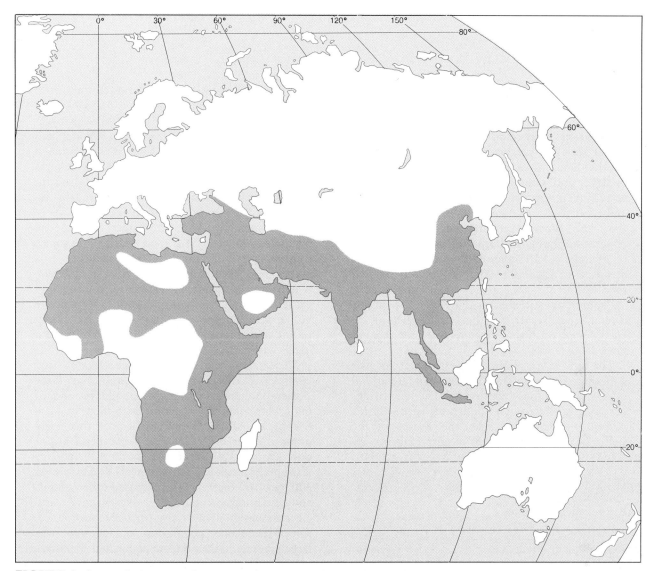

FIGURE 2–2 Population Distribution in Pleistocene The ice sheets of the Pleistocene era limited the area of the world inhabited by humans.

gaged in hunting and gathering must constantly be involved in the process of food gathering. Once such a group developed a system that allowed them to live in their habitat, they had little incentive to change. Sauer suggested that in order for a group to begin to domesticate plants and animals they had to have a food surplus, leisure time, and a tradition of manipulating plants and animals to improve their utility. The constant need for food collection prevented leisure time and advancement of technology or innovation.

Sauer postulated that the warming trend that began at the end of the last Ice Age some 20,000 years ago provided the opportunity for domestication. Inhabitants living at the interface between large rivers of the subtropics and the ocean could obtain a food surplus and had the opportunity to observe rapid evolution of plants as new ecological niches were created by rising sea levels and the alternately flooded and exposed flood plains of rivers carrying the summer water from the glaciers. Sauer believed that fishermen in Southeast

CULTURAL DIMENSIONS
Pleistocene Population

The world's population at the end of the Pleistocene era (about 14,000 years ago) is generally estimated to have been about 10 million total. Calculations of the actual numbers reflect estimates based on presumed carrying capacities of various landmasses and archaeological evidence of dates and rates of migration and economic activity. Theoretical models based on hunting and gathering societies that still exist suggest that the earth's carrying capacity for pre-agricultural human groups was in the range of 1 to 2 people per 10 square kilometers of habitable area or 3 to 5 per 100 square kilometers of total area.

Population estimates for individual continents do not always reach the theoretical carrying capacity level since some regions were occupied at later dates (Table 2–A).

The lowest population is estimated for the Americas, reflecting the hypothesized late arrival of mankind to those continents. The land bridge between northeast Asia and northwest Alaska, currently submerged under the waters of the Bering Strait, was the point of entry. During the last Ice Age, lower ocean levels uncovered this land bridge, and early humans entered the Americas during one (or all) of three periods when geologists believe the land bridge was exposed: 35,000 to 30,000 B.C., 25,000 to 20,000 B.C., and briefly around 10,000 B.C. The small estimated American population of 10,000 years ago reflects the later arrival date. This small population increased rapidly, however, as the abundant resources of the Americas were utilized.

The continents with the largest Pleistocene population were Asia and Africa. Europe's smaller population during the Pleistocene was due to the expansion of the great ice sheets, which limited the area available for hunting and gathering. Highly reliant on hunting, Pleistocene humans in Europe were not only a relatively recent arrival in much of the continent, but faced harsh winters and short, cool summers.

The Pleistocene population in Oceania (the islands of the Pacific, Australia, and New Zealand) inhabited an area where plants and animals were markedly different from those in Asia, Europe, or Africa. It is hypothesized that early humans entered Oceania from Indonesia beginning about 70,000 years ago as the last Ice Age ultimately resulted in ocean levels dropping over 300 feet (100 meters). This exposed other islands and land bridges, making the voyage from Asia to Oceania easier than at any time before or since. The Pleistocene population of Oceania was largely concentrated on the connected landmass of New Guinea and Australia. Adjacent island groups such as the Solomons, New Hebrides, and New Caledonia were also inhabited, but New Zealand and the islands of the Pacific were not occupied until after the Pleistocene era.

TABLE 2–A
Pleistocene Population

Europe	500,000
Asia	5,500,000
Africa	3,500,000
Americas	100,000
Oceania	250,000

Asia with abundant food from the coastal zone and rivers were manipulating vegetation to make nets or extract poison to stun fish, and that the mudflats of the river flood plains were areas where plant invasion could result in rapid evolution.

Sauer's hypothesis is known as the thesis of the "alert fisherman" because fishing provided a reliable and abundant source of protein, supplemented by hunting and gathering. The waste heaps of these individuals included seeds or tubers from plants they had gathered. In the warm climate in the fertile niche created by the wastes, plants could evolve rapidly, creating new varieties. Because these fishing societies had a food surplus, they had leisure time and were more *sedentary* (settled); consequently, they were able to observe plants and recognize new varieties that had desirable attributes. When they recognized a plant with characteristics better than the wild varieties they gathered, they would cultivate and protect it. Once the idea of systematically caring for specific plants had been adopted, the process of domestication was quickly applied to other plants and animals.

The *riverine thesis* is associated with V. Gorden Childe, an archaeologist. Childe also believes that environmental changes associated with the last interglacial period led to plant and animal domestication as North Africa and the Middle East became drier. Increased aridity forced the animals into river valleys, such as the Nile. As the animals the people hunted were concentrated near water sources, the hunters recognized the social nature of the herds. Building on their own social nature, these hunters and gathers began to domesticate some of these wild animals.

The *archaeological record* indicates that plant and animal domestication took place in the foot-hills of the Middle East. Archaeological evidence found here includes tools, permanent settlements, and fossil evidence of some of the earliest plant and animal domesticates. Domestication is believed to have occurred in this region during the last interglacial period. Hunting and gathering peoples were forced into the foothills to find wild grains and grass seeds as the river valleys became too arid to produce sufficient wild food. Proponents of this thesis rely heavily on archaeological evidence found in the *Fertile Crescent*, which extends along the foothills of eastern Iraq and western Iran, north and northwest through Syria and Turkey, and south through Lebanon (Figure 2–3). An abundance of archaeological evidence shows that hunters and gatherers were collecting wild grain in this region. With little change in tools, these groups rather suddenly began growing domesticated varieties. It appears that the residents of this region recognized a hybrid variety that had characteristics that made it more suitable for their use, and began consciously favoring it at the expense of other wild plants and animals (Figure 2–4).

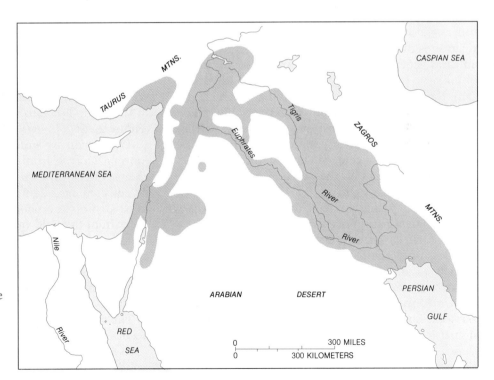

FIGURE 2–3 The Fertile Crescent This broad region of the Middle East is accepted as one of the earliest sites of agriculture.

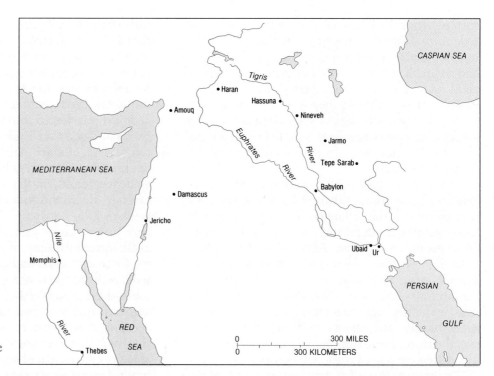

FIGURE 2–4 Early Neolithic Sites in the Middle East Some of the earliest evidence for permanent settlements are found in the Middle East.

Domestication: Changing the World

The human modification of the natural environment made possible by plant and animal domestication is one of the most important elements of cultural geography. From the earliest beginnings of villages, to the cities of the Industrial Revolution, the reliability and surplus of food associated with domestication made *sedentary* life the norm. The central questions associated with plant and animal domestication are when, where, how, and why it occurred, how it diffused and what impact it had on the cultural geography of the world.

The Middle East is one of the earliest hearths where domestication occurred. The occupants of oases, such as Jericho, harvested a wild wheat that was only one of a variety of grasses growing in the area. A new variety of wild wheat may have originated when the original wild wheat crossed with a form of goat grass to produce a hybrid, *Einkorn*. Einkorn was easily recognizable because it had a larger head and produced more wheat than either of its wild ancestors. Over time Einkorn may have hybridized with a second form

of goat grass to produce *Emmer*. Emmer is clearly a domesticated wheat. Whereas Einkorn was plumper and larger than its wild ancestors, it still had many characteristics of wild grasses. If the Einkorn stalk was struck when it was ripe, the seeds separated easily and scattered in the wind. Emmer was larger still and more productive, and when its stalk was struck, the seeds fell directly to the ground or into baskets used for harvesting (Figure 2–5).

The process by which the hybrids Einkorn and Emmer evolved from wild wheat illustrates the changes necessary for domestication. In order to be selected as a domesticate, a plant or animal must exhibit morphological characteristics that differ from the wild varieties and are beneficial to mankind. The domesticate must also rely directly or indirectly on humans for its propagation and be useful and valued by a human population. The hybrid Einkorn clearly meets all of these requirements.

The domestication of wheat in the Middle East occurred between 9,000 and 11,000 years ago. By 7000 B.C. the cultivation and use of the domesticated Einkorn had largely replaced the gathering

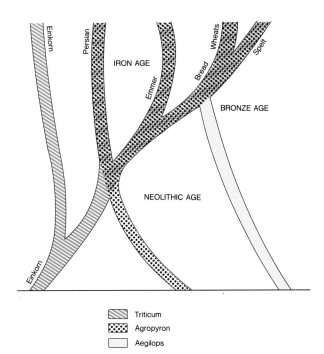

FIGURE 2–5 Diagram of Changes in Wheat The morphological changes in wheat over time illustrate how the process of domestication developed traits in plants and animals desired by humans.

Source: Adapted from Edgar Anderson, *Plants, Man, and Life.* (Little, Brown, Boston: 1952).

of the wild ancestors of wheat. During the same time period, barley was domesticated, providing an alternative crop. Barley became especially important as the idea of domestication diffused beyond the Middle East. Sometime during this same period, domestication of animals began. The earliest animal domesticates included the dog, goats, and sheep. The hearth of domestication for these three animals was in the Middle East also (Figure 2–6).

Western scholars conclude that the spread of grain-producing peoples east and southeast into India and China led to the adoption of rice as a *secondary domesticate* by at least 3000 B.C. Chinese scholars maintain that it was first domesticated in China, but evidence is inconclusive concerning the actual location. Rice replaced wheat because the yield of wheat is less reliable in hot, humid areas where diseases restrict its growth (Table 2–2).

Other secondary domesticates include rye and oats, which were both domesticated about 3,000 years ago in western and northern Europe. Rye and oats will grow in areas where the climate is cool and damp, thus the two crops replaced wheat in areas of northern and western Europe.

The traditional view of the Agricultural Revolution occurring in the Fertile Crescent and then being diffused to the rest of the world is increasingly being challenged. Recent evidence gives dates for domestication between 7,000 and 11,000 years ago in regions distant from one another, and with very different environments; this suggests that the Neolithic Revolution occurred almost simultaneously over a large portion of the world. Early dates for domestication include about 9,000 years ago for northern Nigeria, 8,000 years ago for southeastern Europe, 9,000 years ago for sites in Thailand and highland New Guinea, and 7,000 B.P. for northeastern India, China, and Peru.

This does not eliminate the possibility of diffusion of ideas, but it does suggest that there was not one single center of domestication in the Middle East from which all other domestication attempts derived. Rather it suggests that hunters and gatherers in a variety of places recognized the advantages of domesticated varieties and that awareness of specific domesticates diffused between and among various groups.

Major Plant Domesticates

Millets and sorghums are assumed by some to be secondary domesticates developed through stimulus diffusion. Millets are seed cereals that have a small grain and include several specific species. Sorghums are another group of domesticated grasses that include several varieties. Some sorghum varieties are grown primarily for their sugar content, while others are used for grain. Millets and grain sorghum emerged in the European and Asian periphery of the Middle Eastern hearth of domestication. Millets continue to be a major food crop in areas with harsh environments such as parts of Africa, northern China, the Soviet Union, and India.

Other important domesticates include the legumes, fruits, and vegetables. Legumes are important because they provide a high protein content. The earliest domesticated legumes were peas, which appeared in the Middle East by 6,000 years ago.

The strongest evidence suggests that the initial hearth of domestication for fruit was also the

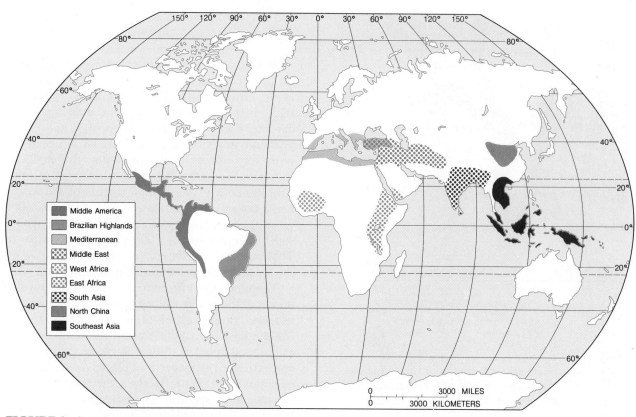

FIGURE 2–6 Culture Hearths of Plant Domestication The major hearths of plant domestication illustrate that domestication occurred in many places.

Middle East. Domestication of grapes (and their use for fermented drinks) is evident by at least 4000 B.C. By 2500 to 3000 B.C., vineyards were widespread in the valleys of the Tigris and Euphrates rivers. Figs, dates, apples, and pears also appear to have been domesticated in the Middle East. All are mentioned in historical documents or are found in the archaeological record by at least 4,000 B.C., although they were probably domesticated earlier. The peach and apricot were originally domesticated in China and spread westward to the Middle East (see Figure 2–6).

An important nonfood domesticate is cotton, which was apparently domesticated for its fibers. Early domestication of cotton in the Old World occurred in India and in Ethiopia, from which it spread to China and to the countries of the Middle East, respectively.

Domestication of plants in the Americas occurred in Middle America and the highlands of South America. The earliest evidence for domestication in the Americas dates from 5500 B.C. The major crops domesticated in the Americas were corn (maize), potatoes (both "Irish" and sweet), peanuts, dry beans (lima, pinto, white, and others), tomatoes, red peppers, cacao (cocoa), squash, pumpkins, tobacco, manioc, and pineapple (see Table 2–1). The earliest indication of maize in the archaeological record of Mexico is between 7,000 and 5,000 years ago. Pollen evidence gives an even earlier date. Other domesticates from Mexico include the common bean, tomato, squashes, and peppers.

The Andes (especially Peru) were the hearth for the domestication of the Irish and sweet potatoes as well as peanuts; all these were root crops that grew well in this area. New World domesticates

TABLE 2–1
Location of Domestication of Common Plants

Mesoamerica	Highland South America	South America	North America	Europe
Avocado	Chile pepper	Cacao	Artichoke	Buckwheat
Common bean	Common pepper	Cassava	Blueberry	Cabbage
Cotton	Cotton	Cotton	Cranberry	Grapes
Maize	Grain amaranth	Pineapple	Sunflower	Oats
Pumpkin	Lima bean	Sweet potato	Tepary bean	Olives
Scarlet runner bean	Quinoa (grain)	Tobacco	Tobacco	Raspberry
Sieva bean	White or "Irish" potato	Peanut		Rye
Squash				Sugarbeet
Sweet potato				
Tomato				

Middle East	Africa	India	Southeast Asia	China and Central Asia	South Pacific
Alfalfa	Coffee	Black Pepper	Banana	Alfalfa	Breadfruit
Apple	Cotton	Citrus	Citrus	Apricot	Coconut
Barley	Cowpea	Cotton	Coconut	Cabbage	Sugarcane
Chick pea	Finger millet	Cucumber	Mango	Cherry	Taro
Date	Muskmelon	Eggplant	Rice	Common millet	
Fig	Watermelon	Mango	Sugarcane	Foxtail millet	
Flax	Yam	Pigeon pea	Yam	Melon	
Grapes		Sesame		Onion	
Lentil				Peach	
Lettuce				Rice	
Onion				Soybean	
Pea					
Pear					
Wheat					

were to be highly significant because Irish potatoes and maize along with wheat and rice provide the bulk of the food consumed by mankind (see the Cultural Dimensions boxes on pages 56–57).

Animal Domesticates
Animal domestication had its origin in the Old World where the dog was first domesticated. Other early animal domesticates found in the Old World include sheep and goats, domesticated in the Fertile Crescent area about 7,000 B.C. The goat was the first herd animal that provided a reliable source of protein and related products through its meat and milk. Sheep were domesticated in the same area as goats, presumably as a source of wool for clothing (Table 2–2).

Present-day domestic cattle are descendants of the ancestral strain *bos primigenius*, the wild urus. The urus originally ranged over a large area of Europe and Asia from the Pacific to the Atlantic, and are believed to have been first domesticated in Europe or the steppes of Russia. The pig is found from the Pacific margins of Asia westward through Europe and northern Africa. The earliest evidence for the domestication of the pig is found in the Middle East in 6000 to 7000 B.C. From there it spread to India, China, and Europe.

The donkey was the earliest domesticate (3000 B.C.) that could effectively carry large loads and be used as a draft animal. It was domesticated in North Africa, where it was used for a variety of purposes. The most important advantage of the donkey was its ability to bear burdens and to survive in drier climates than cattle, which served as beasts of burdens in more humid areas.

CULTURAL DIMENSIONS
Major Foods of the World

It is surprising to find how few foods actually compose the majority of our diets. Wheat, rice, corn (maize), and potatoes are each at least twice as important as the next nearest crop, barley (Table 2—B).

Wheat is the single most important crop produced on the earth. It has specific requirements for optimal production, but can be grown in most middle-latitude climates. It does not do well in hot tropical climates because of a disease called smut, which affects it under damp conditions. In practice, wheat tends to be forced from the best lands because it will produce well in less optimal conditions than crops like corn and soybeans. Wheat is concentrated in the steppe climates of the world and, to a lesser extent, in the humid continental, marine west coast, and Mediterranean climatic zones. As a crop, wheat has unique attributes that make it suitable for human use. It matures relatively quickly; a crop of wheat can be grown in most regions with a growing season of sixty days or longer. A longer growing season will result in higher yields, but wheat will yield even under shorter growing seasons and under conditions of limited moisture. Wheat stores well after harvest and is preferred as a foodstuff because it contains gluten, which allows bread dough to trap the gases generated by yeast, resulting in a lighter bread. Wheat is the only major cereal with this characteristic. It is also useful as an animal feed. Much of the wheat produced in the world is consumed by animals.

CROPS	PRODUCTION	COMMENTS
	(Millions of Metric Tons)	
Wheat	505	Mostly human consumption
Rice	475	Mostly human consumption
Maize (corn)	488	Much consumed by livestock
Potatoes	300	Mostly human consumption
Barley	175	Part fed to livestock
Manioc	100	Local consumption
Soybeans	100	Much fed to livestock
Oats	50	Much fed to livestock
Cane sugar	76	Low nutritive value
Beet sugar	46	Low nutritive value
Citrus	55	High in vitamins
Beans and peas	41	High nutritive value
Peanuts	20	High oil and protein
Rye	33	Much fed to livestock
Bananas	40	Much local consumption
Tomatoes	60	High in water
Millets	25	Mostly human food and beer
Sesame	23	High oil content
Palm oil	10	High caloric value
Sweet potatoes and yams	10	Local consumption
Cocoa	2	Cash crop

continued

Rice needs a longer growing season than wheat and is concentrated in the Mediterranean, subtropical, and tropical environments. The primary advantage of rice as a food grain is that under optimal conditions it will yield more calories per acre than any other grain crop. It is the common food throughout most of the less-developed regions of Asia, Africa, and portions of Latin America.

The potato is an important crop because it will grow in cool, damp climates. A tuber, it is unique among root crops in having a fairly high protein content. It yields a very high tonnage per acre, but much of the potato is simply water and carbohydrates. It is difficult to store and must be kept cool and dry to prevent decay.

Maize (corn) will produce under a wide variety of climatic settings. It does best in a hot, humid environment with fertile soil, and highest yields are obtained in the humid continental climates. Unlike the other major food crops, the majority of corn produced in developed regions is consumed by livestock. It is used in poor areas for human consumption because it has higher yields than most other grains; it also stores well and can be kept for a long period of time without significant deterioration.

The other major food crops of barley, sweet potatoes, manioc (cassava), soybeans, millet, and yams—have varying qualities that make them suitable for either developed or less-developed regions. Barley is produced primarily in climates characterized by drought. It will mature rapidly, so it is grown in the margins of steppe lands, in Mediterranean lands, or in climates where the growing season is short. It is widely grown in the developed regions of the world where it is used as livestock feed. Barley is also used as a survival food in some countries, such as Ethiopia.

Manioc is a tropical tuber that will grow in poor soils; it is concentrated in the tropics and savanna regions. Manioc yields a high volume per land unit, but consists mainly of carbohydrates with little protein or mineral nutrients. It does not store well when removed from the ground,

but does provide food over a long period because it can be left in the ground for years without deteriorating. Manioc is primarily consumed by the farmers who grow it rather than entering the worldwide market exchange system.

Oats are similar to barley in that they will grow in climates with a short growing season; they do best in cool, damp climates. Oats are primarily used for animal feed, and the producing areas are concentrated in the higher latitudes in the developed countries. Sorghum and millet are grown for cereal grain, and some species of sorghum are produced for their sugar content. The grain sorghums and millet serve as emergency food sources in the savanna lands and the margins of desert regions of the underdeveloped countries. Some varieties of millet are known as "hungry millet" because, although yielding a very low volume, their short maturation period enables them to produce some yield even during persistent drought.

Soybeans are an important crop because of their high protein content; they have growth requirements similar to corn. Major producing areas are the United States, Brazil, and China, but much of the high protein soybeans produced are consumed by animals or exported to the wealthy industrial countries. Processed for their high oil content, soybeans form the basis for cooking oils, margarines, and other processed foodstuffs in the industrial countries.

Beans and peas represent a relatively small proportion of total food in the world, but they are of great importance. Beans, peas, lentils, and certain other crops in this family have a high protein content. In less-developed countries these small bean crops are an important source of essential proteins not found in the staple foodstuffs of rice, potatoes, or manioc. Beans and peas store well after harvest and can be used in a variety of dishes. Certain varieties produce not only the bean but also leaves for green salads; some can be eaten before they are mature and are suitable for production of food year-round in areas of long growing seasons.

TABLE 2–2
Location of Domestication of Common Animals

Latin America	North America, Near East and the Mediterranean Regions	Interior Eurasia and Central Asia	South, Southeast Asia, and China
Alpaca	Cat	Bactrian camel	Cat
Dog	Cattle	Cattle	Chicken
Guinea pig	Dog	Goat	Dog
Llama	Donkey	Horse	Duck
Muscovy duck	Dromedary	Reindeer	Peacock
Turkey	Duck	Sheep	Pheasant
	Goat	Yak	Pig
	Goose		Water buffalo
	Guinea fowl		Zebu cattle
	Horse		
	Rabbit		
	Sheep		
	Swine		

The donkey was particularly important for nomadic sheep-herding groups who used it to carry women and children, as well as tents, food, and other material possessions.

The bactrian (two-humped) camel allowed the movement of peoples into even more arid regions of Asia and North Africa; it was probably domesticated about 4,000 or 5,000 years ago in Central Asia. The dromedary (single-humped) camel was domesticated later in North Africa and the Middle East. The dromedary is more tolerant of heat and aridity and is still important in this region.

Evidence for domestication of the horse dates from about 3000 B.C. in the steppes of Russia east of the Caspian Sea. The horse appears to be a secondary domesticate used to perform the labor that had earlier been performed by cattle or bactrian camels. (Figure 2–7).

There are a few animal domesticates in the New World. Most important are the llama, the alpaca, the guinea pig, and the turkey. The llama and alpaca were first domesticated in Peru and are used for carrying burdens and providing milk and fibers.

In terms of human history, the process of domestication appeared rather suddenly and involved only limited species of plants and animals. Through diffusion or independent invention, the

The two-humped camel has been an important beast of burden in certain areas of the Middle East.

idea of domestication (or specific domesticates) spread to all areas of the world. After the voyage of Columbus, diffusion of specific domesticates was rapid. Eastward diffusion of maize, potatoes, peanuts, squash, tobacco, beans, and manioc was paralleled by the westward diffusion of most of the Old World domesticates.

Today's pattern of crop and livestock production reflects the importance of the early domesti-

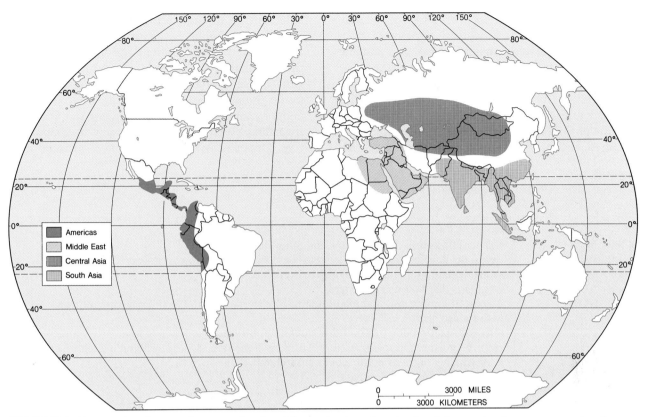

FIGURE 2–7 Hearths of Animal Domestication The major hearths of animal domestication were associated with specific domesticates.

The one-humped camel serves the same purposes as the two-humped camel.

cates. New hybrids of old domesticates and recent domesticates (kiwi fruit, grapefruit, and so forth) have increased the variety of cultivated plants and animals, but the world's occupants still rely on a relatively few crops and animals for the bulk of their food (see Table 2–B on page 56).

The Impact of the Neolithic Revolution: Increasing the Pace of Change

Changes associated with domestication have had a profound impact on the world's cultural geography. The first change was the emergence of permanent villages such as Jericho, or Jarmo, in the Middle East (Figure 2–4). In these villages people with a sedentary life-style were able to develop labor specialization because a food surplus freed some individuals from the daily chores of gathering and preparing food. Populations in hunting and gathering societies are limited since a large area is needed to provide food. Domestication increased the carrying capacity of the land and led to larger populations. Increasing populations, labor specialization, and sedentary life-styles are associated with major technological changes in the centuries that followed domestication.

One of the earliest technological advances that affected the human use of the earth was the substitution of metals for stone, wood, or bone implements (Table 2–3). Not long after the beginning of the first permanent agricultural communities, inhabitants of Middle Eastern villages began to use *native copper*. Native copper is the name given to naturally occurring, relatively pure copper metal found in small quantities in some locations. Relatively soft and easy to shape, native copper provided simple copper vessels, jew-

TABLE 2–3
Dates of Representative Technological and Scientific Advances

5000–4001 B.C.	Egyptian Calendar (360 day year, twelve 30-day months)
4000–3501 B.C.	Copper alloys used by Egyptians and Sumerians.
3500–3001 B.C.	Egyptians develop numbering system. First date in Mayan chronology (3392)
3000–2501 B.C.	Probable date of manufacture of first iron objects.
2500–2001 B.C.	Egyptians discover use of papyrus.
1500–1000 B.C.	Advanced knowledge of shipbuilding in Mediterranean and Scandinavian countries.
1000– 901 B.C.	Earliest use of iron in Greece.
900– 801 B.C.	Iron and steel production in Indo-Caucasian culture.
600– 501 B.C.	First recorded circumnavigation of Africa by Phoenicians—ordered by Pharaoh Necho.
250– 201 B.C.	Eratosthenes suggests earth moves around sun.
271 A.D.	Chinese develop first compass
527 A.D.	First paddle–wheel boats
600 A.D.	Book printing in China
975 A.D.	Arithmetic brought to Europe by Arabs
1000 A.D.	Leif Ericson discovers Nova Scotia
1492 A.D.	Colombus discovers America
1530 A.D.	System for measuring longitude using time suggested
1543 A.D.	Design for Steamboat
1569 A.D.	Mercator: Cosmographia "Map of World for Navigational Use"
1600 A.D.	Telescope invented
1683 A.D.	Newton—Gravity
1718 A.D.	Silk manufacturing machine
1758 A.D.	Ribbing machine for making stockings
1773 A.D.	First iron bridge—Coalbrookdale, Shropshire, England
1782 A.D.	Hot air balloon
1784 A.D.	Threshing machine
1793 A.D.	Cotton gin
1800 A.D.	High pressure steam engine
1803 A.D.	Steamboat
1812 A.D.	Flax spinner
1814 A.D.	First practical steam locomotive
1833 A.D.	Telegraph
1845 A.D.	Machine for combining cotton and wool
1851 A.D.	Sewing machine
1876 A.D.	Telephone
1893 A.D.	Ford builds first car
1903 A.D.	Wright's successfully fly airplane
1914 A.D.	Farm tractor

elry and a few simple tools. A major advance occurred when, in Persia and Afghanistan, the copper ore malachite was heated, allowing relatively abundant amounts of copper to be produced. Copper has a low melting temperature and could be molded, drawn, hammered, or cast. It was made into tools, ornaments, or vessels. Unfortunately, copper is a soft metal unable to retain a sharp edge for cutting. The change in the cultural ecology of groups using smelting techniques to obtain copper marked an important modification in their perception and use of the environment. Until this time, tools had been made from wood, bone, stone, or native copper that only required shaping. The ability to extract metal from ores expanded the resources available to a group.

About 5,000 years ago, a new metal was developed in the Middle East by combining tin and copper. Both tin and copper are soft metals, but when melted together they become an alloy, bronze. Bronze is a relatively hard metal that can be sharpened and will retain an edge. The development of bronze occurred in the Middle East where copper ores that contained tin were heated, creating the natural bronze alloy. The advantages of bronze ensured a demand for more copper and tin than could be produced from naturally occurring ores that combined the two. Discovery of tin in the Danube River basin and on the coast of western Europe led to transportation systems to bring tin to the Middle East. Bronze making diffused rapidly because of bronze's advantages over copper. The use of bronze for artistic purposes reached its peak in China, and numerous outstanding Chinese artifacts of bronze still exist. Bronze was an expensive metal, however, and it is unlikely that the majority of the population had access to it.

In the next great metallurgical achievement, the principle of using fire to alloy metals was applied to iron. It is more difficult to extract iron from iron ore, and the actual development of iron smelting did not occur until about 2500 B.C. The strongest evidence for the earliest smelting of iron indicates a hearth near the Black Sea in eastern Turkey or Armenia.

The development of iron made available a relatively cheap and abundant metal used in a

Bronze artifacts from early Chinese dynasties demonstrate their facility with the use of metal.

variety of products from plows to wheels. Bronze was always limited to the wealthy because of its high cost, but the greater availability of iron ore as compared to tin led to the wide diffusion of iron products. Iron tools greatly increased the pace of environmental change. Iron axes allowed selective cutting of trees instead of wide-scale burning, and the shaping of boards from the trees. With iron tools, houses and ships, temples and aqueducts could be more rapidly constructed. Iron provided metal for plows to expand cultivation into grasslands that could not be cultivated with implements of wood or stone. Iron could also be used for weapons, changing the nature and impact of conflict. Not until iron was alloyed to create steel, did the iron age give way to the modern age of steel.

Perhaps the greatest advance that diffused from the Middle East *culture hearth* was the alphabet. The alphabet was developed in the area of modern Syria and Turkey. From there it diffused rapidly across the Old World and quickly replaced the pictorial forms of writing that had been used previously, including the hieroglyphics of Egypt and the cuneiform system of Sumeria, the two great civilizations of the time in the Middle East. Easily learned, the alphabet was rapidly adaptable to other languages and replaced all but the Chinese pictorial system in the Old World.

Technology and the Industrial Revolution

The technological advances in the centuries following the domestication of plants and animals culminated in the *Industrial Revolution* (also known as the *Technological Revolution*) in the last 200 years. The Industrial Revolution refers to the period in which technological change led to (1) substitution of mechanical devices for human or animal labor, (2) mass production, and (3) the change from agricultural occupations to manufacturing and service occupations. The great increase in the pace of change and the proliferation of technological advances associated with that increased pace epitomize the age of industrialization. To give some idea of the increase in the rate of technological change, of 26,000 patents taken out in England during the 1700s, one-half were in the period 1785–1800.

The basis of the Industrial Revolution was the increased dissemination of knowledge and increased interconnections of the world that began with the Renaissance in Europe. (The Renaissance, from the fourteenth through the sixteenth centuries, was characterized by renewed interest in education, religion, art, literature, music, mathematical science, and geography.) Trade expanded as technological advances in sailing ships allowed more predictable and rapid voyages. Gutenberg's invention of metal type cast in molds, which allowed widespread use of printing and the printing press, was a landmark achievement that enabled information to be preserved

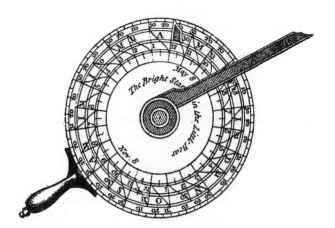

The compass was a major technological advance which revolutionized navigation.

and distributed more easily; by 1482 the first pocketbooks were printed in Italy by Manutius.

The most spectacular results of technological advance were improvements in transportation. The invention of the compass (one of the first achievements of western Europeans not borrowed from the classical civilizations around the Mediterranean) and advances in sailing ship design, which allowed ships to tack into the wind, were among the most important. Mills using wind or water power and clocks with complex gears and levers paved the way for the advances that led to this modern Industrial and Technological Revolution.

Prince Henry the Navigator of Portugal, who was interested in learning about the rest of the world, played a key role in this increase in transportation technology. Calling a council of individuals with information and knowledge about sailing, Prince Henry began a process that culminated in the European *Age of Discovery* during the 1400s. The Age of Discovery is best known for the discovery of the New World by Christopher Columbus, but earlier voyages around Africa were equally important.

As a result of the discovery of the New World, the explanation of the earth-sun relationship was questioned. In 1543, Polish-born Nicolaus Copernicus demonstrated that the sun was at the center of the solar system. Copernicus helped revise the calendar to compensate for the differences between the 24-hour day and 365-day year and the actual rotation of the earth around the sun. Copernicus studied the relationship of the earth to the sun for a long period before he finally published his landmark *De Revolutionibus Orbium Coelestium* ("The Revolution of the Heavenly Orbs") in 1543. Concern over the impact of the then heretical theory of a heliocentric, or sun-centered, rather than earth-centered universe delayed publication of the book till he was nearly seventy.

Nearly forty years later (1582), Pope Gregory XIII introduced the Gregorian calendar now in use throughout most parts of the world. The Gregorian calendar reflects the revisions of Copernicus and other scholars to the Julian calendar introduced during the time of Julius Caesar.

Just twenty-one years after the death of Copernicus, Galileo Galilei was born in Italy. Galileo improved upon an early telescope invented in the Netherlands. Widely used by Venetian trading firms to see ships while they were still far at sea, his improved telescope provided trading firms who used it with a decided advantage. Using longer telescopes, Galileo discovered new planets and concluded that in fact the Copernican model of a sun-centered solar system was correct (Figure 2–8).

Growing trade and colonial relationships required sailing long distances, increasing the need for better navigation equipment. Technological advances based on an understanding of the rota-

tion of the earth around the sun were developed in northern and western Europe. Acceptance of the Copernican view of the universe allowed sailors to find their location north or south of the equator (latitude) by knowing the relationship of the earth to the stars. Isaac Newton contributed to the problem of navigation by calculating the earth's rotational period and developing a mathematical formula for locating distance east or west of a specified point (longitude). Newton understood that since it takes 24 hours for one rotation of the earth, each of the 360° of longitude occupies 4 minutes of time. Sailors could compare *solar noon* (when the sun was directly over the ship) with noon on a clock at a known location (known because a clock on board was set to that time) and then calculate how far east or west of the known location they were. Attempts to develop accurate clocks for determining longitude led to technological advances in clock making that provided the basis for the instruments and machines of the Technological and Industrial Revolution.

The Industrial Revolution: Changing the Human-Land Relationship

Development of a clock that could keep accurate time on a swaying ship seems far removed from the modern world of the late twentieth century, but that achievement was a major factor in the emergence of a new relationship between man-

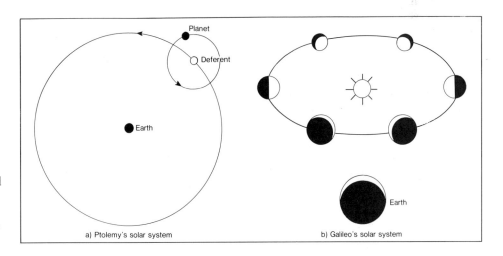

FIGURE 2–8
Reproduction of Galileo's and Ptolemy's Solar System This reproduction of the solar system as envisioned by Ptolemy and Galileo demonstrates the accuracy of their understanding of earth-sun relationships.

kind and the environment. In the middle of the seventeenth century, people's lives were only marginally different from the lives of the Greeks 5,000 years earlier. A relatively few people lived in cities, where even fewer individuals were privileged to be scholars. Limited numbers were involved in trade and commerce, but the vast majority were still engaged in agriculture. The human-land relationship consisted primarily of planting, harvesting, and fallow followed by planting and harvesting once again. The interest in mechanical devices stimulated by attempts to develop a reliable and workable clock led to a transformation in this relationship.

The manufactured items at the time were made by craftsmen: watchmakers, blacksmiths, seamstresses, or tailors. The crafts themselves had been perfected over thousands of years, but still required hand labor. In the Industrial Revolution inanimate power sources were substituted for animate sources. The beginnings of the Industrial Revolution are found in saw and flour mills using water or wind power (Figure 2–9). Then the complex mechanical devices developed from clock making allowed this power to be transformed to a variety of other manufacturing activities.

The first great advances that led to the Industrial Revolution occurred in the textile industry, beginning with the water-powered spinning jenny invented by James Hargreaves. The spinning jenny spun thread so rapidly that each could

FIGURE 2–9 Early Mechanical Devices of the Industrial Revolution These early mechanical devices were designed to allow machinery to substitute for human or animal labor.

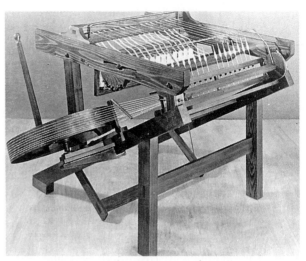

The Spinning Jenny was one of the first inventions of the Industrial Revolution.

supply twelve weavers. In turn, the role of weavers decreased when Richard Arkwright invented a spinning frame powered by water in 1769. Another major advance occurred when James Watt developed a steam engine that replaced water power in the textile industry. By the end of the 1700s, steam engines were being used in mines for lifting and drawing small railroad cars along a track. In 1829 George Stephenson developed a steam locomotive that could move a 13-ton load at 29 miles per hour over a set of rails.

Manufacturing was transformed from individual craftsmen making single items to factories using mass assembly in only fifty years. The use of steam-driven machines to operate textile mills paralleled the development of the railroad for transportation. The dramatic changes in productivity, life-style, and environmental impact associated with the Industrial Revolution are comparable in effect to those of the Neolithic Revolution 10,000 years earlier.

Early mass industrialization was concentrated in the British Isles, reflecting the role of the United Kingdom as the leading manufacturing nation of the time. With numerous colonies producing raw materials and relying upon them for manufactured items, England and Scotland were primary manufacturers of clothing, buckets, barrels, metal implements, wagons, and pots and pans. The impact of the increased productivity of the Industrial Revolution was widespread in En-

The Spinning Frame combined with the Spinning Jenny to allow mass production of textiles.

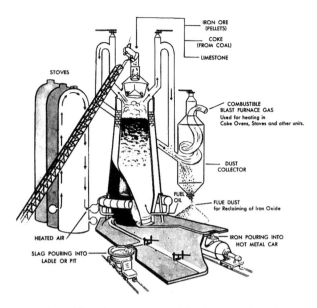

A modern blast furnace is used for large scale refining of iron ore to make pig iron.

gland. Transportation prior to 1750 had consisted of slow wagons, canal boats, or sailing ships, now the steam engine allowed transportation overland at speeds surpassing anything experienced before. The relative ease of laying rails freed manufacturing and other activities from coastal locations, and raw materials could be accumulated to greatly increase the scale of manufacturing.

Mass production signaled the essential demise of the craftsman. Major advances in the technology of making iron and steel greatly increased productivity. Coal largely replaced charcoal for smelting iron ore by the middle of the eighteenth century, greatly expanding the quantity of iron that could be produced. At the end of the eighteenth century, a small amount of high-quality steel was produced. Later, the invention of the Bessemer furnace (1856) and the open hearth furnace (1857) provided the means for making large quantities of high-quality steel (Figure 2–10).

The Industrial Revolution changed the landscape and the pattern of population distribution in its hearth area in the British Isles and in Western Europe and North America where it diffused. The great productivity of industrial plants led to the development of a society based on mass production and consumption. The larger scale of manufacturing associated with the factory system led to concentration of production in

FIGURE 2–10 The Bessemer Furnace The Bessemer and Open Hearth Furnaces revolutionized iron and steel production by vastly increasing the scale of production.

large factories. Large numbers of people moved to the cities to provide labor for the newly emerging industries, increasing both the number and the size of urban areas.

Prior to the Industrial Revolution, the cities of the world were relatively small. Their size was dependent upon their ability to obtain food and to produce the manufactured handicrafts required by their inhabitants. London had only 150,000 inhabitants in 1600, yet was one of the largest cities in the world. Bristol, the second largest city in England in 1600, had only about 20,000 people.

Few of the world's settlements had populations of more than 10,000 people. The majority of the population, except for royalty, the privileged elite, and the craftsmen, lived in abject poverty. The growth of cities during the Industrial Revolution is illustrated by London, which reached 865,000 in 1800, and 2,362,000 in 1850. The growth of other industrial cities was equally impressive. Thus the Industrial Revolution transformed economies from agricultural systems derived from the Neolithic Revolution to urban-based systems (Figure 2–11).

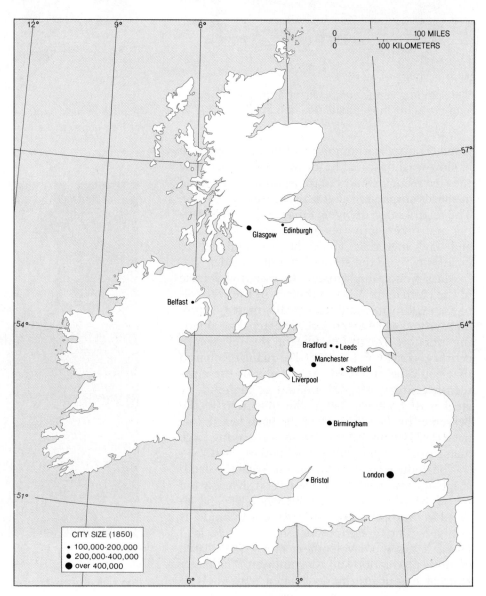

FIGURE 2–11 Location and Site of Early Cities during the Industrial Revolution

CITY SIZE (1850)
• 100,000–200,000
● 200,000–400,000
⬤ over 400,000

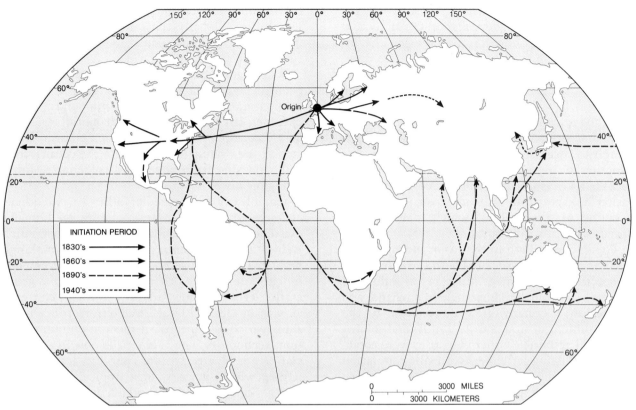

FIGURE 2–12 Map Showing Diffusion through Time of the Industrial Revolution Diffusion of the industrial revolution spread its environmental impact to much of the world.

A massive increase in the circulation and movement of goods and people was associated with the increasing city size and industrial productivity. Prior to the Industrial Revolution, travel required large amounts of time and money. Figuratively speaking, the Industrial Revolution made the earth smaller by providing an ever-expanding transportation network based on the railroad and steam ships. The Industrial Revolution was also responsible for a social revolution that changed the nature and quality of life. The life of the laborer in the textile and steel mills of the Industrial Revolution was harsh by today's standard, even worse than the life of rural farmers and their families. Over time, however, the demands of society fostered laws that improved social conditions in industrial centers. Limits on hours and conditions of work, improved salaries, and advances in medicine, education, and the general society were associated with the emer-

gence of the urban middle class in Europe and North America.

The increased productivity of the Industrial Revolution benefited the entire United Kingdom. The increased exports and associated wealth of the Industrial Revolution fueled greater industrialization in England and financed expansion of its colonial empire. The success of the Industrial Revolution in improving England's place on the world stage led other countries to industrialize. Diffusion of the ideas and technology of the Industrial Revolution involved elements of both expansion and relocation diffusion.

The countries closest to the British Isles were the first to adopt the ideas of the Industrial Revolution through expansion and diffusion. Industrialization spread from England to Belgium and then to northern France and Germany (Figure 2–12). The Industrial Revolution also diffused to the northeast coast of the United States, an example of relocation diffusion. These five areas (United Kingdom, Belgium, France, Germany, and the northeastern United States) became the industrial core lands of the world by the late nineteenth century.

In the twentieth century, the ideas of the Industrial Revolution diffused along the longer trade and transportation routes to western Russia, Poland, Czechoslovakia, northern Italy, and ultimately to Japan, Australia, and New Zealand. The adoption of the Industrial Revolution and its technology was facilitated by increased transportation linkages from railroad expansion and the adoption of steam ships, which increased the speed of transoceanic voyages.

Population growth in the major industrial cities of the world is associated with the emergence of specialized settlement functions in industrial centers. Cities such as Birmingham, England, and Dusseldorf, Germany, grew in response to the establishment of iron and steel mills and related industrial activities. Other industries were established in existing settlements, causing population growth. By the third decade of the twentieth century, more people lived in the cities than on farms or in rural villages in the industrialized countries (Figure 2–13). The growth of the cities of the industrialized countries combined with the manufacturing role of the industrialized core of the world to foster increased international trade and interdependence.

Greatly improved knowledge of the nutritional requirements of our species and new practices in public health and medicine, including the pharmaceutical revolution, have resulted in a remarkable change in population dynamics in the modern industrial phase. As a consequence, the human population is now doubling at the rate of once every 35 to 40 years, in contrast to the overall rate of once in 1,500 years during the early farming and early urban phases.

Another outstanding feature of the modern industrial phase, and one that determines many of its other characteristics, has been the introduction on a massive scale of machines and manufacturing processes powered by inanimate energy. In highly industrialized countries like the United States, the amount of power used on a per capita basis is now thirty times greater than it was prior

to the diffusion of the ideas of the Industrial Revolution. Globally, the amount of energy flowing through human society is doubling about twice as fast as the population.

A significant change of a qualitative nature has been the synthesis, often in very large amounts, of thousands of new chemical compounds, many of which have potent effects on biotic systems. Vast quantities of these synthetic chemical substances are finding their way into the oceans, the soil, and the atmosphere and, of course, into living organisms.

Complexity and unpredictability are two other characteristics of the present phase of human existence. Modern technology and the economic web of interdependence of countries are making problems of land use and resource management much more complex than they were even a generation ago. Machines are now available, for example, for clearing large areas of forest in a very short period of time. With modern communications technology, a decision made now in a financial center in a temperate zone can lead to the felling, within a week, of a tropical forest tree 10,000 kilometers away, which has taken one hundred years to grow.

If certain widely accepted predictions about future increases in the level of energy use turn out to be correct, by the year 2050 mankind will be using about as much energy as is used by all other animals and plants put together. One does not have to be a specialist to see that this kind of growth in energy use, and the concomitant increase in the use of resources and the outpouring of wastes, cannot go on indefinitely.

Industrialization in Great Britain and Western Europe was associated with colonial empires. The European country provided manufactured goods for the colonies and extracted food products or raw materials for industrial use at home. The reliance of England and Western Europe on colonies for grain, tropical products such as coffee or bananas, and markets for manufactured items such as transportation equipment, textiles or other goods created interdependence. The less industrialized countries of Latin America, Africa, and Asia relied on Europe's technical and manufacturing expertise, Europe, and to a lesser extent North America, obtained raw materials from the

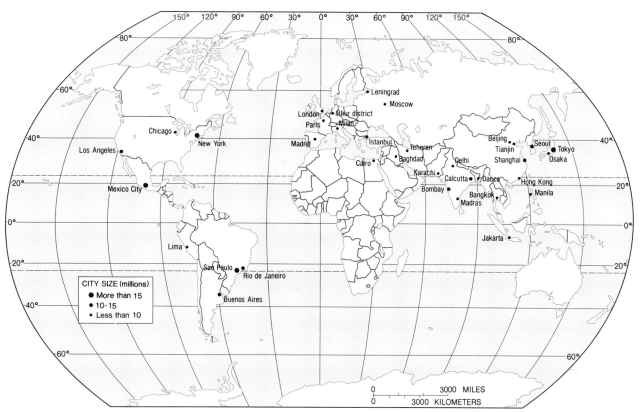

FIGURE 2–13 City Distribution in 20th Century The distribution of the world's largest cities at the present time is related to their role as centers for industrial and commercial activity.

less industrialized countries. This relationship was asymmetrical since the European and North American countries manufactured high-value goods and received low-value raw materials in return, thus creating an interdependence that favored the industrialized nations at the expense of the less industrialized nations (Figure 2–14).

The interrelationships developed in the early decades of the twentieth century changed dramatically after World War II. European colonial empires were destroyed as the individual countries of Asia and Africa gained their independence. Politically, these countries are now self-governed, but many still rely on Europe for manufactured goods and high technology items. This continued reliance on former colonial relationships, which effectively limits the economic development of a former colony, creates *neocolonialism.*

The real, or perceived, continued economic domination by European nations through neoco-lonialism remains a major issue in many countries. The result is an uneasy dichotomy in which Western influence remains important, but some of the individuals and governments in the newly independent countries are hostile toward former European colonial powers and the United States, which are perceived as neocolonial powers because of their global economic and political roles.

The distribution of industrial production in existence at the end of World War II is changing rapidly. Some less industrialized countries of Asia, and to a lesser extent Africa and Latin America, have adopted the ideas and technology of the Industrial Revolution and are now competing with nations of the industrial core who are often their former colonial masters. Increasing wage costs in Europe caused the earliest industries of the Industrial Revolution (textiles, boots and shoes, clothing) to move to the developing countries in the 1960s and 1970s. Cameras, tele-

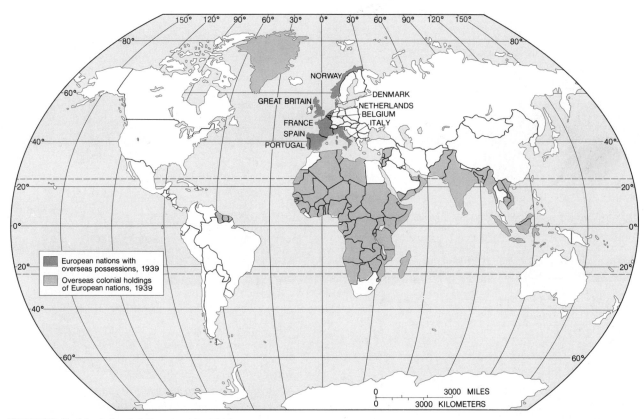

FIGURE 2–14 Colonial Empires in 1939

visions and electronic goods, and automobiles were being mass-produced in countries like South Korea, Taiwan, China, Hong Kong, Singapore, and Brazil by the 1980s. Iron and steel, the basis of the European Industrial Revolution, is increasingly produced in newly industrialized countries such as South Korea and Brazil.

Instead of the simple interconnections in which the less industrialized colonies provided raw materials and the industrialized countries provided manufactured goods, in the late twentieth century a system has emerged in which the industrialized countries produce high technology items (such as computers or sophisticated transportation systems), while the developing countries in the early stages of industrialization concentrate on those items that require a high labor element and a lower level of technology (Figure 2–15).

The changing nature of the international interconnections of the twentieth century represents an important change in the world's cultural geography. Imperialistic domination by European powers was almost completely obliterated in only fifty years as nations of Africa and Asia became independent. The results of European domination, however, are evident in the cultural geography of former colonies in place names, political organization, laws, language, religion, and commercial activity. European influence is reinforced by the process of "modernization" in which less industrialized countries consciously adopt elements of Western culture. Modernization efforts of these countries emphasize technology for industry and agriculture or transportation and communication systems, but also include such diverse elements as Western-style dress and fast food outlets.

The diffusion and dominance of Western culture has led to *cultural convergence* throughout the world. Examples of cultural convergence range from the diffusion of selected culture ele-

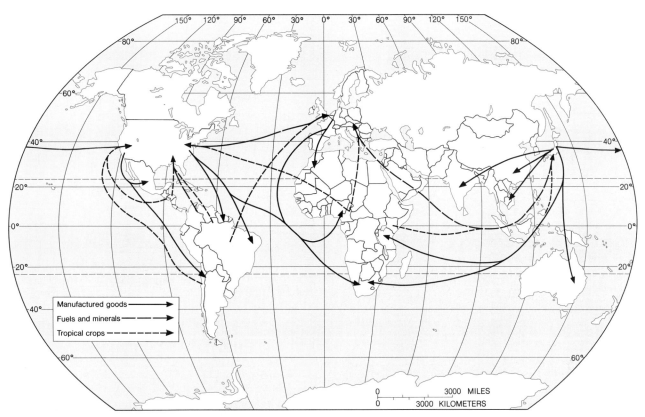

FIGURE 2–15 Map of Commodities and Manufacturing Flow The modern world is linked together by the flow of raw materials and manufactured goods.

The diffusion of western firms is one of the forces leading to cultural convergence.

ments, such as dress styles, language, and technology, to the architectural styles evident in the downtown section of nearly any large city in the world and the presence of Western firms such as MacDonald's Restaurants, General Motors, and the Nestlé Corporation of Switzerland.

Transforming the Agriculture of the Neolithic Revolution

The dramatic changes that occurred in agriculture during the Industrial Revolution are known as the Second Agricultural Revolution. Prior to the Industrial Revolution agriculture in the world was little different from what it had been thousands of years before. Full-grown cattle in Europe in A.D. 1500, for example, averaged only 600 pounds while full-grown sheep averaged only 30–40 pounds. Grains were simply scattered on the

FEUDAL SYSTEM

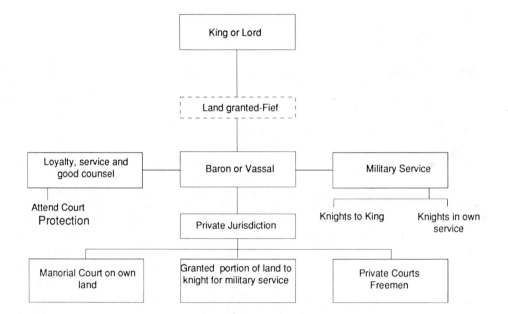

FIGURE 2–16 Diagram of Feudal Social System The roles of individuals related to their position with respect to the monarch. The majority of the population practiced primitive farming on the king's or baron's land.

surface of the soil and then harrowed into the ground, and crop rotation was unknown. Fields were defined as either outfield or infield, with the infield cultivated yearly because of its location near the settlement, while the outfields were cultivated intermittently, allowing alternate fallow periods (Figure 2–16).

The Second Agricultural Revolution had its origin in the Netherlands, Belgium, and England, where selection of better varieties of plants and animals led to higher yields, greater size, and increased productivity. Crop rotation, which involved rotating a legume (timothy) with a root crop (turnips) and grains, further increased yields. The resultant increases in the size of livestock and reliability of the food supply, especially during the winter, allowed urban populations to increase in response to the labor demands of the Industrial Revolution.

During the next century, the use of machines in agriculture in industrial countries allowed fewer farmers to cultivate more land. Combined with advances in plant and animal breeding and the use of chemical fertilizers, mechanization

allowed ever-greater productivity in agriculture. Transportation improvements enabled agricultural products to be moved greater distances to market, facilitating the growth of population in the industrial centers. The result was a pattern of human-land relationships in the industrial countries in which a relatively small farming class provided the basic food stuffs for the majority, who resided in urban areas (Figure 2–17 on page 74).

While the Industrial Revolution and Second Agricultural Revolution were simultaneously changing the landscapes of urban and rural areas in the industrial world, the landscapes of the less industrialized countries of Africa, Asia, and Latin America changed more slowly. In an excellent example of hierarchical diffusion, the technological advances and changes associated with the Industrial Revolution and Western culture spread first to the larger cities of Africa, Asia, and Latin America, while the rural areas were slower to adopt the new ideas, either because they had less contact with the industrial world or, because they had stronger cultural or economic biases against accepting innovations.

CULTURAL DIMENSIONS
The World's Agricultural Resource Base

The land necessary to produce food to support tomorrow's children or to raise the level of living of today's residents is a finite resource. The total land area of the globe (excluding Greenland and Antarctica) is 32.1 billion acres (13 billion hectares). Physical geographical problems of climate, landforms, or soil fertility limit the usefulness of most of this land for crop cultivation. Approximately 6.4 billion acres (2.6 billion hectares) have frost-free seasons of less than 3 months, which effectively prohibits production of most crops. An additional 4.7 billion acres (1.9 billion hectares) have 3 months or less of adequate precipitation to produce crops. The remaining 21 billion acres (8.5 billion hectares) receive sufficient moisture and have an adequate growing season for producing crops, but may have other physical limitations. Mountainous and arid regions with stony or shallow soils comprise 6.4 billion acres (2.6 billion hectares). Deserts total 4.2 billion (1.7 billion hectares), and unusable sandy soils total another 1.7 billion acres (0.7 billion hectares); soils found in tropical areas include 3.46 billion acres (1.4 billion hectares), much of which is unsuitable for cultivation due to soil fertility, waterlogging, or other physical problems (Figure 2–A).

About 24 percent of the earth's land surface excluding Antarctica and Greenland's ice caps can produce crops without irrigation (7.8 billion acres; 3.15 billion hectares) An additional 8 percent of the earth's surface is suitable for crop production with irrigation, providing a total of 10.45 billion acres (4.2 billion hectares) that is suitable for crop cultivation. This 10.45 billion acres is the currently accepted maximum possible limit for cultivation if massive investments of capital are utilized. At present we rely on approximately 3.4 billion acres (1.37 billion hectares) of land to produce food. Perhaps the next great agricultural revolution will be the developments that allow expansion into the two-thirds of the potentially cultivatable land that is now unused.

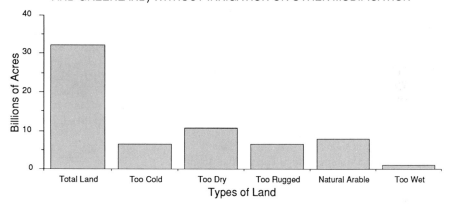

SUITABILITY FOR CULTIVATION OF EARTH'S LAND AREA (EXCLUDING ANTARCTICA AND GREENLAND) WITHOUT IRRIGATION OR OTHER MODIFICATION

FIGURE 2–A Graph of Agricultural Data A graph showing the agricultural resources of the world demonstrates how little of the earth is actually used for crop production.

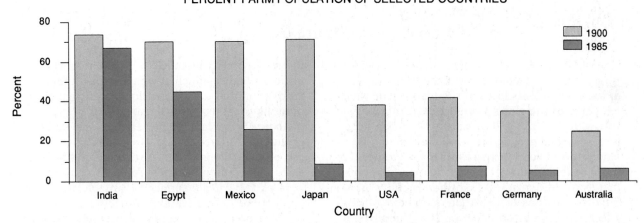

FIGURE 2–17 Graph of Changes in Farm Population in Selected Countries Farm populations are declining in most countries.

Conclusion

The complex mosaic of the world's cultural geography reflects the differential diffusion of ideas, technology, and cultural traits of the last ten millennia. The transformation of the natural environment that began with the first humans rep-resents the changing ability of mankind to use and modify the world. The changes associated with the invention and diffusion of fire, plant and animal domestication, and the scientific and technological advances associated with the Industrial Revolution help explain the culture, cultural landscapes, cultural regions, cultural ecology, and culture history of the world.

QUESTIONS

1. Discuss how fire allowed humans to modify the environment.
2. Define the tripod of culture, and explain how it relates to the human use of the earth.
3. Is there any relationship between climate and the diffusion of humans from Africa? Explain.
4. Describe the three general theories that attempt to explain the hearths of the Neolithic Revolution.
5. What is the Fertile Crescent, and what was its role as a hearth from which diffusion occurred?
6. Discuss the kind of changes that occurred in the world because of domestication.
7. What are secondary domesticates? Give some examples.
8. Identify the major plant and animal domesticates for any region.
9. List at least three major technological advances following the Neolithic Revolution and discuss their role in changing the natural environment.
10. What was the most important change in the human–land relationship that resulted from the Industrial Revolution? Why?

SUGGESTED READINGS

Adams, R. M. *Heartland of Cities: Surveys of Ancient Settle-ment and Land Use on the Central Floodplain of the Euphrates.* Chicago: University of Chicago Press, 1981.

———. "Strategies of Maximization, Stability, and Resilience in Mesopotamian Society, Settlement, and Agriculture." *Proceedings of the American Philosophical Society* 122 (1978): pp. 229–35.

Adams, R. M., and H. J. Nissen. *The Uruk Countryside: The Natural Setting of Urban Socieites.* Chicago: University of Chicago Press, 1972.

Ashton, Thomas S. *The Industrial Revolution*. New York: Oxford University Press, 1964.

———. *The Industrial Revolution: 1760–1830*. London: Oxford University Press, 1948.

Bobek, H., "The Main Stages in Socioeconomic Evolution from a Geographic Point of View," in P. L. Wagner and M. W. Mikesell, eds. *Readings in Cultural Geography*. Chicago: Unviersity of Chicago Press, 1962.

Braudel, F. *Civilization and Capitalism, 15th–18th Century,* vol. 1, *The Structures of Everyday Life*. Translated by S. Reynolds. New York: Harper & Row, 1981.

———. *Civilization and Capitalism, 15th–18th Century*, vol. 2, *The Wheels of Commerce*. New York: Harper & Row, 1982.

Bronowski, J. *The Ascent of Man*. Boston: Little, Brown, 1973.

Brotchie, John F., Peter Hall, and Peter W. Newton, eds. *The Spatial Impact of Technological Change*. London: Croom Helm, 1987.

Butzer, K. W. *Early Hydraulic Civilization in Egypt*. Chicago: University of Chicago press, 1976.

———. *Archaeology as Human Ecology*. Cambridge: Cambridge University Press, 1982.

Carter, George F. *Earlier Than You Think: A Personal View of Man in America*. College Station, Tex.: Texas A & M University Press, 1980.

Cohen, Mark Nathan. *The Food Crisis in Prehistory: Overpopulation and the Origins of Agriculture*. New Haven, Conn.: Yale University Press, 1977.

Darby, H. C., ed. *A New Historical Geography of England*. Cambridge: Cambridge University Press, 1973.

Deane, Phyllis. *The First Industrial Revolution,* 2d ed. New York: Cambridge University Press, 1979.

Denevan, W. M. *The Native Population of the Americas in 1492*. Madison, Wis.: University of Wisconsin Press, 1976.

Duby, G. *Rural Economy and Country Life in Medieval Europe*. Colombia S.C.: University of South Carolina Press, 1968.

Gregory, D. *Regional Transformation and Industrial Revolution*. Berkeley, Calif.: University of California Press, 1982.

Grigg, D. B. *The Agricultural Systems of the World: An Evolutionary Appraoch*. New York: Cambridge University Press, 1974.

Harris, D. R. "Alternative Pathways Towards Agriculture," in C. A. Reed, ed., *Origins of Agriculture*. The Hague: Mouton, 1977.

———. ed. *Human Ecology in Savanna Environments*. New York: Academic Press, 1980.

Hindle, Brooke, and Steven Luban. *Engines of Change: The American Industrial Revolution, 1790–1860*. Washington, D.C.: Smithsonian Institution Press, 1986.

Hole, F., and K. V. Flannery. "The Prehistory of South-Western Iran: A Preliminary Report." *Proceedings of Prehistoric Society* 33 (1960).

Hole, Frank, Kent V. Flannery, and James A. Neeley. *Prehistory and Human Ecology of the Deh Furan Plain: An Early Village Sequence from Khirzistan, Iran*. Ann Arbor, Mich.: University of Michigan, 1969.

Isaac, E. *Geography of Domestication*. Englewood Cliffs, N.J.: Prentice-Hall, 1970.

Kramer, S. N. *The Sumerians*. Chicago: University of Chicago Press, 1963.

Lamb, H. H. *Climate, History and the Modern World*. London: Methuen, 1982.

Langer, W. L. "The Black Death." *Scientific American* 231 (1964): 114–21.

Lee, Richard B., and Irven DeVore, eds. *Man the Hunter*. Chicago: Aldine Publishing Co., 1968.

Pohl, Mary, ed. *Prehistoric Lowland Maya Environment and Subsistence Economy*. Cambridge, Mass.: Harvard University Press, 1985.

Pounds, N. J. G. *An Historical Geography of Europe, 1500 to 1840*. Cambridge: Cambridge University Press, 1979.

———. "The Urbanization of the Classical World." *Annals of the Association of American Geographers* 59 (1969): 135–50.

Pred, A. *Urban Growth and City Systems in the United States*. Cambridge, Mass.: Harvard University Press, 1980.

Rorig, F. *The Medieval Town*. Berkeley, Calif.: University of California Press, 1971.

Sauer, C. O. *Agricultural Origins and Dispersals*. New York: American Geographical Society, 1952.

Siemens, A. H. "Wetland Agriculture in Pre-Hispanic Mesoamerica." *Geographical Review* 73 (1983): 166–81.

Turner, B. L., R. Q. Hanham, and A. V. Portarero. "Population Pressure and Agricultural Intensity." *Annals of the Association of American Geographers* 67 (1977): 384–96.

Ucko, P. U., and G. W. Dimbleby, eds. *The Domestication and Exploitation of Plants and Animals*. London: Duckworth, 1971.

Ucko, P. U., R. Tringham, and G. W. Dimbleby, eds. *Man, Settlement, and Urbanism*. London: Duckworth, 1972.

Vance, J. E., Jr. *This Scene of Man: The Role and Structure of the City in the Geography of Western Civilization*. New York: Harper & Row, 1977.

Vavilov, Nikolai I. *The Origin, Variation, Immunity and Breeding of Cultivated Plants*. Translated by K. Starr Chester. *Chronica Botanica* 13 (1951): 20–43.

Wheatley, P. *The Pivot of the Four Quarters*. Chicago: Aldine, 1971.

White, L. T., Jr. *Medieval Technologoy and Social Change*. Oxford: Clarendon Press, 1962.

Cultural Ecology: Variations in the Human Use of the Earth

The living conditions of individuals in a particular society are governed by the interaction of their work with natural processes to create a distinctively modified environment which reflects their culture, social organization, and techniques, their relations with certain wild or domestic organisms, and the character of the land itself.   P. WAGNER

THEMES

Types of agricultural systems and their cultural ecology.
Regionalization of the production of major crops.
Division of the world by level of economic development.
The relationship of carrying capacity to cultural ecology.
The impact of the Modern Agricultural Revolution.

CONCEPTS

Subsistence farmers
Shifting agriculture
Intercropping
Collective ownership
Horizontal nomadism
Vertical nomadism
Transhumance
Desertification
Peasant society
Periodic market
Sawah
Exotic streams
Green Revolution
Dual societies
Monoculture
Gross national product

Cultural geography is useful as a tool in delimiting regions based on the cultural ecology of individual groups. The level of technology and industrialization of a group provides important insights concerning their use and abuse of the environment. Economic development is also related to the group's standard of living, rates of population growth, and political and military significance. The cultural ecology represented by environmental change, economic systems, population, and society ranges from pre-Neolithic to modern industrial groups (Figure 3–1).

Isolated Peoples

The human use of the earth has changed little over thousands of years in areas with limited interaction with the rest of the world. The least technologically advanced peoples still engage in hunting and gathering as a way of life. These groups are characterized by stone tools and lack of domesticated plants and animals. Such peoples are found in isolated areas where distance, harsh climate, or rugged topography may minimize their interaction with more technologically advanced groups. A few hunting and gathering populations remain in the Amazon Basin of South America, the Congo Basin of Africa, and isolated areas of the mountainous regions of Southeast Asia (Figure 3–2). Other hunting and gathering populations inhabit the desert and savanna lands

The tools and technology of hunters and gatherers in the Amazon rain forest are relatively simple.

of Australia and Southwest Africa. The total population of hunting and gathering groups can only be estimated. Their global numbers probably total between 300,000 and 500,000.

Hunting and gathering societies depend on wild plants and animals. Some rely heavily on one or two basic plants or animals for their food; for example, seals and fish were the primary foodstuffs for the Inuit of Alaska and Canada until contact with more technologically advanced groups destroyed their traditional hunting and gathering life-style. Indian tribes of the Amazon Basin are unspecialized, hunting and gathering a wide variety of plants and animals. Limited trade

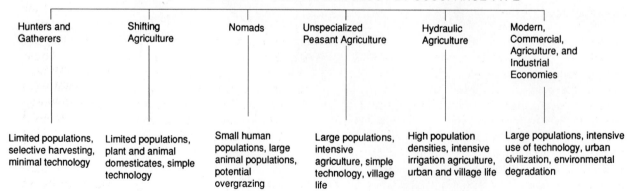

SELECTED ELEMENTS OF CULTURAL ECOLOGY BY OCCUPANCE TYPE

Hunters and Gatherers	Shifting Agriculture	Nomads	Unspecialized Peasant Agriculture	Hydraulic Agriculture	Modern, Commercial, Agriculture, and Industrial Economies
Limited populations, selective harvesting, minimal technology	Limited populations, plant and animal domesticates, simple technology	Small human populations, large animal populations, potential overgrazing	Large populations, intensive agriculture, simple technology, village life	High population densities, intensive irrigation agriculture, urban and village life	Large populations, intensive use of technology, urban civilization, environmental degradation

FIGURE 3–1 The Relationship Between Cultural Ecology and Occupance Type

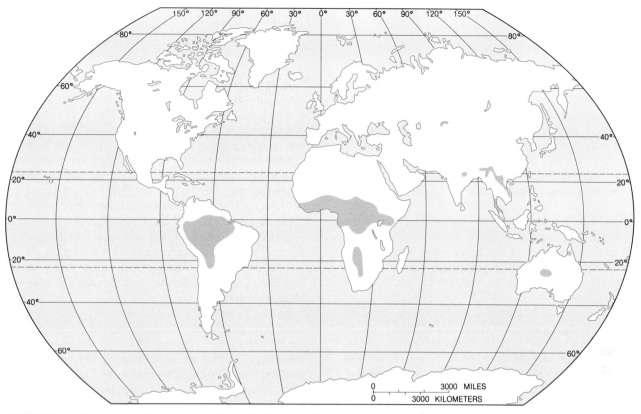

FIGURE 3–2 Location of Hunting and Gathering Peoples Hunting and gathering peoples today occupy areas with harsh environmental settings.

provides metal tools and other items, but their life-style is in stark contrast to the industrial world. Thus, when valuable natural resources or other commodities are discovered on their lands, the small populations of hunting and gathering peoples are at the mercy of more technologically advanced populations.

Increasing global interaction has brought elements from more technologically advanced regions to the hunting and gathering societies that survived into the twentieth century. The result is evident in such peoples as the Indian tribes of the Amazon. They still hunt and gather a variety of wild plants and animals, but now they use implements like steel knives, hatchets, and occasionally even shotguns adopted from neighboring groups with a higher level of technology (see the Cultural Dimensions box on page 80).

An important global issue for humankind is how to prevent hunting and gathering societies

from being completely eliminated by peoples with more advanced technology. Too often the effect of contacts between groups at different technological levels has been to alter or destroy the cultural mores and values of the primitive cultural group, without replacing them with a valid system within the technologically advanced level. Groups to whom this has happened include the Inuit, American Indian peoples, Australian aborigines, and the isolated peoples of the tropical rain forests of the Amazon and Congo. Those members of a hunting and gathering society who adopt the values and life-style of the group with advanced technology generally become alienated from both societies. Unwilling to live at the level of their traditional society, they are often rejected by the group with the advanced technology and become strangers in their own lands. Such individuals and groups face an uncertain future as they try to determine their place in life. Some

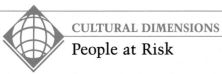

CULTURAL DIMENSIONS

People at Risk

The Penan of Sarawak, nomadic jungle dwellers of Borneo's rain forest, are one of the indigenous groups on this the largest island in Southeast Asia.* They are having to adjust rapidly to a new and very different life.

As recently as ten years ago, a typical Penan family roamed the rain forest, staying in one spot only as long as game and wild vegetation were abundant. They built shelters of palm leaves and bamboo bound with rattan and ate the core of the wild sago palm as their staple starch. They slashed the bark of the tacem tree, collected its white sap, and made a poison for their blowpipe darts strong enough to kill a wild pig in three minutes.

Now Mulu National Park has been created, incorporating much of the forest. Today the government encourages the Penan to settle outside, rather than wander in the park. To foster this, the government gave the people roofing material, seed, and basic instructions on planting and cultivating hill rice and tapioca (cassava).

Next door to the Penan settlement is the park where they may find jobs as laborers or porters. They can sell their crafts to tourists. Most important, they can hunt in the park—blowpipes only—and in the process retain many of their traditional values.

In the upper reaches of another river basin in Sarawak, another group of Penan are adjusting to harsher realities. They have no park or refuge since the area is being logged. The once-clear river is murky with runoff. On one side of the river, machinery plows thin tropical topsoil, making or widening access roads. On the other side are two small shelters, built from milled lumber and roofed with corrugated sheet metal for the native people. The people say they did not know of plans to fell trees on their land until the lumber company sent word asking for employees.

The Penan came here not so much for employment as to claim compensation for the company's use of their land. Meanwhile, some work as laborers. Others trade their handicrafts for supplies and food.

At the moment, wild plants such as the sago palm are abundant. Hunting, however, has already decreased. Fish in the Wat River, once abundantly varied, are now limited to two species adapted to murky waters: catfish and carp. If the pace continues, Sarawak's rain forests may be logged out in as little as twenty years.

The Penan people are at risk of losing their culture and life-style, just as North American Indians and other native peoples have done. Pressures from groups that want to utilize the resources of the land inhabited by the Penan threaten their very existence. Unless the government intervenes, the rain forest, the resource base of the Penan, will be destroyed and with it, their way of life.

*Adapted from the Christian Science Monitor, April 11, 1989, p. 14.

countries, such as Brazil, have adopted programs to protect hunting and gathering groups, but with limited success.

Simple Farming Systems

A second broad type of cultural ecology relies on farming systems only marginally changed since the earliest agriculture. Such groups are classified as *subsistence farmers.* Subsistence farmers produce crops or livestock primarily for their own consumption. Surplus production is incidental, as the system is designed to ensure only that the farmers have adequate food for their own consumption from year to year. Subsistence farmers may produce a crop or livestock for commercial purposes, but such commercial production is secondary. Several types of subsistence farming systems exist. Some emphasize livestock production, while others rely upon a staple crop such as rice to provide their basic caloric intake. Generally, subsistence farmers are associated with other groups with a more advanced technological level. The subsistence farmers exchange their surplus with more advanced farming societies to obtain items they are unable to produce.

Shifting Agriculture

One of the primary types of subsistence agriculture is commonly referred to as *shifting agriculture.* It is characterized by the rotation of land rather than crops. Shifting cultivation systems are known by a variety of terms including *slash and burn,* *ladang* (Indonesia), *milpa* (Central America), *roca* (Brazil), and *chitenmene* (Central Africa). Technically, shifting agriculture occurs when the land is cultivated for a few years and then left fallow for a longer period of time. Slash and burn systems (also known as bush or forest

fallow) allow the land to revert to a secondary vegetation of brush or trees that are cut and burned when the farmers return to farm it again. Burning retains the nutrients from the charred plants in the ashes to be used by the planted crops and destroys weeds, insects, and diseases. Shifting agriculturalists may not always use burning, but allowing the land to lie fallow to restore soil nutrients and fertility is an integral part of the system.

A shifting cultivation system typically consists of either a small village with its adjacent farmlands or isolated, dispersed individual farmsteads. The specific plots cultivated vary from year to year, but the village itself is rarely moved. The actual amount of land cultivated varies, but is generally between 5 and 10 percent of the land (Figure 3–3). The cultural ecology consists of burning the brush or trees and then planting the

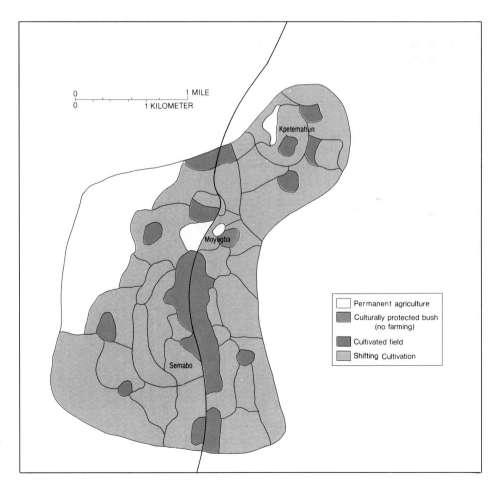

FIGURE 3–3 Shifting Cultivation among the Moyogba A typical village practicing shifting agriculture.

The cultural landscape in shifting agricultural systems includes simple structures of wood or other local materials and permanent tree crops as shown here.

crops in the ashes. The shifting cultivator often practices *intercropping*, the practice of growing several types of crops intermixed in one field. The land is cropped for one to three years, after which the nutrients are largely exhausted, crop yields decline, and the land is allowed to revert back to brush or forest.

The shifting cultivator often maintains a plot near the village where waste products are returned to the soil allowing permanent cultivation. This system of shifting cultivation is analogous to the infield and outfield system practiced in Europe prior to the sixteenth century. The total number of individuals engaged in shifting cultivation is estimated at between 250 and 400 million.

Shifting cultivation is concentrated in the tropics, especially in Africa (Figure 3–4). Tropical climates with heavy rainfall have soils in which the nutrients are affected by *leaching,* or the percolation of nutrients beyond the root zone. Slash and burn systems are a rational means of cultivation in such climates since nutrients are accumulated in the vegetation in the brush and forest. When there is sufficient vegetation, which requires ten to twenty years of fallow, burning follows, providing nutrients for a few years of cropping. Thus shifting agriculture with its mixed cropping and periodic cultivation repre-

sents a type of culture ecology that allows the use of tropical environments in the absence of industrial technologies appropriate to tropical lands.

Cultivation of land in a shifting system begins with the selection of an area for clearing by the men of the village. At the beginning of the dry season the trees are cut down or simply girdled so that they dry, and smaller brush and plants are cut with machetes or axes. The brush and wood are allowed to dry and are burned just prior to the beginning of the rains. Each family uses part of the cleared land for cultivation. Women do much of the planting, intermingling cuttings and seeds of different plants together. The actual planting is done by making a hole with a pointed stick (called a "dibble stick") or hoe (Figure 3–5). Equipment other than the hoe or dibble stick is unsuited for the slash and burn procedure because of tree stumps and fallen, incompletely burned logs.

Intercropping of plants protects the soil from erosion and allows harvesting during different seasons as the crops mature at different times. Interspersing a variety of plant types together may seem chaotic, but in practice it replicates the tropical environment on a miniature scale.

Crops typically grown in shifting agricultural systems include root and tuber crops such as manioc (also called cassava) and yams; grains such as maize (corn), rice, and millet or sorghum; and beans and other vegetables. Some shifting agricultural systems also include pseudo–tree crops such as bananas, papaya, coconuts, or rubber, which can be sold or traded to nearby groups. Maize, manioc, sweet potatoes, bananas, and plantains are the staple crops in the tropics of South America; rice in Southeast Asia; and maize, millets, yams, sorghums, and manioc in Africa. The rice produced in shifting agricultural systems is generally a dry or upland rice variety, which does not need continuous flooding.

Shifting agriculture requires large areas of land to allow a sufficiently long fallow period for vegetation to regrow. Population growth within shifting agricultural societies shortens the period of fallow and prevents the accumulation of sufficient nutrients to restore soil fertility. With fewer nutrients, yields decrease, leading to even shorter fallow periods. The vicious cycle of increased population, followed by decreased fallow and

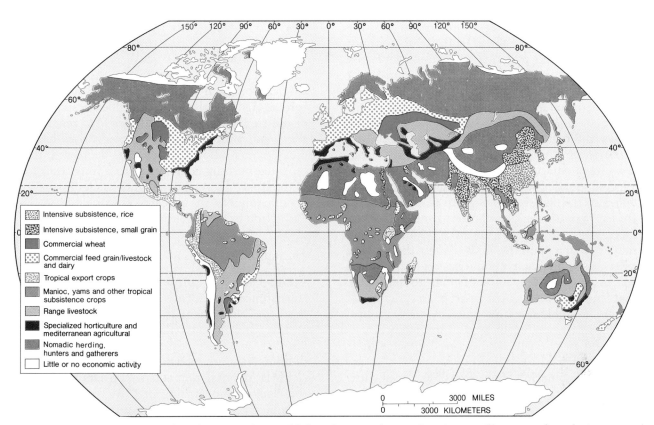

FIGURE 3–4 World Agricultural Types The world distribution of agricultural types illustrates the relative amount of land devoted to each type.

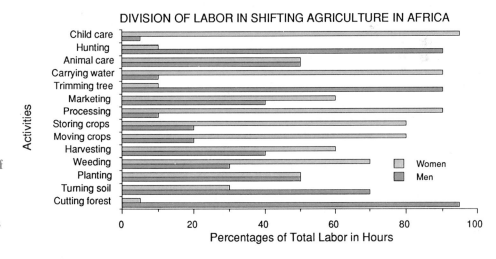

FIGURE 3–5 Division of Labor in Shifting Agriculture in Africa The distribution of labor in most agricultural societies is gender based.

yields, leads to attempts to transform shifting agriculture into permanent cultivation without an extended fallow period. Adoption of permanent cultivation, however, may lead to damage or destruction of tropical environments and may force the former residents to migrate.

Issues in Shifting Agriculture

Pressure on shifting agricultural systems from growing populations and reduced time for fallow are compounded when a shifting agricultural system also has farmers engaged in permanent sedentary agriculture. Population growth is even more rapid among the sedentary farmers, and they encroach upon the lands used by the shifting cultivators. In many countries of Africa and Southeast Asia, governments fail to recognize that land in fallow is in fact part of an agricultural system, a problem compounded by many groups' collective ownership of land. Consequently, fallow land is given to landless peasants for their cultivation.

This combination of problems has forced shifting cultivators into a precarious existence. The agricultural system of the shifting cultivators is not merely a farming method; it is a way of life. Their farming practices from selecting the site for cultivation to clearing, planting, and harvesting are closely linked to religious values and rituals. Group actions and the approval or appeasement of spirits are required at each critical state. Thus pressures on the shifting cultivators threaten not only an economic system, but also the cultural ecology that supports a group.

Government agencies in Africa, Latin America, and Southeast Asia are attempting to provide alternative economic activities for shifting cultivators, but in so doing may threaten their lifestyle. In Indonesia, for example, shifting cultivators are encouraged to plant rubber trees on part of their land to provide a permanent crop. Although this provides some cash income, it takes more land out of the fallow cultivation cycle. In Nigeria some shifting cultivators have been encouraged to plant oil palm trees with the same result. Shifting cultivators face the same problem all groups interacting with a culture group at a higher level of technology and larger population

face. The group with the lower technology is rarely able to sustain its traditional life-style.

Some observers view slash and burn agriculture as an important factor in the loss of the tropical forest on a global scale. Shifting agriculture as traditionally practiced does destroy parts of the tropical forest, but allows it to regrow. As the population of shifting agriculturists has increased, and fallow periods have decreased, their repeated burning contributes to a long-term loss of the tropical forest. Even when such destruction is not complete, the repeated burning may lead to extinction of slower growing species.

Nomadic Peoples

Another form of cultural ecology is the type of subsistence agriculture known as *pastoral nomadism*. Nomadism is the practice of herding animals and includes the movement of animals, families, and their possessions. Two broad types of nomadism are recognized: *horizontal* and *vertical*. Horizontal nomadism consists of moving the flocks from grazing area to grazing area in search of vegetation and water (Figure 3–6). Vertical nomadism responds to the climatic differences of differing elevations. During the summer families move their herds to high moutain pastures, and during the winter they return them to valleys to graze. (Figure 3–7). Vertical nomadism is often referred to as *transhumance*. Transhumance is also practiced by sedentary cultivators who simply truck or drive their animals to mountain pastures during the summer.

Nomadism generally involves reliance upon either sheep, goats, camels, or cattle as the primary means of livelihood. Worldwide, sheep and goats are the most numerous livestock among pastoral nomads because both can live in dry areas and reproduce rapidly. Nomadic herding is an offshoot of early Neolithic agricultural settlements and often emphasizes grazing livestock in areas too dry for other cropping techniques. It is primarily confined to Africa and Asia, in areas with fewer than 35 inches (900 mm) of annual precipitation, such as the steppes and savannas bordering the Sahara and eastern and southern Africa. Nomadism is also practiced in drylands of

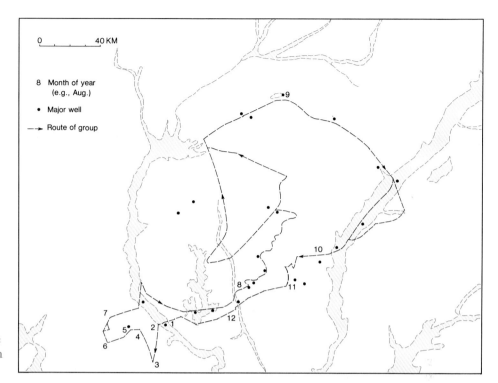

FIGURE 3–6 Range of Movement of Horizontal Nomads Map of the range of horizontal nomadism in Africa.

the Middle East and portions of Mongolia and China (see Figure 3–4). Examples of nomadic societies include the Tibu of the south central Sahara, the Fulani of the savanna lands of West Africa, the Nuer of southern Sudan and western Ethiopia, the Baggara who herd cattle from the White Nile to Lake Chad, the Masai of Kenya, and the Bakhtiari of Iran and Iraq.

Estimates of the present number of nomads range from less than 50 million to 200 million, depending on the definition.

The critical factor affecting nomadic peoples is their reliance upon an ecological niche that is unsuited or marginal for field crops and has great seasonal variability of precipitation or temperature, so that the carrying capacity of the land varies. In addition, a large amount of land is needed to graze one animal—as much as 50 acres (20 hectares) a head. Because nomadism is a semisubsistence economy, cash incomes are low. Nomadic groups rarely average more than a few hundred dollars in cash income per person in a typical year.

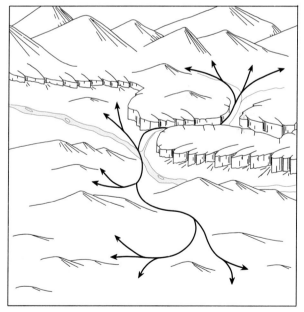

FIGURE 3–7 Range of Vertical Nomadism The range of a group practicing vertical nomadism in the Middle East.

Nomadism has a distinct life-style. Nomadic peoples normally use areas adjacent to permanent cultivators, from whom they are culturally distinct. Livestock are commonly viewed as a status symbol as well as an economic investment. The more livestock a man has, regardless of their quality, the higher his status. Animals are sold primarily to provide essential needs rather than to accumulate capital for development purposes. They are also used for rituals and ceremonies (see the Cultural Dimensions box). As a result, increasing numbers of livestock cause overgrazing and periodic catastrophic reduction of livestock herds, particularly when drought strikes. The repeated droughts of the Sahel region of Africa, just south of the Sahara are the most recent example of this phenomenon, (Figure 3–8). Drought in the early 1970s and again in the first half of the 1980s led to the drastic depletion of vegetation, the deaths of uncounted thousands of animals, and the deaths of an estimated one-half million people. As a result, the people of the Sahel migrated southward, searching for water and pasture for their surviving animals, a move that disrupted the economy and life-style of their cultural ecology.

Drought in the Sahel was especially devastating among nomads for whom wells had been

Nomadic peoples are characterized by their use of tools and structures that can be moved.

drilled after the earlier droughts. The wells allowed the nomads to be more sedentary, concentrating the impact of overgrazing in the region surrounding the wells. Even in the absence of drought, overgrazing associated with large livestock herds causes *desertification*, or the transformation of already sparse grasslands into desert-like landscapes.

The cultural ecology of nomadism also threatens wildlife, especially in Africa. Certain wildlife carry diseases that threaten the nomads' live-

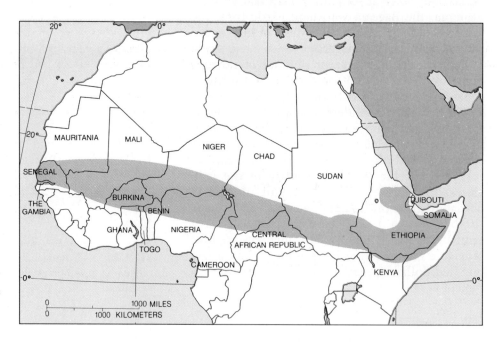

FIGURE 3–8 The Sahel Region The Sahel is a region of aridity along the southern border of the Sahara desert.

CULTURAL DIMENSIONS

Cattle of the Sebei: Life and Wealth among Nomads

Cattle and livestock play a vital role in the society of the Sebei nomads of Africa's Sudan.* A man's herd is structured in the same manner as his household, reflecting the pattern of relationships between the man, his wives, and his children. Cattle play three important social roles—compensation for damage to others, bride price and inheritance—as well as supplying meat, milk, blood, and dung. As compensation, cattle were traditionally given to repay the murder of an individual, providing both economic relief for the family of the victim and preventing additional bloodshed by meeting the demands of justice. Compensation for a murder with cattle suggests that each person has a value calculable in livestock and that individuals and cattle are directly interchangeable.

Cattle are transferred from father to son in a process that Westerners might call inheritance, but it is far more, suggesting the complex intertwining of cattle and life among nomads. Each son has an ultimate right to some of the livestock in his father's herds, which guarantees each son a livelihood. (If there are no sons, the cattle will go to the owner's next younger brother). The sons do not wait until the father's death to receive their inheritance, but receive it at different times. First, the son receives enough animals to acquire his first bride. Later, a son will receive additional cattle whenever the father decides to distribute his wealth. Remaining livestock go to the youngest son, who remains with the father, caring for him, his wives, and the herds till the death of the father.

Cattle are paid to a girl's family as a bride price. Traditionally, marriages are arranged, and the number of cattle is agreed upon in a meeting between the mother, father, uncles, and aunts of the bride and groom to be. Cattle are paid to the family losing the daughter to repay them for her loss, since she goes to live with the extended family of her new husband. Bride prices are fairly uniform, being between three and five cattle plus a small amount of cash (two to five dollars) to pay

for expenses incidental to the meeting and the wedding.

Each animal in a herd belongs to a lineage, just as does each individual of the Sebei people. A lineage is established when an animal enters the herd, and generations later its offspring are still traced to the original progenitor and have the same legal status in the herd and society. Thus if a cow was obtained as payment for damages to a family, descendants of the family have permanent rights relating to the offspring of the cow. In consequence, the Sebei see a herd of cattle differently than Westerners do. Where Americans see a herd of grazing cattle as representing a certain value or producing a certain amount of beef for market, a Sebei man sees much more.

First of all, he is aware that some of the animals are his alone, to do with as he wishes without consulting anyone, while others belong to his several wives. He can use these animals only if the wife agrees: these too are destined to be given to the sons of the wife in question. As a Sebei looks at his cattle, he sees families of cows, each a distinct line like the men of a clan, originating with some particular event in the past. Thus a cow that came from the herd of his father (actually the daughter of a daughter) links him to ancestors long before his time. Each animal has its own history, its own ties with the past.

He also knows each animal by name, and whatever its source, whatever its family line, whatever rights he has in it or has given to his wife, he knows it as an individual animal with its own particular habits, virtues, and faults. The herd is a complex organization of individuals, mirroring the organization of society: his herd reflects the structure of his family, clan, and people, representing the network of social relationships that bind the group together. The herd as a whole stands for what he is not by virtue of its size alone—though a large herd is a measure of his success in the past and the nearest thing he can get to an assurance of his future—but by what it represents as a set of particular social relationships. For the Sebei, his herd is his autobiography and his monument; through it he will gain such immortality as may come to a Sebei herdsman.

*Adapted from Walter W. Goldschmidt, *Culture and Behavior of the Sebei* (Los Angeles: University of California Press, 1976), pp. 136–45.

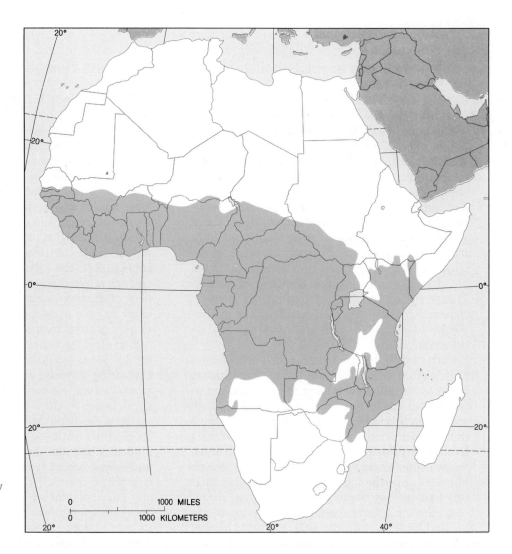

FIGURE 3–9 Areas Infested by the Tsetse Fly Significant portions of Africa are effected by the tsetse fly.

stock; in particular, n'gana spread by the tsetse fly is a problem. To prevent the spread of n'gana, livestock herders may destroy native African wildlife, which they perceive as the source of the hazard to their herds (Figure 3–9).

Traditionally, nomadic tribesmen have relied on trade with sedentary agricultural groups for grain or other items they do not produce. In return, the nomadic peoples sold animal products to the sedentary farmers or provided services such as protection or pasturing their livestock on cropland during the fallow season to fertilize the soil.

Like shifting cultivators, nomadic peoples and their societies are threatened by the expansion of sedentary agriculturists. Ever-growing popula-

tions of sedentary farmers are expanding their cultivated land into areas formerly used by nomadic herders. This reflects the fact that nomadism, which originally developed in a symbiotic relationship with permanent settled farmers, is no longer essential to sedentary peoples. In addition, the recent independence of many African countries, in combination with growing nationalism in the Middle East and Asia, causes border disputes that effectively prevent the movement of nomadic peoples with their flocks. Deprived of their former grazing lands, the nomads are increasingly adopting sedentary farming methods to ensure forage for their animals. They cultivate farmlands during part of the season and turn their animals in to graze on the stubble of the crops after the harvest. Thus, the traditional nomadic life-style is being threatened by pressure from sedentary agricultural villages and political developments as well as by government efforts to establish permanent settlements to facilitate education and medical care. Population pressure and technological change are also forcing nomads to abandon or modify their traditional life-styles.

The Peasant Farmer

Another group of farming peoples are peasant farmers, who make up the majority of the world's farm population. They are characterized by a semisubsistence sedentary village life-style, in which crops rather than fields are normally rotated. The cultural ecology of peasant farming systems varies, as their level of technological

advance may include hydraulic systems (irrigation), draft animals for plowing, small-scale mechanization (including small harvesters and small garden-type tractors), production of commercial crops, or the adoption of new and improved crop varieties. Although peasant agricultural systems produce primarily for their own consumption and only secondarily for a marketable surplus, they generally exchange part of their production rather than farming for total self-sufficiency.

The crops peasants produce emphasize cereal grains that are the staple human energy crops, supplemented by vegetables, fruits, and animals. Concentrated in Asia, Latin America, Africa and the Middle East, their agricultural systems directly support over half the total population of the world (Figure 3–10).

The term *peasant society* is sometimes used to refer to these agricultural systems. The peasant

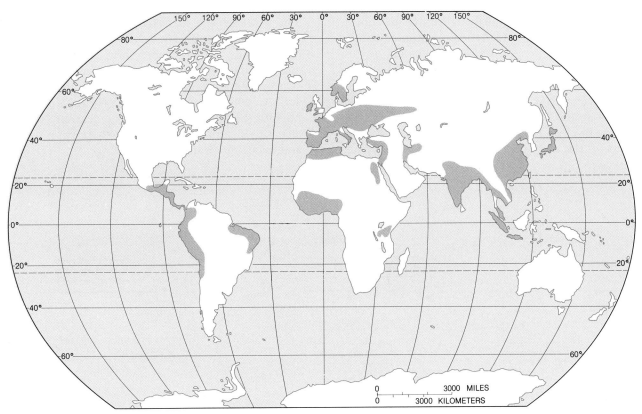

FIGURE 3–10 Major Peasant Sedentary Agriculture Regions of the World The distribution of peasant agriculture systems in Europe is part of an increasingly commercial system.

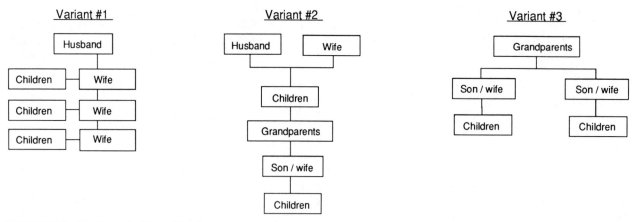

This photo of an extended family illustrates the number of people in such a family unit.

farmers generally farm lands that have been in their families for generations. Nonfarming members of the village may specialize in production of specific items such as pots or services such as carpentry. These various roles in the villages are often a function of birth rather than choice, especially in Asia and Africa. Men and women have specific gender-based, designated duties. Peasant societies are characterized by a high reliance on human and animal labor rather than machinery. The extended family (grandparents, children and their families, and other relatives) living in close proximity is much more common than in the industrial world (Figure 3–11). Since

the majority of the population relies directly on agriculture for their livelihood, the larger family unit can provide additional labor during planting and harvest time. The extended family also helps ensure continuity from one generation to another of roles in the society, and provides care for both dependent children and the elderly.

Another characteristic of the peasant economy is the practice of exchanging goods and services rather than using a cash exchange system. The farmer may exchange his or her crop surplus for the services of a blacksmith or other specialist or for crops not produced on the individual farm. The central feature of the marketing system is a street market, where many individuals sell relatively small amounts of specialized products. The majority of these merchants are women, since roles in peasant systems are highly gender specific. Many markets are *periodic*, operating at some regular interval rather than daily. The peasants involved in the market system have multiple roles in buying, selling, or trading commodities. The limited cash exchanges that take place are used only to accumulate capital for investment in other goods, not for other forms of investment such as land. Entrepreneurs in the peasant market system often engage in several different economic activities, including farming, preparation of products for sale, and marketing. The peasant market also provides an opportunity for social interaction and exchange of information.

DIAGRAM OF EXTENDED FAMILIES

Variant #1

Husband

Children — Wife

Children — Wife

Children — Wife

Variant #2

Husband   Wife

Children

Grandparents

Son / wife

Children

Variant #3

Grandparents

Son / wife   Son / wife

Children   Children

FIGURE 3–11 Extended Families in Peasant Societies

Peasant markets are characterized by large numbers of vendors, each of whom have only a small supply of goods.

This general description of peasant agricultural systems obscures many of their important quantitative and qualitative differences. Peasant agriculture ranges from sophisticated systems, where farmers concentrate on producing a very specific crop under contract to a large firm, to quasi-subsistence economies where only small surpluses are sold. Most peasant societies fall somewhere between these two extremes.

The population density supported by peasant systems varies greatly. In portions of Asia, particularly China, India, Bangladesh, and Indonesia, population densities are among the highest in the world. Key elements in the cultural ecology of these peasant systems that allow such high densities are the use of irrigation and the production of a high-yield staple crop such as wheat or rice (Table 3–1 and Figure 3–12). Peasant systems in Africa and Latin America generally support lower population densities because irrigation is less common and the natural environment less well suited to widespread production of a staple grain. The farms in peasant systems are generally small, averaging between 1 and 2 *hectares* (2–5 acres). Ownership of more than 5 hectares is the exception and is generally associated with a wealthy class of farmers. Because many generations have occupied the same lands, holdings are often fragmented. An individual's 1 or 2 hectares might be spread over five individual, tiny parcels. (Figure 3–13). These small farms, with their multiple plots, reduce the effectiveness of large-scale machinery. The plots are so small that generally a

TABLE 3–1
Comparision of Selected Rural Population Densities, 1989

	AREA (Sq.Mi.)	ESTIMATED RURAL AREA	% RURAL POPULATION	ESTIMATED RURAL DENSITY
Australia	2,967,909	2,879,000	13	1.3
Canada	3,831,033	3,716,000	24	1.7
France	211,208	200,648	23	64.0
U.S.	3,679,245	3,532,075	24	17.0
Japan	145,834	138,542	23	204.0
USSR	8,600,383	8,256,383	31	10.8
Brazil	3,286,487	3,190,892	29	13.3
Argentina	1,068,301	1,025,569	15	4.8
Taiwan	13,900	13,334	34	523.1
Bangladesh	55,598	53,930	84	1,745.0
Ethiopia	472,434	467,710	90	93.0
Nigeria	356,669	353,103	72	229.5
Mauritania	397,955	393,976	65	3.6
Chad	495,755	490,798	73	7.4
Peru	496,224	486,374	31	13.6
Mexico	761,604	746,344	30	34.2

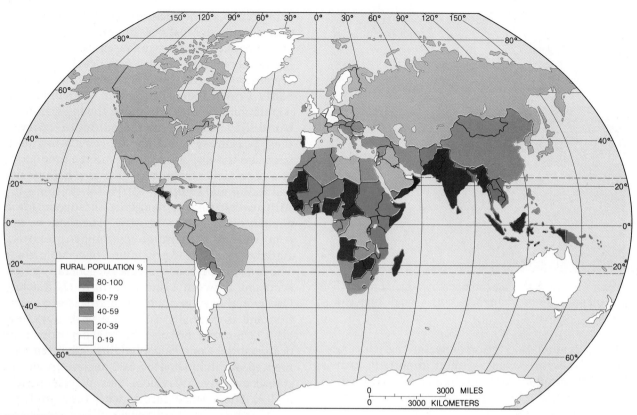

FIGURE 3–12 Rural Population as a Percent of Total Population

garden-type agriculture is more efficient than the large-scale mechanical cultivation typical of North America or western Europe (see the Cultural Dimensions box on page 94).

Consequently, peasant systems rely on tools like the hoe and the plow. The hoe is used widely in Asia, Africa, and Latin America. The plow system had its hearth in the Mediterranean lands of North Africa and the Middle East and later spread to Europe, where it evolved into mechanical systems. Animal-powered plow agriculture is still practiced by peasant farming systems in most areas of the world. Many in peasant farming use the plow only for initial preparation of the land, however, and turn to a wide array of small equipment, often human powered, for leveling, planting, and cultivating. Typical crops grown in areas of plow agriculture are wheat, rice, barley, corn, sorghum, and other staple foodstuffs.

Irrigated Peasant Farming Systems

Another system of peasant agriculture is irrigated rice agriculture, or *sawah* cultivation, which also uses the plow. *Sawah* is the Malay-Indonesian word for a field capable of being flooded for the cultivation of rice. Westerners also refer to irrigated rice cultivation as *paddy* (padi) from the Malay-Indonesian word *padie*, which means unhusked rice. Wet rice (sawah) cultivation is one of the most important agricultural techniques ever invented. Due to irrigation, water can be applied carefully and precisely, and the same field can be used year after year. In addition, when the rice is flooded, the irrigation water brings nutrients to the rice field. The fields are not subject to serious erosion, and by returning the rice stubble and animal and human wastes to the fields, soil fertility has been maintained for as long as 4,000

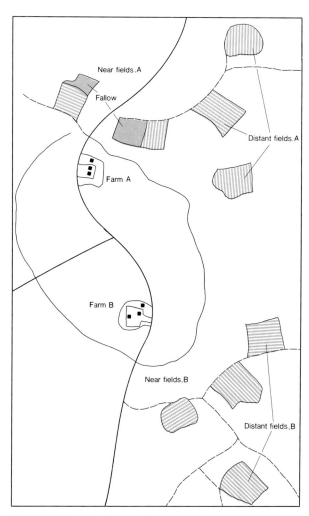

FIGURE 3–13 A Typical Peasant Farm Holding
Land in a typical peasant farming system is
fragmented among many small plots.

ized by terraces, paddies, levees to protect against river floods, reservoirs to provide a reliable water supply, canals, ditches, and water-lifting devices. The cultural landscape in regions of irrigated rice cultivation predates mechanization of agriculture, and the labor requirements of the system prevented farm families from cultivating a large acreage. The social organization required to control and allocate irrigation water and the use of suitable land for levees combine to concentrate farm families in small towns and villages that dot the cultural landscape.

Where climate permits, peasant rice cultivators grow two or more crops in succession in the same year. Where the growing season is shorter, rice may be followed by a vegetable crop, winter wheat, or a fodder crop. Farm animals are generally chickens, pigs, and cattle. Chickens and pigs require little space and efficiently convert grains or other vegetative materials into eggs and meat. Cattle are important as the primary source of power for plowing and transportation. They also provide milk, leather, dung, and meat. Horses are rare in the rice economy. Fish are frequently raised as an adjunct in aquaculture.

Water requirements for paddies limit the amount of land that can be converted to wet rice cultivation. Throughout East and Southeast Asia, where wet rice cultivation is the basis of life, other agricultural systems occupy land unsuited for paddies. These systems include upland rice cultivation, using varieties that do not need constant flooding and rely only on rainfall for their growth, and the production of small grains, maize, potatoes, manioc, sorghum, and millets. The bulk of the land in most of Asia is so rugged that wet rice cultivation is not feasible, and the population density is low (Figure 3–15).

Seasonal demand for labor in intensive wet rice systems is immense. The rice is first planted in small seedbeds, where it is allowed to grow until it is about 6 to 12 inches high. The plants are then pulled up with their roots, tied into small bundles, and taken to larger paddies where they are transplanted one plant at a time. Two to three weeks may be required to transplant the seedlings of a single farm family. Subsequent labor demands include maintaining the rice paddies, controlling

years in areas of southern China and over 1,000 years in Southeast Asia, India, and Indonesia. The practice of sawah cultivation creates distinct agricultural regions in each of these four areas (Figure 3–14).

The importance of the cultural ecology of irrigated rice cultivation lies both in the continuous cultivation of the land for millennia and in the dense population supported in rural areas. Wet rice cultivation supports more *nonurban* residents per unit area than any other peasant system. A distinctive cultural landscape has been created in areas where it is practiced, character-

CULTURAL DIMENSIONS

Changing Peasant Agriculture in Senegal (Excerpts from the Journal of Ted Mooney, a Peace Corps Volunteer in Senegal, Africa, 1974–1977)

July 19, 1974: The rains began today, al'Humdu'illah.* (Praise be to God). Moussa said tomorrow we go to the fields and plant. The rats have left plenty of seeds, he joked, so there should be a few days of hard work.

July 20, 1974: What was a brown desert floor yesterday is a green carpet of spiny grass today. How could it have grown so quickly? These village planting techniques are definitely not FAO [Food and Agriculture Organization] approved. A man wields a scythe-like shovel that scoops a ragged hole in the ground. He drops into it melon, millet and white bean seeds, then covers it over with his foot. When I asked Moussa why they planted this way, he said "This is the way we've always done it. All my fathers have planted like this."

July 21, 1974: Moussa designated a small plot in his field just for me. He even gave me seeds and Soulayman, his 12-year-old son, to work with me. Maybe he feels that I can't do less than what Yalla (God) wills. Whatever that is.

August 20, 1974: Apparently the rains this year are the best in the last 12 years, according to the Voice of America. I believe it. My millet and beans are coming up fast. The melons look a bit iffy but at least most of them have sprouted. When Soulayman and I went out to weed today we ended up chasing away the rats which look like brown guinea pigs. Neither Moussa or Alioune or any of the other farmers seem to do much tending of their fields. I asked Moussa if he would go out and weed his fields and chase away rats and the birds as Soulayman and I have done. "Yalla makes the crops grow. He will decide what kind of harvest we will have."

October 14, 1974: Harvest is a bitch. My back is sore, my legs are sore, my hands look like I've been dead for a month. Soulayman and I loaded the millet into the five sacks Moussa gave us and we still need at least another five. When Moussa saw all the millet we'd reaped from our little plot he said Yalla must like me.

I inspected his field and Alioune's field. The rats had downed hundreds of millet stalks. The melons were scattered on the ground and I could see that where they thrived, the beans and often the millet suffered. Catching up to Moussa I had to draw his attention to this. "See what planting those seeds together does?" I asked. "My beans and melons turned out much better than these." "Yes indeed," he said, "Yalla must like you."

October 21, 1974: Maybe next year, just as the rains come and we're all going out into the fields, I'll hold a class. We'll separate the seeds, designate areas for each type of crop. Then we could . . . .

November 2, 1975: Mahdy Kahn came to my compound with three other women and asked if I would show them how to grow vegetables. They had seen okra, turnips and cabbage in Dagana and wanted to grow them in the village. She told me there were at least fifty women interested. That sounds absurd. The women already do all the work around the compounds. They've got no time. Mahdy promised all would be there tomorrow for lesson number one.

November 3, 1975: Sixty-two. I'm amazed. Sixty-two village women, organized and ready to grow vegetables.

[Ted Mooney completed his tour in Senegal in 1977, then became chief of field operations for the Environmental Impact Assessment of the Senegal River Development Program.]

March 22, 1979: I had to visit my village before leaving Africa. Everyone remembered me. It was quite a time. I didn't have enough sugar and tea for everyone so I gave my bags to Soulayman and let him play Solomon. . . . Mahdy Kahn made her way through the throng of well-wishers and presented two egg-plants and two turnips from the women's vegetable garden. My god, it may even be going when I return next time.

---

*Adapted from Ted Mooney, *"A View From the Field,"* as published in *Development Education Exchange,* a Returned Peace Corps Volunteer publication. Not all Peace Corps or other efforts to change traditional agricultural practices are this successful.

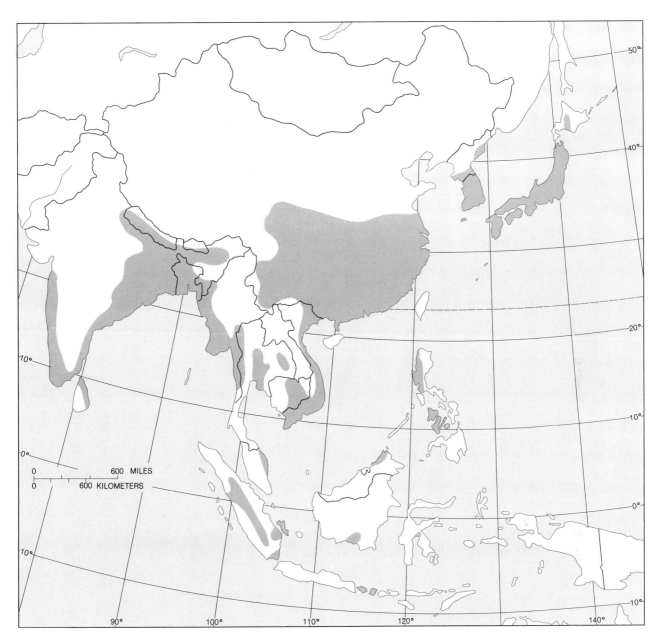

FIGURE 3–14 Regions of Irrigated Rice Agriculture Areas in which land devoted to irrigated rice agriculture predominates.

water levels, weeding, and harvesting. Harvesting is generally done with hand implements, and the grain is often still threshed by hand. Although small, mechanized garden tillers are used in several areas, most sawah systems rely on hand labor. The seasons are identified by the labor demands for sowing, weeding, harvesting, threshing, and marketing the rice crops.

The major areas of intensive irrigation in the Middle East are along the Nile, Tigris, and Euphrates rivers. Wheat grown in irrigated fields has supported civilizations along these rivers for thousands of years. Just as the paddies of sawah systems have been able to sustain constant rice cultivation for millennia, the irrigated fields of the Nile in Egypt and those of the Tigris and

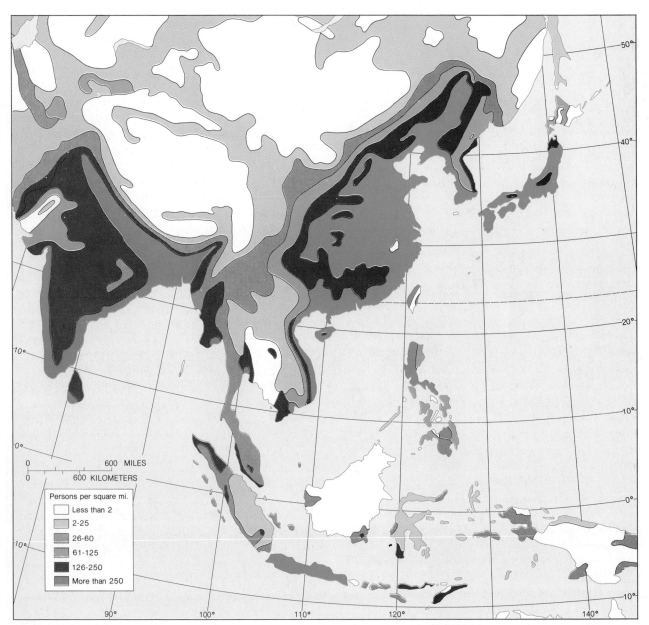

FIGURE 3–15 Population Density of Asia This map illustrates how population density varies greatly from region to region.

Euphrates in Iraq and Syria relied on repeated flooding by the river to deposit silt to provide and maintain soil fertility. Over time, these systems have evolved and adopted other crops including corn, cotton, dates, and alfalfa. In some areas of the Middle East, rice is also an important crop. The cultural landscape is one that is tied to irrigation. Lands that cannot be irrigated are chiefly used for pastoral nomadism.

Egypt epitomizes the importance of irrigation in the cultural ecology of the people of the Middle East. Herodotus said "Egypt is an acquired country, the gift of the Nile." The agriculture of Egypt is found along the Nile and in a very few scattered

Rice cultivation, threshing or its preparation for consumption involves a significant portion of the population of many Asian countries.

oases. The river originates far to the south in more humid areas. The Nile is technically an *exotic stream*, flowing from humid through dry areas to the sea. The annual flooding that deposited silt on the fields was the basis for agriculture in Egypt for millennia. The growing population of Egypt in more recent centuries prompted construction of low dams on the Nile to provide irrigation after the flood season, and completion of the Aswan High Dam in 1970 ended the periodic flooding of Egypt's Nile valley. Agricultural production in the irrigated lands along the Nile has increased with perennial irrigation from the Aswan High Dam and now concentrates on production of winter wheat followed by cotton and maize in the summer. The Aswan High Dam represented a major change in the cultural ecology of Egypt. While providing more reliable irrigation, it eliminated the annual deposit of rich silt, necessitating greater use of artificial fertilizer. The silt that was formerly deposited on the fields now accumulates in Lake Nassar (created by the dam), thus shortening its useful life.

Changes in the agriculture of the Egyptians typify the pressures for change affecting the irrigation systems of the world. Growing populations are associated with increasing farm fragmentation and inability to provide sufficient food. Attempts to solve this problem include consolida-

tion of farm holdings to allow mechanization, distribution of land to peasant farmers or landless laborers, and intensified land use.

Agricultural experimentation in Mexico, the Philippines, and China in the late 1950s and early 1960s resulted in the creation of hybrid grains, a development known as the *Green Revolution.* Widespread adoption of these grains in Asia and the Middle East has resulted in significantly larger crop yields. Previously, inadequate food and repeated famines had been met by world attempts to provide grain from the surplus food producers of Europe, North America, and Australia. With the adoption of high-yielding hybrids, India has been able to meet its grain needs for the past decade.

Adoption of the hybrid seed varieties of the Green Revolution has not been without controversy. The new varieties represent a major modification of a cultural ecology that had developed over centuries of human-environment interaction. The old varieties had been selected because they would provide some harvest even in years when irrigation water was limited. The new varieties are more vulnerable to disease and insects and require more water and more reliable water supplies, as well as heavy fertilization. Large landowners have been best able to afford the increased costs associated with the hybrid varieties of the Green Revolution. Some observers have argued that the hybrid grains make it economically attractive to change to a mechanized system of agriculture, with the result that the large farms are becoming larger at the expense of the small farmers. Some recent studies suggest that the advantages of the hybrid grains of the Green Revolution may now be benefiting the individual small farmers, with resultant increases in yields and a slightly higher standard of living. Most observers still conclude that the benefits of the Green Revolution have had little beneficial impact on peasant farmers.

Nonirrigated Peasant Farming Systems

Peasant systems in Latin America, much of Africa, parts of the Middle East, and Asia are reliant on intensive sedentary agriculture without irrigation. In Latin America, these systems rely heavily on maize or potatoes as their staple product, whereas in Africa they rely on tuber crops, such as manioc, plus millets and sorghum, wheat, and peanuts. These areas are characterized by the same fragmented farms, high labor demands during the crop-growing season, and limited surplus production of other peasant agricultural systems. Peasant agricultural systems in the Mediterranean lands of North Africa (Morocco, Algeria, Libya) and southern Europe (Spain, Portugal, France) are based on wheat, barley, grapes, olives, and figs in conjunction with livestock herding of sheep and goats. Although this area also has irrigation agriculture, traditional nonirrigated peasant agriculture uses much of the land.

Peasant agriculture can also still be found in portions of northern and western Europe. Peasant agriculture in Ireland, parts of Norway and Finland, portions of Poland, and other countries of eastern Europe produces barley, rye, oats, wheat and fodder crops, and livestock.

Peasant agriculture in Europe, however, is only part of the broader commercial agricultural systems that dominate the region. Viewing peasant agriculture on a continuum suggests that the peasant farms of Africa south of the Sahara have been least changed by the technological innovations of the twentieth century, while those of Europe have been the most transformed. This continuum reflects the levels of mechanization and amount of human labor required as well. The remaining farms of the peasant sector in Europe represent the midpoint on a longer continuum that embraces commercial farming, which is markedly different from peasant systems in its cultural ecology and cultural landscape.

Modern Commercial Farming

The commercial farming systems of the world differ in many ways from peasant systems. They generally rely on large acreage and mechanization to produce high yields and food surpluses with limited labor. The overwhelming majority of the production is designed for sale. Individual farms specialize in specific products for market. Unlike peasant farmers, commercial farmers do not even produce the bulk of their own food on their farms. Commercial agriculture is found in Anglo America, Europe, the Soviet Union, Australia and New Zealand, and portions of Latin America, such as the Argentine Pampas and southern Brazil. Areas of Europe associated with Mediterranean agriculture are transitional between peasant and commerical agriculture. They are experiencing a change in their cultural ecology as commercial agriculture, with its crop specialization, rapidly replaces the traditional peasant agricultural system. Where the peasant system is the norm, there are usually some commercial operations, giving rise to a *dual society*. Where commercial agriculture is dominant, there are rarely any major regions devoted to peasant agriculture, although there may be scattered farms as relic cultural landscape features from an earlier peasant system.

The fields in peasant agricultural systems are generally unsuited to mechanized cultivation.

Commercial agriculture (as in this photo from New Zealand) is characterized by large scale and monoculture.

The commercial agricultural regions of the world are interconnected to one another through interchange of crops. Commercial agricultural regions are also interconnected with the regions of grazing, shifting, and peasant agriculture because the commercial regions provide surpluses to meet local food shortages and purchase tropical products that are not grown in the mid-latitudes where much of the commercial agriculture is found.

The modern commercial agricultural landscape is a part of the market exchange economy of the industrial world. The primary motivation of farmers in the commercial system is maximum return on their investment. Unlike the peasant farming systems, commercial farming is a business rather than a way of life. Successful farmers in the commercial agricultural regions of the world are part of the broader industrial society. They rely upon science, technology, and capital investment to increase productivity. Large farms, a high degree of mechanization, high productivity per farmer, relatively little human labor, high capital investment requirements, concentration of fodder for animals, and processing of products away from the farm are characteristic of the cultural ecology of commercial farming systems.

Commercial agriculture ranges from intensive production of vegetables and fruit to semi-extensive agricultural activities associated with grain production or dairy or beef production in grazing activities. Crops produced in commercial agriculture regions include wheat, rice, corn, and potatoes as the most important, as well as a host of related crops such as barley, rye, oats, sugar beets, cotton, and vegetables and fruit crops. Livestock production emphasizes cattle, pigs, and chickens. Farms are characterized by intensive use of the land; a high level of capital inputs in the form of machinery, fertilizers, or structural modification of the soil through leveling or irrigation to make it more productive; and high yields per acre. Intensive production of vegetables and fruits is sometimes referred to as *market garden, truck farming,* or *intensive commercial horticulture.* It is characterized by concentration of large amounts of labor, capital, and equipment in a small area. Intensive agricultural production of this type is normally found either near large market areas, such as cities, or in areas having a unique geographical location, such as a distinctive climate allowing the production of citrus fruit.

The location of market gardening around cities reflects markets and their accessibility as demonstrated by the von Thünen model of spatial interaction discussed earlier. The high perishability of vegetables generally means that they will occupy land close to the market. The major market gardening regions of the world are associated with the large urban populations of the northeastern United States, western Europe, and Japan. An exception to von Thünen's model of location of crop production is related to climate and technology. Vegetables produced during the summer for urban markets will not meet winter demand; consequently, specialty truck farming areas have developed in such places as the Mediterranean lands of Europe, Florida, and southern California. Areas in the Southern Hemisphere, such as Chile, provide fruits and vegetables during the winter season to the industrial areas of the world. Thus, in some cases, climate and the technological advances associated with refrigeration and speedier transport rather than proximity to markets are the principal locational factors.

Von Thünen hypothesized that in areas removed from markets, extensive agricultural land uses would develop. Examination of the agricultural regions of any major industrial nation indi-

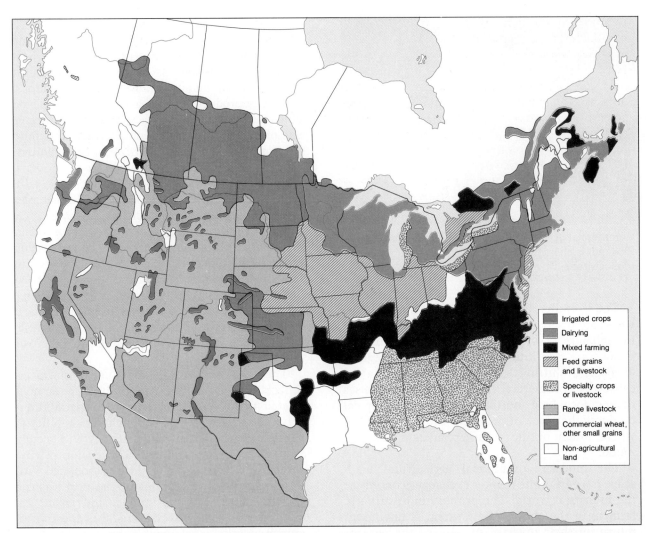

FIGURE 3–16 Agricultural Regions of the United States and Canada The specific crop and livestock combination in an individual location creates distinctive agricultural regions in the United States and Canada.

cates that this is true (Figure 3–16). Farther from major markets, land is used predominantly for dairying/mixed farming, semi-extensive grain farming, and livestock grazing. These farms are characterized by increased acreage, intensive use of machinery, and high productivity. Extensive commercial farming often involves *monoculture*, as in the wheat belt of North America where farmers concentrate on the production of one crop. The specialization of farmers in the industrial world has led to overproduction and surpluses of a variety of agricultural products.

High productivity poses a number of problems for those nations engaged in modern commercial agriculture. The costs of production in countries involved in commercial agriculture dictate the price farmers must receive if they are to continue farming. Because farmers must harvest their crop or receive nothing for their labor, farmers have been faced historically with the necessity of taking whatever price the market offers. High prices were offset by limited quantities in years of low yields. In years of high yield, the quantity of crops available was high, but the price was low. In response to these fluctuations, most industrial countries have adopted price support systems to guarantee crop prices. Consequently, since price guarantees ensure a profit, farmers typically produce as much as they can. Because crop demand does not change dramatically from year to year, governments are left with crop surpluses. Governments must either sell surplus commodities for less than they paid or pay for their storage (see the Cultural Dimensions box on pages 102–103).

The dilemma of a crop surplus is particularly acute for the United States and western Europe in wheat, butter, dry milk, and other staple crops (Figure 3–17). If the countries sell agricultural commodities cheaply or give them to countries with a food deficit, they unwittingly contribute to the food problem in countries with a peasant or subsistence agricultural base. When cheap grain or other agricultural products are exported to these countries, the market prices for locally produced crops are driven down, causing local farmers to cut production or cease market farming entirely.

One of the main characteristics of the commercial agricultural system has been the continued decline in both the number of farms and the number of farmers (Figure 3–18). Most countries of the industrial world (and some in the less industrialized world) have changed from primarily rural agricultural societies to urban societies in the twentieth century. The U.S. census of 1910 revealed that slightly more than 50 percent of the population lived in areas defined as rural. (The U.S. census definition of "rural" is anyone living

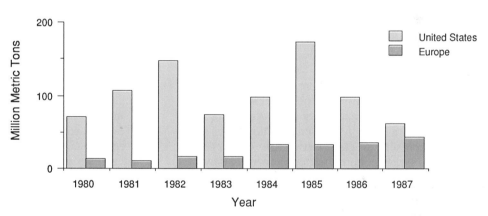

FIGURE 3–17 **Growth in Surplus Cereal Crops in the United States and Europe (End of Season Stock)** The total amount of surplus crops in the world varies from year to year.

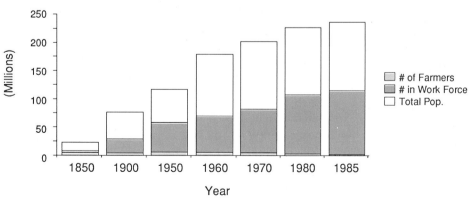

FIGURE 3–18 **Changes in the Farm Populations of the United States** The total number of farmers in the United States has declined sharply, illustrating the role of machinery in agriculture in industrialized countries.

CULTURAL DIMENSIONS
Modern Commercial Farming

Mark Hancock and Franz Schmidt are farmers in the United States and Germany, respectively. Both are part of the modern industrial world, both operate "family farms," and both produce for urban markets. Both rely on machinery to allow them to cultivate large (by world standards) acreage with only one or two hired laborers. Examination of their farming systems illustrates the nature of the cultural ecology in modern agriculture and suggests the problems facing the ever smaller numbers engaged in agriculture in the industrial world.

Farmer Schmidt's farm is part of the picture postcard landscape of Bavaria in southern Germany. The average farm here is only 35 acres, and nearly half are between 2.5 and 25 acres. Approximately one-half of the farmers work at full-time jobs in nearby towns and cities. Full-time farmers like Franz rent land from the part-timers. Franz cultivates 125 acres, 78 of them rented from neighbors. He grows potatoes, wheat, and barley, the latter two are fed to the cattle he produces for beef. Schmidt's farm has two tractors, a small truck, storage facilities for the wheat and barley, a variety of small machinery for cultivating his crops, and a grain harvester. His buildings and fields represent the cumulative effect of 900 years of cultivation. Schmidts have farmed Franz's acres for over 300 years.

Falling birthrates and the attraction of city jobs make it unlikely that family farms will continue in the same family here. In some farm villages not a single family has an heir for their farm. The number of farms in Bavaria, in consequence, fell from nearly 400,000 in the 1960s to less than 250,000 in 1989. German farmers, like other commercial farmers in Europe, also fear the competition from "factory farms" in the United States, Australia, and Canada.

European farmers are highly reliant on government subsidies (nearly $30 billion dollars in 1988 for farmers in the European community). High subsidies translate into high yields, and some German farmers have wheat yields exceeding those in the United States. Increasing costs of farm subsidies have prompted politicians to suggest lowering them. Such comments alienate farmers like Franz, who feel their contributions to the country's economy justifies continued support.

Farmer Hancock of the American Midwest faces similar challenges. Subsidies for his crops also mean the difference between his success or failure in agriculture. The United States pays its farmers nearly as much in subsidies annually as does the European community. Their elimination, farmers claim, would destroy the "family farm," lead to monopolies on production by "corporate farms," and result in high food prices.

The scale of Hancock's operation make European farmers question whether it is really a "family farm." He cultivates 1,200 acres of rich midwestern soil, but even the average American farm size (430 acres) dwarfs European farms. Hancock's farm is big business. He inherited part of his land, bought more, and rents an additional 560 acres. If he farmed only the 240 acres he received from his father, it would keep him busy only a few weeks each year because of the level of mechanization on the farm.

Hancock has two large tractors, a smaller one, two large trucks, a pickup, a combine with grain and corn heads, and a fleet of plows, discs, planters, cultivators, and other equipment. It would cost nearly half a million dollars to replace his machinery and the modern steel building he built to house it. Hancock raises about 600 acres of soybeans and 600 acres of corn each year. Storage facilities for 50,000 bushels of corn and 25,000 bushels of soybeans cost an additional $50,000 when they were constructed. Hancock also raises approximately 2,000 hogs each year, and the hog barn and equipment added another $75,000 to his investment.

Hancock had nearly $2 million invested in his farm operation at the end of the 1980s, and the cash flow related to the sale of his soybeans (over $100,000 per year), the one-half of his corn crop not fed to his hogs, and the sale of his hogs gives him a gross income of over $400,000 per year.

continued

Unfortunately, after paying for seed, fertilizer, pesticides, equipment repair and replacement, interest, wages for one full-time and several seasonal laborers, and taxes, farmer Hancock feels lucky to receive an average annual net income of $30,000 to $50,000. Hancock's return on his investment is so small that a few years of bad weather could threaten his operation.

Hancock is a farmer, but unlike the hundreds of thousands of American farmers who have been forced off the land, he is also a savvy business-man. He keeps good records, and he and his wife spend part of each day on accounts. Drought, insects, disease, and economic conditions are constant threats, but like any businessman he attempts to minimize their worst impact. Hancock buys crop insurance, produces both grain and meat, and closely monitors his agribusiness. He is constantly aware of world events that might impact his farm, from petroleum prices to drought and crop yields in other countries that might affect his operation.

---

in places having fewer than 2,500 people.) By 1988 slightly more than 20 percent of the population were classified as rural, and the number of people actually engaged in farming had declined to less than 5 percent of the total population. Where commercial agriculture is the main element of the cultural ecology, consolidation of small farms into larger economic units and the substitution of machinery for human labor have reduced the number of farms and farmers.

Plantation Agriculture: An Intrusive Commercial Agriculture

Plantation agriculture is a type of commercial agriculture combining characteristics of the intensive commercial agriculture of industrial countries with elements of the peasant agriculture of the less industrialized countries. Plantation agriculture is associated with the production of crops in tropical and subtropical regions. It is characterized by large operations concentrating on cash crops such as rubber, sugar cane, bananas, tea, or coconuts. The plantation system generally represents the transfer of the commercial goals of one culture into an existing agricultural economy. Most of the plantation regions of the world provide cash crops for sale in cities of industrialized areas. (Figure 3–19).

Historically, the plantation emphasized large-scale production of a single crop for export to the markets of Europe or North America. Wherever the plantation was found, there was normally a social distinction between the people who did the labor and the European or North American managers. The impact of plantation agriculture on those areas where it was practiced was profound and included changes in crops, land ownership patterns, density and distribution of the population, and even the political system. One of the most serious consequences of the plantation system is the shortage of domestic foodstuffs, which leads to chronic malnourishment and increased disease.

Europeans achieved the original goal of Columbus—to obtain access to the spices of Asia—by simply acquiring spices produced as a cash crop in subsistence and peasant agricultural systems. Increased European demand led to changes in the agriculture system and the introduction of new crops. Sugar cane, coconuts, rubber, coffee, cacao, pineapples, and bananas all became plantation crops produced by peasant farmers. Originally from Latin America, cacao and rubber diffused to other tropical areas and, with coffee and sugar, found a ready market in Europe.

In addition to the diffusion of tropical crops from Latin America, plantation agriculture changed the ethnic makeup of some areas. The need for additional labor in Southeast Asia, where uncultivated land was planted in rubber or other

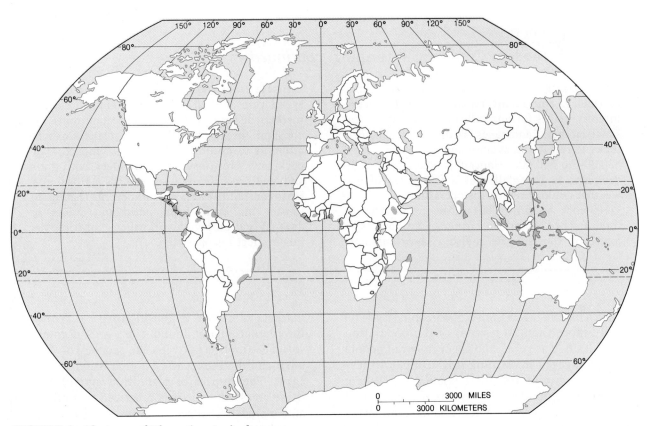

FIGURE 3–19 Areas of Plantation Agriculture

plantation crops, led to the importation of Chinese and Indian contract workers (Figure 3–20). There had been small groups of Chinese in Southeast Asia before the European plantations, but the large numbers of Chinese in Southeast Asian countries today dates largely from nineteenth-century population movements associated with labor for plantations. Approximately 32 million individuals had migrated to Southeast Asia to provide labor by the effective end of the colonial period in Asia (1945). Although many laborers were temporary workers, many Indians in Malaysia and Burma and the Chinese in Malaysia, Vietnam, Thailand, Singapore and Indonesia stayed as permanent residents. Indians also remained in Mauritius, New Caledonia, Kenya, and South Africa as a consequence of the plantation economy.

Plantation agriculture in the New World was established in the southern United States, the Caribbean Islands, and Central and South America. Plantation agriculture in the southern United States initially concentrated on such products as indigo, rice, and tobacco. The invention of the cotton gin during the Industrial Revolution led to cotton becoming dominant. Plantation agriculture in the United States led to the removal of Indians from their lands and the importation of Africans as slave labor. Before the Civil War (1860–1865) approximately 8 to 10 million African slaves were brought to the United States, both directly from Africa and from plantations also caused serious soil depletion and erosion with accompanying silting or stream channels and harbors in the United States.

The Spanish and Portuguese in Latin America created an agricultural system distinct from the plantation. The Spanish granted the communally held Indians lands to a grantee under the *encomienda* system, which did not technically

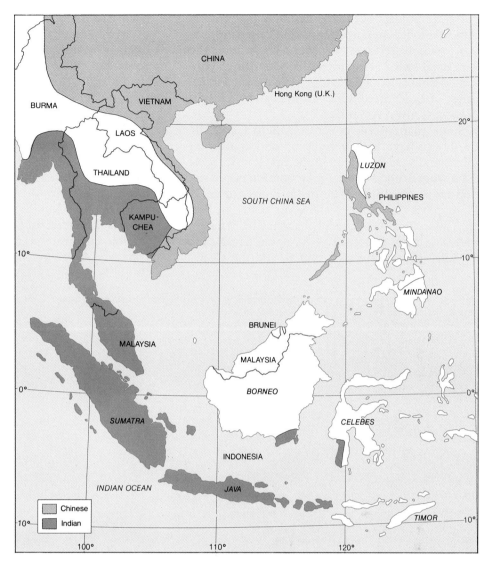

FIGURE 3–20 Population Movements Associated with Plantation Agriculture in Southeast Asia The movement of Chinese and Indian population from plantation agriculture in Southeast Asia.

grant land ownership to the Spaniard, but did give him the right to use Indian labor. Plantation agriculture did develop in northeastern Brazil where the Dutch attempted sugar cane production using Indian labor. Growing demand for sugar in Europe led to the expansion of sugar cane production in northeast Brazil and in the Caribbean Islands after the Portuguese forced the Dutch out of Brazil. British and French sugar cane plantations were later established on other islands in the Caribbean. When disease decimated the Indian laborers, they were replaced by African slaves. In time, cotton, tobacco, cacao, coffee, and bananas were added as plantation crops in Latin America.

The impact of plantations on Latin America included the decimation of Indian populations by European diseases; ethnic changes as Africans replaced Indian populations on the islands of the Caribbean and were introduced to the northeastern portions of South America; development of roads, railroads, and ports to facilitate the export of plantation crops; and consolidation of land ownership in the hands of a few wealthy landowners (Figure 3–21). Large plantations often continued to be dominated by North American or

European powers even after individual countries achieved independence. The term "banana republics," which referred to countries where the American-based United Fruit Company controlled the banana industry, reflects the dominance of the United States in the banana plantations of Central America.

Some plantations in Latin America and the Caribbean have been divided up and granted or sold to individual farmers in the past few decades. Even where this is the case, production of plantation crops continues. The major crops grown in these quasi plantation systems are coffee in Brazil, Colombia, and Ecuador; bananas in Ecuador, the Caribbean, and Central America; sugar cane in Cuba, the northeastern Brazil, and the Carib-

bean Islands; and the fiber henequen in the Yucatán Peninsula of Mexico. Although actual production is now in the hands of small farmers, control of marketing by foreign firms still effectively limits the choices open to the farmers.

A modern example of how outside markets affect crops grown in Latin America is the drug trade. A market for drugs in North America and Europe has created a new type of export crop, whose marketing and profits are removed from the peasant farmers (see the Cultural Dimensions box on pages 108–109).

European colonists came late to Africa, with most of the continent coming under European domination only near the end of the nineteenth century. Formerly, Africa was important to Europeans pri-

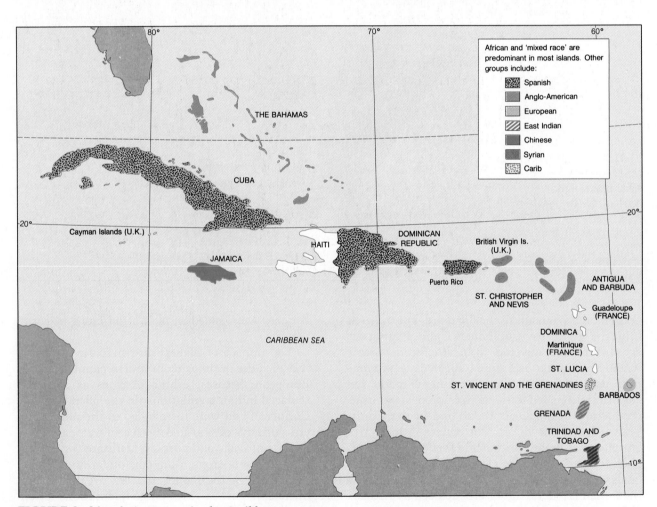

FIGURE 3–21 Ethnic Groups in the Caribbean

Plantation agriculture involves both the production of the crop and selected elements of market preparation as in this banana plantation.

marily for slaves. European demand for slaves reoriented trade with Europe from North Africa to the coastal slave ports of the empires south of the Sahara. The European exploration of the interior of the African continent began in the mid-nineteenth century. The motivations for these expeditions of exploration were financial gain, expansion of the geographical knowledge of the world, world status as European countries attempted to gain colonies, and (to a lesser extent) Christian missionary work.

European colonization of Africa led to establishing plantations in several areas: West Africa, the highlands of East Africa, and South Africa. West African plantation crops include palm oil, cacao, rubber, and peanuts. Except for rubber, plantations are not the dominant producers. In West Africa, Liberia still has rubber plantations operated by the Firestone Rubber Corporation. East Africa plantation agriculture emphasizes coffee in highland areas, and the fiber crop sisal in the lowlands of Kenya. South African sugar cane and other crops are produced on large commercial estates owned by descendants of European colonists in a commercial agriculture system with some characteristics of plantation agriculture.

The impact of plantation agriculture on the cultural geography of the world is greater than the territorial extent of present plantation agricul-

ture. Relics of plantation agriculture persist in the form of European or other foreign domination of markets for cash crops, dependency status to industrial markets for countries where an export crop inherited from earlier plantation systems is the basis of agriculture, and changed ethnic and cultural patterns associated with migration of people to staff plantation agriculture.

An important aspect of the cultural ecology of plantation agriculture is associated with land ownership and is among the most unpleasant heritages of plantation agriculture. Frequently, a few wealthy landowners own large tracts of land while the majority of the population are landless laborers or occupy tiny farms too small to support their families. Even when plantations have been divided up and land provided to individual farmers, the individual peasant farmer must continue to market his or her crop through the old plantation infrastructure. Marketing bananas, coffee, cacao, or other crops in Europe or Anglo America requires greater quantities than the individual peasant farmer can produce. Thus, simply dividing the lands among the former plantation laborers still leaves them at the mercy of the old plantation management, creating a marketing system described as *neo plantation*. Land ownership issues can be a major source of conflict in countries with a strong colonial heritage. The relic plantation systems with their large land holdings and landless laborers create an environment that can contribute to social unrest and revolution. Plantation and associated land ownership problems have been a contributing factor in revolutions in Cuba, Nicaragua, and other areas.

Crops and People: The World Pattern

One of the most important impacts of cultural ecology has been the change in the plant geography of the world since the time of the Neolithic Revolution. The process of human diffusion to most of the earth's surface has been accompanied by transformation of the world's plants and animals. This has meant a substitution of cultivated crops and domestic animals for the original wild plants and animals in a region. The combination of plants and animals in a given place can be used

CULTURAL DIMENSIONS
Drugs: A New Plantation Economy in Latin America?

Production of illegal drugs is spreading across parts of Latin America.* The prospect of easy money has led hundreds of farmers to abandon traditional crops in favor of the lucrative cultivation of coca. This poses a major problem for the region's agricultural economy, which every day loses acres of its best land, formerly used for food production.

According to a report by the United States Department of State, between 15,000 and 17,000 hectares in Colombia and 5,000 hectares in Bolivia are currently used to grow coca. They yield an annual crop of between 52 and 60 metric tons of cocaine destined for the United States, where between 54 and 71 tons are consumed each year. Consumption in western Europe is estimated at 10 to 20 metric tons a year. The farmers grow coca because it pays $2,000 to $4,000 per acre per year, far above what any other crop would bring.

As for marijuana, an estimated 10,000 hectares are under cultivation in Colombia, yielding a total of 12,000 metric tons a year. Additional thousands of hectares are devoted to marijuana in Mexico, Guatemala, and Jamaica (Figure 3—A).

Farmers and Crime

The farmers' solution to their economic problems is temporary, since by involving themselves in a criminal business, where the rules are laid down not by society but by the current "boss," they expose their families to threats from a rival eager to acquire their harvests. They are also likely to become enmeshed in one of the common vendettas, which the press never hears about owing to the remoteness of the jungle areas where the crops are grown.

The drug traffic is controlled by a few families. According to a document prepared by the Colombian government after the murder of a prominent newspaper editor, the lucrative business is run by

*Adapted from *Development Forum* (June 1987): p 6.

254 people, many of them related, in a country of 30 million inhabitants.

And it's not only coca growing that keeps the countryside at work. Tens of thousands of people are employed in transporting the chemicals, kerosene, acetate, and ether necessary to process coca leaves into paste. Along certain highways, flatbed trucks brim with teenage boys in knee-high rubber boots on their way to crush troughs of chemical-soaked coca leaves into paste. A stomper can earn $10 to $12 a day, nearly the monthly minimum wage, for a 10-hour shift of stomping coca leaves.

The illicit profits do not return to Latin America. A study by the Institute for Latin American Studies of the University of St. Gallen in Switzerland maintains that, of the profits from the illegal trade in drugs and traditional agricultural products outside Colombia, which rose from $US 6 billion to $US 9 billion between 1980 and 1984, no more than 10 to 20 percent returned to the country. The balance was deposited in banks in the United States, the United Kingdom, Panama, and Switzerland and placed in showy investments in well-known high-growth industries, including some in Colombia.

The challenge to U.S. authorities in the face of this serious problem is demonstrated in an action carried out by the Drug Enforcement Agency in Miami, known as "Operation Swordfish." More than 90 Colombians were prosecuted in this case, and at least 4 of them received sentences of more than 10 years and were fined millions of dollars. Yet, no bank managers or presidents were put in prison, and the few fines levied on them did not exceed $US 500,000, since it is difficult to prove they knew the money deposited was drug money.

As long as foreign banks continue to accept drug money as indifferently as they do funds from legal sources, any efforts by the Latin American countries to stop drug production will go unsupported. This is all the more true when one considers that the United States and the Federal

continued

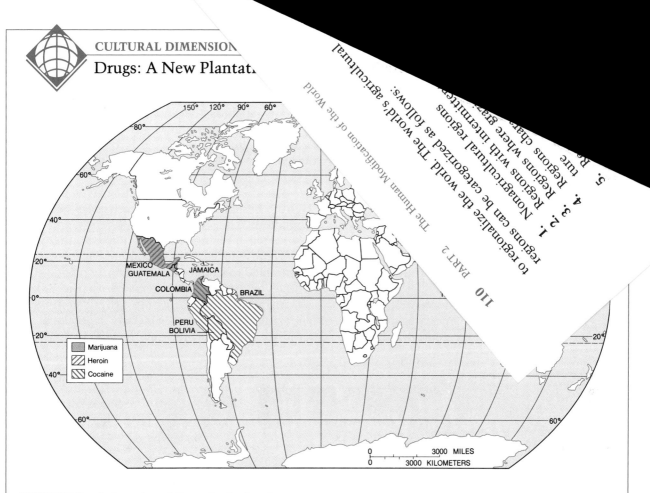

FIGURE 3—A Location of Drug-Producing Areas

Republic of Germany continue to be the leading exporters of ether, a substance essential to the processing of cocaine, to Latin America.

The jungle of the illicit trade in drugs prompts us to think too about the responsibility of the present generations for those that will follow them. If, to a greater or lesser degree, our political system, administrative organization, and productive apparatus are the legacy of our forebears, we cannot escape the fact that the receipt of these benefits implies a similar obligation on our part vis-à-vis those to come. In principle, then, it is incumbent upon political and economic leaders as well as the trade unions and other lobbies,

which generally do not look beyond their own interests, to take up the fight against obvious social injustices, without which crime and corruption could not flourish.

If the farmer finds ways of putting his products on the market, if the state guarantees that they will be sold, so that it is the farmer, and not the middleman who ends up with the profits, and if the farmer's children enjoy access to education, health, and subsequent opportunities for decent and well-paid work, then the drug traffickers will have lost an ally, and Latin American countries will have earned a second chance for their future generations.

...t agriculture
...ng dominates
...acterized by shifting agricul-

...gions with peasant crop agricultural systems

6. Regions dominated by commercial crop and livestock production.

The first two regions consist, respectively, of one in which there is truly no agriculture and another in which agriculture is temporary or occupies only a small fraction of the total land (Figure 3–22). These two regions are associated with areas having harsh environmental settings, such as the Arctic and subarctic regions, high mountain ranges, deserts, or tropical rainforest;

such areas are characterized by seasonally extreme cold, extreme aridity, or tropical conditions with infertile soils and plant and animal diseases. Only in Antarctica and Greenland is there no agricultural activity at all. Reindeer-herding groups such as the Suomi (Lapps) in the Arctic still use the seasonal vegetation. Agriculture in desert regions is restricted to oases where water supplies allow crop cultivation or to livestock herding by nomads.

Another agricultural landscape is associated with the extensive grazing activities that cover much of Asia, Africa, Latin America, and North America. Grazing activities may be either generally subsistence with a commercial component (intended to supplement the family production of food) or strictly commercial with the animals being produced for sale only. Strictly commercial grazing activities are generally associated with North America and the Soviet Union, with

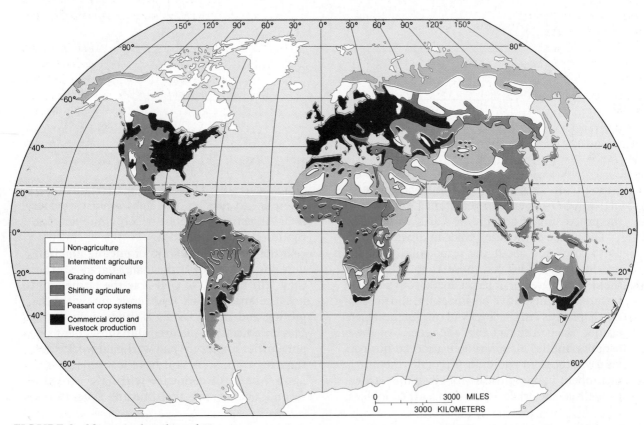

Legend:
- Non-agriculture
- Intermittent agriculture
- Grazing dominant
- Shifting agriculture
- Peasant crop systems
- Commercial crop and livestock production

FIGURE 3–22 Agricultural Land Use

smaller areas in western Europe, Australia, and Latin America. Subsistence or semisubsistence production of livestock is carried out across much of Eurasia, Africa, and portions of Latin America.

Regions that reflect the cultural ecology associated with shifting agriculture also occupy large areas. The subsistence economics of shifting agricultural societies support only low population densities (Figure 3–23). Much higher population densities are associated with regions of peasant agriculture. Peasant agricultural systems use large amounts of labor and have relatively low inputs of capital in the form of machinery or fertilizer. The regions devoted to commercial crop agriculture occupy a relatively small area. They are of immense importance to the world, however, because they provide surplus food for use in many countries. Characterized by high carrying capacity because of their technology, these regions emphasize monoculture in cropping, as in the American corn or wheat belts. Production of commercial crops is often associated with livestock production, as in the hay and dairy belt of the United States.

Categorizing the world's agricultural systems is complicated because the difference between commercial agriculture and peasant agriculture is somewhat blurred. The two represent points upon a continuum that begins with technologically primitive subsistence shifting agriculture and ends with highly technical commercial agriculture. Dividing the world into regions on the basis of this continuum is difficult because peasant agricultural systems contain examples of commercial agriculture, and commercial agricultural regions include examples of peasant systems. Examples of commercial agriculture in predominantly peasant systems include the large landowners along the Nile River who produce cotton for use in Egypt's textile industry or export to Europe; carnation production in Costa Rica for export to North American consumers; and coffee, cotton, or sugar cane production in Colombia or southern Brazil. In spite of this difficulty, the general pattern of agricultural systems indicates the type of cultural ecology and associated cultural landscape found in broad areas of the world.

The Industrial Revolution and World Regions

The countries of the world can be grouped according to whether their economies emphasize the commercial and manufacturing activities of the Industrial Revolution or are more oriented toward the peasant, shifting, or nomadic agricultural system. On the basis of the level of industrialization and standard of living, nations can be arranged on a continuum with wealthy, industrialized countries at one end and poorer, less industrialized countries at the other. One such division contains four broad groups: industrialized nations, newly industrialized nations, surplus capital oil exporting nations, and the least industrialized countries (Figure 3–24). The industrialized nations include Canada and the United States, all of Europe (excluding Albania), the Soviet Union, Japan, Australia, and New Zealand. These are countries with a limited population engaged in agriculture. Incomes are high as defined by per capita *gross national product*, and literacy rates and life expectancies are high.

The newly industrialized countries have an important industrial sector, but normally have a much larger percentage of their population engaged in agriculture. Newly industrialized countries include Taiwan, Hong Kong, South Korea, Singapore, Malaysia, Colombia, Brazil, Venezuela, Argentina, Chile, Mexico, and others that are making the transition from an agricultural to an industrial economy.

The surplus capital oil exporting nations are countries that until the 1970s were clearly part of the least industrialized realm, but experienced enormous transfers of capital from the industrial countries when the price of petroleum soared in the late 1970s and early 1980s. Most of the people in these countries are still engaged in peasant agricultural activities, but there is also a small elite class with tremendous wealth and opportunities for travel, education, and material goods. The surplus capital exporting countries include Saudi Arabia, Kuwait, Oman, and others that have changed abruptly from peasant societies to dual societies. Recall that in a dual society one distinctive sector enjoys a standard of living much higher than the majority who are

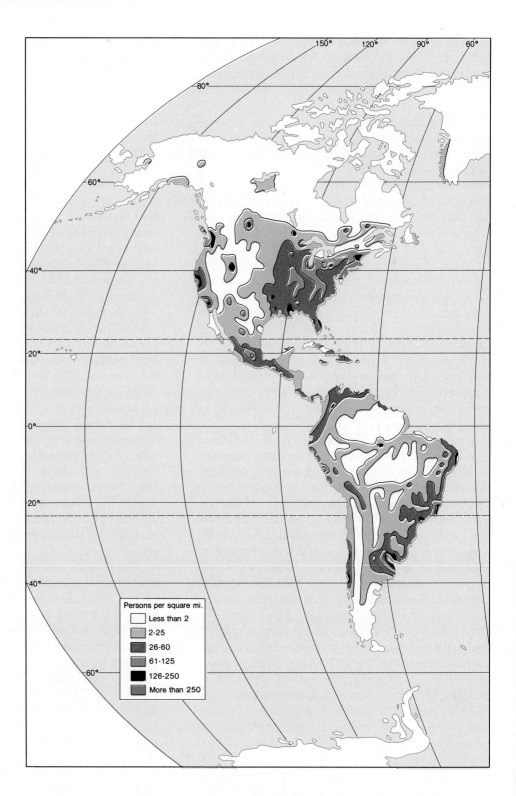

FIGURE 3–23 World Population Density

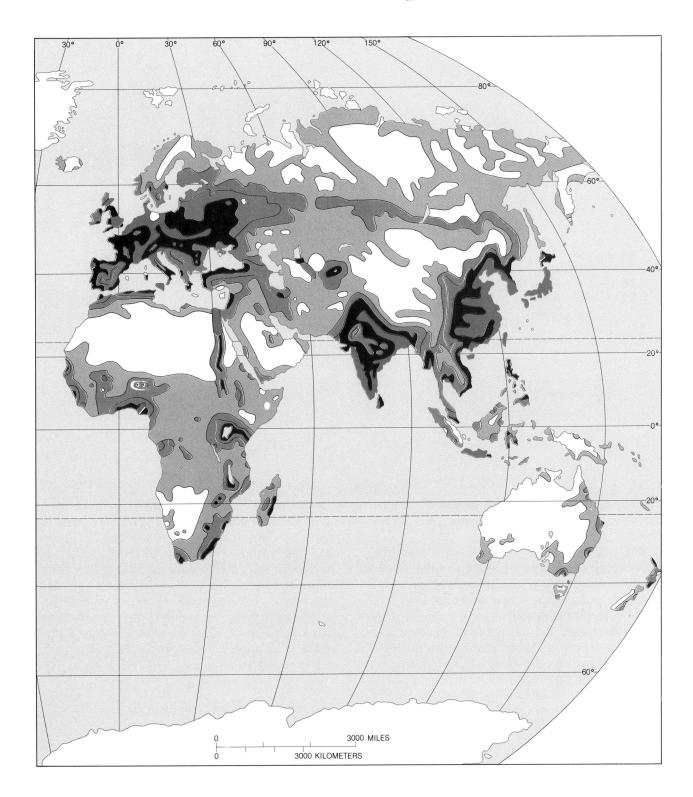

0 — 3000 MILES

0 — 3000 KILOMETERS

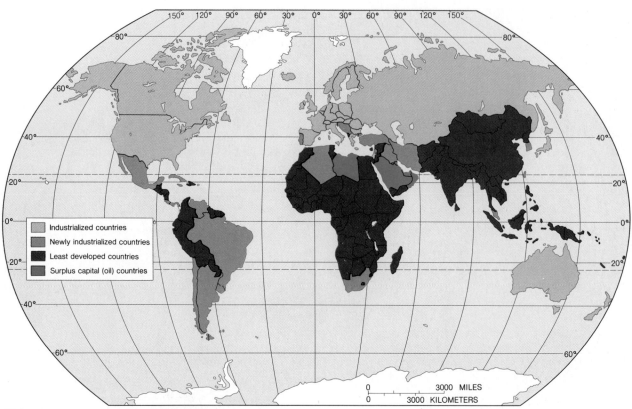

FIGURE 3–24 Division of the World on the Basis of Level of Industrialization

still engaged in peasant agriculture and continue to have high illiteracy rates and high death rates.

The fourth group of countries includes the least industrialized nations of the world. In these countries the majority of the population is engaged in agriculture, which includes intensive peasant agriculture as well as subsistence agriculture, nomadism, and examples of hunting and gathering. Infant mortality, malnutrition, illiteracy, underemployment, periodic food shortages, and inadequate housing continue to plague a large segment of the society (Figure 3–25). Highly reliant upon an agricultural economy, these countries are approaching the end of the twentieth century with severe contrasts between the wealthy few and the impoverished majority.

The majority of the world is part of the region of least industrialized countries. Two countries, India and China, contain 40 percent of the world's population. These two countries typify the least industrialized countries, which are attempting to move to the level of newly industrialized countries. They have a *dual economy* in which there is an industrial base, producing everything from bicycles to jet aircraft, but industrial activity employs a minority of the country's population. The majority of the population is involved in agriculture, and production or acquisition of adequate food supplies occupies a critical part of their daily lives.

Carrying Capacity: Production for Humans

From the time of the earliest human inhabitants, the basic issue of life on earth has been providing sufficient food to maintain the population. The carrying capacity of a specific place is a function

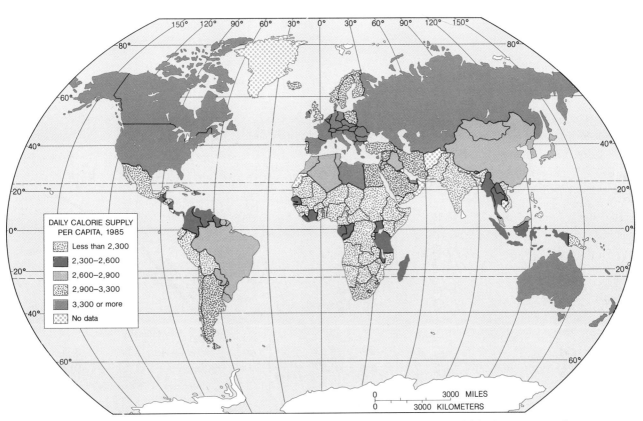

FIGURE 3–25 The Quality of Life Worldwide Maps related to issues of quality of life demonstrate the contrast between the industrialized and less industrialized regions of the world.

of the technological level of the people. Prior to the Neolithic Revolution, the number of people that could be supported by hunting and gathering was very small. Use of domesticated plants and animals changed the carrying capacity of the earth by increasing the productivity per unit of land, thus increasing the number of people the earth could support. The increased carrying capacity that came from the Agricultural Revolution allowed the global population to grow dramatically. Nevertheless, a fundamental question remains. How many people can be supported by the earth's food production?

Increasing agricultural production in the less industrialized world is still largely a matter of human labor. In the industrialized world, changing technology and changing crop and livestock patterns create food surpluses with the labor of few people. These surpluses are possible only because of large energy subsidies and the use of artificial fertilizers and pesticides (Figure 3–26). Measured in terms of yield per unit of energy, the cultural ecology of peasant agricultural systems produces greater yields than those from commercial agriculture in the industrial world. The relative crop yields found in different agricultural systems are important because they affect the carrying capacity of the world. The high yields associated with commercial agriculture, obtained by energy subsidies and use of artificial pesticides and fertilizers, are also important because they may cause permanent damage to the environment.

The pressure on existing agricultural systems to produce more food to feed growing numbers of people is causing changes in agricultural systems and techniques (Figure 3–27). Foremost among these changes has been the adoption of hybrid grains (Figure 3–28). Increase in yields by this method has not been accomplished without cost, however. Consumption of artificial fertilizer has

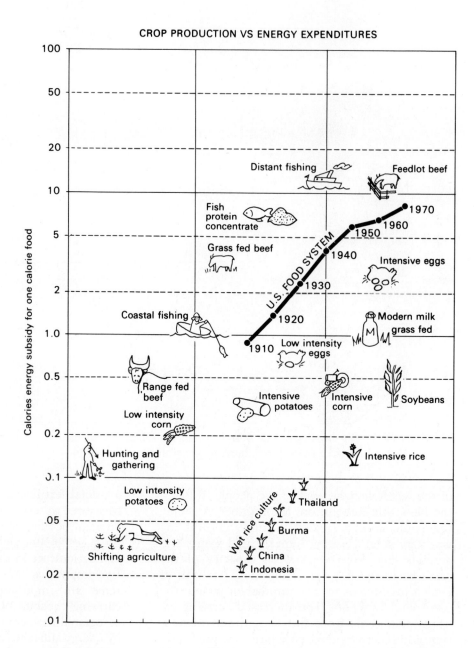

FIGURE 3–26 Energy Costs of Food Production
The energy costs for food production are lowest in shifting agriculture and highest in commercial agriculture.

Source: Carol E. and John S. Steinhart, *Energy Sources, Use, and Role in Human Affairs* (Wadsworth Publishing Company, Inc., Belmount, Calif.: 1974) p. 84.

increased geometrically where these grains have been adopted. Fertilizer consumption increased more than tenfold in India in the period from 1955 to 1985. Not all of the fertilizer is consumed by the plants, but is carried away by water runoff to rivers, lakes, and ponds where it also results in a highly fertile medium for the production of algae and other plant life. The resulting changes in rivers and streams also affect the production of fish and other wildlife.

In addition, in areas where hybrid grains have been adopted, landlords have replaced tenant farmers with mechanized large-scale farming. Critics conclude that although there is more food as a result of the Green Revolution, the human dislocations involved in destroying the age-old tenant systems may be greater than the benefits of additional food.

Closely related to the efforts to increase agricultural production through better varieties of

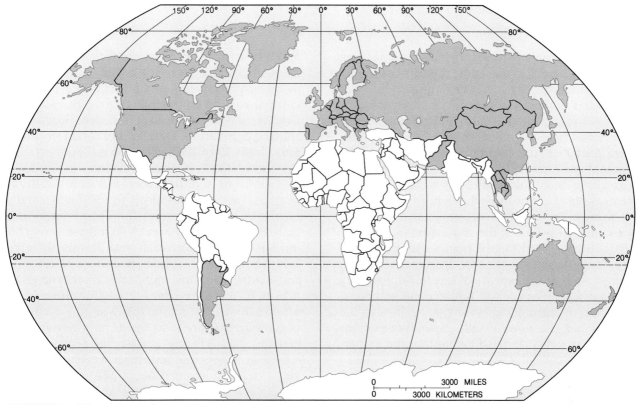

FIGURE 3–27 Food Surplus Nations of the World

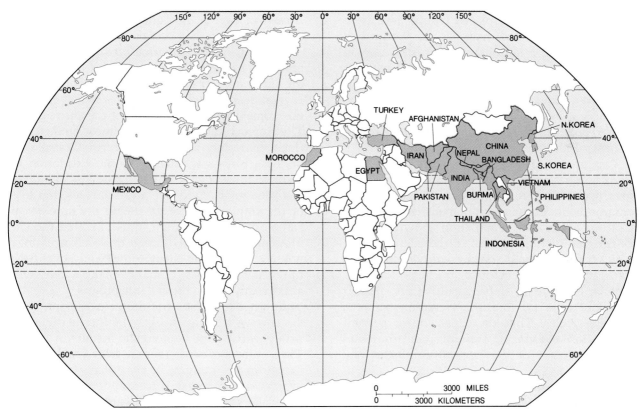

FIGURE 3–28 Areas That Have Adopted Hybrid Grain Countries in which the hybrid grains of the Green Revolution are important are concentrated in Asia.

crops are attempts to increase yields by expanding cultivated land. Large dam projects have expanded irrigation by increasing the reliability of the water supply and provided hydroelectricity for modernization and industrialization. The issues associated with such expansion of agriculture center on the environmental effects of the dams themselves and on flooded river valleys that may have been cultivated for generations, as in the case of the Aswan High Dam.

Related to the problems associated with expanding irrigation is the expansion of peasant agricultural systems in the tropics and subtropics. The territorial expansion of these farming systems comes at the expense of less technologically advanced peoples such as hunters and gatherers, nomads, or shifting cultivators, as well as at the expense of the forest itself. Expanding peasant farming systems are pushing hunters and gatherers, nomadic peoples, and shifting cultivators into smaller and smaller areas. Because nomads, hunting and gathering peoples, and shifting agriculturalists rarely have individual title to the lands they use, they have little recourse against the encroachers.

Life-styles of the displaced groups have been destroyed as their lands have become too limited to support their traditional ways of life. Many of the members of such groups are forced to leave their ancestral lands and relocate in the broader societies around them where they are untrained to function and consequently lose their culture or turn to prostitution, gambling, alcoholism, drugs, or other undesirable activities.

Pressure from increasing populations, both from competing agricultural systems and their own population growth, leads to damage to the environments occupied by nomads, gatherers, or shifting cultivators. Population growth among shifting cultivators has led to the adoption of a new term, *"terminal shifting cultivation."* This refers to the problem created when increasing populations among shifting cultivators prevent adequate time for forest rejuvenation between clearing cycles. Thus the expansion of peasant cultivation and grazing activities contributes to the destruction of the tropical rainforest and subtropical forest.

The issues surrounding the loss of the rainforest reflect the lack of definite figures on how much is actually lost and the lack of agreement on how to define loss. To the biologist, loss may mean either conversion of primary forest to agricultural pasture or tree plantations, or modification of the forest through selective logging or shifting cultivation. Estimates of total forest loss range from 50 million acres (20 million hectares) per year (the U.S. National Academy of Sciences, 1980) to 13 to 16 million acres (5.3 to 10.5 million hectares) per year based on the Wildlife Fund's estimate of 25 to 50 acres lost each minute; the United Nations estimates 18.75 million acres (7.5 million hectares) lost each year. If only the area lost to permanent peasant agriculture practices is considered, the total amount is in the range of 6.25 to 6.75 million acres (2.5 to 3 million hectares) each year. Much of this loss to permanent peasant agriculture is in the form of pastures for grazing. This grazing activity is primarily designed to benefit the industrial world. For example, a tremendous increase in pasture in Central America is in response to the need for cheap low-fat beef for use in fast-food outlets in the United States. The animals grazed on the grass are very lean, and although unprocessed Central American beef is not allowed into the United States, the processed beef patties are.

The expansion of agriculture presents other problems unrelated to the impact it has on shifting cultivators or hunting and gathering peoples. A great deal of evidence indicates that the loss of the tropical rainforest will affect the entire globe. The tropical rainforests are the most biologically rich resource on the face of the earth. Destruction of these forests is leading to the extinction of plant and insect species and to the loss of biological gene pools. Evidence also indicates that loss of the tropical rainforest affects the amount of carbon dioxide in the air, thus affecting the climate of the earth. Although these changes may not have a dramatic effect immediately, over time they may result in climatic changes that will affect many of the earth's inhabitants.

Another form of agricultural expansion with destructive consequences is the increasing numbers of cattle kept by nomadic herders in Africa. The numbers of cattle that are grazed in the Sahel region on the margins of the Sahara have led to overgrazing and desertification. The margins of the Sahara have crept outward into less arid regions as a result of overgrazing that destroyed the plant cover and allowed the soil to be moved by the winds. Consequently, nomadic peoples moved to more humid regions where food might be available, and the traditional nomadic society broke down. Thus several issues are associated with nomadic peoples:

1. The loss of traditional lands used for grazing as peasant agriculturalists expand
2. Overgrazing and resulting desertification associated with increasing numbers of livestock
3. Whether the government should force nomadic peoples to adopt a sedentary lifestyle.

There are no easy answers to these questions, but they represent important issues related to expanding crop or animal production.

A final issue associated with the cultural ecology used by groups as they occupied and changed the earth relates to the increasing industrialization

and urbanization of the world. By the beginning of the twenty-first century, a majority of the world's population will reside in urban places. The world will shift from being basically agricultural to being basically urban in the half century between 1950 and 2000. Although there will still be countries in which a majority of the population live in villages and are engaged in agriculture, the global trend is to urban life. Even in the least industrialized countries, the most rapidly growing places are the cities even though rural populations continue to increase also (Figure 3–29). Associated with the growth in urban population are issues of housing, employment, education, and an adequate life for urban residents. Closely allied with these issues are problems of industrialization as the growing urban populations turn to the production of textiles, boots, and shoes or the assembly of electronic goods. The spread of the Industrial Revolution to the traditionally agricultural countries of the world is causing environmental problems as well. On a global scale, issues such as acid rain, air and water

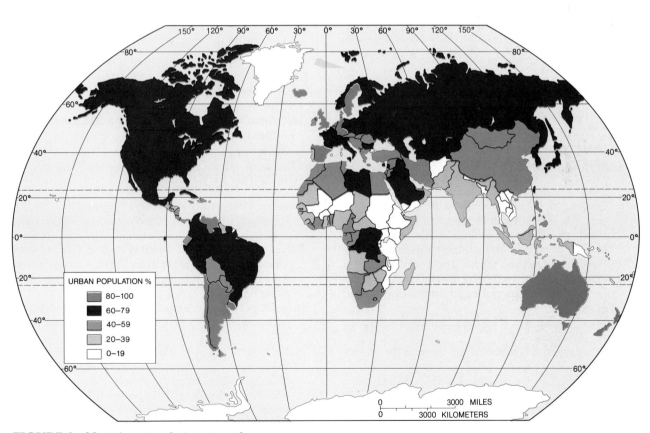

FIGURE 3–29 Urban Population Growth

quality, and related environmental problems are indicative of the environmental impact of human use of the earth.

The freeing of some individuals from agriculture and the labor specialization associated with the earliest villages of Jarmo or Jericho created their own challenges, just as the urbanization and industrialization of large regions of the world are doing today. Changes in quality and quantity of production and resource use are associated with environmental change in each case. The primary issue for the human world today is how to ensure that the continued use of the earth does not result in the impairment of its carrying capacity. The earth has been able to support a population that has increased geometrically numerous times, but a fundamental issue for tomorrow is the extent to which it can continue to increase that carrying capacity.

Conclusion

The development of the distinct cultural ecology of each group of people, and their resultant cultural landscapes, has created the global mosaic of cultural geography. The changes brought by advances in agriculture, in combination with scientific and technological advances since the Neolithic Revolution, have created the cultures and landscape phenomena of the world. In the process of creating the cultural geography of the world, the environment has been transformed. Some of these changes have been destructive, leading to extinction of plant and animal species and threatening the groups who occupy some areas, such as nomads in arid regions. Other destructive changes are more widespread, but have not been as immediately devastating to the occupants of the world. In spite of the negative impact of some aspects of the human occupation of the world, the changes in agriculture and technology have allowed the world's population to increase dramatically. Examination of the growth of the world population indicates its future growth potential and the challenges the world will face if continued population growth is associated with the aspects of culture geography that have been environmentally destructive.

QUESTIONS

1. Describe the location and cultural ecology of hunting and gathering peoples of the world.
2. Where are nomad societies found today? What aspects of their cultural ecology are unique?
3. Discuss the general characteristics of peasant farming, indicating three major types.
4. What are the main differences in the cultural ecology of peasant agricultural systems and commercial agricultural systems?
5. Discuss how the von Thünen model applies to the location of commercial agriculture in the United States.
6. What impact did plantation agriculture have on the Americas?
7. What types of land tenure are there in the world? What issues are associated with each?
8. Describe five types of rural landscapes recognizable in the world.
9. What are some generalizations that can be made about population densities and agriculture?
10. Describe the divisions of the world based on the Industrial Revolution.

SUGGESTED READINGS

Ashmore, Wendy, ed. *Lowland Maya Settlement Patterns*. Albuquerque, New Mexico: University of New Mexico Press, 1981.

Bassett, Thomas J. "The Political Ecology of Peasant-Herder Conflicts in the Northern Ivory Coast." *Annals of the Association of American Geographers* 78 (September 1988): 453.

Butzer, Karl W. *Early Hydraulic Civilization in Egypt: A Study in Cultural Ecology*. Chicago: University of Chicago Press, 1976.

Butzer, Karl W., Juan F. Mateu, Elisabeth K. Butzer, and Pavel Kraus. "Irrigation Agrosystems in Eastern Spain: Roman or Islamic Origins?" *Annals of the Association of American Geographers* 75 (December 1985): 479–509.

Carneiro, Robert. "Slash and Burn Cultivation among the Kuikuru and Its Implications for Cultural Development in the Amazon Basin,'' in J. Wilbert, ed., *The Evolution of Horticultural Systems in Native South America: Causes and Consequences, A Symposium.* Caracas: Editorial Sucre, 1961.

Carr, Claudia J. *Pastoralism in Crisis: The Dasanetch and Their Ethiopian Lands.* Research Paper no. 180, Department of Geography, University of Chicago, 1977.

Chakravarti, Aninda K. ''Diet and Disease: Some Cultural Aspects of Food Use in India,'' in Allen G. Noble and Ashok K. Dutt, eds., *India: Cultural Patterns and Processes,* pp. 301–23. Boulder, Col.: Westview Press, 1982.

———. ''Green Revolution in India.'' *Annals of the Association of American Geographers* 63 (1973): 319–30.

Chisholm, Michael. *Rural Settlement and Land Use: An Essay in Location,* 3d ed. London: Methuen, 1979.

Cole, Donald P. *Nomads of the Nomads: The Al Murrah Bedouin of the Empty Quarter.* Chicago: Aldine Publishing Co., 1975.

Dove, Michael R. *Swidden Agriculture in Indonesia: The Subsistence Strategies of the Kalimantan Kantu.* Amsterdam: Mouton, 1985.

Duckham, A. N., and G. B. Masefield. *Farming Systems of the World.* New York: Praeger, 1970.

Evans, E. Estyn. ''The Ecology of Peasant Life in Western Europe,'' in William L. Thomas, ed., *Man's Role in Changing the Face of the Earth,* pp. 217–39. Chicago: University of Chicago Press, 1956.

Frank, R., and B. Chasin. *Seeds of Famine: Ecological Destruction and the Development Dilemma in the West African Sahel.* Montclair/New York: Allenheld, Osmun/Universe Books, 1980.

Gregor, Howard F. *Geography of Agriculture: Themes in Research.* Englewood Cliffs, N.J.: Prentice-Hall, 1970.

Grigg, David. ''Agricultural Geography.'' *Progress in Human Geography,* 5 (1981): 268–76; 7 (1983): 255–60.

———. ''The Agricultural Regions of the World: Review and Reflections.'' *Economic Geography* 45 (1969): 95–132.

Grossman, Larry. ''The Cultural Ecology of Economic Development.''*Annals of the Association of American Geographers* 71 (June 1981): 220–36.

Grove, A. T. ''Population Densities and Agriculture in Northern Nigeria,'' in K. M. Barbour and R. M. Prothero, eds., *Essays on African Population,* pp. 115–36. London: Routledge & Kegan Paul, 1971.

Hale, Gerry A. ''The Origin, Nature, and Distribution of Agricultural Terracing.'' *Pacific Viewpoint* 3 (1961): 1–40.

Hart, John Fraser. ''Cropland Concentrations in the South.'' *Annals of the Association of American Geographers* 68 (December 1978).

Horvath, Ronald J. ''Von Thünen's Isolated State and the Area around Addis Ababa, Ethiopia.'' *Annals of the Association of American Geographers* 59 (1969): 308–23.

Huang, Philip C. C. *The Peasant Economy and Social Change in North China.* Stanford, Calif.: Stanford University Press, 1985.

Isaac, Erich. *Geography of Domestication.* Englewood Cliffs, N.J.: Prentice-Hall, 1970.

Johannessen, Carl L. ''The Domestication Processes in Trees Reproduced by Seed: The Pejibaye Palm in Costa Rica.'' *Geographical Review* 56 (1966): 363–76.

Johnson, Douglas L. *The Nature of Nomadism: A Comparative Study of Pastoral Migrations in Southwestern Asia and Northern Africa.* Chicago: Research Paper No. 118, Department of Geography, University of Chicago, 1969.

Jordan, Terry G. ''Early Northeast Texas and the Evolution of Western Ranching.'' *Annals of the Association of American Geographers* 67 (March 1977): 66–87.

King, F. H., *Farmers of Forty Centuries.* Emmaus, Pa.: Organic Gardening Press, 1927.

Lappe, Frances Moore, and Joseph Collins. *Food-First: Beyond the Myth of Scarcity.* Boston: Houghton Mifflin, 1977.

Lee, Yuk (with Richard Stevens). ''A Spatial Analysis of Agricultural Intensity in a Basotho Village of Southern Africa.'' *The Professional Geographer* 31, (May, 1979): 177–83.

Masuda, Shozo, Izumi Shimada, and Craig Morris, eds. *Andean Ecology and Civilization. An Interdisciplinary Perspective on Andean Ecological Complementarity.* Toyko: University of Tokyo Press, 1985.

Monod, Theodore, ed. *Pastoralism in Tropical Africa.* London: International African Institute, Oxford University Press, 1975.

Pannell, Clifton. ''Recent Chinese Agriculture.'' *Geographical Review* 75 (April 1985): 170–85.

Roberts, Bryan. *Cities of Peasants.* London: Arnold, 1978.

Sauer, Carol O. *Agricultural Origins and Dispersals,* 2d ed. Cambridge, Mass.: M.I.T. Press, 1969.

Stanislawski, Dan. *Landscapes of Bacchus: The Vine in Portugal.* Austin: University of Texas Press, 1970.

Steward, Norman R., Jim Belote, and Linda Belote. ''Transhumance in the Central Andes.'' *Annals of the Association of American Geographers* 66 (September 1976): 377–97.

Symons, Leslie. *Agricultural Geography,* 2d ed. Boulder, Col.: Westview Press, 1979.

Turner, B. L., Robert Q. Hanjam, and Anthony V. Portaararo. ''Population Pressure and Agriculture Intensity.'' *Annals of the Association of American Geographers* 67 (September 1977): 384–96.

Udo, K. R. *The Human Geography of Tropical Africa.* London: Heinemann, 1982.

Uyanga, Joseph T. *A Geography of Rural Development in Nigeria.* Washington, D.C.: University Press of America, 1980.

Vermeer, Donald E. "Collision of Climate, Cattle, and Culture in Mauritania during the 1970s." *Geographical Review* 71 (1981): 281–97.

Von Thünen, Johann Heinrich. *Von Thünen's Isolated State: An English Edition of Der Isolierte Staat.* Translated by Carla M. Wartenberg. Elmsford, N.Y.: Pergamon Press, 1966.

Wittlesey, Derwent S. "Major Agricultural Regions of the Earth." *Annals of the Association of American Geographers* 26 (1936): 199–240.

Wilkensen, J. C. *Water and Tribal Settlement in South-East Arabia: A Study of the Aflaj of Oman.* Oxford: Clarendon Press, 1977.

Yappa, Lakshman S. "The Green Revolution: A Diffusion Model." *Annals of the Association of American Geographers* 67 (September 1977): 350–59.

Yelling, Y. A. *Common Field and Enclosure in England, 1450–1850.* Hamden, Conn.: Archon Book, 1977.

The Geography of Population: Patterns and Problems

Population growth does not provide the drama of financial crisis or political upheaval, but its significance for shaping the world of our children and grandchildren is at least as great. . . . Failure to act now to slow growth is likely to mean a lower quality of life for millions of people."   A. W. CLAUSEN

THEMES

Population growth in the world.
Explanations of the demographic transition.
The role of culture in population growth.
Regional variations in population characteristics.
The changing distribution of population.

CONCEPTS

Population growth
Birth rates
Life expectancy
Population doubling time
Death rates
Demographic transition model
Population pyramids
Dependency ratio
Zero population growth
Malthusian equation
Positive checks
Preventive checks
Completed family size
Biological and cultural resistance
Infant mortality rate
Lifeboat earth
Physical quality of life
Sickle-cell anemia
Primogeniture

The population geography of the earth is related to the earth's carrying capacity. The number of people inhabiting the earth at any given time is a reflection of their ability to obtain food and other needed resources, which in turn is based on the knowledge and technology of the specific population. The population at the time of the Neolithic Revolution is estimated to have been between 5 and 10 million. In the last 10,000 to 12,000 years, the population has soared to over 5 billion (Figure 4–1). The rapid increase in the world's population during the last ten millennia is due in part to technological changes that have increased the earth's carrying capacity. Many observers worry that we are now outstripping global carrying capacity and are creating a "population bomb" in which the earth's food-producing capability will be overwhelmed by ever greater populations. While total global food production has increased, in some regions population is growing more rapidly than food production. This is particularly true in Africa, where the amount of food produced per person has actually declined since 1960 (Figure 4–2).

Per capita grain production is declining in 40 of the world's 160 countries. Declining per capita grain production usually leads to a decline in the standard of living. "Absolute poverty" is a threshold nutritional level set at 2,250 calories per day. An estimated 800 million people in the world now fall below this level. The population living in absolute poverty is one indicator that leads some observers to conclude that the world is facing *overpopulation*. Overpopulation is defined as the condition that exists when the number of people exceed the available food resources.

Statistics on food-deficit countries are, at best, estimates since in countries where food deficits are a problem, much of the food production is not marketed. Total food production in Africa, for example, has increased 20 percent in the last decade, but per capita production in Africa decreased 11 percent in that same period (Figure 4–3).

An old riddle asks when a lily pond will be half covered by lilies if the pond has one lily on the first day, the number of plants doubles each day, and the pond is full on the thirtieth day. The answer is the twenty-ninth day. Many observers believe that the present world population is sim-

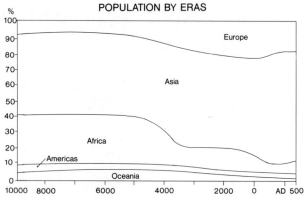

FIGURE 4–1 Population by Eras: Early History Global population between 10,000 B.C. and A.D. 500 In the era from 10,000 B.C. to 5000 B.C. Africa and Asia combined for over 80 percent of total population, but Africa's proportion of the earth's population declined and Europe's increased between 5000 B.C. and A.D. 0.

ilar to the lilies on a pond. On the twenty-ninth day the lily pond is only half covered, but on the thirtieth day it will be completely covered with lilies, obscuring the water and beginning a process of deterioration in the pond. If the earth's population is indeed analogous to the lily pond, the next generation will consume all the resources the earth can provide.

Those who maintain that world population growth is a crisis argue that millions of people in Africa, Latin America, and Asia are forced by circumstances beyond their control to destroy the very resources from which they eke out their living.

The dilemma for the world posed by the destruction of resources is analogous to a person with a savings account of $100,000. If the account earns 10 percent interest per year, the owner can spend $10,000 each year without affecting the principal. If the individual spends more than $10,000 per year, not only will the principal decline but future years' interest will also decline. When population exceeds the carrying capacity of the earth, the residents must consume principal from their "bank account" by expanding agriculture into marginal lands, eliminating fallow periods, increasing animal numbers, or otherwise engaging in activities that impact the environ-

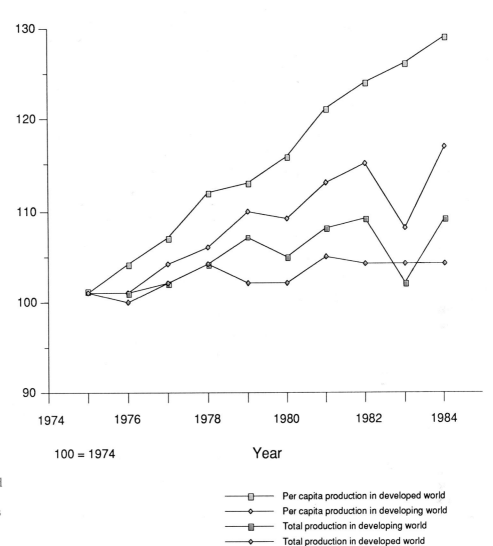

FIGURE 4–2
Agricultural Production
The rate of increase in
both total and per capita
food production is higher
in the more industrialized
"developed" countries of
the world than in the less
industrialized "develop-
ing" countries.

ment. Thus there is a basis for the concern that
the rapidly increasing number of people constitutes a population crisis. If the world is indeed in
the demographic twenty-ninth day, the population issue is a worldwide concern.

The Population Crisis

The significance of the population issue is evident in the simple arithmetic of population

growth. Approximately 143 million children were
born and 51 million people died from old age,
disease, accidents, or malnutrition in 1989. This
net growth (92 million) exceeds the population of
Mexico or the population of any Western European country. Two years of such growth would
exceed the population of Japan, and three years'
growth would equal the populations of the United
States and Canada combined. The net daily increase in population is approximately 225,000
people. During the time it has taken you to read

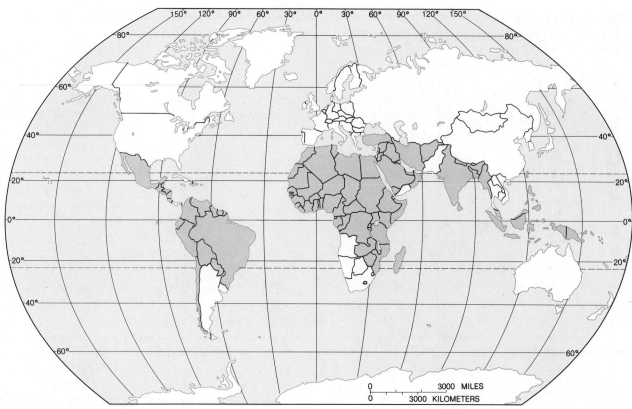

FIGURE 4–3 Food-Deficit Countries of the World Food deficit countries of the world are found in both less-industrialized countries and industrialized countries, but wealthy industrialized countries face fewer problems in purchasing additional food from surplus producers.

this paragraph, 175 people have been added to the earth's population—nearly 3 persons each second.

Growing global population places greater demands upon resources. To provide each new resident of the earth a single glass of milk and a one-pound loaf of bread daily would require increased production equivalent to adding 1,500 dairy cows and 45 acres of land for wheat each day, or more than 16,000 acres of wheat and nearly 550,000 cows each year.

Population through Time: A Theme in Cultural History

A related population issue of interest to cultural geographers is the theme of cultural history. Changing patterns of population distribution, regional population patterns, and the spread of

humans are of central importance in understanding the human occupance of the world. The study of world populations through time illustrates the problems faced by cultural geographers in reconstructing cultural history. Census records would provide the best source of information, but unfortunately they do not extend very far into the past and are unavailable for many areas.

The earliest census records date from 2000 B.C. when some areas in China and the Middle East tried to count their populations periodically. These attempts to calculate population were not comprehensive, nor did the data survive intact. The Romans began census taking in the sixth century B.C. and continued to count the inhabitants of Roman lands at regular intervals for hundreds of years. They used the census information as a basis for levying taxes, recruiting armies, and commandeering labor for public works.

Growing populations present difficult challenges to countries trying to provide education, jobs, and shelter for their youth.

Early modern censuses include French records from their Quebec colony in 1665 and Swedish records from 1749. Most census information dates from the eighteenth and nineteenth centuries in Europe, North America, and Japan. The United States began its first national census in 1790. The United Kingdom and France began comprehensive census enumeration in 1801. Census taking has become a worldwide activity in the twentieth century; over 95 percent of the population of the world resides in countries where regular official censuses are taken (see the Cultural Dimensions box).

Attempts to develop information about population numbers, distribution, and change for periods before census records produce only crude estimates. Numbers given for the earliest ancestors of humans one million years ago are based on attempts to reconstruct the carrying capacity of the natural environment. Such estimates suggest a total population for *Australpithecus* of 125,000 (Figure 4–4). People occupied essentially all of the earth's surface except Antarctica, the ice caps of Greenland, and the highest mountain areas by the time of the Neolithic Revolution.

Population increased at least tenfold in the first 4,000 years after the Neolithic Revolution increased the earth's carrying capacity. It is estimated that 2,000 years ago the total world population was between 125 and 150 million (Table 4–1). The largest populations of the world at this time were in South and East Asia: an estimated 40 million people lived in today's India, Pakistan, and Bangladesh; 45 million in East Asia, including China; and 15 million in the Middle East, primarily in what is now Egypt. Europe's population was about 30 million people, primarily around the Mediterranean controlled by the Roman Empire. Other areas of the world were more sparsely inhabited, with important population clusters in Southeast Asia, the Sudan of Africa, and Central and South America. Worldwide, av-

FIGURE 4–4 Population Growth of the World
Global population increased slowly until the last few hundred years as births and deaths were nearly equal. The rapid growth in recent centuries is caused by lowering death rates.

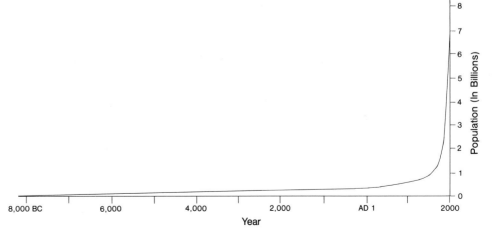

CULTURAL DIMENSIONS

A Census Is More Than Counting People

The census-taking process is much more than simply counting the numbers of people in a place. According to the United Nations, a census is the "total process of collecting, compiling, and publishing demographic, economic, and social data pertaining, at a specified time or times, to all persons in a country or delineated territory."* (UN 1958:3). This definition indicates the great variety of information that may be included in a census. Modern census information for the industrialized countries of the world may include such things as place of birth; race and ethnicity; standards in housing (including availability of running water, electricity, and plumbing); place of residence; number of persons in the household; occupation; and movement since the last census.

Early censuses and those taken in the less industrialized world are extremely important to geographers and other social scientists because they include a wide variety of information about the life and culture of the people themselves. Questions may include distance to the nearest water source, number of people per room in the habitation, amount of production of various types of crops, number of years of education, number of births in the family in the past year, or the number of wives per husband.

Simple things, such as age profiles, tend to reflect culture. In the past the process of recording births and birth rates in most less industrialized countries was less formal than in industrial countries. Birth rates and ages given to census enumerators were sometimes rounded off rather than precise. The average person might have said that she was forty or fifty years old rather than forty-three, forty-seven, or fifty-four. Such generalized ages were particularly in evidence in societies where certain years are important for religious or political events. Many individuals born near the time of some particularly important cultural event claimed that date as their year of birth.

The census provides information far beyond simple numbers. The great enumeration of the

Iſd Giſlebt teñ *HADESORE*.7 Walter *IN CLENT HVND.*
gener ei de eo. Bricſmar tenuit teiñ regis.E.Ibi . II . hidæ
In dñio ſunt . II . car.7 II . uilt 7 VIII . bord 7 IIII . cotmanni
cū . II . car.7 tcia poſſet .ee. ibi. Ibi . IIII . bouarii.7 VII . ſa
linæ reddt . CXI . mittas ſalis.
T.R.E . ualb . LX . ſolid. Modo . XLV . ſolid.

.XXI. **D**TERRA DROGONIS FILII PONZ. *IN DODINTRET HD.*
DROGO filius ponz teñ de rege *HOLIM*. Vlmar tenuit
7 potuit ire quo uoluit. Ibi . I . hida geld.7 una car poſſet .ee. ibi.
Waſta . ē 7 uuaſta fuit.T.R.E . ualb . V . ſolid.
Iſd Drogo teñ *STILLEDVNE*. Vlchet tenuit.7 ñ poterat
diſcedere á dño ſuo Vlmaro. Ibi dimid hida geld . Tra . ē
II . car. Valuit . V . ſolid. Modo. ē Waſta.
Iſd Drogo teñ *GLESE*. Vlmar tenuit 7 potuit ire
quo uoluit. Ibi . I . hida geld. In dño dimid car.7 I . uilt
7 III . bord cū . I . car.7 alia poſſet ibi. ee. Ibi molin de
IIII . ſot 7 VIII . denar. Valuit . xx . ſot. Modo . x . ſolid.
Iſd Drogo teñ unā v in *MERLIE* ᛗ regis.7 geldat.
Ibi hr . I . Radman redd . VI . ſolid p annū . Ernuin tenuit.

.XXII. **H**TERRA HERALDI FILII RADVLFI COMIT.
HERALD fili Radulfi teñ de rege . I . hidā in *WICH*.
7 ibi hr . XX . burgſes . cū . VII . ſalinis . reddt L . mittas ſalis.
Valuit 7 ualet . XL . ſolid.

XXIII. **W**TERRA WILLI FILII ANSCVLFI. *IN CAME HVND.*
WILLS filius Anſculfi teñ de rege *ESCELIE*.7 Wibt
de eo. Vluuin tenuit. Ibi ptin una Bereuuiche *BER*
CHELAI. Int tot . IIII . hidæ.   In dño . ē dimid car.
7 II . uilti 7 IX . bord cū . IIII . car. Nem . I . leuua lg.
T.R.E . ualuit . c . ſot. Modo . LX . ſolid.
Hoc ᛗ emit iſd Wluuin T.R.E . de eƥo ceſtrenſi ad
ætatē triū hōum. Qui cū infirmat ad finē uitæ ueniſſet.

FIGURE 4—A A Page from the Domesday Book
This page from the Domesday book records the population, land, farm animals, crops and equipment of one small part of England.

British Isles undertaken under the direction of William the Conqueror in A.D. 1085–86 (the Domesday Book) counted farms, farm animals, farm buildings, equipment, types of cultivation, villages, and economic activities, so that William could have a better idea of what he had conquered (Figure 4—A). The Domesday Book recorded not only the population numbers, but the total way of life, providing insights to scholars centuries later.

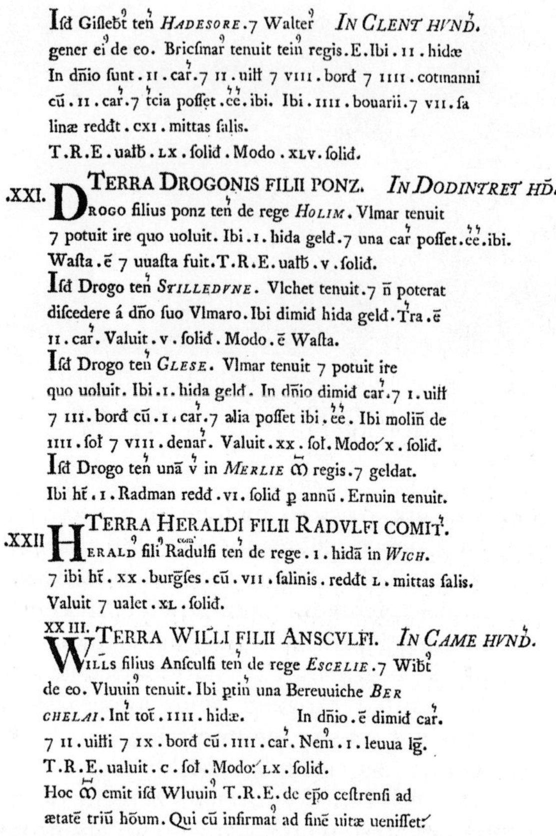

TABLE 4–1
Population of the World 2,000 Years Ago (in millions)

South and Southeast Asia	40
East Asia	45
Middle East	15
Europe	30
Americas	4
Oceania	1
Africa	17
Total	152

Source: Colin McEvedy and Richard Jones, *Atlas of World Population History* (New York: Facts on File, 1978).

erage life expectancy is estimated to have been between 30 and 40 years. Average *life expectancy* remained constant and low until the advent of scientific medicine in western Europe in recent times.

The population of the world increased slowly to approximately 450 million by A.D. 1500 (Figure 4–5). World population doubled between 1500 and 1800, reaching one billion by 1830. This rapid population increase was due largely to the later Agricultural Revolution, which increased reliability of food production, and to medical and social advances that increased life expectancy.

European Population Growth: Regional Population Dynamics

The population of Europe initially grew slowly as cultivated crops diffused from the Middle East about 7,000 years ago. Europe's population reached its pre-Renaissance peak in A.D. 200, due to the stability provided by the Roman Empire (Figure 4–6). This level was not reached again until A.D. 1000–1350 when the population more than doubled, reaching an estimated 80 million people. This population growth was due to the expansion of agriculture in northern and western Europe as the forests of Germany, England, southern Sweden and Finland, and European Russia's plains were brought under cultivation. Population growth slowed after A.D. 1300 as less new land was cultivated. The absolute number of people in Europe declined precipitously between 1347 and 1353 primarily because of the bubonic plague (caused by *Pasteurella pestis*). The bubonic plague diffused to the Crimea from Mongolia where the plague was *endemic*. One of the caravans traveling the ancient silk route brought the plague to the ports of the Black Sea in 1347, beginning a process of hierarchical diffusion to the major ports of Europe (Figure 4–7).

Plague is a disease that affects rodents, fleas, and humans. The ships of the Middle Ages were ideal agents for transmitting the disease, and the major ports of the Mediterranean received the infestation in a few months. Trade in agricultural products and handicraft goods between the Mediterranean and northern Europe ensured its spread across the highly populated French countryside to the rest of Europe. The disease appeared in both continental Europe and the British Isles by 1348.

FIGURE 4–5 World Population by Region in A.D. 1500 Asia, Europe and Africa continued to have the largest populations, but the Americas and the countries around today's Near East together were home to over one half million people.

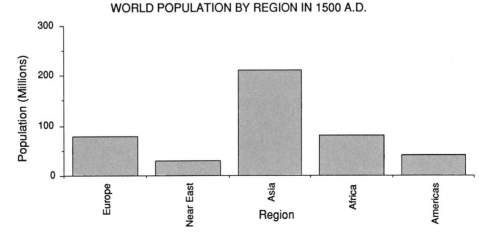

WORLD POPULATION BY REGION IN 1500 A.D.

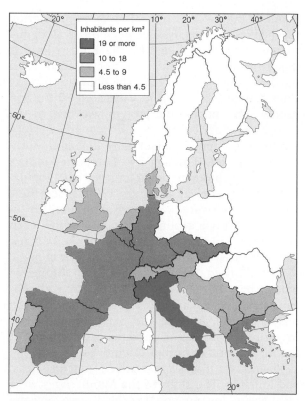

FIGURE 4–6 Population of Europe in A.D. 200 The southern areas of Europe had the highest population densities, suggesting the importance of the Mediterranean region at this time.

It spread northward into Scandinavia and then eastward into Poland and Russia at a slower rate since the population densities were lower.

Between a quarter and a third of Europe's population died in the great plague of 1347–1353 that we know as the "Black Death." Subsequent outbreaks of the plague and other diseases reduced Europe's population to the level of two hundred years earlier. The growing numbers of people who were resistant to the plague ultimately led to its disappearance as a threat to the population of Europe, and after A.D. 1400 the population once again began to grow.

Population growth rates in western Europe increased dramatically after 1700 as social changes associated with industrial, technological, and agricultural advances occurred. Improvements in technology and productivity were paralleled by improvements in plant and animal vari-

eties and changes in agricultural systems. Hybrid plants and improved animal husbandry provided more food, and agricultural practices incorporated the more scientific methods of the Modern Agricultural Revolution, including use of fertilizer and crop rotation.

Industrialization brought improvements in the productivity of labor, transportation, and energy use that gave people freedom and opportunity on a scale never before dreamed possible. Advances in hygiene, sanitation, and medicine improved the life expectancy allowing an increased population growth rate. Rapid population growth in Europe after 1700 fueled the spread of European culture to the rest of the world.

Between 1500 and 1900, several major population migrations occurred, most of them movements of Europeans to less industrialized countries (Figure 4–8):

1. Northern Europeans to North America.
2. People from the Mediterranean countries of Europe to Middle and South America.
3. People from the United Kingdom to Australia, New Zealand, India, and Africa.
4. African slaves to America, Arab and Turkish areas.
5. Russians eastward into Siberia.

Over 5 million people left Europe between 1500 and 1845, and at least 50 million more migrated between 1845 and the beginning of World War I (1914). Russia and the British Isles were each the source of roughly 10 million migrants, while Germany, Italy, and the Austro-Hungarian Empire added over 5 million each. Scandinavia, with its relatively small population, contributed 2.5 million. Thirty million of these immigrants came to the United States, with the balance going to Canada, Latin America, Australia, New Zealand, and to Russian Siberia (Table 4–2).

As they migrated, Europeans spread the awareness of hygiene and medical care to other areas, leading to an increase in the population growth rate throughout the world. Ever-increasing growth rates after 1700 are reflected in average annual growth rates by time period and in population *doubling times* (Figure 4–9). The first global population of one billion was reached after

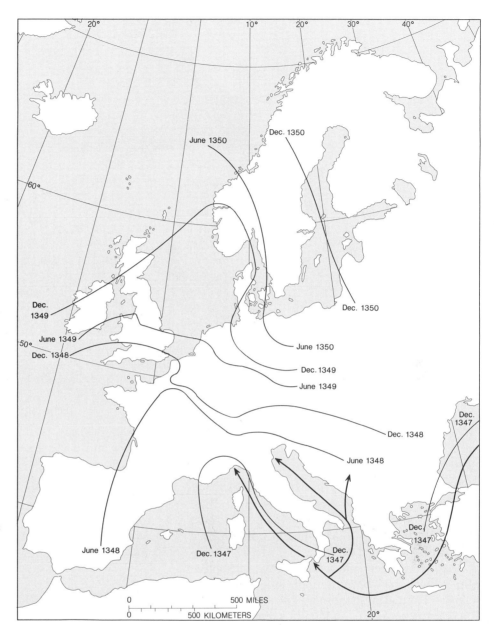

FIGURE 4–7 The Spread of the Bubonic Plague The Bubonic Plague diffused to Europe from Asia via ship, and then spread across Europe, killing between one-quarter and one-third of the people.

TABLE 4–2
Migration of Europeans to major destination countries and regions, 1866–1915

REGIONS	1866–1875	1876–1885	1886–1895	1896–1905	1906–1915
Canada	99,722	277,097	325,653	480,350	2,069,206
United States	1,939,153	3,315,934	4,691,508	4,887,155	6,475,241
Latin America	196,684	757,108	2,047,993	2,029,190	3,747,956
Asia	190	1,478	4,144	14,316	45,660
Oceania	232,139	462,218	286,609	286,816	805,856
Africa	51,916	199,158	397,282	693,212	669,400

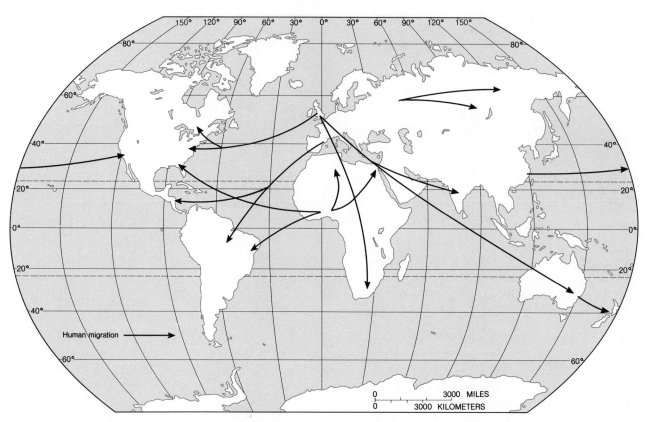

FIGURE 4–8 Migrations (A.D. 1500–1800) The great migrations during this period were from Europe to North and South America, (with lesser numbers to Australia, New Zealand, and South Africa); from European Russia to Siberia, and from Africa.

millennia of population growth. A second billion was added in only one hundred years. The most recent increase from four to five billion required only twelve years.

As we have seen, the history of world population growth has been one of long periods of relative population stability followed by relatively short periods of sudden upsurge in population. The first major increase in population occurred when the use of fire as a tool and the subsequent development of sophisticated blade tools expanded the earth's carrying capacity. The Neolithic Revolution began another period of rapid growth, and the Industrial and Technological Revolutions of the eighteenth and nineteenth centuries provided another upward jump in total numbers. Diffusion of scientific, agricultural, and medical advances to most of the world in the twentieth century prompted an even greater in-

crease in global population. The rapid increase in total numbers in the last century masks important regional differences in population numbers and rates of change.

The Demographic Transition

Population change (whether growth or loss) in an individual country is a function of the number of births per thousand population (*birth rate*), the number of deaths per thousand population (*death rate*), the number of emigrants leaving a country per thousand population (*emigration rate*), and the number of immigrants into a country per thousand population (*immigration rate*). The population change of a country is obtained by calculating the actual number of births, deaths, emigrants, and immigrants and subtracting the

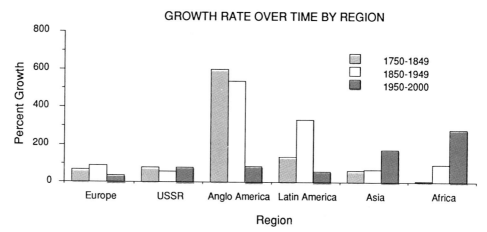

FIGURE 4–9 Population Growth Rate Over Time by Region The growth rates of the major world regions have changed dramatically over time. The highest rates of growth today are found in less industrialized countries.

number of deaths and emigrants from the sum of births and immigrants. Examination of the resulting rate of population change reveals great differences between individual countries. Some countries' populations are growing at rates of over 4 percent each year, a rate that could double their population in less than twenty years if maintained. Other countries have stable populations, with growth rates that are near zero or negative (Table 4–3).

The difficulty of estimating future global growth rates lies in determining how growth rates will change in individual countries over time. *Population projections* for future numbers and rates of change involve calculating the future population based on some specified rate of change.

Historic changes in population provide one basis for estimating future population numbers and rates of change. Examination of the population history of European, Asian, Latin American, and African countries shows that those countries of Europe that were the first to experience rapid population growth now have the lowest population growth. The rate of population growth in these countries seems to have followed a pattern of rapid increase with subsequent decline to a slow growth rate (Figure 4–10). This pattern is referred to as the *demographic transition* (Figure 4–11).

The demographic transition divides the growth rates of the world into four broad stages. In the first stage, both birth and death rates are high, and the resulting growth rate is low. Stage one repre-

TABLE 4–3
Population Growth Rates of Selected Countries

COUNTRY	GROWTH RATE (Percent)	YEARS TO DOUBLE
Brazil	2.0	34
Cuba	1.0	69
Egypt	2.8	24
Fiji	2.3	31
Guatemala	3.2	21
Haiti	2.8	25
India	2.0	35
Israel	1.6	42
Japan	0.5	133
Jordon	3.6	19
Kenya	4.1	17
Libya	3.1	22
Mali	2.9	24
Singapore	1.0	71
Spain	0.4	154
Sweden	0.1	673
Switzerland	0.3	277
Togo	3.3	21
United States	0.7	99
Zambia	3.7	19

Source: Population Reference Bureau, *1988 World Population Data Sheet.*

sents the characteristics of population growth for most of human history. Stage two is characterized by continued high birth rates, but declining death rates associated with better medical care and sanitation. During this period the growth rate is high, and a "population explosion" occurs. Stage three occurs when birth rates drop to levels that

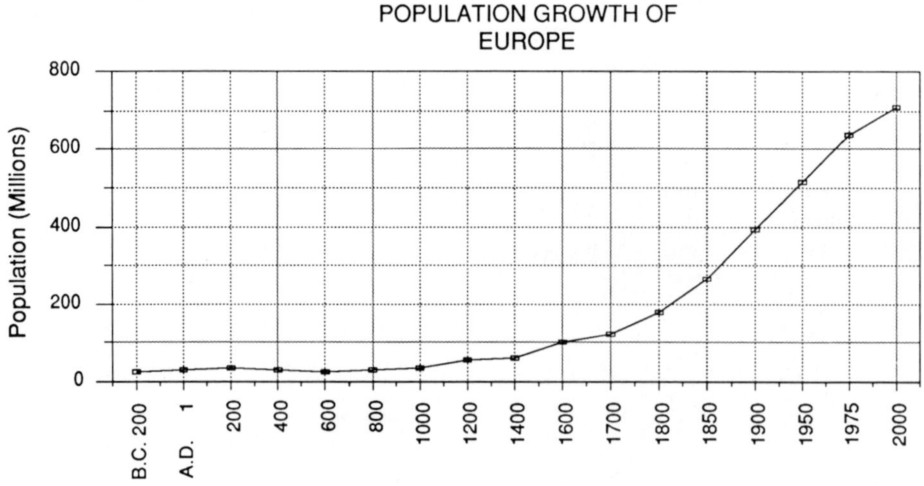

POPULATION GROWTH OF
EUROPE

FIGURE 4–10
Population Growth of Europe The population of Europe increased slowly prior to A. D. 1400, but growth since that time demonstrates the effect of exponential increase.

approach the death rates, and population growth is slow. Stage four occurs when population growth approaches zero or a negative rate averaged over many decades.

Both death and birth rates have been high for most of human history, with birth rates hovering around forty per thousand and death rates only marginally different. Good harvest years might result in increased births and fewer deaths, allowing an increase in population, but the earth's population remained small since average life expectancy was low. Populations grew relatively slowly even though birth rates were very high.

Western Europe clearly entered stage two of the demographic transition in about 1700. Adoption of better hygiene and better medical care that allowed more infants to live to maturity was an important factor allowing the change from stage one to stage two. The countries of the less industrialized world generally entered the second stage in the twentieth century, and their populations continue to increase dramatically.

Factors that seem to determine the movement from stage two to stage three are not as easily identifiable as those accounting for movement from stage one to stage two. Western Europe and other industrialized regions seem to have moved to stage three or four in part because of the actual and perceived value and benefit of children as opposed to their disadvantages and costs. The impact of a child born to a family in the industrialized world is far different from that of one born

in a less industrialized country, as are the relative advantage and utility of each child. The benefits of children may include their labor, the security they provide for parents in their old age, personal satisfaction through seeing the continuation of the family name, or such miscellaneous benefits as tax deductions or government subsidies. The costs of children may include expenses associated with child rearing (including education, food, clothing, and shelter); the restricted options for the parents caused by the responsibility of caring for children; the loss of the mother's labor as a result of pregnancy and child care; and miscellaneous disadvantages such as mental stress on the parents.

The relative benefits and costs of children vary between countries with an industrial economy and those with an agricultural economy, and may also vary significantly between individual families. A child in less industrialized countries may provide the only form of security in old age available to an average set of parents since governments rarely provide programs to assist the elderly. The dominance of agriculture in the economies of less industrialized countries makes children economically useful at an early age. Children can weed the garden, tend poultry, collect firewood, fetch water, or other tasks by the time they are six. Social status associated with large families, particularly with male children, also acts as an incentive for having more children.

The cost of additional children in less industrialized countries is less than in industrialized

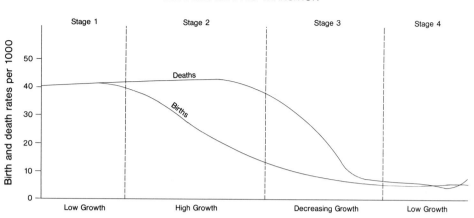

THE DEMOGRAPHIC TRANSITION

FIGURE 4–11
The Demographic Transition The demographic transition model suggests that populations change over time from high birth and high death rates with low growth rates; to high birth and low death rates with high growth rates; to low birth and low death rates with low growth rates.

countries. Educational opportunities are limited; food, clothing, and medical care are available to a much lesser degree, and hence the cost of rearing a child is lower. The mother in less industrialized countries can more readily return to work and can continue to care for the infant in the field or the kitchen. Women are unable to bring young children to the office or factory in industrialized countries. Thus parents in less industrialized countries are less restricted by additional children since additional children are not viewed as adding significantly to their poverty.

The benefits and costs of children are directly opposite in industrialized countries. Governments provide much of the care for the elderly. There are numerous ways for an individual or couple to derive satisfaction other than child bearing. Legal and social restrictions prevent children from working until maturity. The costs of providing education and consumer items such as televisions, clothing, and automobiles are much higher. Restriction on the actions of the parents are greater because families in the industrial world tend to consist of only the mother and father and their immediate offspring. In less industrialized countries, the family often includes the grandparents or other members of the extended family, which gives parents the option of having home care for their children. The smaller the family in industrialized countries, the greater the freedom of choice of both material goods and individual activities. The family may buy a larger home or otherwise spend its discretionary income

by limiting the number of children. Whereas children help earn the living and support the parents when they are old in less industrialized societies, in an industrial society children may be viewed as an economic liability.

The Question of Numbers

Birth and death rates for representative countries vary dramatically. The average number of children born per woman (known as the total fertility rate or TFR) ranges from 8 in Kenya to only 1 in West Germany (Figure 4–12). On a regional basis, Europe has the lowest average TFR with 1.9

The constant need to provide for children in less industrialized countries falls largely on their mothers.

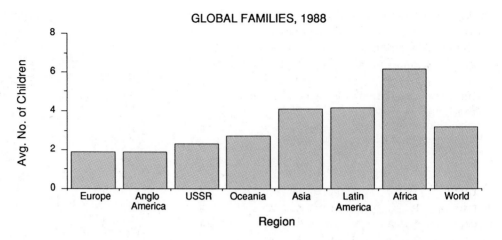

FIGURE 4–12 Global Families, 1988: Average Number of Children by Region The average number of children born to a woman during her reproductive years varies greatly from region to region.

children born to each woman, while Africa has the highest (6.5 births per woman). The global TFR in 1988 was 3.9 children, more than 50 percent greater than the 2.2 children per woman needed to maintain a stable global population.

Regionally, the rapid growth rates of Africa cause concern about the continent's carrying capacity. In the industrialized regions of North America and Europe, slow growth rates associated with small families cause concern that their populations will level off and actually decline. Examination of the projected population doubling time for selected nations indicates the problems that this might create. Kenya, with 21 million people at present, could reach 42 million within 17 years if present growth rates continue. The

United Kingdom, with 56 million people in 1988, would not reach the 100 million mark until approximately A.D. 2600 if present growth rates continue. Such vast differences in population change cause regionally different problems.

Population Pyramids

One way to examine population is through *population pyramids*. A population pyramid consists of tiered graphs that show the number of male and female population by age ranges. Normally each tier represents either a five- or ten-year interval. The male population is shown on the left and the female population on the right. It is referred to as a population pyramid, because it frequently take a pyramidal shape (Figure 4–13).

FIGURE 4–13
Population Pyramids of Industrialized and Less Industrialized Countries
Population pyramids illustrate the impact of the demographic transition. Industrialized countries are generally in the last of stage 3 or stage 4, with lower numbers of people in the youngest age groups. Less industrialized countries with high growth rates, have more people in the youngest ages.

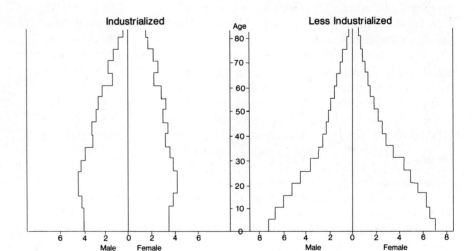

Population pyramids of countries in stage three or four of the demographic transition show a relatively uniform number of residents in all age groups until late middle age. Population pyramids in the countries of stage two where the population is expanding rapidly have many more young than old people. Lower age groups in high-growth countries comprise a larger percentage of the total population. Countries in stage one of the demographic transition have a population pyramid with a shape similar to that of countries in stage three, but with lower numbers of elderly. Countries in stage four have fewer people in the youngest age categories than in the middle-aged groups.

A concern of countries in both rapid- and slow-growing situations is the *dependency ratio*, a ratio of the young and old who must be economically supported to those who are working. Children fifteen years of age or younger or elderly over retirement age (normally 62–65) are counted as dependents. The dependency ratio in less industrialized countries is generally between 40 and 50 percent and is dominated by children who have not yet reached the age of fifteen. The dependency rate in industrialized countries is between 20 and 30 percent, with a higher proportion of elderly in the dependent group.

Children constitute as much as half the population in less industrialized countries, posing the challenge of providing them with education, jobs, housing, and other social services. The less indus-

In industrialized societies the elderly comprise an ever larger proportion of the dependent population.

trialized countries face the problem of creating an estimated additional 500 million jobs by the year 2000 simply to provide jobs for children living in 1988. If present trends continue, by the year 2025, the working age population in the less industrialized countries alone will be larger than the population of the entire world in 1988. The higher proportion of elderly in industrialized countries presents a different challenge. Most of these elderly will never enter the economic sector again, and costs of medical and other services increase as their health declines. The larger number of elderly in the industrialized countries creates more jobs in the health care industry, but may place a serious strain on the social systems that provide health care.

Population projections suggest that the industrialized regions of the world will reach approximately stable populations between 1995 and 2020. The less industrialized regions of the world are projected to reach a level of stability sometime between 2020 and 2045 (Figure 4–14). Pop-

Children are dependent on their parents and society to provide education and job skills. These schoolgirls are from Spain.

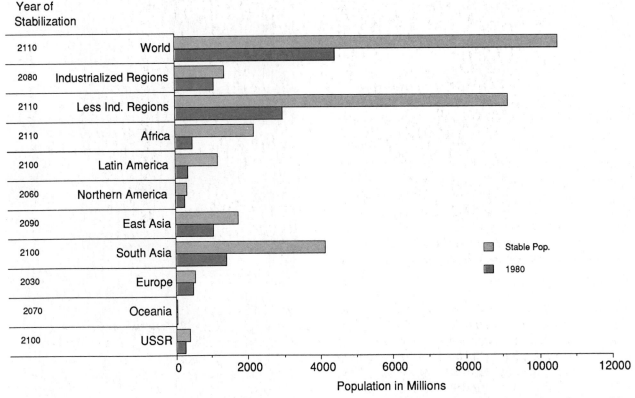

DATE OF REACHING ZERO POPULATION GROWTH AT 1980 GROWTH RATES

FIGURE 4–14 **Date of Reaching Zero Population Growth at 1980 Growth Rates**   Earlier dates of reaching stable populations are found in industrialized countries.

ulation stability occurs when *zero population growth* is reached. Zero population growth occurs when net births and immigration are canceled out by net deaths and emigration. A few decades' delay in reaching zero population growth would have a dramatic demographic impact. Achieving zero population growth for the world in the year 2045 versus 2005 would equate to five billion more global residents, equal to the world population in 1987.

The Malthusian Equation

Some observers believe that the rapidly growing world population is a problem of crisis proportions. Other observers argue that population growth is not a threat; they maintain that increased numbers of people in the last few millennia made possible the development of civilization

as we know it. The increased population of Europe after A.D. 1500 prompted the Industrial Revolution of the eighteenth and nineteenth centuries and improved the standard of living to a level previously unknown in the world. Such divergent views on the impact of population growth reflect individual and group values and goals. During periods of rapid population growth such as that in less industrialized countries from 1940–1990, concern for overpopulation is often manifest.

The rapid increase in population that began at the time of the Industrial Revolution also aroused fears of overpopulation. An English theologian (and amateur economist), Thomas Malthus, published *An Essay On the Principle of Population* in 1798. He concluded that the population of the world increases more rapidly than food production. His thesis, called the *Malthusian equation*, emphasizes the importance of food or other resource limitations as controls on population growth. The basic premise of his theory is that population increases geometrically while food supply increases only arithmetically (Table 4–4).

Malthus maintained that individuals, particularly those from lower socioeconomic groups, could continue to reproduce more rapidly than food could be provided for them. He concluded that the human need for food and desire for procreation were powerful, uncontrollable, and unchanging and would lead to ultimate misery for most of the human race. He believed that the only controls to population growth were *positive checks*, which destroy human life (such as disease [pestilence], famine, and war); or *preventive checks*, which limit births (delayed marriage or sexual abstinence). He hypothesized that the preventive checks would not exist except in conditions of extreme deprivation; consequently extreme misery was the inevitable lot of humans. For Malthus, preventive checks would occur only in conjunction with or after famine, disease, or war.

The experience of Malthus with famine, war, and disease reflected European experience. The last outbreak of bubonic plague in England, the Great Plague of 1665, was widely remembered as the last occurrence of the Black Death which had stalked Europe for three hundred years. For some observers today the repeated droughts and famines that have afflicted Africa's rapidly growing population are modern evidence of the validity of the Malthusian equation (see the Cultural Dimensions box on page 140).

In the nearly two centuries since Malthus' essay was published, the world's population has increased more than fivefold. Although there have been great wars and periodic famines, deaths on a scale projected by Malthus have not occurred. Some observers maintain that Malthus was correct, and his predictions will come true in the future. These neo-Malthusians maintain that migration to less inhabited areas in the nineteenth and twentieth centuries relieved population pressure in Europe, but that his ideas are being borne out today (see the Models in Geography box on pages 142–143).

Others, however, view Malthus's theory, as simply invalid. One of the earliest writers to question the Malthusian thesis was Karl Marx.

TABLE 4–4
Basic Calculations of the Malthusian Equation

POPULATION	FOOD
2	2
4	3
8	4
16	5
32	6
64	7
128	8
256	9
512	10
1,024	11
2,048	12
4,096	13
8,192	14
16,384	15
32,768	16
65,536	17
131,072	18
262,144	19
524,288	20
1,048,876	21

Marx argued that the basis of human poverty and misery was not population growth, but rather the capitalist system, which allowed a few to enjoy lives of luxury while the masses were paid low wages for their labor. Marx maintained that in an ideal society population would not be an issue since the resources would be equitably distributed among all people, eliminating poverty and overpopulation.

A recent critic of Malthus has argued that increasing population growth can actually improve the human condition. Esther Boserup maintains that peasant societies can produce more food through population growth. Her thesis is that additional inputs of labor on a given area of land can increase output. She concludes that in the absence of political or economic forces that decrease the price of local food production, increased population leads to greater food production in agrarian societies. Boserup concludes that both Malthus and Marx were wrong because they failed to recognize that population can grow and living conditions simultaneously improve over a long period; indeed global living standards have

CULTURAL DIMENSIONS
Drought in the Sahel

The wiping out of a herd of livestock is a personal catastrophe for the nomad, whose life is geared to the raising of animals. The drought has hit the nomads hard, many have lost their herds and come into the villages and cities for help. Mopti, a city of Mali with a normal population of about 55,000, now has 110,000 people living in and around it. Nouakchott, the sandswept capital of Mauritania, has an estimated population of 120,000. Its normal population is 40,000. Outside of Nouakchott a number of camps have been set up for the destitute nomads.

Ehel Lemeounek is a nomad who pitched his tent in one of the camps. There are fifteen people in his family. He once owned two hundred camels. Now he has none. "My whole life is herding camels," he says. "I know no other life. I want to go back to camel herding. I would work as a guard if I could, but I would rather take care of five camels than anything else in the world." Mr. Lemeounek sits under his sheep's wool tent with his family most of the hot, windy days. What do they do? He shrugs his shoulders. "We meditate," he says.

In another of the nomad camps that ring the city, Ehel Ahmedouval has brought his family, which also numbers fifteen. He once owned twenty cows and sixty sheep. He still has two calves, a donkey, and a few sheep. Ehel Idi had never had contact with city people before. His world was the edge of the desert. His life was herding cattle, and until a few years ago he owned three hundred head. He was considered a prosperous, lucky man. Now, all he has left is ten sheep and an emaciated goat. His family of six wants to go back to the life they know.

"But I am forced to stay here," Mr. Idi says in Hassinya, an Arabic dialect. "I tried to survive. We had finally thirty-five cattle left. I cut down trees for them to eat the leaves and keep them alive. But it was no use. They all died and this is what I have left. I didn't want to come to the city even then, but others told me I must live. I could stay alive and find food for the family if we came here."

Adapted from the *Christian Science Monitor*, Feb. 11, 1987, p. 11.

improved in the last two centuries in spite of the tremendous increase in population (Figure 4–15).

Culture and Population Growth

Population growth in a country or region can be indicated as a growth rate for some specified time period or as the total fertility rate. The fertility rate is also used to determine the average *completed family size* for a group of women in their reproductive years. The two figures are essentially the same for forecasting purposes in that

FIGURE 4–15 Boserup Theory of Population and Technological Changes The Boserup model of population growth and standard of living hypothesizes that population growth causes greater agricultural productivity.

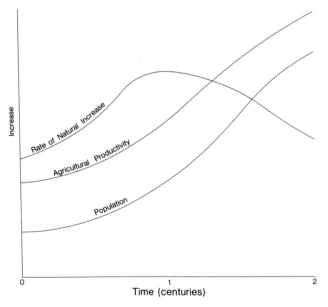

BOSERUP THEORY OF POPULATION AND TECHNOLOGICAL CHANGES

they refer to total fertility, but the average completed family size allows for easy comparison between countries because it is readily understood (Figure 4–16).

The importance of average completed family size lies in its dramatic impact on total population growth. A baby girl who grows up to have 3 children instead of 6 would have 27 great-grandchildren rather than 216 assuming subsequent generations have the same number of children.

Decisions to have larger families create major differences in population. Nigeria, where women at present indicate they hope to have more than seven children, is projected to be the third most populous country in the world, with 500 million people by the year 2100 at present growth rates. India would have 1.7 billion and China 1.5 billion people at that time if present growth rates continue. The birth rate, fertility rate, and average completed family size have been decreasing for most of the world (Figure 4–17). Yet women in some countries, such as Kenya, which has the most rapid population growth in the world, continue to bear large numbers of children. Asked why they continue to have large families, Kenyan women gave the following answers:

1. Among many children there are not only likely to be rogues and robbers but also professionals such as doctors, lawyers, and engineers.

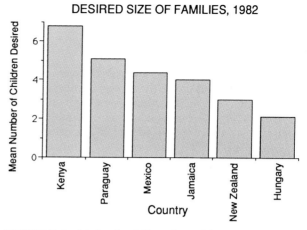

FIGURE 4–16 **Desired Size of Families, 1982** The desired family size as measured in number of children differs markedly between countries in the industrialized versus the less industrialized world.

2. A wife continues to bear children to prove her worth and to prevent her husband from taking another wife.
3. Where polygamy is practiced, wives may compete to have the largest number of children.
4. Children provide labor for fetching firewood and water.
5. Children provide security for aged parents.
6. The husband decides how many children the family will have.

FIGURE 4–17 **Changes in Fertility Rate and Family Size** The great difference between fertility rates and family size in the less industrialized and the industrialized world indicates why some countries face serious problems with growing populations.

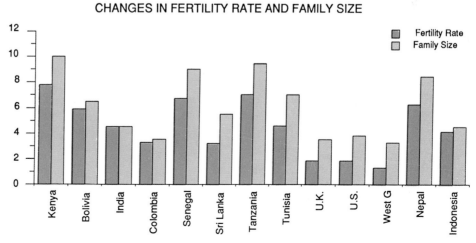

MODELS IN GEOGRAPHY

Population Growth

The Malthusian model of population growth was one of the earliest attempts to model population growth, yet many of its fundamental principles are still accepted. Malthus's model is called an *exponential model* because it describes a simplified situation in which growth (or decline) in population is constant. Exponential models conclude that the amount of growth is related to the size of the population; the larger a population, the faster it grows. In its simplest form, an exponential model demonstrates that even a small rate of increase (or decrease) maintained over an extended time will cause substantial change. Populations growing at 1 percent per year will double every 70 years, for example. Exponential models also show the importance of even small changes in growth rates. If the rate is halved (.5 percent per year), the population requires 140 years to double.

When graphed, exponential models of population growth have a curve that steepens until it is nearly vertical. The graph of historic global population growth is representative of such curves (see Figure 4–4). All exponential models include some mechanism for preventing unchecked growth. At some point any population will reach the carrying capacity of its environment. Incorporating the concept of carrying capacity into exponential models suggests three possible general models of the change in growth rate as the carrying capacity is reached (Figure 4—B). Hypothetically, there could be an instantaneous adjustment to exceeding the carrying capacity, but this is unlikely. More likely, as carrying capacity is reached, population growth and resultant populations will adjust over time to the

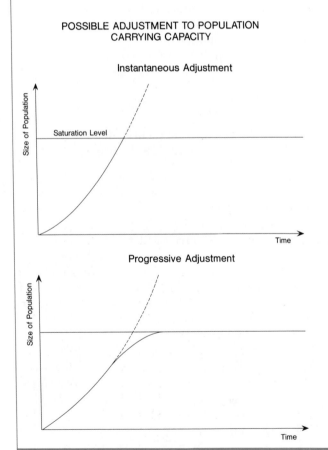

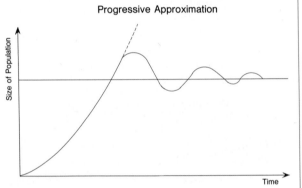

FIGURE 4—B Possible Adjustments to a Population Reaching Carrying Capacity Possible adjustments to a population reaching carrying capacity suggest three possibilities. Theoretically the population growth could end *instantaneously* when the capacity of the environment is reached; or population growth may slow as it nears capacity in a *progressive adjustment*; or the population may *progressively approximate* carrying capacity, exceeding it at first, but falling over time to the level that can be supported.

continued

MODELS IN GEOGRAPHY
Population Growth *continued*

actual carrying capacity. These adjustments may take the form of decreasing population size (through migration, famine, disease, or conflict) or increasing the carrying capacity by technological or social changes. Historically, events such as the domestication of plants and animals, or more recently the development of hybrid plants and animals, increased the earth's carrying capacity.

Since it is essentially impossible to predict what actual adjustments will occur as a population nears or exceeds its carrying capacity, models of future population growth are used mainly to illustrate what will occur if present population trends continue or to make assumptions about future adjustments that will affect population growth rates. Projections made in the 1950s and 1960s concluded that world population would be between 4.5 and 7.6 billion in the year 2000. The lower projections have already been exceeded, and it now appears that global population in A.D. 2000 will be near 6 billion.

Models of population growth that incorporate the demographic transition conclude that at some point, zero population growth (ZPG) will be attained. Several industrialized countries are now experiencing zero (or negative) population growth, and world growth rates are declining. Models of population growth make assumptions about how quickly global growth rates will slow, resulting in major differences about when global population will stabilize. One important aspect associated with the differences in conclusions is the degree of pessimism or optimism about the future condition of the world's population. Observers who see growth rates slowing more rapidly are more optimistic than those who believe it will take longer to reach ZPG.

Declining fertility rates, family size, and population growth rates occur for a number of reasons. Biologically a woman could conceivably bear some thirty children if she had one child each year during her reproductive years. Few actually reach this potential because of *biological* and *cultural resistances.* Biological resistances to having large families include sterility, malnutrition, breast feeding (which suppresses ovulation), and other physical problems that inhibit conception. Natural sterility results from genetic disorders, disease, or other problems in both men and women. Decreased reproduction may be associated with inadequate food supplies, poor health, and related difficulties. Historically, food supplies may have been the primary regulator of population growth.

Cultural resistances that affect birth rates involve a wide variety of techniques and practices designed to limit population growth. A period of abstinence from sexual relationships is prescribed in many societies after the birth of a child. Other societies place a high premium on celibacy and delaying marriage.

Industrializing countries of Europe during the nineteenth century would have had even higher birth rates and population growth had it not been for the practice of late marriage. The average age at marriage in Belgium during the years 1846–1856 for women was 28.6 and for men 30.3. It is estimated that 15 percent of the women and 18 percent of the men in Belgium during this period never married or had children. Other European countries exhibited similar patterns, thus preventing growth rates from equaling those found today in Kenya (see the Cultural Dimensions box).

Attitudes towards age at marriage affect the birth rate and family size because of the length of time a female may bear children. Marriage in many countries in the less industrialized world is almost universally early. Sixteen was the average age for marriage in Korea in 1925, fourteen in India in 1950, and eighteen in Kenya today. Attitudes towards marriage age change over time. The age at marriage in the United States declined rapidly between 1890 and 1950, but began increasing again after 1960 (Figure 4–18). Some countries, including India, have adopted mini-

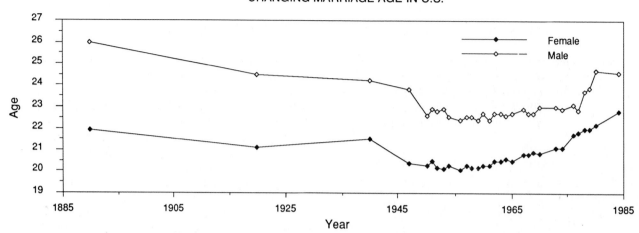

FIGURE 4–18 Changing Marriage Age in the United States The average age at marriage in the United States dropped rapidly between 1945 and 1965, but has increased markedly in the last decade.

mum ages for marriage to slow the population growth rate.

Other means of slowing the birth rate and lowering average family size are various types of contraception. On a world basis, contraceptives have been used for millennia. The number of people using contraceptives has been increasing, growing from 8 percent of couples in India in 1969 to 25 percent in 1980; from .5 percent of couples in Indonesia in 1970 to 35 percent in 1980; and from less than 10 percent of couples in Thailand in 1972 to nearly 66 percent in 1985.

A Question of Values: Dealing with Increased Population

A number of alternatives have been suggested concerning possible world response to the population explosion in less industrialized countries. Three of the most widely discussed are the lifeboat earth concept, triage, and the small planet view.

The *lifeboat* earth view of population is based on the Malthusian theory. Proponents of this view assume that the earth has limited amounts of resources, just as a lifeboat has only limited resources of food and water. According to this scenario, when a lifeboat is surrounded by other drowning passengers, pulling more survivors into

Many countries rely on public campaigns to change attitudes concerning family size, as in this poster encouraging small families in Bombay, India.

the lifeboat is not in the best interests of those occupying the boat. Additional survivors will only decrease the amount of time the limited food and water in the lifeboat will last, and if too many survivors are pulled into the boat, none may reach the land safely. Underlying this view is the belief that the industrialized wealthy countries should concentrate on the their own well-being and let the poorer, less industrialized countries, the passengers in the ocean, sink or swim on their own. Such a view, if accepted, might well condemn those countries with the most rapidly growing populations to everlasting poverty, misery, and ultimate overpopulation.

CULTURAL DIMENSIONS
Sweden, An Early Example of Population Change

Sweden provides a good case study for population change because of the length of its recorded history. Population statistics, available from 1749, reveal Sweden's change from stage one to stage four of the demographic transition.

The Early Period

Sweden in 1750 was typical of agricultural countries in Europe. The population was chiefly engaged in farming in a society characterized by a rigid, hierarchical class system. The income of the majority of the population was low, approximating that of less industrialized countries today. The wealthy elite had more material goods, but their health and life expectancy was little different than that of the masses. The average life expectancy for men was only 34.2 years, and that for women was 37.6.* The population in 1750 totaled 1.7 million people, who suffered from repeated infectious diseases such as whooping cough, smallpox, measles, and related infections. The country was subject to periodic famine and the infant mortality was high. Birth rates exceeded forty per thousand, but death rates were equally high.

Sweden began the transition to stage two of the demographic transition in the period 1750–1850. The life expectancy for Swedish men had increased to over 39 and to over 43 for women by 1850. During the same period the population doubled to nearly 3.5 million.

The Population Explosion in Sweden: Examination of Culture History

Advances in medicine after 1850 (including vaccination against smallpox and antiseptic and hygienic conditions in hospitals) combined with improved sanitation and sewage systems to decrease the death rate precipitously. Increased life expectancy led to a population boom that aroused fears of overpopulation in Sweden because the rapid population growth placed tremendous pressure on the agrarian society, resulting in the emigration of 20 percent of the total population (mostly to the United States).

Attitudes toward population increase had changed by 1930, and birth rates declined, while average life expectancy improved. Life was still difficult for many, with unsanitary and crowded housing and inadequate hygienic conditions for the poor. Contrasts between the wealthy and the poor were reflected in birth and death rates and life expectancy. *Infant mortality rates* ranged from forty-nine per thousand births in low-income families to only fourteen per thousand in high-income families. Sweden introduced some forms of government social security programs between 1930 and 1940 including national health insurance, employer medical insurance, maternity welfare, housing allowances, nursery schools, children's health services, free school meals, and information on nutrition and health. The population of Sweden grew to 6.4 million by 1940.

The Present Population Situation in Sweden

Sweden's population increased from 6.4 million to 8.3 million between 1940 and 1985. The trends begun during the earlier part of the twentieth century led to decreasing growth rates, making immigration responsible for a significant proportion of the population increase from 1940 to 1985. Fertility rates continued to decline, with the average woman bearing only 2.22 children by 1955 and only 1.55 by 1985. Sweden will reach zero population growth if this average total fertility rate continues, and the number of Swedes may actually decline.

Concern for the slow growth of Sweden's population is reflected in a national survey in 1982. The results of this survey indicate that Swedish women have small families for the following reasons:

1. More women are attending school longer and are therefore postponing childbearing.

---

*Royal Ministry of Foreign Affairs, *The Biography of a People* (Stockholm-Government Printing Office, 1974).

continued

CULTURAL DIMENSIONS

Sweden, An Early Example of Population Change *continued*

2. Highly educated women have fewer children.
3. A larger percentage of women are in the labor force and have more education and higher status jobs.
4. More unmarried couples are living together and are less likely to have children than married couples.
5. Effective methods of contraception are being widely used.
6. Working women have difficulty finding adequate day care centers.
7. The burdens of housework, caring for the children, and cooking and shopping are being borne primarily by women whether they work or not.

The survey revealed that almost all women considered having children one of the meaningful things in life, but the problems presented by children in Sweden's industrial society indicate that family size will continue to decrease. Sweden's population grew at a rate of 0.01 percent between 1980 and 1985, and since 1985 it has actually declined. Life expectancy in Sweden is one of the highest in the world (75.8 years average), and infant mortality (seven per thousand) is also among the lowest.

Sweden's experience in moving from stage one to stage four of the demographic transition has been repeated in other industrial countries and is currently occurring in some less industrialized countries. Sweden is unique in having long-term population statistics, but the changes in its birth, death, and growth rates as the country moved from stage one to stage three of the demographic transition have been mirrored in other industrial countries. The experience of Sweden and other industrialized countries has prompted some observers to conclude that if a country industrializes, its population growth rate will also decline.

Triage is a medical term that had its origin in warfare. After any battle, there are generally more wounded than the available medical personnel can handle. In order to use its limited medical facilities to best assist the survivors, the military divides casualties into three groups:

1. Those who are slightly wounded and will recover without any additional care.
2. The seriously wounded, whose prospects for medical survival will be greatly improved by medical attention.
3. Those who are gravely wounded, for whom additional medical care will do little to save their lives.

The most severely injured require large amounts of medical attention with limited results. The first and third groups are ignored in battle conditions if a choice must be made, and attention is given to the second group, who should benefit most from medical care.

Proponents of the triage concept applied to world population maintain that only those less industrialized countries with smaller and slower growing populations should receive assistance from the wealthy industrial countries. Countries with overwhelming population problems are seen as beyond help, and aid would only delay their ultimate disaster, while not benefiting the general world population. Countries with the potential to increase their standard of living through industrialization are the logical targets for assistance from the wealthy countries in the triage view.

Observers favoring the *small planet* view maintain that both the lifeboat and triage views are not only wrong but immoral, and represent only the rationalizations of the wealthy trying to justify their continued dominance of the world's resources and wealth. Small planet proponents insist that all humans inhabit the same earth and are equally entitled to its resources. The population problem that must be solved if the inequity in distribution of the world's wealth is to be overcome is a world problem, not a regional one. The solution does not involve arbitrarily abandoning large segments of the earth's population to misery and human degradation. It involves international cooperation to ensure harmony between the demand for resource and population in each region. Proponents of this view conclude that residents of the wealthy industrialized countries must recognize that they demand a disproportionate share of the world's resources because of their high consumption of fuels, minerals, and foods; and they must therefore be willing to share their wealth to assist less fortunate countries.

Which of these views of the world population is the right one? The answer to this question reflects individual and group values since the issues of population revolve around the culture and values of those who observe the problem.

The rapidly increasing population in many countries of Latin America, Asia, and Africa are the basis for concern about global population growth. A few countries, because of their large base population, represent a high percentage of the total annual population increase. The countries of Pakistan, India, Bangladesh, Indonesia, and the People's Republic of China are home to nearly one-half of the world's population. At the present time, almost half of the world's population lives in four countries (India, People's Republic of China, Soviet Union, and the United States), and another 25 percent live in the fifteen next most populous countries. Ninety percent of the world's population lives in the fifty most populous countries.

Examination of the most populous countries indicates why issues relating to population growth, particularly the question of overpopulation, are culturally based. The population density of the Soviet Union is only 33 per square mile (13 per square kilometer), while Japan's is 841 per square mile (322 per square kilometer), yet Japan has a higher standard of living. The People's Republic of China has a population density of 296 people per square mile (114 per square kilometer), while Mauritania has only 5.3 per square mile (2 per square kilometer), yet both are part of the less industrialized world. Japan's high population densities are noteworthy because only a small frac-

tion of the nation is suitable for agriculture, an indication of the significance of technological level in understanding population problems. The majority of Japan's population are employed in manufacturing and service activities and have high standards of living in crowded urban areas. The majority of Mauritania's population are employed in agricultural-related activities and have low standards of living in rural areas. Determining whether either Japan or Mauritania is "overpopulated" is impossible for outside observers. Only the residents of the country involved can decide whether additional population growth will be a blessing or a curse. Population growth reflects the overall impact of each individual couple's decisions about family size. Such individual decisions are based, consciously or unconsciously, on the values of the group, but these values may change over time. Thirty-two of the fifty most populous countries, containing over three-fourths of the world's total population, have experienced declining population growth rates in the last decade. Changes in growth rates in these particular countries have a dramatic impact on total global growth rates, and changing attitudes towards population in India and China in particular have slowed the world growth rate.

India: Population Problems in an Agrarian, Democratic Society

India is a country that commonly prompts images of overpopulation, starvation, poverty, and deprivation in the minds of uninformed Western observers. Cultural values in India are often misunderstood in the West, resulting in erroneous conclusions concerning the so-called "sacred cow" and its relationship to Indian poverty. The immense size of India's population ensures that it will attract the attention of outside observers, but simplistic reporting masks as much information as it reveals. India's population growth, its causes, implications, and the Indian government's response exemplify the problems of population. The absolute annual population increase in India at present (1989) is greater than that in any other country in the world, a net increase of over 15 million persons per year. The birth rate has declined in the last thirty years, but the population continues to increase because the death rate has declined even more (Figure 4–19). There were only 250 million people in India in 1920. In 1989 there were an estimated 835 million. India's population increases at the rate of 1.2 million per month, or 40,000 each day. The present annual increase of 15 million people is essentially equal to the population of Australia. There will be one billion people in India at the turn of the century if growth rates continue at the present rate. By the year 2020 India would become the most populous country in the world.

India's attempts to resolve its population issues illustrate both the challenges and dilemmas faced by countries with rapidly increasing population. India's initial programs were modest. Gandhi, the architect of India's independence from Britain in 1947, believed that India could develop a society in which people's needs could be met informally without the great variation in wealth found in most countries. Escalating population led to the adoption of a family planning program as part of India's first five-year plan in 1952; birth control clinics were established, and information on family planning was provided to all who had access to the clinic.

There was a total of 4,165 clinics by 1961, but the number was inadequate since the majority of Indians lived in the estimated 580,000 villages of the country. In the 1960s India adopted a mass education program encouraging people to have

India's population growth contributes to problems of housing and poverty, epitomized by the homeless as seen here in Calcutta.

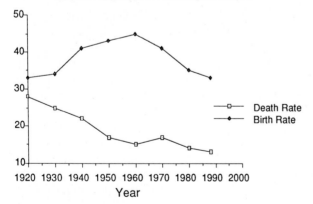

BIRTH AND DEATH RATES IN INDIA

FIGURE 4–19 Birth and Death Rates in India Both birth and death rates have fallen in India, but the growth rate remains high.

smaller families. Billboards showing families with two children maintained that smaller families lead to prosperity and happiness, while larger families create problems for the entire country. As part of the education program, voluntary sterilization programs were established in many of the states of India, and as many as five million male Indians were sterilized. The minimum legal age for marriage for females was raised to 15, but the population continued to increase rapidly.

A new national population policy was adopted in 1976. Its major elements included increasing the minimum legal age for marriage to 18 for females and 21 years for males, linking federal financial assistance to state governments to performance in limiting births, providing sex education in schools, increasing compensation for voluntary sterilization, and providing incentives to encourage couples to limit the size of their family. These incentives included both *negative* incentives in some of India's states (such as loss of retirement income, jobs, or promotions for civil servants who have had more than two children) and *positive* incentives (such as payments or gifts for sterilization and the establishment of retirement accounts, which were increased each year, for women who went a prescribed number of years without having children).

The 1976 population policy was greeted with hostility by many Indians, and the nation's political leaders were forced out of office in the 1976 elections. Some observers maintain that the defeat of the ruling party came about because of the harsh penalties for large families. The new law's proposals were particularly frightening because of widespread reports that local officials had forced men to be sterilized or had otherwise abused their power under the old law.

The new national leaders instituted a major change in 1977, making participation in the population program voluntary. The 1977 program still provided free sterilization for men and women and maintained the increased marriage ages. Only 8 percent of federal financial aid to individual states was based on success in family planning. The new national government set a target birth rate of 25 per thousand by 1984 and proposed assigning a community health worker for each of the county's villages. This target was not reached, and by 1989 the birth rate was still 32 per thousand. The majority of the Indian population still rely on agriculture for their livelihood. It is difficult for farmers to limit family size to benefit the nation. The individual farm family still perceives additional children as a benefit, reflecting the importance of additional labor in their culture ecology.

China: Population Change in a Communist State

China, with the world's largest population, has a much different population program. The Communist party established the People's Republic of China in 1949. The population of China in 1949 was estimated at 540 million, but by 1989, it had increased to nearly 1.1 billion. The growth in population might have been higher had it not been for government's attempts to limit family

size, especially after 1970 when programs encouraging small families were widely adopted.

Widespread adoption of delayed marriage, sterilization, and contraceptives cut the Chinese population growth rate dramatically during the decade of the 1970s, but even with a reduced growth rate of 1.3 percent, nearly 13 million people were added to China's population yearly, more than the population of over one-half of the world's countries. The rapid population growth of the 1960s and 1970s meant even greater numbers of females in their childbearing years during the 1980s and 1990s, further compounding the population problem.

China adopted a new program of population control in 1980. It called for reducing the annual growth rate to .5 percent by 1990 and to zero by the end of this century. To achieve such a rapid demographic transition, China embarked on a program of positive economic incentives and social pressure designed to encourage compliance.

Urban couples who have only one child and agree to take measures to prevent additional children are given a "one-child certificate." Certification entitles them to a cash grant every month until the child reaches age 14. A single-child family is entitled to living space equal to a two-child family and receives preferential treatment in obtaining housing. Children of single-child families are given priority in admission to schools and in job selection. When the parents of a single-

China's population program uses social and financial pressures to encourage people to have small families.

child family retire, they are entitled to larger pensions than they would otherwise receive.

One-child families in rural areas can receive additional monthly work points (the basis for payment on the agricultural communes) until their child reaches age 14, or they can receive more land to farm under a contract system where they keep the profits over the basic rental requirements. Rural one-child couples are entitled to the same grain ration as two-child couples, and all couples regardless of family size receive the same amount of land for private cultivation, plus the same size lot for their home. Should the child of a single-child family die or become disabled, the family is permitted to have another child and continue to enjoy the same benefits.

To ensure that people do not unfairly use these economic incentives, penalties are imposed for those who break their agreements. For example, a third child born after January 1, 1980, is not eligible to participate in the worker's family medical coverage. Grain rations for a third child are obtained at a higher price than those obtained under the rationing system for a single-child family. Bonuses and incentives given to a certified single-child family that has a second child must be returned.

To counter the common fear among rural societies that the lack of a large family may leave no

Even with a greatly reduced population growth rate, China still faces the challenge of caring for the large numbers born before the population rate declined rapidly.

one to care for the parents when they are old, the Chinese government has adopted the "five guarantees system," the five guarantees being food, clothing, shelter, medical care, and a decent funeral. Every individual is entitled to a minimum grain ration of 500 pounds (225 kilograms) of grain per year plus a survival ration of other basic foods. Subsidies are provided for individuals who need help. Farms, communes, and factories have "respecting the old houses" to care for the elderly and those supported by the public welfare fund. An elderly person or couple whose single child dies before they do can stay in their own house and receive aid from relatives or neighbors through the five guarantees.

To counter the problem of a single-child family in which the bride goes to the household of the groom (as is traditional in China and other less industrialized countries), the government is encouraging the daughter of a family without sons to bring her husband to her parents' household upon marriage. Chinese law now requires that relatives not interfere with the bride and groom and their choice of residence. The farm production or factory team to which the bride belongs may not discriminate against the new groom in job assignments, wages, housing, or admission of their children to school or work. The groom is expected to care for the bride's parents when they become old and inherits their property.

China's "one-child family" program of the 1980s has substantially decreased birth and growth rates. Pressure from the Communist party for single-child families was extreme, apparently leading to forced abortions and female infanticide in some cases. (It is unclear how frequent such events were, but the Chinese government officially denounces them.) The combination of social pressure and economic incentives prompted wide adherence to the standards of the population program in the first half of the 1980s. Economic changes in China since 1985 are associated with a slight upturn in birth and growth rates, especially in rural areas where families are now allowed to have larger families. Still, China's move from stage two to stage three of the demographic transition represents one of the most rapid demographic changes the world has known. The birth rate had declined to 21 per thousand in 1989, and

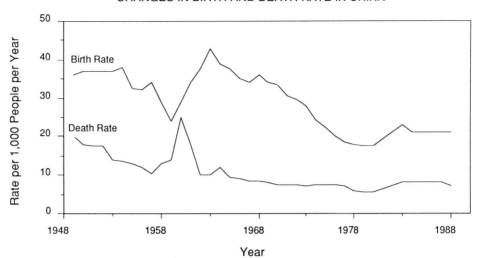

CHANGES IN BIRTH AND DEATH RATE IN CHINA

FIGURE 4–20 Changes in Birth and Death Rates in China Birth and death rates in China are low for a less-industrialized country, resulting in a slow rate of population growth.

the annual growth rate had declined from 2.34 percent in 1971 to 1.4 percent (Figure 4–20).

China's growth rate is still higher than its final goal, but reduction in the growth rate has resulted in a much smaller population than would have occurred had the growth rates of the 1960s continued. China was more successful than India in slowing population growth because the control of the Communist party facilitated widespread enforcement of the elements of the population program.

Population and Quality of Life

Rapid population growth in much of Africa, Asia, and Latin America affects the quality of life. It is difficult to measure the respective quality of life in countries as diverse as Burkina Faso and the United States, but comparison of basic standards such as per capita gross national product, *infant mortality rate* (the percentage of children who die during their first year), availability of food, access to drinking water, and related issues, such as literacy, provides some basis for comparing the quality of life in different areas. Early attempts to measure or compare the relative standard of living used a simple comparison of average gross national product (GNP). Such an index is misleading because GNP is very unevenly divided in many countries. An alternative measure of stan-

dard of living is the *physical quality of life* index. The physical quality of life index uses adult literacy, per capita GNP, birth rate, and infant mortality rate to calculate a quality of life for individual countries (Figure 4–21).

Quality of life data must be used cautiously, however, since the factors being considered are not uniform. A GNP figure of $200 per capita per year in a country in which the majority are engaged in agriculture and produce their own food provides a higher quality of life than $200 in an industrialized country. Nevertheless, the magnitude of differences between a per capita GNP of less than $500 (which characterizes much of the less industrialized world) and one of over $5,000 (which characterizes much of the industrialized world) indicates major differences in quality of life. Differences in literacy rates of 90 percent versus 20 or 30 percent also suggest a great gulf in living standards.

Distribution: Patterns and Problems

Global population is distributed in a series of clusters of densely populated areas separated by large areas of low density. If the people living on the earth in 1989 were distributed evenly over the land area of the planet (including Antarctica), there would be approximately 87 persons per square mile (34 per square kilometer). Physical,

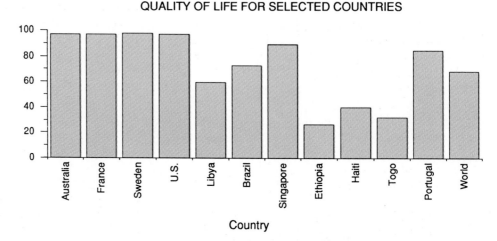

QUALITY OF LIFE FOR SELECTED COUNTRIES

FIGURE 4–21 Quality of Life for Selected Countries
The quality of life index ranges from 0 to 100. Less industrialized countries generally have a lower quality of life as measured by literacy, GNP, birth rates and infant mortality.

technological, social, political, and economic factors interact over time to prevent such uniform distribution.

Seventy-six percent of the world's population live in Europe and Asia (Eurasia). North America has 9 percent of the world's population, Africa has 10 percent, 5 percent reside in South America, and the Pacific Islands and Australia/New Zealand have less than .5 percent (See Figure 3–23, pages 112–113). The greatest number of people are found in Asia, which has more than one-half of the world's population. Europe with its smaller land area has a greater population density than Asia. The highest population densities in Europe are found in the Netherlands and the industrialized portions of the British Isles, France, and Germany (Figure 4–22).

The world is commonly divided on the basis of population density into densely settled regions with 250 or more persons per square mile (100 or more per square kilometer); moderately densely inhabited areas with 60 to 250 persons per square mile (25 to 100 per square kilometer); sparsely settled areas with 2 to 60 persons per square mile (1 to 25 per square kilometer); and essentially unpopulated areas having fewer than 2 persons per square mile (less than 1 per square kilometer). The areas with the highest population densities are associated with areas of the world that have especially favorable locations such as coasts or river valleys, or climatic conditions suitable for agriculture. Areas with low population densities reflect harsh environmental condi-

tions. Deserts in the interior of Australia and western China, areas with cold temperatures and short growing seasons in the high latitudes of the Soviet Union and Canada, and some areas in the tropical climates of central Africa and the Amazon Basin of Latin America are examples of low-density regions.

As we have seen, high population density is not directly associated with poor countries with rapid population growth rates. Nor does population density alone reveal a great deal about the quality of life that people enjoy. A density of over 250 per

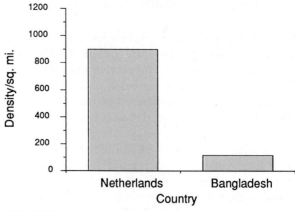

COMPARISON OF POPULATION DENSITY

FIGURE 4–22 Comparison of Population Density
Comparison of the population densities of individual countries reveals that some European countries have higher population densities than densely populated Asian countries.

square mile (100 per square kilometer) does not represent a very dense population in a city, but the same density in a rural area is very high. Consequently, regions based on population density are primarily important for the information they reveal on where the population is clustered rather than for any information they provide about the quality of life of the residents, the changes taking place in the population, or other issues that are of concern to cultural geographers.

Explaining the Pattern of Distribution

Factors affecting population distribution can be divided broadly into environmental and cultural factors. Environmental factors include climate, landforms, accessibility, availability of water, resource availability, and soil type. Historically, climate was the most important locational factor as early humans subsisted from hunting and gathering. Humans were able to migrate into other areas with harsher climates with the advent of fire and advances in tool use and technology. After the Neolithic Revolution, the human population increased rapidly in those areas that were most suitable to early agriculture. Diffusion of agriculture and the subsequent Industrial and Technological Revolutions have not diminished the importance of climate in explaining global population distribution.

Comparison of population distribution patterns with elevation reveals a pattern of human populations clustered at lower elevations. This reflects both the difficulty of access of higher elevations, which tend to be associated with rougher topography, and their cooler climates, which limit crop production. An exception is the equatorial areas of Africa and Latin America where populations are concentrated in the higher elevations because the climate there is cooler and more similar to the subtropical or mid-latitude humid climates. The importance of the ocean or navigable rivers is also clearly evident. Clusters of dense population are located along suitable coasts and rivers in each continent. This pattern of population distribution is in part related to the importance of location; it is also partially a cultural phenomenon, as these areas were settled by migrants from Europe who established communities and settlements along coasts and rivers in North and South America.

An indirect environmental factor affecting population distribution is disease. Disease is related to population distribution because it strikes both humans and their crops and livestock. One example of the impact of disease on population distribution is Africa, which has a range of disease-carrying pests. Major ones include mosquito vectors carrying malaria, a variety of parasitic worms, and the tsetse fly carrying trypanosomiasis (sleeping sickness). Malaria is endemic in tropical Africa, since mosquitoes thrive in the standing water of the swamps.

Although malaria is rarely fatal, it does affect people's long-term well-being. The genetic makeup of the population of tropical Africa often includes an abnormal hemoglobin associated with *sickle-cell anemia*. The sickle-cell hemoglobin is found in as many as one quarter of the African people in malaria-infested areas, as it offers some protection from death from malaria, although not from its other impacts.

Mosquitoes also transmit yellow fever (which is fatal but vaccines are available); a variety of small parasitic worms (filariae); and dengue, the so-called "break-bone" fever because of the intense pain it produces. The tsetse fly carries sleeping sickness to humans and the destructive disease *n'gana* to cattle and horses. Much of Africa is also plagued by parasitic worms, particularly the hook worm, which is nearly endemic among the peoples of tropical Africa. Diseases caused by parasitic worms, which are transmitted by the black fly, include *onchocerciasis* (river blindness), which afflicts millions in Africa; and *oschistosomiasis*, or liver flukes. Both of these diseases seem to be increasing as irrigation projects and other developments create areas of standing water that provide breeding locations for mosquitoes. Africa also suffers from periodic invasions of locusts, as in 1986 and 1987 (see the Cultural Dimensions box).

The combination of climate and disease found in tropical regions such as those in Africa result in low population densities. Groups avoid such areas or may migrate when faced with disease problems.

Technology has decreased the need for individuals to produce their own food. Consequently,

CULTURAL DIMENSIONS

A Plague of Locusts

An outbreak of desert locusts that began in Ethiopia in 1985 has expanded into the world's worst locust plague for a quarter of a century; it is damaging crops throughout a swath of northern Africa and posing a threat to sixty-five countries as far afield as India and the West Indies. The United Nations Food and Agriculture Organization (FAO) expects it will rage at least until 1990 before being brought under control.

This is the worst outbreak since the great fourteen-year infestation between 1949 and 1963, itself the worst on record. The pests are threatening famine-plagued sub-Saharan nations with new food shortages. A swarm of 5 billion locusts covers 60 square miles and eats as much as 80,000 tons of vegetation a night. The new attack found Africa unprepared. In between plagues the locusts can fade away for years, thereby encouraging the mainly poor countries on which they prey to relax their guard. In this case a generation of Africans had never seen a locust, and many of the old experts had retired. Out of three organizations set up in Africa to control locusts during earlier plagues, only one still operates.

Locusts are Dr. Jekylls. In between plagues they become harmless grasshoppers. Then, when conditions are right, they change their color and size and start to migrate in crop-destroying swarms. In 1985 the conditions for these changes were ideal. After years of drought heavy rains fell in Ethiopia, part of the desert locust's favorite breeding ground along the Red Sea. This provided damp soil suitable for egg laying and shrubbery to feed the hatchlings.

Attempts to confine the outbreak to Ethiopia were frustrated by that country's civil wars. Helped by unusually good rains, the locusts went northwest and reached the Atlantic coast in 1987. Some have now flown on to the Cape Verde Islands in the Atlantic, while other swarms have doubled back to the east. For a few hours in September 1988, they blotted out the sun over Khartoum; later they poured into Saudi Arabia over a 400-mile front. The FAO warned Kenya and Iraq in 1989 that they risked invasion for the first time since the previous great plague.

Wherever they go the pests are voracious eaters. During migration a desert locust can consume the equivalent of its own weight every day. A single square kilometer of swarm (swarms may be hundreds of times bigger) contains about 50 million locusts. They can devour as much food in a day as a village of 500 people will eat in a year.

The plague has done immense damage already. Although fears that it could wipe out virtually all the crops of countries such as Morocco and Tunisia have not been realized, the danger is not over. Most governments have been too busy protecting their crops to spend much time in halting the locusts' migration or attacking the remote places where they breed. In the arid Sahel, where a bumper harvest has followed a drought, the locusts have revived fears of another famine.

Destroying the locusts by attacking them while they are on the ground is a formidable undertaking. Early in October of 1988, it was estimated that 7 million hectares (17.3 million acres) of Africa were infested with infant locusts at the crawling stage. Killing them before they could fly and start to breed would have required a ten-day operation by a fleet of 700 aircraft dumping 200,000 gallons of insecticide. The FAO feels that a different insecticide would also help. Dieldrin, the poison that was once the mainstay of locust control, has been widely banned because it lingers in the soil, but the FAO wants it brought back. The war against locusts is an age-old one, however, and the pesticide issue represents only one incident in the ongoing relationship between humans and the land in African peasant farming systems.

---

*Adapted from *The Economist*, October 22, 1988.

people now have a greater range of choice in living sites, and the most important locational feature has become accessibility. The present distribution of population largely reflects the situational relationships of places that emerged during the Industrial Revolution as manufacturing centers. Of primary importance to these centers was accessibility by water, which allowed the agglomeration of the raw materials for manufacturing and the subsequent dispersal of the manufactured items to markets. With advances in transportation technology, rail and motor transport supplemented the oceans as the most important transportation forms.

The early locations of cities in Europe, North America, Asia, Latin America, and Africa remain the focus of high population densities. Regions where high population densities occurred at an early date tend to maintain high densities because of inertia. Once population clusters develop, barring catastrophic environmental change or major changes in cultural values, the population tends to remain in those locations. Early population centers reflected the environmental advantages of an individual site, and these remain important today.

Cultural Variables Affecting Population Distribution

The distribution of the world's population is affected by a variety of cultural factors, including religion, legal systems, marriage habits, and personal residential preference. Religion affects population since it often involves traditions concerning age at marriage, number of wives, attitudes toward population control, and the importance of large families. Historically, many religions emphasized the importance of marriage as the "normal" situation for men and women and often indirectly encouraged early marriage, thus providing the potential for more births. The church in a specific region sometimes had regulations or taboos affecting the use of contraceptives or abortion as means of controlling population. Historically, the Catholic church, for example, has opposed any "artificial" contraception methods, and the larger families associated with Latin America and southern Europe have in part reflected this doctrine. Some critics maintain, however, that the later adoption of industrialization also contributed to the persistence of high birth rates in southern Europe.

Religion and population distribution do not always correspond. France, for instance, which is largely Catholic, has traditionally had one of the lowest population growth rates in Europe. The French population grew slower than that of Germany, the United Kingdom, the Netherlands, or Belgium after 1550 even though those countries were Protestant. The slow growth of France's population indicates that although the Catholic church may have played an important role in the life of the people, the public did not accept the official church view on birth control. Instead French inheritance laws requiring land to be divided among all children caused couples to desire small families.

Population growth rates in Germany and England during the late nineteenth and early twentieth centuries reflect another cultural variable, politics. Political variables can affect population in a variety of ways, and in the case of Germany, the perception that military strength and political power were based on population size led the government to encourage large families. Slower growth rates in France allowed the German population to become the largest in Europe. Germany's large population and rapid growth led to the migration of Germans from the German hearth along the Rhine River to areas of eastern Europe and to North and South America. The growing population of Germany led to the adoption of the concept of *Lebensraum* (literally, "living room") as a justification for territorial expansion. Past political views thus help to explain present population distribution and density.

Another political factor affecting population relates to selective policies affecting individuals or individual groups in a particular country. Germany's laws and attitudes in respect to its Jewish population were an example of this. The ethnocentric cultural views of Adolf Hitler led to the persecution, forced removal, and death of the majority of the Jewish population in Germany and the European lands where Germans had migrated. (Table 4–5).

TABLE 4–5
Changes in the Jewish Population of Europe: 1939 to 1980

	JEWISH POPULATION 1939	JEWISH POPULATION 1980	CHANGE IN JEWISH POPULATION 1939–1980
Albania	204	300	+96
Austria	No data	13,000	
Belgium	60,000	41,000	−19,000
Bulgaria	48,398	7,000	−41,938
Czechoslovakia	356,830	12,000	−342,830
Denmark	5,690	7,500	+1,810
Finland	1,000	1,755	+755
France	140,000	650,000	+410,000
Germany	240,000	38,000	−202,000
Greece	72,791	6,000	−66,741
Hungary	444,567	80,000	−364,567
Ireland	3,986	1,900	−1,786
Italy	47,825	41,000	−6,825
Luxembourg	3,144	1,000	−2,144
Netherlands	156,817	30,000	−126,817
Norway	1,359	900	−459
Poland	3,113,900	6,000	−3,107,900
Portugal	1,200	600	−600
Spain	4,000	12,000	+8,000
Sweden	6,653	17,000	+10,347
Switzerland	17,973	21,000	−3,027
Romania	900,000	45,000	−855,000
United Kingdom	300,000	410,000	+110,000
Yugoslavia	68,405	5,500	−62,905
Soviet Union	3,020,141	2,630,000	−390,141

Another political action affecting population is the imposition of barriers to the migration of people from one place to another. The Berlin Wall between West Germany (the German Federal Republic) and East Germany (the German Democratic Republic) was constructed in the 1950s in response to the movement of population from East to West Germany. The wall has maintained the population of East Germany at a higher level than if free migration had been allowed to occur.

Another physical political barrier affects the population distribution in India. The country of Bangladesh has one of the world's highest population densities (Figure 4–23). Adjoining Bangladesh to the north is the Indian state of Assam, which has a low population density. Population pressure in Bangladesh prompted 2 million people to migrate across the border into India's Assam between 1960 and 1987. Resentful of the settlers from Bangladesh, some Indians attacked them in the early 1980s, leading to hundreds of deaths. The Indian government responded to the concerns of its citizens by attempting to build a wall to prevent the immigration. India's attempt to block immigration failed because of the tremendous population of Bangladesh (120 million people living in an area slightly smaller than the state of Iowa) and its rapid increase (7,700 people per day). Continued population pressure has forced Bangladesh to negotiate with India concerning the proposed barrier to movement, but the problem remains unresolved.

Another type of political issue affecting population density and distribution involves conflict.

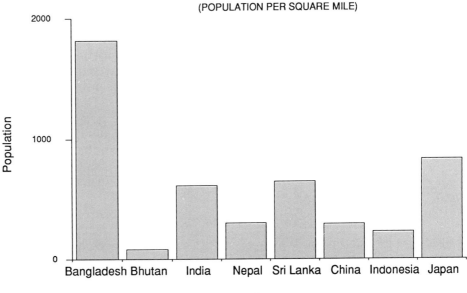

POPULATION DENSITIES IN SELECTED ASIAN COUNTRIES, 1989.
(POPULATION PER SQUARE MILE)

FIGURE 4–23
Population Densities in Selected Asian Countries, 1989 (Population per Square Mile) The population densities of South Asian countries illustrate that total numbers alone can be misleading. Bangladesh has only about one-tenth as many people as China, but its density is much greater.

Wars force people to leave. One-third of the population in Afghanistan (5 to 6 million) fled as the result of the conflict with the Soviet Union between 1977 and 1989. It is estimated that nearly 10 percent of the population of Afghanistan died during this migration. Similar conflicts in a host of sites around the world also affect population by destroying communities, farms, and societies. Most noteworthy are the conflicts in the Middle East that have affected Palestinians; the destruction of much of the Armenian society in the early 1920s; the campaign against the Kurds by Iraqi and Iranian forces; the killing of an estimated one-third of the population of Kampuchea (Cambodia) by the Pol Pot regime in the late 1970s and early 1980s; the flight of hundreds of thousands of refugees from Vietnam in the late 1970s and early 1980s; and the flight of Cubans to the United States since Castro's ascent to power.

Political upheavals in Sudan, Ethiopia, Uganda, and South Africa have also changed population distribution. In the cases of South Africa and Ethiopia, these have taken the form of specific government actions to relocate populations. Elsewhere, people have simply left what they viewed as an intolerable situation.

The range of impacts of political decisions on population does not end with population movement. Political decisions such as those in China that affect the birth rate also have a tremendous impact on the population of a region. The increased recognition by less industrialized countries that rapid population growth handicaps attempts to increase the standard of living creates another political impact on population growth. Consequently, the governments' efforts to affect population growth rates will continue to have an important impact on population numbers, density, and distribution.

Economic conditions based on political decisions also affect the distribution and density of population. Mass migration to North America in the last three centuries reflects economic conditions and the perception of economic possibilities in other areas (Figure 4–24). Migration to the United States from Mexico, the Caribbean, and Asia today represents attempts by residents of countries with rapidly growing populations and limited economic opportunities to acquire citizenship in a region where economic conditions are perceived as being better. Today's migration includes both legal and illegal migrants.

IMMIGRATION TO UNITED STATES: 1820-1986

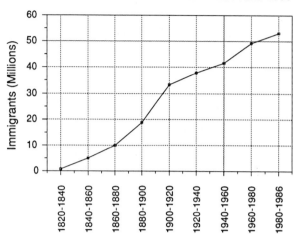

FIGURE 4–24 Immigration into the United States: 1820–1986 Immigration into the United States has been a major factor in population growth in this country.

Another cultural factor affecting population is the tradition of a specific group regarding family size. In some societies the number of children born in a family (particularly male children) is perceived to bring status. This is often related to economic conditions as well, because it is associated with rural societies where children, particularly males, can assist with the farm work. The impact of such attitudes is sometimes overcome only when population pressure leads to excessive fragmentation of farm property.

Legal restrictions on the inheritance of property also affect population, as in the case of France. Inheritance laws in Europe are divided between those that require property to be given to the eldest son intact (*primogeniture*) and those that require all property to be divided among all the children. The area of Europe that was dominated by the Roman Empire emphasizes division of the property among all children, whereas the area dominated by the Germanic peoples emphasizes primogeniture. Where primogeniture is common, the tendency has been for people to move from the farms to the cities, to migrate to other lands, or to have smaller families. The initial movement in Germany was to the United States where more land was available, followed by movement to Germany's cities as the Industrial

Revolution began, and culminating in the move to smaller family size in the twentieth century.

The changing environmental perception of a group can also affect variables that influence population distributions. Historically, groups viewed the environment in terms of its ability to provide a livelihood based on small-scale agriculture. Changes in agriculture in the nineteenth and twentieth centuries led to the expansion of population to the frontiers of North America, Australia, Latin America, and the Soviet Union to develop farms based on larger scale monoculture. The perception of place in the twentieth century has led to the migration of population from the farms to the urban areas of the less industrialized world in response to a perception of economic opportunity. In industrial countries, people have moved to the suburbs or areas perceived to have a favorable climate, such as Florida or the Mediterranean lands of Europe.

Conclusion

All of the factors that affect population growth, change, density, and distribution combine to create the mosaic of population that characterizes the cultural geography of the globe. Population trends divide the world into geographic regions according to the rate of population growth (Figure 4–25). Five broad regions categorized as slow

New York City. Deteriorating housing stock in many American cities is paralleled by suburban growth.

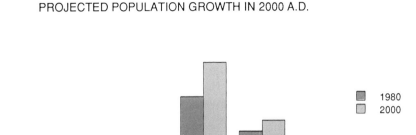

PROJECTED POPULATION GROWTH IN 2000 A.D.

FIGURE 4–25 Projected Population Growth in A.D. 2000 The projection of population to the year A.D. 2000 demonstrates the impact of continued high population growth rates in the less industrialized world.

growth contain 2.4 billion people. Western Europe, Anglo America, Eastern Europe and the Soviet Union, Australia and New Zealand, and China and Japan combine for an average growth rate of only .08 percent per year. Rising living standards and low growth rates in these countries reinforce one another. Another five broad geographic regions are in the rapid population growth group, which contains nearly 2.7 million people and averages 2.5 percent growth yearly: Southeast Asia, Latin America, South Asia (India, Bangladesh, and Pakistan); the Middle East, and Africa. Many of these countries face both rapid population growth and falling incomes that result in lower standards of living. A few of the countries in this broad group are anomalies in that their population growth rate slowed decades ago (such as Argentina, Uruguay, and Cuba).

This global division based on slow or rapid growth represents the broadest cultural regionalization of the world that has explanatory powers. Categorizing a country as part of the slow-growth or high-growth region also characterizes it in terms of other aspects relating to life-style and environmental problems. The richest countries generally have the lowest growth rate, while the poorest countries have the highest. The central question regarding population is whether the carrying capacity can be increased so the world will be able to increase the living standard of today's population and provide for those yet unborn. If

the carrying capacity cannot be increased, or if it cannot be increased rapidly enough, the standard of living of large areas of the world may decline. The greatest challenges for the world are in the less industrialized countries, which are growing at a rate three times faster than that found in the industrialized countries.

Rapid population growth in these countries creates challenges and problems, but it also represents one of the greatest achievements in hu-

Rapid population growth is associated with high unemployment in less industrialized countries, while technological change may cause unemployment in industrialized countries.

man history. Life expectancy in the less industrialized countries increased by 20 percent between 1961 and 1986, ensuring that most children born can anticipate living to adulthood. This achievement is sometimes obscured by concern for population growth. It is also significant that growth rates in the less industrialized countries are beginning to slow and that on a world scale the growth rate is actually declining.

Issues related to resource allocation, population distribution between wealthy industrialized countries and less industrialized countries, and further slowing of population growth will be of central interest to the coming decades, but given the great advances that people have made since the emergence of *Homo Sapiens*, these problems seem resolvable. Solutions may involve diffusion of technology from the industrialized to the less industrialized world, adoption of methods to slow population growth rates, or movement of surplus population between countries, but whatever the solution, the world's population distribution and density will change in the course of the next decades.

QUESTIONS

1. Describe the changes affecting world population that cause some observers to believe a "population crisis" is imminent.
2. Discuss the development of formal censuses of population, giving examples.
3. What are five major population migrations that have occurred since the sixteenth century?
4. Describe Europe's population growth, including the major events that affected the growth.
5. What is the demographic transition? Why does it occur?
6. Describe and give an example of the three broad stages of the demographic transition.
7. Compare and contrast the role of children in less industrialized and industrialized societies.
8. Describe the major elements in the Malthusian equation and indicate how they apply to the world population at present.
9. Compare and contrast India's and China's population growth and population policies. Which seems to be the most effective? Why?
10. Describe the world regions with high population density. What geographic factors help to explain these concentrations of population?

SUGGESTED READINGS

Bequjeu-Garnier, Jacqueline. *Geography of Population*, 2d ed. S. H. Beaver, trans. London: Longmans, 1978.

Bennett, D. Gordon. *World Population Problems: An Introduction to Population Geography*. Champaign, Ill.: Park Press, 1984.

Berry, Brian J. L. "Challenges of the Numbers Game." *The Professional Geographer* 1981, 33 (May): 161–62.

Birdsall, Stephen S. "Analysis of Population Age Balance." *The Professional Geographer* 1980, 32 (November): 467–70.

Boserup, Esther. *Population and Technical Change: A Study of Long-Term Trends*. Chicago: University of Chicago Press, 1981.

Bouvier, Leon F., "Africa and Its Population Growth." *Population Bulletin* 30 (February 1975).

_____ . "Planet Earth 1984–2034: A Demographic Vision." *Population Bulletin* 39 (February 1984).

Bouvier, Leon, Elinore Atlee, and Frank McVeigh. "The Elderly in America." *Population Bulletin* 30 (September 1975).

Bowden, Martyn, et al. "The Effect of Climate Fluctuations on Human Populations: Two Hypotheses," in T. M. L. Wigley, M. J. Ingram, and G. Farmer, eds. *Climate and History: Studies in Past Climates and Their Impact on Man*, pp. 479–513. Cambridge: Cambridge University Press, 1981.

Brown, Lester. *The Twenty-ninth Day: Accommodating Human Needs and Numbers to the Earth*. New York: Norton, 1974.

_____ . "World Food Resources and Population: The Narrowing Margin." *Population Bulletin* 36 (December 1981).

Brown, Lester R., and Jodi L. Jacobson. "Our Demographically Divided World." Worldwatch Paper 74. Washington, D.C.: Worldwatch Institute, December 1986.

Clark, Colin. *Population Growth and Land Use*, 2d ed. London: Macmillan, 1977.

Clarke, John I. *Geography and Population: Approaches and Applications*. Oxford and New York: Pergamon Press, 1984.

Coale, Ansley J. "The History of Human Population." *Scientific American* 231 (September 1984): 40–51.

Coleman, David, and Roger Schofield, eds. *The State of Population Theory: Forward from Malthus*. Oxford and New York: Basil Blackwell, 1986.

Day, Lincoln H. "What Will a Zero-Population Growth Society Be Like?" *Population Bulletin* 33 3 (June 1978).

Du Paquier, J., and A. Fauve-Chamoux, eds. *Malthus Past and Present*. New York: Academic Press, 1983.

Espenshade, Thomas J. "The Value and Cost of Children." *Population Bulletin* 32 (April 1977).

Feshbach, Murray. "The Soviet Union: Population Trends and Dilemmas." *Population Bulletin* 37 (August 1982).

Freedman, Ronald, and Bernard Berelson. "The Human Population." *Scientific American* 231 (September 1974): 30–39.

Gendell, Murray. "Sweden Faces Zero Population Growth." *Population Bulletin* 35 (1980).

Gould, W. T. S., and R. Lawton, eds. *Planning for Population Change*. Totowa, N.J.: Barnes and Noble Books, 1986.

Grigg, David. "Modern Population Growth in Historical Perspective." *Geography* 67 (April 1982): 97–108.

_____ . "The Growth of World Food Output and Population 1950–1980." *Geography* 68 (October 1983): 301–6.

Hornby, William F. *An Introduction to Population Geography*. New York: Cambridge University Press, 1980.

Hsu, Mei-Ling. "Growth and Control of Population in China: The Urban-Rural Contrast." *Annals of the Association of American Geographers* 75 (June 1985): 241–57.

Jones, Huw R. *A Population of Europe: A Geographical Perspective*. Harlow, England: Longman, 1970.

Jowett, A. J. "China: Land of the Thousand Million." *Geography* 69 (June 1984): 252–57.

Keyfitz, Nathan. "The Population of China." *Scientific American* 250 (1984): 38–47.

Kleinman, David S. *Human Adaptations and Population Growth: A Non-Malthusian Perspective*. Montclair, N.J.: Allanheld, Osmun, 1980.

Lee, David (with Ronald Schultz). "Regional Patterns of Female Status in the United States." *The Professional Geographer* 34 (February 1982): 32–41.

Lo, G. P., and R. Welch. "Chinese Urban Population Estimates." *Annals of the Association of American Geographers* 2 (June 1977):246–53.

Malthus, Thomas Robert. *An Essay on the Principle of Population*, Phillip Appelmen, ed. New York: W. W. Norton, 1976.

Meadows, Donella H., et al. *The Limits of Growth: A Report for the Club of Rome's Project on the Predicament of Mankind*. New York: Universe Books, 1972.

Menken, Jane, ed. *World Population and U.S. Policy*. New York: W. W. Norton, 1986.

Merrick, Thomas W. Population Pressures in Latin America." *Population Bulletin* 40 (July 1986).

_____ . "World Population in Transition." *Population Bulletin* 41 (April 1986).

Momsen, Janet Henshall, and Janet G. Townsend, eds. *Geography and Gender in the Third World*. Albany, N.Y.: State University of New York Press, 1987.

Newman, James L. *Malthus Past and Present*, J. Du Paquier, ed. London: Academic Press, 1983.

Newman, James L., and Gordon E. Matzke, *Population Patterns, Dynamics, and Prospects*. Englewood Cliffs, N.J.: Prentice-Hall, 1984.

Nortman, Dorothy. "Changing Contraceptive Patterns: A Global Perspective." *Population Bulletin* 32 (August 1977).

Nostrand, Richard L., A. J. Jaffe, Ruth M. Cullen, and Thomas D. Boswell. *The Changing Demography of Spanish Americans*. New York: Academic Press, Studies in Population Series, 1980.

Peters, Gary L., and Robert P. Larkin. Population Geography: Problems, Concepts, and Prospects. (Second edition) Dubuque, Iowa: Kendall-Hunt, 1983.

Population Growth and Policies in Sub-Saharan Africa. Washington, D.C.: World Bank, 1986.

Scientific American. *The Human Population*. San Francisco: W. H. Freeman, 1974.

Tien, H. Yuan, "China: Demographic Billionaire." *Population Bulletin* 38 (April 1983).

Van de Kaa, Dirk J., "Europe's Second Demographic Transition." *Population Bulletin* 42 (March 1987).

Van der Walle, Etienne, and John Knodel. "Europe's Fertility Transition." *Population Bulletin* 34 (February 1980).

Visaria, Pravin, and Leela Visaria. "India's Population: Second and Growing." *Population Bulletin* 36 (October 1981).

Warns, Anthony M., ed. *Geographical Perspectives on the Elderly*. New York: John Wiley & Sons, 1982.

Woods, Robert. *Population Analysis in Geography*. New York: Longman, 1979.

_____ . "Population Studies." *Progress in Human Geography* 7 (1983): 261–66.

_____ . *Theoretical Population Geography*. London and New York: Longman, 1979.

Wrigley, E. A. *Population and History*. New York: World University Library, 1969.

Yinger, Nancy, Richard Osborn, David Salikever, and Ismail Sirageldin. "Third World Family Planning Programs: Measuring the Costs." *Population Bulletin* 38 (1983).

Migration: Population Movement in Cultural Geography

The motives for immigration . . . have been very similar from first to last;
they have always been a mixture of yearnings—for riches, for land, for
change, for tranquility, for freedom, and for something not definable in
words.   M. W. JONES

THEMES

Types and effects of population
 mobility.
Barriers to migration.
Distance and intervening opportunities
 in migration.
Character and characteristics of
 migrants.
Process and patterns of world
 migration.

CONCEPTS

Mobility
Circulation
Migration
Immigration
Emigration
Pioneers
Chain migration
Push-pull factors
Refugees
Natural hazards
Gravity model
Social distance
Economic distance
Intervening opportunities
Return migration
Selectivity
Ideological differences

Humans are restless and constantly on the move. Individuals and cultural groups commute from home to work, travel from place to place, flee from environmental or societal disasters, and move their hearth and home in search of a better life-style. The movement of humans over time has shaped the cultural geography of the earth. Place-names are borrowed from ancestral homes. Neighborhoods and regions reflect the language, religion, foods, dress, or architecture of groups who moved from one place to another.

Repeated waves of immigrants to new places help create the character of those areas. Movement of Spanish and Asian peoples to the United States in the past two decades has had a marked and easily recognizable impact on the cultural geography of towns and cities from Florida to California. Earlier movements to the United States by peoples from Europe, Africa, Asia, and Latin America created the fabric of the country's cultural geography (Figure 5–1).

The movement of population includes both temporary and permanent changes in location. The degree of population movement indicates the *mobility* of a group. A population's mobility refers to the nature and frequency of both short-

and long-term movement. Daily or weekly movements of people from home to work or school, vacations, and cyclical and other short-term changes in location are called population *circulation*. The circulation of people is manifest in the cultural geography in many ways, including commuter traffic, vacation resorts, and the location of schools and shopping areas (Figure 5–2).

Long-term (generally longer than one year) or permanent population movement is called *migration*. Migration implies that individuals or groups have relocated their place of residence. Such movement may be permanent or temporary, voluntary or forced. Migration may involve movement to a new country or to a new location within the same country. It may involve an individual, a family, an entire neighborhood or community, or an entire social or ethnic group. Migrants may choose to move to start a new life, to obtain amenities in their place of residence or work, or to be safe from war, famine, or other social or environmental stresses.

It is important to distinguish between *immigration* and *emigration*. Emigration is migration *out* of a country, while immigration is migration *into* a country. The terms emigrant and immi-

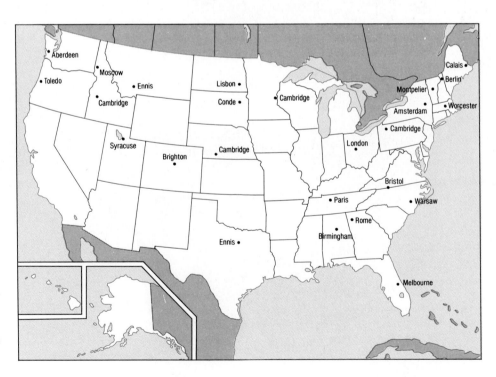

FIGURE 5–1 Selected European Place-Names in the United States An important element in the landscape is the name applied to places, which often provides clues to the origins of the residents.

Causes of Migration

Each individual or group moves in response to factors that attract them to a new destination or repel them from their present location. People perceive places as having negative, positive, or neutral attributes that affect their decisions concerning migration. The intuitive rationality of the existence of push and pull factors is obvious. *Push-pull* factors can be divided into a number of types, including economic, political, social, and environmental factors.

Economic Factors

Economics is often regarded as the primary push-pull factor affecting migration. Economic push factors include such things as high rates of unemployment, low salaries, or a declining economy. Even in locations where there is a high rate of employment, perceptions that the wage rates are low relative to other places may impel people to migrate. Population growth and associated pressure on land resources push rural people to migrate to urban settings. Limited opportunities for highly trained people in less industrialized countries may be a push factor contributing to the "brain drain," in which highly trained people from the more agrarian world move to countries of the industrial world. Economic push factors are often the basis of migration, particularly in combination with economic pull factors.

Economic pull factors include labor shortages, higher wage rates, land availability or cost, or lower cost of living. Each of these acts in concert with push factors to attract individuals and groups. Much of the immigration to the United States has been the result of economic pull factors. Abundant low-cost land, an industrializing society, and relatively high wage rates attracted migrants to the United States (Figure 5–4). During the eighteenth, nineteenth, and early twentieth centuries, migrants from the United Kingdom, Scandinavia, Germany, Italy, and Russia came in anticipation of cheap land or jobs in the emerging industries of U.S. and Canadian cities. Migrants from Korea, Mexico, the Caribbean Is-

lands, and elsewhere came to the United States in the 1980s because they perceived job opportunities and a higher standard of living (Figure 5–5). These economic pull factors continue to make the United States most attractive to migrants.

Economic conditions in the countries of Kuwait, Saudi Arabia, Qatar, and other oil-rich countries of the Middle East have attracted numerous migrants in the last two decades. As many as half of the total populations of Kuwait and Qatar are nonnationals employed in the petroleum industry, in construction work, or in service positions in restaurants, laundries, and other low-status jobs. Rapidly expanding economies in Europe in the 1960s and 1970s coincided with declining European population growth rates to attract numerous *guest workers* from Algeria, Turkey, Yugoslavia, Pakistan, and the British and French Caribbean Islands (Table 5–3). These examples illustrate the broad range of economic factors that affect migration. Economic factors related to migration are often affected by political factors as well.

Political Factors

Political factors affecting migration are related to forms of government and the laws and regulations that they foster. One key political element affecting migration seems to be personal freedom. Lack of

TABLE 5–3
Foreign Workers in Selected Countries

West Germany	2,000,000
United Kingdom	1,600,000
France	1,600,000
Greece	750,000
Italy	400,000
Libya	280
Oman	249,400
Kuwait	396,200
Saudi Arabia	258,000
Bahrain	81,200
United Arab Emirates	464,000

Source: The World Fact Book, 1987 (Washington, D.C.: Central Intelligence Agency, 1987).

FIGURE 5–4
Advertisement of Cheap Land Advertisements such as this attracted many migrants to the United States and to specific states. Land in the northern states of the Midwest attracted many Scandinavians in the period from 1860–1890.

UNITED STATES IMMIGRATION IN SELECTED YEARS

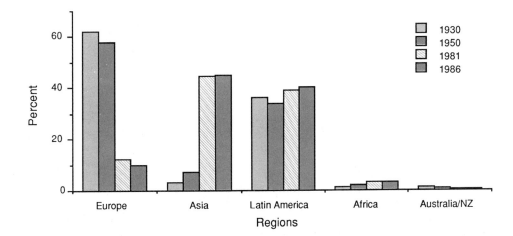

FIGURE 5–5 United States Immigration in Selected Years

personal freedom often causes large migrations. Over three million people emigrated from Southeast Asia between 1976 and 1985. Many of these emigrants came to the United States where they thought they would have greater individual freedom.

Another politically induced migration occurred in Afghanistan when the country was occupied by forces from the Soviet Union in 1978. Over four million people fled to Pakistan or Iran to escape the conflict between the government forces supported by the Soviet Union and Afghan

Turkish Workers. Guest workers often prompt the development of services to provide elements of their culture in the host country.

guerilla forces. The overthrow of the shah of Iran in 1978 and the rise to political power of Ayatollah Khomeini and fundamentalist religious leaders caused tens of thousands of Iranians to emigrate to other countries. Push factors related to political activity are a recurrent phenomenon in the earth's history, ranging from the migration of Jews to the New World in the eighteenth century when the French government proclaimed that the entire country was to be Catholic to the mass movement of population between India and Pakistan when the Indian subcontinent was partitioned between the two countries in 1947 (Figure 5–6).

Individuals or groups who leave a country because of political situations (including wars) are known as *refugees*. The United Nations defines a political refugee as someone who cannot return to his or her home country for fear of persecution. Political refugees may or may not be permanent migrants depending on whether they are allowed to take up residence in a new country. Some refugees are in the unclear class of stateless persons since they reside in refugee camps but are not accepted as permanent migrants by the country where the camp is located. This is the case with Palestinian refugees in Lebanon, African refugees in a variety of countries, many people from Central America trying to obtain residency in the United States, and the Afghan people who

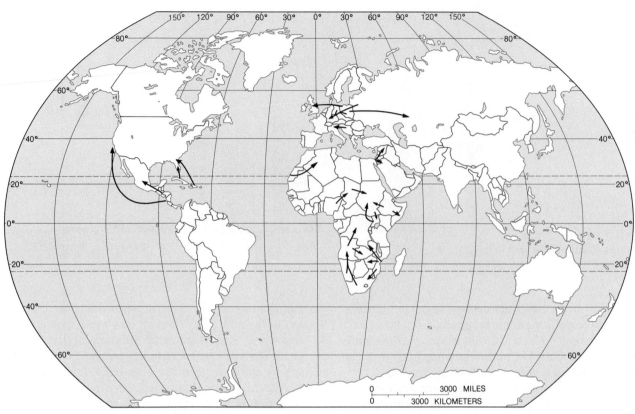

FIGURE 5–6 Migration Resulting from Political Factors

Refugee Camp, Korem, Ethiopia. Refugees are often the most impoverished migrants.

have fled to Pakistan or Iran. Although technically refugees may not be classed as permanent migrants, their inability to return to their homes over extended periods of time makes them de facto migrants (Figure 5–7).

Political pull factors are often associated with countries perceived as being more democratic. Many of the migrants to the United States came because they perceived the country as a democratic haven in which individual freedoms and economic actions were protected. Migration of people to countries having perceived political advantages is common. The existence of the Berlin Wall in Europe was a forcible reminder that residents of the German Democratic Republic (East Germany) were unable to move freely to the Federal Republic of Germany (West Germany) until the government of the GDR opened it in November of 1989. The movement of peoples from Nicaragua and El Salvador to adjacent countries or to the United States and the migration

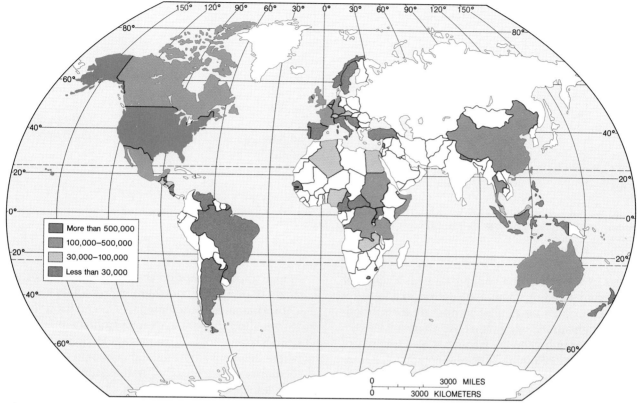

FIGURE 5–7 World Refugees—countries receiving refugees from other countries or from internal relocation of selected groups.

from Cuba to the United States and other Latin American states are examples of population movement resulting from political push and pull factors. Political factors affecting emigration are commonly intertwined with economic factors, making it difficult to designate the migration of a group as solely politically motivated.

Environmental Forces

Environmental factors related to migration include environmental quality and environmental hazards. Natural disasters often serve as a push factor affecting the migration of the population. In the 1840s, Ireland's population declined from 8 million to less than 5 million largely as a result of a potato blight. Famine associated with the potato blight forced many Irish from their homeland to England or North America (Table 5–4). Other environmental push factors include droughts and

The Berlin Wall epitomized the efforts of some countries to prevent emigration.

floods. The drought in the Sahel region of Africa, which forced the relocation of millions because of loss of pasture for animals, is paralleled in other

Table 5–4
Population Changes in Ireland

IRELAND		NORTHERN IRELAND		REPUBLIC OF IRELAND	
1821	6,802,000	1921	1,275,000	1926	2,972,000
1831	7,767,000	1937	1,280,000	1936	2,968,000
1841	8,175,000	1951	1,371,000	1951	2,961,000
1851	6,552,000	1961	1,425,000	1961	2,818,000
1861	5,799,000	1966	1,485,000	1966	2,884,000
1871	5,412,000	1971	1,536,000	1971	2,978,000
1881	5,175,000				
1891	4,705,000				
1901	4,459,000				
1911	4,390,000				

Source: Whitaker's Almanack, 1989. (London: William Clowes Limited, 1989).

areas by the relocation of population as a result of recurrent flooding. Where an option exists to relocate, people may choose to migrate from flood- or drought-prone areas to those that are less affected, depending on the frequency and severity of the drought or flooding. Other secondary environmental hazards affecting migration include severe storms (hurricanes, tornadoes, blizzards), earthquakes, landslides, volcanic activity, soil erosion, and disease. Perception of the severity of such hazards, awareness of alternative places to live, and the ability to relocate enter into the actual decision to move away from hazardous locales.

Virtually every place on earth is subject to some *natural hazard*, and local residents tend to minimize the danger to justify their living there. When the natural hazard is perceived as too threatening, they move. Given the same level of hazard, some people will be more apt to move than others, reflecting their perception of both the hazard and the real or perceived opportunities to move.

Some environmental factors are perceived as attractive, creating pull factors. Environmental pull factors include climate, access to recreational amenities, and the possibility of more fertile land or other perceived resources. Internal migration within the United States reflects environmental conditions in the so-called Sunbelt (Figure 5–8). There has been a great increase in migrants—particularly of retirees—to Florida, California, Arizona, and Texas since 1970 in re-

sponse to the perception that the climate is better there than in the northern states (Table 5–5). California alone increased its population from 15.7 million in 1960 to over 28 million in 1990. A similar phenomenon is taking place in Europe, where residents of England, Germany, and the Netherlands are purchasing retirement homes in southern Spain and in France.

Social Factors Affecting Migration

Social factors also affect the decisions of individuals or groups to migrate. Social push factors are difficult to separate from political or economic factors since the three are closely interrelated. Social push factors may include customs or traditions of a culture that result in social stratification of the society, persecution or discrimination against specific groups, or inadequate access to jobs, education, or medical care. The perception or reality of lack of discrimination, persecution, social stratification or the availability of social amenities serves to attract people to other locations.

The movement of black Americans from the South after World War II, although primarily the result of economic factors, in part represented the real or perceived threat of persecution in the South and social and economic opportunities in the North. The movement of Chinese from Southeast Asia to the United States or Europe reflects anti-Chinese sentiments in Southeast Asia that have induced them to leave.

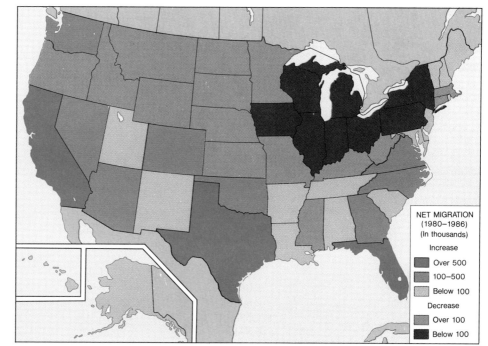

FIGURE 5–8 State Population Change The migration of people from one region to another is in part responsible for the rapid increase in the population of the southern and western states.

NET MIGRATION
(1980–1986)
(In thousands)

Increase

Over 500

100–500

Below 100

Decrease

Over 100

Below 100

TABLE 5–5
Population Change in the United States

STATE	1970 (in thousands)	1986 (in thousands)	PERCENTAGE CHANGE
United States	**203,302**	**241,078**	**18.6**
New England	**11,848**	**12,737**	**7.5**
Maine	994	1,173	18.0
New Hampshire	738	1,027	39.2
Vermont	445	541	21.6
Massachusetts	5,689	5,832	2.5
Rhode Island	950	975	2.6
Connecticut	3,032	3,189	5.2
Middle Atlantic	**37,213**	**37,280**	**0.2**
New York	18,241	17,772	−2.6
New Jersey	7,171	7,619	6.2
Pennsylvania	11,801	11,888	0.7
East North Central	**40,262**	**41,737**	**3.7**
Ohio	10,657	10,752	0.9
Indiana	5,195	5,504	5.9
Illinois	11,110	11,552	4.0
Michigan	8,882	9,145	3.0
Wisconsin	4,418	4,785	8.3

TABLE 5–5
Continued

STATE	1970 (in thousands)	1986 (in thousands)	PERCENTAGE CHANGE
West North Central	**16,327**	**17,576**	**7.6**
Minnesota	3,806	4,214	10.7
Iowa	2,825	2,851	0.9
Missouri	4,678	5,066	8.3
North Dakota	618	679	9.9
South Dakota	666	708	6.3
Nebraska	1,485	1,598	7.6
Kansas	2,249	2,460	9.4
South Atlantic	**30,678**	**40,916**	**33.4**
Delaware	548	633	15.5
Maryland	3,924	4,463	13.7
District of Columbia	757	626	−17.3
Virginia	4,651	5,787	24.4
West Virginia	1,744	1,918	10.0
North Carolina	5,084	6,333	24.6
South Carolina	2,591	3,377	30.3
Georgia	4,588	6,104	33.0
Florida	6,791	11,675	71.9
East South Central	**12,808**	**15,209**	**18.7**
Kentucky	3,221	3,729	15.8
Tennessee	3,926	4,803	22.3
Alabama	3,444	4,052	17.7
Mississippi	2,217	2,625	18.4
West South Central	**19,326**	**26,864**	**39.0**
Arkansas	1,923	2,372	23.3
Louisiana	3,645	4,501	23.5
Oklahoma	2,559	3,305	29.2
Texas	11,199	16,685	49.0
Mountain	**8,289**	**13,023**	**57.0**
Montana	694	819	18.0
Idaho	713	1,002	40.5
Wyoming	332	507	52.7
Colorado	2,210	3,267	47.8
New Mexico	1,017	1,479	45.4
Arizona	1,775	3,319	87.0
Utah	1,059	1,665	57.2
Nevada	489	963	97.0
Pacific	**26,549**	**35,737**	**34.6**
Washington	3,413	4,462	30.7
Oregon	2,092	2,698	29.0
California	19,971	26,981	35.1
Alaska	303	534	76.2
Hawaii	770	1,062	37.9

Source: U.S. Statistical Abstract (Washington, D.C.: U.S. Government Printing Office, 1988).

Barriers to Migration

Barriers to migration may be environmental, economic, political, or social. Environmental barriers inhibit the migration of population to the Arctic, Antarctic, highlands, and deserts. The Berlin Wall in East Germany was constructed for both economic and political reasons to prevent migration into West Germany.

Rural to urban migration affecting major cities of the Soviet Union such as Moscow, Leningrad, and Kiev has resulted in limiting further migration to these cities for social reasons. The government accepts the thesis that cities should not become too large because of the resultant increases in crime, difficulty of commuting to work, and higher costs of providing urban services (Figure 5–9). Migrants are officially allowed into such closed cities only if they have a skill that is needed (although political connections allow many to circumvent the law).

Individual countries have migration regulations to prevent them from being overwhelmed by immigrants. The individual regulations may take the form of restrictions on specific classes of refugees, ethnic groups, or total numbers of migrants. Most countries practice selective immigration control, limiting the entry of persons with certain backgrounds (such as those with criminal records or diseases). South Africa has a whites-only migrant policy that requires proof of European descent.

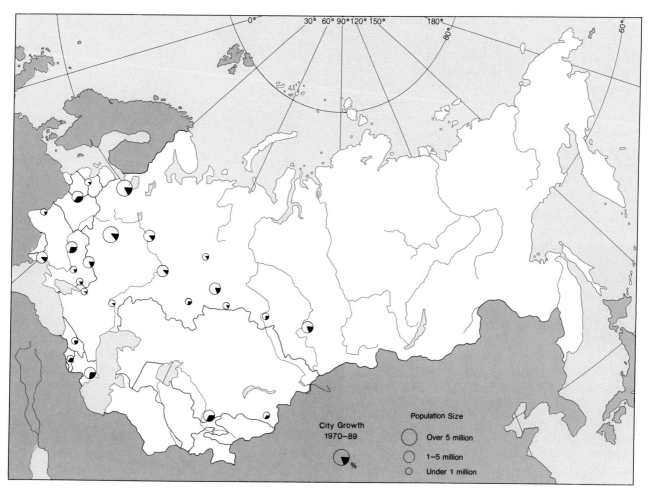

FIGURE 5–9 Population Changes in Soviet Cities Population growth in cities of the Soviet Union is caused in part by rural to urban migration.

Restrictions on migration fall into two classes, those preventing entry and those preventing exit. Exit restrictions are imposed in countries of the Soviet Bloc, China, and Vietnam. The degree of permeability of these political borders for potential emigrants varies from country to country. Motivation for controls on exit are generally economic, but may involve political reasons as well. The Soviet Union and the East European bloc countries limited emigration because of slow population growth and high labor demand and the negative image created by mass migration from societies nominally dedicated to championing the working class. The migration of highly trained and highly skilled people was perceived as a major ideological problem for these countries. Many of these restrictions were removed in 1989.

Other Communist countries, particularly Yugoslavia, began lifting some of their restrictions on emigration earlier. Elderly migrants in East Germany who were no longer useful in the labor force were allowed to migrate to the West before 1989, but had to leave their pensions behind. Political pressure from the West, particularly the United States, periodically resulted in larger numbers of Jews being allowed to emigrate from the Soviet Union before 1989 (Table 5–6).

Factors designed to prevent or limit migration are rarely successful in stopping migration for more than short periods. The human propensity to move, particularly when faced with economic, political, environmental, or social factors that are perceived to be life threatening, overwhelms all but barriers that are themselves life threatening.

Distance and Intervening Opportunities

Two other variables that affect migrants and migrations are distance and intervening opportunities. The greater the distance between places, the lower the volume of migration. The *gravity model* of spatial interaction postulates that places interact with each other in relation to their size and the distance between them (Figure 5–10). Two places that are large will have a greater amount of interaction than two places that are small if all are an equal distance apart.

TABLE 5–6

Jewish Emigrants from the Soviet Union, 1959–1985

YEAR	NUMBER OF EMIGRANTS
Total	271,506
1959	7
1960	102
1961	128
1962	182
1963	388
1964	539
1965	1,444
1966	1,892
1967	1,162
1968	229
1969	2,979
1970	1,027
1971	13,022
1972	31,681
1973	34,733
1974	20,628
1975	13,221
1976	14,261
1977	16,736
1978	28,865
1979	51,333
1980	21,472
1981	9,448
1982	2,683
1983	1,320
1984	883
1985	1,141

Source: Israel Yearbook on Human Rights (Tel Aviv: Tel Aviv University, 1986).z

Distance is more than a simple straight line, since it may involve *social* or *economic* "distances" as well, a distance that represents economic or social difference in the community. A family moving from the suburb of a large city may move a greater distance to a suburb on the other side of a large city than to the nearer central city. The decision to move farther to avoid the central city reflects the social and economic distance or difference between the residents of the central city and those of the suburbs.

Intervening opportunities also affect migrants and migrations. An intervening opportunity (sometimes referred to as an intervening obstacle) is an opportunity that exists between two places.

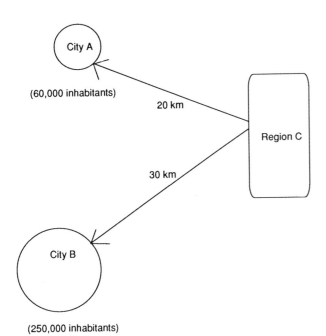

GRAVITY MODEL

City A

(60,000 inhabitants)

20 km

Region C

30 km

City B

(250,000 inhabitants)

FIGURE 5–10 Gravity Model of Migration Flow
Larger cities are attractive even if they are farther
away than smaller cities.

For example, a migrant from Mexico to the
United States may anticipate migrating to New
York City because he or she perceives economic
opportunity there. Any large city between Mexico
and New York City that provided job opportunities would be an economic intervening opportunity. Intervening opportunities can be recognized

in environmental, economic, political, or social
contexts. The number of intervening opportunities between two places reduces the number of
migrants who actually arrive at their original
destination. Many migrants began with the idea
of moving to the frontier for free land in the
historic development of the United States, but
ended up staying in the large cities of the East
because of the availability of jobs in the industrializing society, the presence of friends or family in
the eastern cities, or the lack of funds to move on
to the frontier (Figure 5–11). Sometimes the intervening situation is an obstacle rather than an
opportunity. The Sahara is an intervening obstacle that prevents migrants from the drought-
stricken Sahel from moving north into the coastal
areas of Africa. Laws and regulations restricting
migration from individual countries are political
obstacles.

The Nature of Migrations and Migrants

Neither the actual process of migration nor the
participants are uniform. Examination of the nature of migrants and migration helps to explain
present global patterns of population and culture.

One characteristic of migration in the industrial world is known as *step migration*. Early
observers noted that migrants generally did not go
directly from the farm to the largest city in a
region, but moved from the farm to a small town,
from the small town to a larger city, and from the
larger city to a metropolitan area. This pattern of

INTERVENING OPPORTUNITY

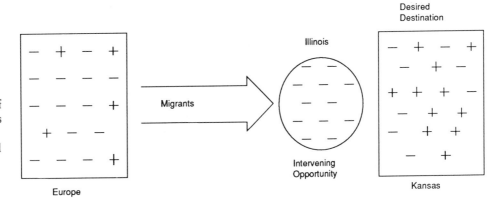

**FIGURE 5–11 Model of
Intervening Opportunities**
Migrants moving to
Kansas for cheap land had
to pass through Illinois,
which also offered fertile
farm land.

Europe

Migrants

Illinois

Intervening
Opportunity

Desired
Destination

Kansas

step migration was the norm in the industrial world through the middle of the twentieth century. Movement from the central city to the suburbs and from the suburbs to a small town and from the small town to a rural area by a small counterstream of migrants is a modern form of step migration.

The model of step migration in the industrial world does not apply to the rapidly urbanizing less industrialized world. The dominance of a few cities in these countries often results in direct movement from the farm to the largest city of the country. Potential migrants' greater awareness of jobs and job opportunities in the dominant city, the focus of roads and railroads on the dominant city, and the relative level of urban amenities (education, medicine, and so forth) make the larger cities more attractive to migrants.

Alongside the migration to the dominant cities of the less industrialized world, there is a counterstream of migrants returning to rural areas. Some of those who moved to the city in the hopes of finding a better environment may instead find themselves without jobs or other means of support and return to their families. If a migrant to the city is successful, the process of pioneer and chain migration may occur as the successful migrant encourages other members of the family to move to the city (Figure 5–12).

Many examples of chain migration could be cited. Migrants from European and Asian countries in the eighteenth and nineteenth centuries came to the United States, established themselves, and then actively encouraged others to migrate. Prior to World War II, much of the cultural diversity of large American cities with their associated ethnic enclaves reflected chain migration. Chain migration continues today in both the industrial and less industrial world. Migrants to the United States from Mexico, for example, maintain contact with family and friends in Mexico, leading to chain migration (Figure 5–13). Through chain migration, contacts are maintained between migrants and their home community for an extended period of time.

Migration and Return Migration

Chain migration is also associated with *return migration*, which occurs when individuals or groups migrate to an area, maintain residence for an extended period in the new locale, but ultimately return to their hearth area. Migrants from the rural South and Appalachia in the period between 1940 and today present an example of return migration. Poor rural residents (both white and black) migrated North to jobs in urban areas like Detroit, Pittsburgh, Cleveland, Gary, or the New York City metropolitan area. Many of these individuals or their descendants have returned to the South since 1970.

The push and pull factors that prompted the original migration from the rural South have changed in the last three decades. Job opportunities in the late 1930s and 1940s were limited in the South, isolation of blacks was the norm, and discrimination against them was common. The North, in contrast, was perceived as a region with less persecution, where employment and education were attainable. These push-pull factors have been reversed in the post-1970s. Loss of employment in the traditional industries of the North

PIONEER OR CHAIN MIGRATION

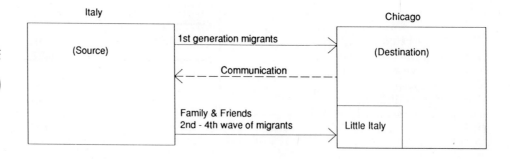

FIGURE 5–12 Pioneer or Chain Migration The pioneer or chain model of migration postulates that initial migrants (pioneers) influence later migrants' destination, as in this example of Italian migrants to Chicago.

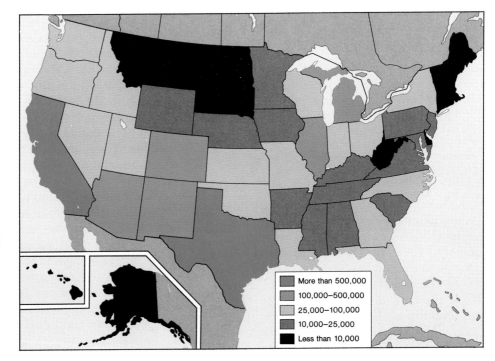

FIGURE 5-13 Mexican Migration to the United States The distribution of Mexicans in the United States provides some indication of their migration in the absence of accurate figures on actual numbers of migrants.

(steel, automobiles) and the movement of industry and manufacturing to the South have combined with social changes to reverse the flow of migrants. Retirement programs (both private and public) allow many of the older migrants from the South to return. This return migration to the South is one of the important trends in migration in the United States, but like other return migrations, it has occurred in much smaller numbers than the original migration flow.

Migrants: Character and Characteristics

Any migration is selective since certain people are more likely to emigrate than others. The *selectivity* associated with migrants includes such factors as age, sex, income, education, and culture.

Age has an important effect on migration. Although emigrants can be of any age, generally they are young adults. Repeated studies demonstrate that in both the industrial and less industrial world the late teens to the early thirties is the primary age of migration. In the industrial world, individuals in this age group are most likely to move as a result of education or career

opportunities, while in the less industrialized world, these individuals are often forced to move from the family farm as population pressures overwhelm its carrying capacity. The preponderance of individuals in the 17-to-35 age group in migrations has a distinct impact upon both the hearth and destination areas of migrants. Regions with a large out-migration of the young are left with an elderly population, as in the midwestern United States, where the average age of farmers has increased yearly as rural youth migrate to cities (Table 5-7).

TABLE 5-7
Age of Migrants, 1985-1986

AGE	NUMBER OF MIGRANTS
1-14	9,925,528
15-19	3,015,528
20-29	14,112,728
30-34	4,597,650
35-44	5,103,756
45-54	2,288,862
55-74	759,645
Over 75	563,886

Source: U.S. Statistical Abstract, 1988 (Washington, D.C.: U.S. Government Printing Office, 1988).

TABLE 5–8
Destination of Moves Made by Elderly Americans

	AGE	
TYPE	64–74	Over 75
Same county	55.7	58.2
Different county—same state	21.7	21.5
Contiguous state	6.4	6.1
Noncontiguous state	15.4	14.2

An exception to the youth versus age migration is found in some industrial countries as a result of retirement. An example is the movement of elderly retirees in the United States to the Sunbelt. This phenomenon is greatly overstated, however, as the vast majority of retirees and elderly do not move any great distance (Table 5–8). The apparent mobility of the elderly occurs because of the geographic concentration of long-distance elderly migrants. The development of retirement centers in Florida, Arizona, and California has resulted in major demographic changes in certain counties in each of these states. More than 20 percent of the population of Florida for example, is above age 65, and certain sections of the state, such as St. Petersburg, have become enclaves for the elderly. Retirement centers are often characterized by restrictions against children, creating a unique cultural landscape.

Some elderly in industrialized countries migrate to areas with perceived amenities, as in this retirement center in Anaheim, California.

Gender is another variable that affects migration, with males generally more apt to migrate than females. Larger numbers of male migrants were the norm in migration to the United States, especially during the period of rapid industrialization in the last half of the nineteenth and early twentieth centuries.

Male dominance in migrations is also notable during periods of rapid population growth in less industrialized countries and among individuals recruited to work in another country or place. Chinese, for example, were recruited to move to the United States to help build the transcontinental railroad in the American West and to work on sugar and pineapple plantations in Hawaii. Single Chinese men were brought in large numbers, but females did not arrive until much later. An unusual phenomenon associated with male migration flows is "picture brides." Absence of Chinese females in California and Hawaii led to Chinese males agreeing to pay for the migration of women, whom they would marry solely on the basis of pictures they had seen. This phenomenon has occurred among other cultural groups as well.

Many other migrations create a demographic imbalance between two areas. The need for men to work in mines, factories, and farms in the southern portions of Africa (South Africa and its tribal homelands, and neighboring states of Lesotho, Mozambique, and Zimbabwe) results in the literal depopulation of the young male component of many towns and villages. The resultant pattern of male migration to the cities for economic opportunities creates a society of females who care for the children and the farms and are visited by their husbands for only a short time each year. Similar gender differences in migration can be found in many of the major migrations that have occurred in the world.

Other variables that affect the characteristics of migrants are income, education, and occupation, all of which influence the real or perceived options of an individual or group. Higher education raises the threshold of expectation of jobs and income, changing the way an individual's home area is perceived. Individuals with higher education and occupational skills, and the income to support a major move, are able to relocate to places that they perceive as offering

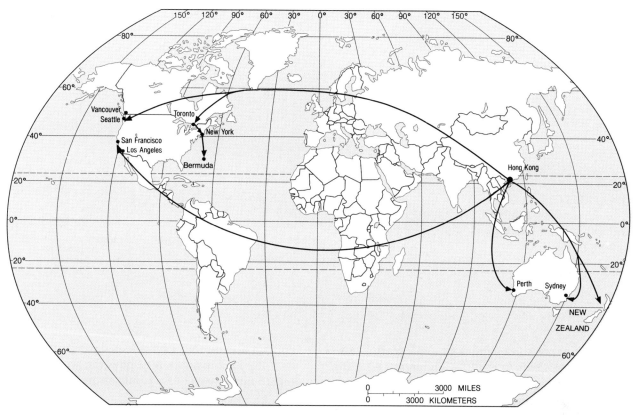

FIGURE 5–14 The Brain Drain from Hong Kong The movement of the educated elite from less industrialized countries is exemplified by the present migration of Chinese from Hong Kong.

greater economic opportunities. The result, in part, is the "brain drain," in which the educated and skilled elite of the less industrialized world are attracted to the industrialized world. Doctors, engineers, and other skilled professionals are disproportionately represented in the international migration stream between the less industrial and industrial worlds and within the internal migration stream of the industrial world (Figure 5–14).

Migration within the less industrialized world presents a different pattern. Internal migrants in less industrialized countries are more often poor, poorly educated, and poorly trained for urban industrial or service jobs. They represent surplus population from rural areas who are unable to find jobs, inherit land, or obtain a formal education. Such migrants are in reality a type of internal refugee. Economic refugees from the dearth of opportunities in their rural homes, they migrate to the cities in the hope of finding some form of

sustenance (Figure 5–15). There they frequently comprise the occupants of the slums, which surround the rapidly growing urban centers of the less industrialized countries.

A component of the industrial world's internal migration is also made up of the less well educated and less well trained. These migrants are economic refugees who for various reasons are unable to provide for themselves in the industrial economy. They occupy niches in the urban settings of large cities in the United States, Western Europe, and to some extent Japan. Sometimes referred to as street people, derelicts, vagrants, or bums, they eke out a living by begging, using public assistance, or engaging in petty theft. Although large American cities are not surrounded by squatter settlements of migrants, as are those of most less industrialized countries, they are surrounded by suburbs, which reflect migration and population movement in the industrial

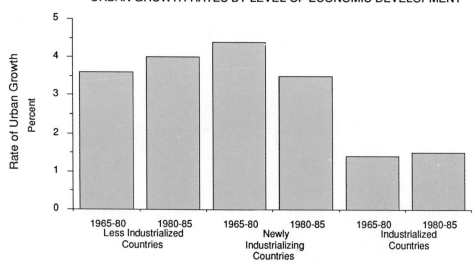

URBAN GROWTH RATES BY LEVEL OF ECONOMIC DEVELOPMENT

FIGURE 5–15 Urban Growth Rates by Level of Economic Development Growth rates of urban areas in industrialized countries is much lower than those in less industrialized countries.

world. The cultural landscape and spatial relationships of cities in both the industrial and less industrial world reflect the movement of their residents.

The combination of push-pull factors, intervening opportunities or obstacles, and distance combine with the individual characteristics of migrants to create the great migrations that have affected the cultural geography of the world. Examination of past migrations reveals the forces of

Caracas, Venezuela. In the less industrialized countries housing of the poor occupies undesirable building sites near the homes of the wealthy.

culture history that have affected past migrations and provides insights into how the world of the future may change as a result of migrations.

World Migrations: Process and Patterns

Information about prehistoric migrations is limited, but by the time plants and animals were domesticated, distinct cultural groups occupied Asia, Africa, Europe, the Americas, Australia, and New Zealand. The European Age of Discovery marked the beginning of revolutionary changes in the world's cultural map associated with migrations of Europeans (and the associated migration of Africans).

European Migration

Among the greatest of historic migrations has been that from Europe (Figure 5–16). European migration began following the discovery of the New World by Columbus and accelerated in numbers and impact over the course of the next four centuries. Prior to 1830, the total emigration from Europe was approximately 3 million. These migrants primarily settled in Canada (French), the United States (British, Scottish, German), and

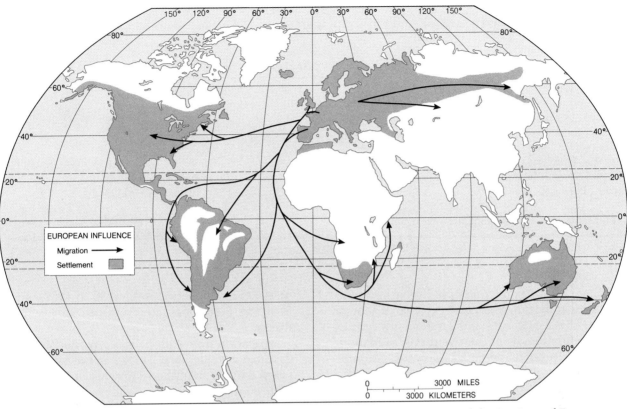

FIGURE 5—16 European Migration and Areas Settled by Europeans Major routes and destinations of European migrants.

Latin America (Spanish and Portuguese). The majority of these early migrants were from the British Isles and Germany.

Following 1830, the number of migrants from Europe increased dramatically, with over 70 million migrating from Europe between 1830 and the Great Depression of the 1930s. The Europeans established numerical dominance in the United States, Canada, New Zealand, Australia, and some countries of South America (Uruguay, Argentina, Chile, and southern Brazil) in this hundred-year period. Even where Europeans were not numerically dominant, they established political and cultural dominance, bringing European culture to most of Africa and Asia.

Over one-half of all European migrants moved to the United States or Canada (Table 5—9). The movement of Europeans to Anglo America and other areas resulted in cultural changes associated with diffusion of languages, Christianity,

legal systems, and technology. Native inhabitants in North and South America, Australia, New Zealand, the Hawaiian Islands, South Africa, and elsewhere were either subjugated, destroyed, or relegated to positions as second-class citizens.

The large number of migrants from Europe reflected Europe's rapid population growth. The early industrialization of England was accompanied by rapid increases in population, providing a source of migrants to North America, Australia, and New Zealand. The later population growth in southern and eastern Europe was associated with migrants to South America and Anglo America. Early migrants to South America and the Caribbean came primarily from Spain and Portugal, while later migrants from Italy, Greece, Turkey, Poland, Czechoslovakia, and Russia came to the United States.

While Europeans were migrating westward to newly discovered lands and colonial possessions,

TABLE 5–9
European Migrants: Percentage of All European Overseas Emigrants Leaving from Selected Countries, 1846–1963

COUNTRIES	PERIOD			
	1846–1890	1891–1920	1921–39	1946–1963
British Isles	47.9	17.7	29.0	27.7
Germany	20.2	3.4	9.8	15.7
Sweden, Denmark, Norway	6.9	3.8	3.8	2.1
France, Switzerland, Netherlands	4.2	1.5	2.5	14.9
Italy	8.2	27.0	18.6	19.0
Austria, Hungary, Czechoslovakia	3.7	15.9	4.1	?
Russia, Poland, Lithuania, Estonia, Finland	2.1	13.0	12.0	?
Spain, Portugal	6.9	15.3	15.0	12.1
Total emigrants from Europe per year (thousands)	376	910	366	585

Source: Calculated from Leszek A. Kosinski, *The Population of Europe: A Geographical Perspective* (London: Longmans, 1970), p. 57.

another group was moving from European Russia eastward to occupy the great expanse of Siberia. During the last decades of the nineteenth century and the first decade of the twentieth, as many as 5 million Russians migrated eastward (voluntarily or as prisoners) to assist in opening and colonizing Siberia. Completion of the Trans-Siberian Railroad in 1892 hastened this movement, and developments since the Communist Revolution of 1917 have caused still more to move east. Between 15 and 20 million Russians migrated eastward in the period from 1930 to 1987, but as many as 5 million later returned to European Russia.

Migration from Europe since World War II has declined considerably. The great exodus of Europeans that began in the 1830s with approximately 100,000 migrants per year and peaked in the first decade of the twentieth century with nearly 1.5 million per year has now dwindled into relative insignificance (Figure 5–17). The decline in European migration reflects changes in political, so-

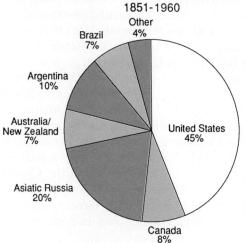

DESTINATIONS OF EUROPEAN MIGRANTS, 1851-1960

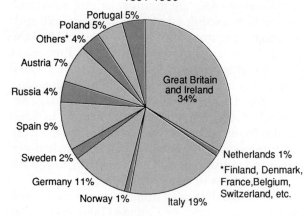

DIFFUSION OF EUROPEANS: ORIGINS OF MIGRANTS, 1851-1960

FIGURE 5–17 Migration from Europe (1851 to 1960) Examination of the origins and destinations of European migrants from 1851 to 1960 demonstrates the importance of the United Kingdom, Germany-Austria, and Italy as sources and the United States and Canada and eastern Russia as destinations.

cial, and economic conditions in Europe and in the rest of the world. Declining population growth rates in Europe and the loss of overseas colonies by European countries are two of the most important factors contributing to the decline in European migration. The low numbers of European migrants today does not change the fact, however, that for four centuries, European migration was a dominant force shaping the cultural geography of the world.

African Migration

A second major international migration since the Age of Discovery was the movement of black slaves from Africa. Unlike the great European migration, the black migration was a forced mi-

gration (Figure 5–18). The total numbers involved may never be known, but estimates suggest that a minimum of 10 to 15 million blacks were taken from Africa between 1500 and 1880 (when slavery legally ended). This was the period of the most intensive capture and removal of Africans for slave labor in the Americas by European slave traders. In addition, during this time, still more slaves were sold for use in the Americas and the Middle East by African and Arab slave traders (who had historically been involved in trading slaves to North Africa and the Middle East).

It is important to note that the total number of slaves going to or through Arab lands was small before A.D. 1500. It has been estimated that fewer than 5,000 per year were moved north to the

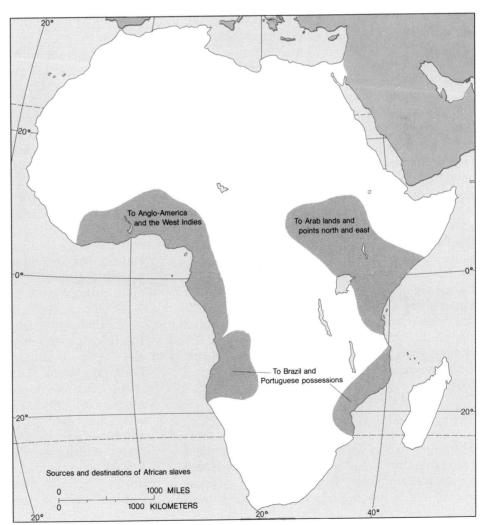

FIGURE 5–18 African Migration Participation in the movement of Africans to other locations was generally involuntary.

major cities of North Africa, Egypt, and Portugal prior to the discovery of the New World. The numbers increased substantially after 1500, however. Between 1.5 and 2.5 million slaves were removed from Africa and taken to the Arab lands of North Africa between 1500 and 1800, and at least 10 million—and possibly as many as 15 million—were moved to the Americas. Between 1500 and 1880 an additional 1.5 million to 2.5 million slaves were moved eastward to the Middle East and South Asia by European and Arab slave traders.

The great majority of African migrants were brought to Brazil, the Caribbean Islands, the Guianas, Venezuela, Colombia, and the United States where they were used as labor. The first great slave trade began with the movement of Africans to the northeast coastal regions of Brazil, then to the Caribbean where they worked in sugar plantations, and then to the United States, where they worked on plantations producing cotton, tobacco, rice, and indigo. Less than 10 percent of Africans came directly to the United States, since most went to the Caribbean or Latin America.

Descendants of African slaves are the dominant population in nearly all Caribbean countries (including Haiti, Barbados, St. Lucia, and Jamaica) and form an important cultural element in other countries such as the United States, Brazil, and the Guianas. The movement of Africans to the Americas effectively handicapped the population growth of Africa until recent times and reflects the sex-specific and age-specific nature of migration. Blacks selected by slavers were predominantly young males. The resultant demographic imbalance combined with an average life expectancy of only seven years for a slave working in the fields limited the growth of the black population in the Americas until the end of slavery. The population of blacks in the United States was just 1 million in 1800 compared to over 30 million in 1990.

Other Colonial Migrations

The largely voluntary migration of Europeans and the involuntary migration of Africans constituted the largest migration flows during the colonial period, but a number of other migrations also

Many Indians were brought to the Caribbean by the British, creating distinct ethnic groups in nations such as Trinidad.

helped shape the cultural geography of the world. Africa's cultural geography has been modified through the addition of migrants from India. Laborers were brought as indentured servants to South Africa and several countries of East Africa by the British who had outlawed slavery. Indian entrepreneurs emerged in these areas of Africa, ultimately leading to local trade and commerce in East Africa being dominated by Indians. The British were also the primary force behind relocating Asians, primarily Indians, to the Caribbean Islands. Countries such as Trinidad and Tobago now have large Indian minority populations, and Guyana has an Indian majority. The Dutch imported laborers from Java (Indonesia), adding another cultural group to the Caribbean (See Figure 3–21 on page 106).

Another major change in cultural geography was related to the migration of Chinese to Southeast Asia during the nineteenth century. European-owned plantations producing tropical crops of rubber, coconut, rice, tea, and coffee used millions of migrant laborers. Over 10 million individuals had participated in migrations to Southeast Asia by the end of the colonial period. Many of these were temporary laborers, such as Indians who worked in Malay or Burma. Many others remained permanently, such as the millions of Chinese who migrated to the region extending from southern Vietnam through Thailand, Singapore, and Java in Indonesia. Eventually, the Chinese emerged as the dominant group in the merchant class of Southeast Asia.

Chinese, at the present time, constitute 5 percent of the population of the Philippines, 14 percent in Thailand, 32 percent in Malaysia, and over 75 percent in Singapore (Table 5–10). Chinese also constituted a significant minority in Cambodia and Vietnam before 1980. The Chinese of Southeast Asia are overwhelmingly urban since native peoples or European colonists own the agricultural land. The Chinese are more influential in finance, trade, and commerce than their numbers would indicate, reflecting their historic role in business (See Figure 3–20 on page 105).

Post-Colonial Migrations

The European colonial period effectively ended with the close of World War II (1945), although it lingered in Africa until the 1960s, but a large number of migrations involving several major ethnic groups have occurred since then. These migrations can be categorized as either continuations of migration routes established during the colonial period, reversals of migrations that occurred during the colonial period, or involuntary migrations. The largest of these post-colonial

migrations has been that to the United States and Canada, the two major destinations for migrants.

Prior to World War II, most voluntary migrants to Anglo America were from Europe, but in the last four decades this has changed. The primary sources of migrants to the United States and Canada in the last decades of the twentieth century are Asia, Latin America, and the Caribbean Islands (See Figure 4–24 on page 158). Prior to the late nineteenth century, most immigrants to Canada and the United States came from northern and western Europe. Between 1890 and 1915, migrants from southern and eastern Europe dominated. Hostility toward migrants with cultural backgrounds other than those from northern and western Europe led the United States to adopt restrictive immigration laws, beginning with a prohibition against immigrants from Asia in 1882.

The National Origins Quota System had an even greater impact on immigration to the United States. Adopted in several stages after 1924, it established a quota system for migration into the United States. The quota system allowed immigration into the United States in proportion to the numbers of each ethnic group already in the United States. Initially based on the 1920 census, quotas were later tied to earlier censuses that further discriminated against migrants from areas other than western Europe. The quotas favored Great Britain, Germany, and Ireland, whose people had migrated earlier, at the expense of southern and eastern Europe, the countries with large numbers of migrants in the twentieth century (Table 5–11). The National Origins Quota System remained in effect until 1965 when the Kennedy-Johnson Act replaced the quotas with a preference system allowing up to 270,000 immigrants each year. The changes in United States immigration laws have allowed large numbers of legal migrants from Asia and Latin America to enter the United States in the last two decades (Table 5–12). At least as many immigrants from Latin America also entered illegally, prompting adoption of a new immigration act in 1986. It allowed illegal migrants already in the country to acquire legal status, but increased the penalties for future illegal migration. (See the Cultural Dimensions box on page 191).

TABLE 5–10
Chinese Population of Southeast Asian Countries (1980, 1981)

COUNTRY	CHINESE POPULATION
Brunei	46,280
Burma	723,320
Indonesia	4,000,000
Kampuchea	400,000
Laos	Several thousand
Malaysia	4,350,392
Philippines	500,000
Singapore	15,558,802
Thailand	4,427,814
Vietnam	2,000,000

Source: The World Factbook 1988 (Washington, D.C.: Central Intelligence Agency, 1988).

TABLE 5–11
Immigration Quotas under the National Origins System

RANK	COUNTRY OR AREA	QUOTA
1	Great Britain and Northern Ireland	65,721
2	Germany	25,927
3	Irish Free State	17,853
4	Poland	6,524
5	Italy	5,802
6	Sweden	3,314
7	Netherlands	3,153
8	France	3,086
9	Czechoslovakia	2,874
10	Russia	2,701
11	Norway	2,377
12	Switzerland	1,707
13	Austria	1,413
14	Belgium	1,304
15	Denmark	1,181
16	Hungary	869
17	Yugoslavia	845
18	Finland	569
19	Portugal	440
20	Lithuania	386
21	Greece	307
22	Rumania	295
23	Spain	252
24	Latvia	236
25	Turkey	226
26	Syria and Lebanon	123
27	Estonia	116
	All others	100

Source: U.S. Department of State, *Admissions of Aliens into the United States* (Washington, D.C.: U.S. Government Printing Office, 1932), pp. 102–4.

The changing origins of migrants to the United States have been paralleled by migration into Europe. Loss of population and diminished births during World War II combined with Europe's move to stage four of the demographic transition to create labor shortages in the old industrial countries of Europe. Consequently, residents of the less industrialized world have migrated into most of the western European countries. Migrants to the United Kingdom from India, Pakistan, and the Caribbean Islands reflect associations with former colonies. The United Kingdom has a special relationship with its former colonies through the British Commonwealth of Nations, which originally allowed residents of individual member countries to migrate between those countries. Consequently, Indians, Pakistanis, and Caribbean Islanders have become large minority group members in the United Kingdom (and Canada, another commonwealth member) (Table 5–13). Hostility to these migrants prompted the United Kingdom to restrict their entry in the 1970s (see the Cultural Dimensions box on page 193).

France has received large numbers of migrants from former colonies in Africa (Algeria and Morocco), from Vietnam, and from other less industrialized countries bordering the Mediterranean. Germany, Switzerland, and Austria have accepted "guest workers" from Mediterranean countries and Southwest Asia. Encouraged to come as "temporary" workers between 1960 and 1980, the children of these guest workers view themselves as citizens of the western European countries where they were born. They comprise small, but important minority groups within these countries.

TABLE 5–12
Immigration into the United States (in thousands)

REGION	1820–1960	1961–1970	1971–1980	1981–1985	1986
Europe	34,574.6	1,238.6	801.3	321.8	62.5
Asia	1,097.8	445.3	1,633.8	1,376.3	268.2
Caribbean	619.8	519.5	759.8	371.6	101.6
Central America	115.6	97.7	132.4	123.1	28.4
South America	234.7	228.3	284.4	184.0	41.9
Africa	47.5	39.3	91.8	77.0	17.5

Source: U.S. Statistical Abstract, 1988 (Washington, D.C.: U.S. Government Printing Office, 1988).

CULTURAL DIMENSIONS
Mexican Migrants to the United States

Many residents of Mexico's rural farming communities want to know how to cross the border safely, find work, and perhaps someday become legalized under the amnesty provision of the U.S. immigration law. Along with more than half of the men in Santiago Apostol, a dirt-poor village of 3,000 Indians in the southern Mexican state of Oaxaca, Mr. Sanchez hopes to slip into the United States to work for the next eight months. Each year at this time, as the planting season approaches in the vast agricultural stretches of the American West, hundreds of thousands of Mexican men prepare for the trek north.

Despite the new immigration law of 1986, which makes it a crime for U.S. employers to hire undocumented aliens, the tide of seasonal laborers continues unabated, according to immigration experts. In some southern villages it even seems to be swelling. Many small farmers of Mexico are overcoming their fear of crossing the border as they realize a stark economic reality: workers on U.S. farms can earn more than $30 a day, whereas rural laborers in Mexico make an average daily wage of 4,000 pesos, or about $1.80.

Although U.S. jobs are a powerful magnet for these impoverished farmers, only a few have remained permanently. Some men say they don't feel comfortable in a land so alien to their traditional culture. Others return for financial reasons, as the money they send back in a season would not go far in the United States, but in Mexico it buys upward social mobility. Returned migrants are easy to tell apart from the older men who wear sombreros, ride donkeys, and tell time by the angle of the sun. Many of these migrants spend part of every year in the United States.

The financial impact of migration is clearly evident in this traditional town, where children herd goats through the dirt streets, women carry bowls of cornmeal on their heads, and most families live in bamboo huts with thatched roofs. But on nearly every block, red-brick houses are springing up as symbols of those who migrate to the United States.

In a few weeks, more than 200 men from the village of Santiago Apostol will head north for the first time. The cost of the week-long trek to Oregon, which is mainly a steep fee for the professional smugglers at the border known as "coyotes," is about $450. Several farms from California and Oregon—impressed by the hard-working men from Santiago—are using villagers who worked for them last year to help round up teams of workers. These recruiters arrange the trip to a specific region (generally to an agricultural area in the American West), promising an eight-month work permit and a job. It seems harsh, but to many Mexican men it is the only way they can find to care for their families.

By selling a "package"—a prearranged trip with a "coyote" and promises of eight-month work permits—a seventeen-year-old who worked in Oregon last year says he will have no problems signing up more than 70 men. Sitting next to his grandfather in the adobe house built with migrant money in 1954, the seventeen-year-old may seem worlds apart in his lavender shirt and multicolored jeans. But the two men are not so different: both have relied on fields in a faraway foreign land to support their traditional way of life.

_____

Adapted from the Christian Science Monitor, February 11, 1988, pp. 26–27.

Another small group of migrants that some see as part of the colonial influence are those who have emigrated to Israel. Much of the territory now occupied by Israel and surrounding states was part of the Turkish empire until the twentieth century. Jewish emigration after 1919 occurred while the area of Palestine was a League of Nations mandate under the direction of the United Kingdom. By 1948 as many as 750,000 Jewish residents were in Palestine. An independent state of Israel was proposed by the United Nations and established by the Jewish settlers in that year in spite of opposition from Palestinians and their allies. (Figure 5–19).

TABLE 5–13
Minorities in the United Kingdom: Origin and
Volume of Immigration, 1970–1984

COUNTRY	IMMIGRANTS (in thousands)
Commonwealth countries*	1,356.0
Australia	314.9
Canada	118.4
New Zealand	108.8
African countries	270.1
India, Pakistan, Sri Lanka	50.8
West Indies	62.8
Other	337.8
Non-Commonwealth Countries	798.7
South Africa	165.3
Latin America	35.5
United States	107.7
Western Europe	215.5
Others	274.4

*The Commonwealth is an association of nearly 50 former
British colonies that recognize the British monarch as head of
the Commonwealth.

Source: Europa Yearbook (London: Europa Publications Limited, 1974,
1977, 1980, 1984, 1986).

Natural increase and subsequent migration of
Jews from Europe, North Africa, the Middle East,
the Soviet Union, and even the United States
have increased the Jewish population in Israel to
3.5 million. The creation of Israel in 1948 immediately created a class of nonvoluntary migrants
as many Palestinians fled to adjoining countries
to await the defeat of the new nation so that they
could return to their homes. Subsequent border
conflicts brought many of these Palestinians
within the borders of Israel, but hundreds of
thousands of Palestinians who had fled Israel
remain stateless.

Refugees

Migrants such as the Palestinians are refugees.
Such peoples have played a major role in postcolonial migrations (Table 5–14).

The total number of refugees historically can
only be estimated. Examination of only the major
groups whose status changed from refugee to
nonrefugee in the first half of the twentieth
century reveals approximately 50 million individ-

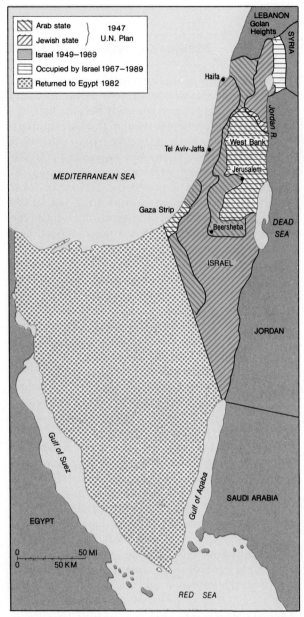

FIGURE 5–19 Changing Boundaries of Israel The
changing boundaries of Israel are the result of modern
migration to Palestine and subsequent conflicts with
Palestinians and other peoples.

uals. Such numbers can only suggest the magnitude of suffering experienced by the refugees as
they were forced to flee their homes (Table 5–15).

Currently, millions of displaced peoples in
Asia, Africa, and Latin America struggle to main-

CULTURAL DIMENSIONS

Minorities in England

Only a few minutes' walk from London's financial district is a school where you have to look long and hard for a white face. At the Mulberry School for Girls in London's East End, the local version of the British school uniform is a tunic and trousers to accommodate the Muslim precept that women's legs should be covered. A conversation with the matronly and very British headmistress is punctuated by wafts of Asian music, cumin, and curry. Welcome to multicultural Britain, a reality that is challenging the educational establishment.

Only fifty years ago Britain was a fairly homogeneous, Anglo-Saxon society. But the prosperity of the postwar years and the crumbling of the British Empire brought a flood of Indians, Pakistanis, Africans, and West Indians into the country. Since the 1970s, strict legislation has stopped further mass immigration, but the nonwhite population has grown to 2.5 million. Nationwide, nonwhites represent only 4 percent of the population, yet their concentration in urban areas makes the question of educating a multicultural student body urgent.

Some English think that schools need to emphasize the English language, English tolerance, English institutions, and Christianity, which they feel are part of the English legacy. In their opinion, a multicultural approach is counterproductive, because it makes children more aware of their differences. This attitude is rejected by the Commission for Racial Equality, a government body set up by the 1976 Race Relations Act. Racism, in the commission's view, does not just mean individual prejudices of teachers or children. It also means discrimination that is built into the system itself, or what is called institutional racism.

The existence of institutional racism was recognized by the 1985 Swann report, an official inquiry into the reasons for underachievement among West Indian pupils. Today, most regional bureaucracies governing state schools concur with the report's conclusions. Out of 110 authorities nationwide, about 70 have issued statements committing themselves to "antiracism" policies. Teacher training colleges, too, are making antiracism an integral part of their courses.

Still, in rural areas, where the population is mostly white, many schools see no need to adopt a multicultural approach. And even in schools that have embraced antiracism, what happens in the classroom varies considerably. Many whites resent what they see as special treatment being received by nonwhites. Recognizing this failure in communication, the commission is planning a public relations counterattack that will replace antiracism with a term like "fair education."

At some schools the headmasters are oblivious to semantics. For them it's all very simple. Antiracism is a full education. Part of a teacher's duty is to educate students to challenge received opinion. At one school in London, whites, at 15 percent, are a minority among children of Bangladeshi origin. An antiracist approach here means, among other things, making sure white students are not isolated in the classroom. Multiculturalism and antiracism permeate the curriculum at such schools, and teachers must show a commitment to this approach before joining the staff.

Britain is having a hard time adjusting to its multicultural identity. What kind of national culture will eventually emerge is unclear, although you can catch glimpses through the Notting Hill West Indian carnival, the biggest street festival in Europe. Britain's ethnic minorities may not be English, but they are British, and they are here to stay.

TABLE 5–14
World Refugees

REGIONS	NUMBER OF REFUGEES
Africa	2,251,000
Asia	613,200
Latin America	268,000 to 400,000
North America (including Mexico)	1,187,000
Middle East	4,637,000
Total	10,032,000

Source: *The New Book of World Rankings 1987* (New York: Facts on File, Inc., 1987).

tain their lives and livelihood while awaiting the final deposition of their lives. Many of the migrants of the postcolonial period began as refugees, especially those from Southeast Asia who fled to the United States, or those from Caribbean nations such as Cuba who fled Communism and were accepted as refugees in the United States. But while some have obtained permanent homes in the United States, Canada, and to a lesser degree Europe or Australia, many refugees remain in refugee camps. These refugee camps form a cultural landscape of their own.

Refugee camps cannot provide more than a minimal level of nutrition, medical care, and accommodation. They are constantly overburdened and incapable of meeting the needs of their residents. Moreover, the countries least able to care for migrants are often those that have received large refugee populations. Pakistan, Thailand, Sudan, Somalia, and other countries of the less industrialized world are faced with millions of individuals who need humane treatment these countries are often ill-equipped to provide. The refugee problem is a global problem. Contributions from the United Nations, individual donor states in Europe, North America, and elsewhere help to supply the necessities of life for these refugees, but do not provide them a home.

Part of the problem faced by migrants classified as refugees is one of definition. The United Nations has a special agency, the High Commission for Refugees, that recognizes a category of people called "displaced persons." Displaced persons are those who have been forced to leave their homes

TABLE 5–15
Historic Refugee Movements

1. 1.2 million Greeks from Greece-Turkey to Greece (1922–1923).
2. 1.2 million Russians to European countries other than the Soviet Union (1918–1922).
3. 1.1 million Poles from Russia to Poland (1918–1925).
4. 1 million Poles from former Russian and Austrian Poland to former German Poland (1918–1921).
5. 700,000 Germans from western Poland to Germany (1918–1925).
6. 6 million Germans from Poland to Germany (1944–1947).
7. 6 million Jews from Germany and German-occupied territory to extermination camps in Poland and elsewhere (1940–1944).
8. 4 million Germans from Soviet-occupied Germany to West Germany (1945–1946).
9. 3 million Poles from Russia to Poland (1945–1947).
10. 2.7 million ethnic Germans from Czechoslovakia to Germany and Austria (1945–1946).
11. 1.8 million Czechs and Slovaks from Czechoslovakia to former German Sudetenland (1945–1947).
12. 1 million ethnic Germans from Polish territory created from Germany (1944–1945).
13. 17 million Hindu and Islamic people moved between India and Pakistan when the two countries become independent from the United Kingdom (1947).
14. 500,000 Koreans from North Korea to South Korea (1950–1953).
15. 700,000 Vietnamese from Vietnam to the United States, France, Canada, and Australia (1976–1987).

and cross an international boundary; as such they are clearly refugees. The High Commission for Refugees assists in their care and keeps some statistical records on their numbers.

Other refugees may be *internal refugees*, who are forced to move from one portion of their homeland to another. For example in the early 1970s and again in the 1980s, hundreds of thousands of residents of the Sudan, Chad, and Ethiopia were forced into temporary dwellings around existing communities as they fled drought and famine in the region known as the Sahel along the southern borders of the Sahara. Since they did not cross an international border, the people were not defined as refugees, making it more difficult for them to get assistance.

Like other migrants, refugees can be classified as temporary or permanent. Temporary refugees would return to their home given the opportunity, while permanent refugees have no intention of returning. Classifying refugees in this manner is sometimes difficult, however, since some refugees may live in an area for decades, but would be willing to return if the forces that caused them to become migrants were changed. At the present time examination of the causes of refugee movement helps to differentiate between temporary and permanent refugees. Among the causes for refugee movement are conflict, environmental problems, geopolitics, and ideology.

Conflict can be the result of either warfare between competing nations or civil war within a nation. Examples of international conflicts and the refugees associated with them include 750,000 Ethiopians residing in Somalia; 600,000 Ethiopians (including 300,000 from the disputed area of Eritrea in Northeast Ethiopia) in Sudan; 3.5 million Afghans in Pakistan; another 1.5 million Afghans who have fled to Iran, France, or other countries; and nearly 300,000 Kampucheans in Thailand. An estimated 100,000 refugees have fled Nicaragua to El Salvador, and tens of thousands more have fled El Salvador to Guatemala, Honduras, and Mexico or Guatemala to Mexico. Other peoples who have been forced to migrate because of conflict in the postcolonial period lost their lives or established permanent homes and are no longer regarded as refugees.

In addition to the millions of refugees resulting from actual conflicts, millions more have become refugees because of *ideological differences*. Ideological differences have been particularly important in the flow of migrants from the Soviet Union to Israel, from Cuba to the United States (1962 to 1987), from East Germany and other Communist Bloc countries to Western Europe or the United States, and from the Republic of China to Hong Kong (1949–present).

In the Middle East ideological differences and conflicts have caused 3 to 4 million Palestinians to become refugees, including those in Jordan (1.3 million), Lebanon, (.33 million), Syria (.25 million), other Arab states (.75 million), and Israel (800,000). The total number of Palestinians is one of the largest groups of refugees in the world.

Environmental hazards also create refugees, particularly in the less industrialized world. Environmental disasters destroy more property in dollar value in the industrial world, but have a much greater impact on the lives of individuals in the less industrialized world, where the reliance of the majority of the population on agriculture for a livelihood means that drought or floods may displace hundreds of thousands of people. The largest regional concentration of environmental refugees is in Africa. The great droughts of the 1970s in the Sahel caused millions to flee to other

Refugees may create problems in the countries where they settle, as in the case of these Palestinians engaged in military training in Lebanon.

parts of their own country or to adjacent countries. Even more severe droughts in the 1980s in the Sahel and south of the equator in countries such as Botswana and Mozambique created even more refugees. It was estimated that there were 5 million refugees in Africa (excluding Palestinians) in 1989. The largest number of these are refugees who have been displaced from their homes, but have not fled their countries.

The largest single cluster of African refugees is in Somalia, where nearly a million refugees from Ethiopia live in refugee camps. During the period of drought in the 1970s when refugees were fleeing the most severely affected regions of Ethiopia, a Marxist government came to power. The new government displaced more Ethiopians in a large (but misguided) resettlement scheme. Then, in the early 1980s, Somalia invaded Ethiopia to aid ethnic Somalians living there. The Somalian forces were turned back by Ethiopians with Cuban and Soviet assistance. Many of the refugees in Somalia represent nomads and seminomads of Somali ethnicity who had been living in Ethiopia but returned to Somalia with the retreating Somalian forces.

The northern part of Ethiopia (once ruled by the Italians as a colony called Eritrea) has tried for years to become independent. Continued attempts after the Marxist revolution in Ethiopia created an additional estimated 1 million refugees who fled to the Sudan (Figure 5–20). Famine and drought from 1983 through 1985 and again in 1987 exacerbated the impact of the conflict to create hundreds of thousands more refugees to the Sudan.

It is impossible to completely separate environmental factors from ideological and conflict-related factors in explaining the refugees of Africa. What began as an environmentally induced refugee movement in Ethiopia was intensified by resulting or existing conflicts and ideological differences within the region to create the largest group of refugees in the world. Continued environmental degradation associated with increasing populations in Africa will result in ever greater numbers of environmentally induced refugee movements in the continent.

The governments of Africa south of the Sahara are generally unstable, reflecting their emergence

Refugees in Ethiopia. Refugees may be fleeing social, political, or environmental hazards.

from colonial domination only in the last few decades. Political conflict in Africa will probably continue to accompany the environmentally induced refugee movements. In countries where civil war does not break out and international assistance is available, many refugees will ultimately be able to return to their homes. Unfortunately, the economy of the region bordering the most environmentally sensitive portion of Africa, the Sahel, is dependent upon nomadic or seminomadic pastoralism. The environmental effect of drought leads to the destruction of the food for the animals upon which these societies rely, making it difficult for refugees to return to their homelands. Thus additional hundreds of thousands of refugees are residents in their homelands, but have been forced from their homes to squalid refugee camps.

The Geographical Impact of Refugees

The geographical impacts of refugees can be divided into the following categories:

1. Impact upon the individual refugees and their families
2. Impact upon the country from which they fled
3. Impact upon the recipient country

The individual or family impact of refugee status is difficult to comprehend. Individuals, families, and entire communities are torn from the geographical setting with which they are familiar and

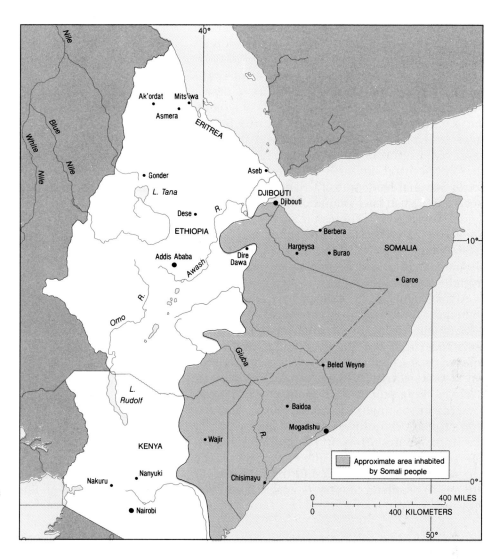

FIGURE 5–20 Ethiopia, Eritrea, and Somalia The boundaries of the countries of Ethiopia, Somalia, and Kenya divide the Somali people among three nations, creating conflict and refugees.

transported to a new location. Whether this location is within their country or in a new country, the problems associated with such a movement and consequent readjustment involve their cultural ecology and associated way of life.

Unlike voluntary migrants who are able to engage in at least some planning for their move, refugees seldom have this privilege. Refugees rarely carry with them more than the clothes on their back and a few personal items. In addition to leaving behind their physical possessions, the refugees are in many cases also afflicted by either poor health, injury, or both. Those fleeing from famine associated with drought often lose family

members and suffer physical debilitation that affects them for life. Those fleeing conflict may also suffer from chronic shortage of food, disease, and injury. The human suffering associated with the refugee movement is immeasurable and can be perceived but dimly by those who have not experienced it.

The impact upon the country from which refugees migrate is equally profound. Communities may be depopulated, farmlands abandoned, and economies thrown into disarray by the loss of large segments of the population. The push factors that cause refugees create additional problems for the country from which they flee.

Drought, war, or ideological conflict may damage the country as severely as the loss of the individuals themselves.

The impact of refugees on the recipient country can be equally great, although refugee movements usually do not affect all regions of a recipient country. In the case of the United States, southern Florida has been culturally changed because of the influx of Cubans and Haitians. Part of the older central city area of Miami (known as Little Havana) has a cultural landscape that reflects the Cuban background of many of its residents. Palestinian refugees in the Middle East (including those in Israeli-occupied territories) create a large, culturally distinct population concentrated in camps with low standards of income, hygiene, and quality of life.

Afghan refugees in Pakistan have imposed an immense burden upon a government that has difficulty providing for its own people. Even with assistance from the United Nations and other relief agencies, countries of the less industrialized world that receive millions of refugees face an overwhelming problem. The influx of Ethiopian refugees into Somalia, a country with a population of only 7.7 million (1987), is equivalent to the United States receiving 100 million refugees. The impact of such population movement on the geography of the local area and the resources of the entire recipient country is apparent to even casual observers.

Internal Migration

The great movement of European peoples to populate North America, portions of Latin America, Australia, New Zealand, and South Africa often obscures the importance of shorter movements within Europe itself. The movement of Europeans within Europe prior to the twentieth century was much greater than that to foreign countries. Internal migration has always involved the greatest number of migrants. Prior to the development of the railroad and large, fast, steam-powered ships, migration out of a country was limited by the size of ships and the cost, dangers, and time involved in travel. Historically, the population of individual countries expanded to gain control of territory. Today, people move as population pressure pushes them to the city from the rural areas. The importance of internal migration is twofold: (1) it determines the general distribution of population within countries and (2) it concentrates population clusters within individual countries in response to push and pull factors.

Migration within Europe

From the thirteenth to the eighteenth century, Europe's slowly growing rural population moved from one rural location to another as the vacant or sparsely settled land suitable for cultivation was populated. Initially, these movements involved very short distances, as was typical of migration before modern transportation. In 1841, 80 percent of Sweden's population lived in rural areas, and 93 percent resided in the county in which they had been born. Eighty-five percent of the inhabitants born in a Danish town in the second half of the 1700s were buried there. Most women married men from their own or a nearby village.

The primary movement during the preindustrial period in Europe was from village to village or from village to urban area or small town. This reflected the quality of life in towns and cities, which was much lower than that found in villages. It is estimated that in A.D. 1300 only 4 percent of Europe's population lived in towns of more than 2,000. London had only 60,000 inhabitants by 1520. Paris, the largest city in preindustrial Europe, had only 250,000 in 1600, but had grown to 500,000 by 1700.

Rapid population growth in Europe after A.D. 1600 forced people to move to cities and to migrate to new lands such as North America. London replaced Paris as the largest city in western Europe, a reflection of the Industrial Revolution and the associated creation of jobs in cities, which caused rapid urban population growth. The growth of large urban centers occurred in spite of repeated outbreaks of disease that decimated their populations. In 1580 30,000 people in Paris died from typhus. In London, 33,000 in 1603, 41,000 in 1625, and 69,000 in 1665 and 1666 lost their lives to the recurrent plague. Continued population growth in rural areas combined with the Industrial Revolution to attract more people to the cities despite their unhealthy conditions.

Migration within China

Similar internal voluntary and involuntary migration was taking place in other countries during the preindustrial period. China, the most heavily populated country of the world in terms of numbers, exemplifies the cultural ecology and cultural landscape associated with internal migration. Chinese migration began several thousands of years ago in an attempt to occupy land devoid of Chinese and to expand the frontiers of Chinese settlement. Only the settlement of the American West and the expansion of czarist Russia into Central Asia and Siberia are comparable in numbers of migrants. Over thousands of years, the internal migration of Chinese peopled an area as large as the continent of Europe.

Most migrations in China were government sponsored. Records indicate that they began in the fifth millennium B.C. or earlier, but detailed records are available only from 1000 B.C. The numbers involved in the Chinese migrations are impressive. Between A.D. 300 and 500 alone, more than 5 million Chinese were moved by government action to populate border areas, for political reasons, or because of famines and population pressure in the central core of Eastern China. In an earlier government-sponsored migration, Emperor Wu moved 725,000 people to a new location in the year 120 B.C. The government provided housing, transportation, food, and clothing to assist people moving to the northern border as colonists. The government levied a special tax on all nobility and issued China's first recorded silver currency to help pay for this movement.

China vividly illustrates the impact of internal migration on the cultural geography of the world. The Chinese are often viewed as an example of a generally homogenous population with a single language whose characters are understood and used by nearly all of the 1.1 billion inhabitants of the country. The land now known as China was much more ethnically diverse 3,000 years ago. Today's dominant ethnic group in China, the Han, did not exist. Their predecessors lived within the North China Plain, which comprises only 10 percent of modern China. The bulk of China was occupied by independent empires with rigid ethnic boundaries reflecting political divisions.

The beginning of wide-scale migration after 1027 B.C. dramatically changed the ethnic map of China. Migrations affected all of the country as people were moved to the north, to the south, and to the west. The intermarriages accompanying migration led to greater cultural homogeneity. The present Chinese writing system emerged as a common binding cultural force in association with the Chou migration nearly 2,200 years ago.

Migration within the United States

The internal migration in China and western Europe that created the patterns of peoples is similar to the experience of other areas. Migrations within the United States have shaped and continue to reshape its cultural geography. The earliest orientation of internal migration in the United States was from east to west. The passing of the American frontier by the end of the nineteenth century led to different forms of internal migration that partially continued the earlier east to west movement, but with additional components. Internal migrations within the United States today can be broadly divided into two flows:

1. From the central city to the suburbs
2. From the East to the West and South

The general movement from city to suburb has been discussed previously and is the migration that involves the greatest numbers of people in the United States (Table 5–16). Examination of

TABLE 5–16
Percent Change in Population by Decades in Central Cities and Suburbs.

	PERCENT	
TIME PERIODS	Central Cities	Suburbs
1900–1910	35.3	27.6
1910–1920	26.7	22.4
1920–1930	23.3	34.2
1930–1940	5.1	13.8
1940–1950	13.9	34.7
1950–1960	10.7	48.6
1960–1970	6.4	26.8
1970–1980	0.6	17.4

Source: Census' of Population, 1960–1980 (Washington: U.S. Government Printing Office, 1963, 1972, 1983).

internal migrants in the United States reveals that the persons most apt to migrate within the country are those between the ages of 18 and 30. This is the period when children leave home to attend school, begin their careers, marry, and begin to form families.

Nearly 15 percent of the U.S. population moves to a new address yearly. These are generally short migrations, with nearly two-thirds of the yearly migrants relocating within the same county. Longer distance internal migration creates important changes in the pattern of population distribution in the United States.

Examination of population change in the United States from 1970 to 1980 reveals that the population of the South and West grew rapidly while that of the old urban Northeast grew slowly or declined (Figure 5–21). California, Florida, Texas, and Arizona led in terms of total number of in-migrants. The movement of people to these states reflects a perception of the environmental, economic, political, or social conditions there as being better than those in the states of origin. The rapid growth of the southern and western states has changed political representation within the United States as California is now the most

populous state and has the most representatives (Table 5–17).

The general pattern of population movement to the South and West in the United States masks an important change related to the ethnic and racial composition of this migration. The period from the 1940s to 1970 was characterized by migration of the black population from the rural South to the industrial cities of the urban Northeast. In 1900 only 10 percent of the black population of the United States lived outside the South, by 1980 almost 50 percent did. Black residents outside the South are predominantly urban residents. Today, 80 percent of rural black Americans live in the South, while in the North over 90 percent of the black population live in metropolitan areas.

The black migration to central cities in the twentieth century parallels the migration of the rural population of Europe to its industrial cities a century earlier. After the Civil War, the rural black population of the United States was faced with the problems of landlessness, high population growth, and little economic opportunity in a region with limited civil rights. The economic and social pressures of the rural South combined

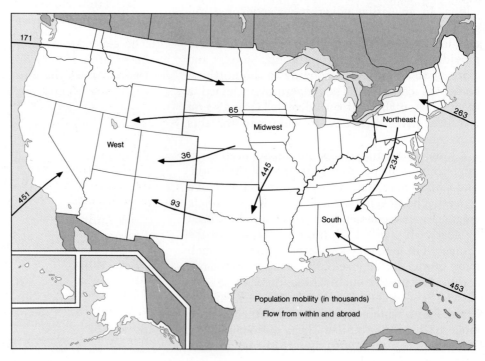

FIGURE 5–21 Migration Flows in the United States Migration flows between the four major regions of the United States in the last decade demonstrate the net gain of the South and West at the expense of the Midwest and Northeast. (Arrows refer to regional movement.)

Population mobility (in thousands)
Flow from within and abroad

TABLE 5–17
Changes in Representation by Region and State,
1981–1987

	1981	1987
New England	25	24
Massachusetts	12	11
East Central	79	71
New York	39	34
New Jersey	15	14
Pennsylvania	25	23
East North Central	86	80
Ohio	13	11
Indiana	11	10
Illinois	24	22
Michigan	19	18
West North Central	35	33
Missouri	10	9
South Dakota	2	1
South Atlantic	65	69
Florida	15	19
East South Central	27	28
Tennessee	8	9
West South Central	42	45
Texas	24	27
Mountain	19	24
Colorado	5	6
New Mexico	2	3
Arizona	4	5
Utah	2	3
Nevada	1	2
Pacific	57	61
Washington	7	8
Oregon	4	5
California	43	45

Source: U.S. Statistical Abstract, 1988 (Washington, D.C.: U.S. Government Printing Office, 1988).

with the growing industrialization in the urban Northeast to attract black migrants. Restrictions on foreign immigrants after the mid-1920s increased this flow, and rapid expansion of industry in the United States after 1940 attracted even more to the jobs of the steel and auto industries of the northeastern and north central United States.

Since 1970 the century-long migration of blacks from the South to the North has reversed. At the present time, more blacks are leaving the North and returning to the South than are going North in response to increased economic opportunities in the South.

Unlike the earlier migration to the North, however, this is not a rural to urban migration. Blacks from northern cities are moving to the urban areas of the South, where they are joining rural black migrants from the South itself. Thus the economic declines of the traditional iron, steel, and automobile industries in the late 1970s and early 1980s in the North have prompted thousands of blacks (along with their white counterparts) to move to the South where both environmental and economic conditions are perceived as more favorable. The combination of perceived or real economic and environmental attraction and the real or perceived improvement in civil rights conditions in the South has made the black migration back to the South one of the important internal migrations in the world today.

Migration within the Less Industrialized World

To some degree, migration within the less industrialized world parallels that which occurred in the industrial world during the eighteenth and nineteenth centuries. Most of that immigration is internal and reflects movement from rural to urban areas. Although there has been some movement out of the less industrialized world to other countries, particularly to the United States, it is a small portion of the total migration. Major migrations in the less industrialized countries include those in Latin America where population has moved from densely populated highland areas to the lowlands of the Pacific Coast. Smaller but important movements from the highlands of western Latin America to the tropical lowlands east of the Andes and from the densely populated southeastern and northeastern regions of Brazil to the Amazon Basin have also occurred. Migration in India includes that to Assam in the Northeast and to drier areas where extension of irrigation allows greater rural population. Migration in Indonesia has been from the island of Java to the island of Sumatra. A modern Chinese migration has been westward to drier, colder environments where reclamation projects have resulted in the resettlement of millions. Many countries of Africa have developed agricultural schemes (many involving irrigation) that have attracted migrants. The rural to rural and urban to rural movements in the less industrialized world reflect efforts to colonize regions that have not been intensely cultivated. These government programs account for only a portion of the migration of the less industrialized world, however, as the majority of migrants perceive urban locations as more desirable.

The major migration flow in the less industrialized world is rural to urban. The pace of urban growth in the less industrialized world has increased dramatically in the last half century. In 1920 only 87 million people lived in urban places (defined as those with 20,000 or more inhabitants) in Asia, Africa, and Latin America, compared to 170 million in the Soviet Union, Europe, and North America. By 1960, however, the urban population of the developing countries was greater than that in the industrialized countries (Table 5–18). Nearly one-half of the population in the less industrialized world will reside in urban places by 1990.

The migration to urban areas in the less industrialized world reflects the push factor of rapidly growing populations in rural areas. Not enough industrial jobs are available for the large numbers of migrants to the cities, and most are forced by their lack of training to take low-level, unskilled jobs and live in neighborhoods characterized by poor housing and inadequate services; nevertheless, population pressures in rural areas encourage continued migration. Thus the rural to urban migration is important in slowing the rate of rural population growth, but has created a deluge of peoples to the cities.

The migration from rural to urban areas in less industrialized regions is similar in some respects to that of Europe a century before. The majority of the migrants move only relatively short distances, and most are young adults between fifteen and thirty years of age. The majority of migrants tend to be better educated than the average of the community they leave, unlike Europeans a century before, and males predominate to a much greater extent than in the European experience.

Conclusion

The rapid growth of cities in the less industrialized world is only the most recent change population movement has caused in the world's cultural geography. The complex pattern of culture regions (whether based on language, ethnic group, race, or other culture variables) is another visible manifestation of population movements. The diffusion of ideas with migrants is reflected in global patterns of livelihood, architectural and dress styles, community form and function, and numerous other aspects of the cultural landscape. Understanding past migrations and present migration trends helps us categorize and explain the cultural landscapes that characterize the earth.

Santiago, Chile. Migration to the cities in the less industrialized world has created teeming slum settlements.

TABLE 5–18
Rural-Urban Population Changes by Region (in millions)

	1920	1930	1940	1950	1960	1970	1980
World							
Total	1,860	2,068.6	2,295.1	2,508.4	2,986	3,610.4	4,374.1
Urban	360	450	570	712.1	1,012.1	1,354.2	1,806.8
Rural	1,500	1,618.6	1,725.1	1,796.3	1,973.9	2,256.2	2,567.3
Africa							
Total	142.9	163.8	191.5	220.2	272.8	351.7	460.9
Urban	10	15	20	31.8	49.5	80.4	133
Rural	132.9	148.8	171.5	188.4	223.3	271.3	327.9
Latin America							
Total	89.5	107.5	129.9	163.2	215.6	283	371.6
Urban	20	30	40	66.3	106.6	162.4	240.6
Rural	69.5	77.5	89.9	96.9	109	120.6	131
North America							
Total	115.7	134.2	144.3	166.1	198.7	226.4	248.8
Urban	60	75	85	106	133.3	159.5	183.3
Rural	55.7	59.2	59.3	60.1	65.4	66.9	65.5
Asia							
Total	1,023.2	1,120.2	1,245.1	1,374	1,634.7	2,028.1	2,514.5
Urban	90	115	160	216.3	341.6	482.4	689.3
Rural	933.3	1,005.2	1,085.1	1,157.7	1,302.1	1,545.7	1,825.2
Europe							
Total	324.9	353.9	378.9	391.9	425.2	459.1	486.5
Urban	150	175	200	216.3	266	318.4	369.3
Rural	174.9	178.9	178.9	175.6	159.2	140.7	117.2
Oceania							
Total	8.5	10	11	12.7	15.7	19.3	23.5
Urban	4	5	6	7.7	10.4	13.7	17.8
Rural	4.5	5	5	5	5.3	5.6	5.7
Soviet Union							
Total	155.3	179	195	180.1	214.3	242.8	268.1
Urban	25	35	60	70.8	104.6	137.6	173.7
Rural	130.3	144	135	109.3	109.7	105.2	94.4

Source: Philip M. Hauser, et al, *Population and the Urban Future* (Albany, New York: State University of New York, 1982), pp. 3–5.

QUESTIONS

1. Describe and give examples of two forms of migration.
2. Indicate four factors that cause migration. Which is most important?
3. Which two of the barriers to migration do you feel is most important? Justify your answer.
4. Compare and contrast migration in industrializing countries and industrialized nations.
5. Did migration from Europe or migration from Africa have the greater impact upon the world? Justify your answer.
6. How has involuntary migration changed the character of the receiving country? Give some examples.
7. Discuss three types of migration that have created serious refugee problems in the world.
8. Describe the geographical impact of refugees.
9. What are the general patterns of migration in the United States? Why do they occur?
10. What problems are presented by the migration patterns in the less industrialized world?

SUGGESTED READINGS

Allen, James P. "Changes in the American Propensity to Migrate." *Annals of the Association of American Geographers* 67 (December 1977): 577–87.

Arreola, Daniel D. "The Chinese Role in Creating the Early Cultural Landscape of the Sacramento–San Joaquin Delta," *California Geographer* 15 (1975): 1–15.

Baines, Dudley. *Migration in a Mature Economy: Emigration and Internal Migration in England and Wales, 1861–1900.* New York: Cambridge University Press, 1986.

Baltensperger, Bradly H. "Agricultural Change among Great Plains Russian Germans." *Annals of the Association of American Geographers* 73 (1983): 75–88.

Bigger, Jeanne C. "The Sunning of America: Migration to the Sunbelt." *Population Bulletin* 34 (1979).

Bouvier, Leon F. *Immigration and Its Impact on U.S. Population Size.* Washington, D.C.: Population Reference Bureau, 1981.

———. "Immigration at the Crossroads." *American Demographics* 5 (October 1981): 17–24.

———. "International Migration: Yesterday, Today, and Tomorrow." *Population Bulletin* 32 (1977).

Bouvier, Leon F., Henry S. Shryock, and Harry W. Henderson. "International Migration: Yesterday, Today, and Tomorrow." *Population Bulletin* 32 (1977).

Bouvier, Leon F., and Robert W. Gardner. "Immigration to the U.S.: The Unfinished Story." *Population Bulletin* 41 (November 1986).

Brown, Lawrence A., and Victoria A. Lawson. "Migration in Third World Settings, Uneven Development, and Conventional Modeling: A Case Study of Costa Rica." *Annals of the Association of American Geographers* 75 (March 1985): 29–47.

Brunn, Stanley D. (with James H. Johnson, Jr.). "Residential Preference Patterns of Afro-American College Students in Three Different States." 32 (February 1980): 37–42.

Clark, Gordon L. *Interregional Migration, National Policy and Social Justice.* Totowa, N.J.: Rowman and Allanheld, 1983.

Clark, Gordon L., and Meric Gertler. "Migration and Capital." *Annals of the Association of American Geographers* 73 (March 1983): 18–34.

Clark, W. A. V. *Human Migration.* Beverly Hills, Calif.: Sage Publications, 1986.

Clarke, J. J., and L. A. Kosinski, eds. *Redistribution of Population in Africa.* London: Heinemann, 1982.

Clayton, Christopher. "Interstate Population Migration Process and Structure in the United States, 1935–1970." *The Professional Geographer* 29 (May 1977): 177–81.

Davis, Cary, Carl Haub, and JoAnne Willette. "U.S. Hispanics: Changing the Face of America." *Population Bulletin* 38 (June 1983).

Davis, Kingsley. "The Migrations of Human Populations." *Scientific American* 231 (September 1974): 95–105.

Demko, George J., and William B. Wood. "International Refugees: A Geographical Perspective." *Journal of Geography* 86 5 (September–October 1987): 225–28.

Dorigo, Guido, and Waldo Tobler. "Push-Pull Migration Laws." *Annals of the Association of American Geographers* 73 (March 1983): 1–17.

Fredrich, Barbara E. "Family Migration History: A Project in Introductory Cultural Geography." *Journal of Geography* (November 1977).

Golledge, Reginald G. "A Behavioral View of Mobility and Migration Research." *The Professional Geographer* 32 (February 1980): 14–21.

Greenwood, Michael. *Migration and Economic Growth in the United States.* New York: Academic Press, 1981.

Harris, Richard S. (with Eric C. Moore). "An Historical Approach to Mobility Research." *The Professional Geographer* 32 (February 1980): 22–29.

Hudson, John C. "Migration to an American Frontier." *Annals of the Association of American Geographers* 66 (1976): 242–65.

Jones, Huw, Nicholas Ford, James Caird, and William Berry. "Counter-urbanization in Societal Context: Long-Distance Migration to the Highlands and Islands of Scotland." *The Professional Geographer* 36 (November 1984): 437–43.

Jones, Richard C. "Undocumented Migration from Mexico: Some Geographical Questions." *Annals of the Association of American Geographers* 72 (March 1982): 77–78.

Khan, Abdullah Al-Mamum. "Rural-Urban Migration and Urbanization in Bangladesh." *Geographical Review* 72 (1982): 379–94.

Kosinski, Leszek A., and R. Mansell Prothero. *People on the Move.* London: Methuen, 1975.

Kritz, M. M., C. B. Keely, and S. M. Tomasi, eds. *Global Trends in Migration: Theory and Research on International Population Movement.* New York: Center for Migration Studies, 1981.

Lai, Chuen-yan David. "An Analysis of Data on Home Journeys by Chinese Immigrants in Canada, 1892–1915." *The Professional Geographer* 29 (November 1977): 359–65.

Lee, Everett. "A Theory of Migration." *Demography* 3 (1966): 47–57.

Lewis, G. J. *Human Migration: A Geographical Perspective.* London: Croom Helm, 1982.

Lowell, Lindsey B. *Scandinavian Exodus: Demographic and Social Development in Nineteenth-Century Rural Communities.* Boulder, Col.: Westview, 1987.

Mascie-Taylor, C. G. N., and G. W. Lasker, eds. *Biological Aspects of Human Migration.* Cambridge: Cambridge University Press, 1988.

McHugh, Kevin E. "Black Migration Reversal in the United States." *Geographical Review* 77 (April 1987): 171–82.

_____. "Explaining Migration Intentions and Destination Selection." *The Professional Geographer* 34 (August 1982): 315–25.

McNeill, W. H., and R. S. Adams, eds. *Human Migrations, Patterns and Policies.* Bloomington, Ind.: Indiana University Press, 1978.

Meyer, Douglas K. "Southern Illinois Migration Fields: The Shawnee Hills in 1850." *The Professional Geographer* 28 (May 1976): 151–60.

Meyer, Judith W., and Alden Speare, Jr. "Distinctively Elderly Mobility: Types and Determinants." *Economic Geography* 61 (1985): 79.

Miller, Kerby A. *Emigrants and Exiles: Ireland and the Irish Exodus to North America.* London: Oxford University Press, 1985.

Mings, Robert C. (with Patricia Gober). "A Geography of Nonpermanent Residence in the U.S." *The Professional Geographer* 28 (May 1982): 164–73.

Mitchell, Robert D. "American Origins and Regional Institutions: The Seventeenth-Century Chesapeake." *Annals of the Association of American Geographers* 73 (September 1983): 404–20.

Morrison, Peter A. *Population Movements: Their Form and Functions in Urbanization and Development.* Liege, Belgium: Ordina Editions for International Union for the Scientific Study of Population, 1983.

Murphy, Elaine M., and Patricia Chancellier. "Immigration: Questions and Answers." Washington, D.C.: Population Reference Bureau, Inc., 1982.

Newland, Kathleen. "International Migration: The Search for Work." Worldwatch Paper 33. Washington, D.C.: Worldwatch Institute, November 1979.

_____. "Refugees: The New International Politics of Displacement." Worldwatch Paper 43. Washington, D.C.: Worldwatch Institute, March 1981.

Ogden, P. E. *Migration and Geographical Change.* New York: Cambridge University Press, 1984.

Ostergren, Robert, "A Community Transplanted: The Formative Experience of a Swedish Immigrant Community in the Upper Middle West." *Journal of Historical Geography* 5 (1979): 189–212.

_____. "Land and Family in Rural Immigrant Communities." *Annals of the Association of American Geographers* 71 (March 1982): 77–87.

Papademetriou, Demetrios G., and Nicholas DiMarzio. *Undocumented Aliens in the New York Metropolitan Area.* New York: Center of Migration Studies of New York, Inc., 1986.

Ravenstein, E. G. "The Laws of Migration." *Journal of the Royal Statistical Society,* 52 (1889): 214–305.

Rogerson, Peter A. "New Directions in the Modelling of Interregional Migration." *Economic Geography* 60 (1984): 111.

Rogge, John R. "A Geography of Refugees: Some Illustrations from Africa." *The Professional Geographer* 29 (May 1977): 186–93.

Roseman, Curtis C. *Changing Migration Patterns within the United States.* Washington, D.C.: Association of American Geographers, 1977.

_____. "Migration as a Spatial and Temporal Process." *Annals of the Association of American Geographers* 61 (September 1971): 589–98.

Rouse, Irving. *Migration in Prehistory.* New Haven, Conn.: Yale University Press, 1986.

Rykiel, Zbigniew. "Intra-metropolitan Migration in the Warsaw Agglomeration." *Economic Geography* 60 (1984): 55.

Simon, Rita J., and Caroline B. Brettell, eds. *International Migration: The Female Experience.* Totowa, N.J.: Rowman and Allanheld, 1986.

Stephenson, George M. *A History of American Immigration.* New York: Russell and Russell, 1964.

Svart, Larry M. "Environmental Preference Migration: A Review." *Geographical Review* 66 (1976): 314–30.

Tabbarah, Riad. "Prospects of International Migration." *International Social Science Journal* 36 (1984): 425–40.

U.S. Bureau of the Census. *Historical Statistics of the United States, Colonial Times to 1957.* Washington, D.C.: U.S. Government Printing Office, 1960.

Ward, David. *Cities and Immigrants: A Geography of Change in Nineteenth-Century America.* New York: Oxford University Press, 1971.

Warren, Robert, and Ellen Percy Kraly. ''The Elusive Exodus: Emigration from the United States.'' Washington, D.C.: Population Reference Bureau, 1985.

White, Paul E., and Robert Woods, eds. *Geographical Perspectives on the Elderly.* New York: John Wiley & Sons, 1982.

White, P., and R. Woods. *The Geographical Impact on Migration.* New York and London: Longman, 1980.

Wiseman, Robert F., and Curtis C. Roseman. ''A Typology of Elderly Migration Based on the Decision Making Process.'' *Economic Geography* 55 (1979).

Wolpert, Julian. ''Behavioral Aspects of the Decision to Migrate.'' *Papers, Regional Science Association* 15 (1965): 159–69.

Cultural Landscapes in Rural Regions

The works of man express themselves in the cultural landscape. There may be a succession of these landscapes with a succession of cultures. They are derived in each case from the natural landscape, man expressing his place in nature as a distinct agent of modification.
CARL O. SAUER, 1925

THEMES

Causes and characteristics of variations in cultural landscapes.
Types of land tenure.
Land division systems and their impact upon the landscape.
Architectural styles and landscape types.
Farming in the modern world.

CONCEPTS

Landscape regions
Land use
Cultural convergence
Traditional landscapes
Environmental modification
Pristine state
Desertification
Land tenure
Mir
Kolkhoz
Sovkoz
Kibbutzim
Commune
Freehold
Sharecropping
Cadastral patterns
Metes and bounds
Fragmented holdings
Compact holdings
Long lot
Centuriation

Cultural landscapes are a visual expression of human relationships. The landscape of any place records the interactions between people and their natural environment, creating the distinctive character of each place on the earth. Analysis of cultural landscapes in terms of age, degree of human modification, or characteristic features assists in explaining the cultural geography of a place. The elements of a landscape reveal information about its inhabitants, ranging from their history to religion, technology, and economy. Recognition that each individual group of people create a distinctive landscape in varying environments is an essential part of explaining the earth's geography.

The cultural geographer delimits specific *landscape elements* as the first step in creating *landscape regions*. Landscape elements are distinct and recognizable features that recur in different locations. One rural landscape in the American Midwest, for example, is composed of landscape elements that include large, rectangular fields and farmsteads with large barns and the farm home, surrounding a town laid out in a rectangular grid. The landscape of New England, by contrast, is characterized by houses gathered together into a village, surrounded by fields with irregular shapes and few houses. (Figure 6–1). Landscape regions occur when a group of landscape elements domi-

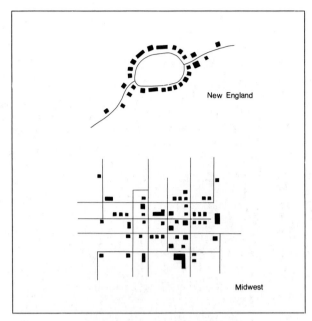

FIGURE 6–1 Village Patterns in the Midwest and New England These patterns illustrate the importance of the village common in New England and the rectangular grid pattern in the Midwest.

nate a recognizable region, such as the rural landscapes of the American Midwest.

Landscape elements and regions vary in scale from the continental to the local. On a continental scale, African landscapes are markedly different from European landscapes. At the local level, metropolitan Los Angeles has a landscape that is distinct from that of the Sierra Nevadas. These landscape differences reflect both the imprint of human actions in the natural environment and differences in the natural environment. The degree to which humans have changed the natural environment and the character and function of the forces for change in landscapes are one focus of cultural geography.

Geographers examine the landscapes of the earth to find those elements of a specific cultural landscape that are characteristic and which may combine to create a broader landscape region. Examination of landscapes and their components helps to explain the dynamics of the human-land relationship in a place and illustrates how landscape classification and analysis contribute to understanding the world in which we live.

An African landscape, illustrating elements of the natural environment. Without specific cultural features, it is difficult to recognize this as unique to Africa.

The city of Koblenz on the Rhine River demonstrates many of the characteristics of the European cultural landscape.

Landscape: A Definition

Landscape means different things in various academic disciplines. Traditionally, landscape has been defined as the fixed and enduring objects of the world we see. In this definition it consists of natural topographic features (mountains, plains, hills, lakes, woods, or even clouds) and the buildings, roads, communities, and monuments built by humans. A broader definition includes what a person sees or senses when he or she is outside. This includes both the natural and human-built enduring elements of the landscape and the smells, sounds, people, wildlife, and experiences that contribute to the individual's reaction to a specific place. This definition is much more in-clusive as it recognizes the role culture and perception play in defining landscape features. The advantage of the traditional definition of landscape, however, is that it can easily be depicted in a snapshot or sketch.

The narrower definition of landscape also reflects the origin and development of the term. The word "landscape" comes from the Anglo-Saxon influences on Western Europe. It was introduced into England by Germanic speaking peoples (Saxons, Jutes, Danes) after the fifth century A.D. and appears in several Old English variations, including *landskipe* and *landscaef*. Modern Germanic languages of Europe retain the term, as in the German *landschaft* and the Dutch *landscap*. Although the term is not always used in the same sense in each language today, originally it simply referred to attractive land. During the Middle Ages the word landscape fell out of use but it was revived by Dutch painters around A.D. 1600 to refer to the pictorial representation of a scene, either as the subject of a painting or as the background to a portrait. Soon landscape came to mean not only the painting but the view itself and became an essential term in the working vocabulary of poets, artists, gardeners, and the educated elite. Landscape in this usage meant the composition of humanmade spaces on the surface of the earth.

Cultural Landscape

Geographers began using the phrase "cultural landscape" in the late nineteenth and early twentieth centuries to describe a specific phenomenon, the surface of the earth as modified by human activities. Geographers study the cultural landscape to reveal the patterns and forms produced by the interaction between individuals or groups and the natural environment. Because culture includes the technology, customs, and beliefs of a group whereas environment incorporates the climatic, geomorphological, and botanical factors, landscape analysis emerged as a synthesis between the physical and cultural approaches in geography. Geographers study landscapes on a variety of scales, ranging from the landscape of an individual farm to that of a village or community, from the landscape of a mountain valley, to the generalized landscape of a

The dominance of the landforms does not detract from the importance of the cultural elements in this landscape of a Swiss mountain village.

region or country. In their studies geographers attempt to recognize and define those elements that combine together to make the appearance of each locality unique.

The first scientific geographic research that evaluated the cultural landscape was conducted by French geographers in the late nineteenth century. The early French studies led to similar landscape studies in the United States where one of the most influential cultural geographers was Carl O. Sauer at the University of California at Berkeley. In his presidential address to the Association of American Geographers in 1925, entitled ''The Morphology of Landscape,'' Sauer described the process by which the natural environment is transformed into the cultural landscape through human modification of the earth's surface (Figure 6–2). Sauer maintained that it was essential for cultural geographers to generalize from local landscape characteristics to a broader regional model. The process of generalization involves ignoring minor landscape variations at the local level in order to see the regional characteristics of the landscape. Thus the generalized cultural landscapes hypothesized by Sauer are achieved by categorizing the general characteristics of a culture region's landscape; these generalized landscapes, in turn, provide a model for examining local variations in those landscapes.

Geographers use three major methodologies to study the cultural landscape. In the first, geographers simply adopt the generalized regional landscapes suggested by Sauer. Studies of this type attempt to describe regions in terms of their landscape, explaining landscapes as the product of interacting historical, cultural, and natural factors. An example of a study of this type would be categorization of countries of the Middle East as a landscape region reflecting the Islamic religion.

Geographers using the second method try to formulate regional patterns and systems of landscape features to provide a systematic classification of landscape types. Such studies might methodically examine rural landscapes, for example, to develop a typology of classification. Although a good deal of effort has been devoted to attempts to classify landscapes in this way, no acceptable scheme of classification has been widely accepted, and many geographers have abandoned the effort.

The third method used in examining cultural landscapes goes beyond mere description of the

Carl Sauer is one of the most important figures in American cultural geography.

SAUER'S MODEL

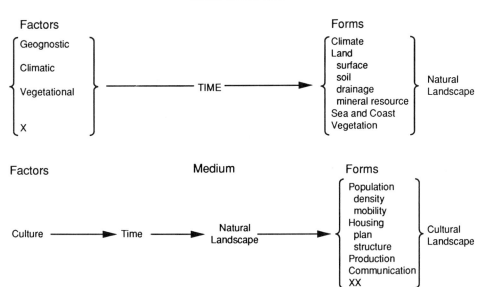

FIGURE 6–2 Sauer's Model Carl O. Sauer's model of the process of creation of the natural and cultural landscape.

landscape and its components and uses landscape as a clue to culture. This view maintains that the cultural landscape is an expression of human values and practices and represents a cultural artifact that is as informative as a book in providing information about the people who occupy that landscape. The American geographer Pierce Lewis expanded Sauer's views of the cultural landscape in his article "Axioms of the Landscape" in 1976. Lewis maintains that all landscapes have cultural meaning and that this meaning can be understood if we learn to "read" the landscapes. (Table 6–1). Landscape meanings are not always self-evident and require the analysis of the peoples who created the landscape, but such study provides understanding of the people, their values, and their culture.

Cultural landscapes are the end products of the expenditure of a considerable amount of time, effort, and resources by their human creators. The elements of the landscape, whether fields or fast-food restaurants, are there because they serve a function and are valued by the inhabitants of that landscape. Whether the landscape is characterized by large fields of single crops as in North America or by the traditional fragmented farms of African subsistence farmers, the resulting landscape reflects the technology, values, goals, and tools of its occupants. The structure of a town or city also reflects the real or perceived needs of those who inhabit it. Landscapes are like great documents spread out around us, providing us in countless ways (both subtle and obvious) with information about the values of our culture. The information in the landscape remains a closed "book" unless we are willing to investigate its elements. Comprehending the information revealed by landscapes opens a whole new sphere of understanding to those who are observant.

The cultural landscape provides us with information about the people who lived there in the past, how their use of the earth changed over time, and the structures involved in their use of the environment. Examination of the geographic patterns in any area (whether field or farm, village or city) provides insight into how a particular landscape has come to assume its present shape and appearance. For the cultural geographer, understanding why there are different house types in a community, why fields are divided a specific way, why villages have abandoned or deteriorating structures, or why certain place-names appear on the land are all part of the process of understanding the world in which we live.

TABLE 6–1
The Axioms of Landscape

1. *Axiom One: The Axiom of Landscape as a Clue to Culture:* The man-made landscape—the ordinary run-of-the-mill things that humans have created and put upon the earth—provides strong evidence of the kind of people we are, and were, and are in process of becoming.
   Corollary A: The investment of time, money, and emotion represented by a landscape prevents it from changing unless a major change occurs in the culture.
   Corollary B: If regions of a country look different, their cultures are also probably different.
2. *Axiom Two: The Axiom of Cultural Unity and Landscape Equality:* Nearly all items in human landscapes reflect culture in some way. There are almost no exceptions. Furthermore, most items in the human landscape are no more and no less important than other items—in terms of their role as clues to culture.
3. *Axiom Three: The Axiom of Common Things:* Common landscapes—however important they may be—are by their nature hard to study by conventional academic means.
4. *Axiom Four: The Historic Axiom:* In trying to unravel the meaning of contemporary landscapes and what they have to "say" about a cultural group, history matters.
5. *Axiom Five: The Geographic (or Ecologic) Axiom:* Elements of a cultural landscape make little cultural sense if they are studied outside their geographic (i.e., locational) context.
6. *Axiom Six: The Axiom of Environmental Control:* Most cultural landscapes are intimately related to physical environment. Thus, the reading of cultural landscape also presupposes some basic knowledge of physical landscape.
7. *Axiom Seven: The Axiom of Landscape Obscurity:* Most objects in the landscape—although they convey all kinds of "messages"—do not convey those messages in any obvious way.

Source: After Pierce F. Lewis, "Axioms for Reading the Landscape" in D. W. Meinig, ed., *The Interpretation of Ordinary Landscapes* (New York: Oxford University Press, 1979), pp. 11–32.

Remember that the term *cultural landscape* as used by geographers is distinct from the term *landscape* as used by artists. For the artist, the landscape is a pleasant scene often divorced from

The Palace at Versailles, France demonstrates the cumulative effect of human use of the environment. The Palace is no longer used by kings, but remains an important element of the French landscape.

the reality of the people who must work within it, but to the geographer, the landscape is more, for it is an assemblage of natural and human alterations that represents the end product of the earth as the home of human beings. The cultural landscape of the geographer emphasizes equally the open spaces of suburban malls and the open space of the formal landscape gardens of the palace Versailles in France. Each landscape is an expression of the culture, values, technology, and goals of its human occupants—the mall for ordinary Americans, Versailles for the Kings of France.

The neon signs of a typical American urban shopping strip may not appear beautiful to the artist, but to the geographer they are an indication of the functional nature of the landscape. The neon signs, competing with one another in garishness and size, attract some Americans by providing cues to the utilitarian nature of the landscape itself. The pastoral farmlands of the Swiss Alps not only reflect the use of the land for growing small grains and grazing animals, but are a utilitarian attempt to maintain a landscape attractive to tourists fleeing the cities for their annual vacation. In each case, the landscape provides cues to observers, but it also provides resources for those who created the cultural landscape through their use of the environment.

Landscape Classification

Because of the wide variety of landscapes, it is difficult to classify the landscapes of the world on a generalized basis. It is possible to use almost any criteria for examining landscape types, ranging from population density to land use, from predominant crop types to housing, from the style of dress of the people to the physical characteristics of the region. Whatever criteria are used, the resulting patterns of landscapes are fragmented since they reflect the great variation in physical and cultural elements that combine to create the cultural landscape.

Despite these difficulties, the landscapes of the world are often divided in terms of general level and type of land use, reflecting levels of economic development. At one end of a continuum of land use is the industrialized world. Landscape characteristics and regions of the industrialized world are broadly similar and reflect the process of *cultural convergence*. Technology, house types, job types, building systems, and even dress styles have diffused from a hearth in Western Europe and the United States to the periphery of the modern industrial world in Australia, New Zealand, and Japan. The landscapes of the industrialized world are characterized by mechanization, movement, and affluence: their typical landscape elements include the following features:

1. Emphasis on transportation functions (railroads, freeways, airports), reflecting the importance of movement of goods and people.
2. Factories and farms dominated by machines; and cities dominated by glistening skyscrapers that epitomize the machine age.
3. Structures in urban, suburban, and rural settings that emphasize the modernity and affluence of the occupants.
4. Contrasts between the recent landscape elements representing the modern world and relic landscape elements from the past.

At the opposite end of the continuum from the modern industrial world is the less industrialized, traditional world, whose landscape elements are based primarily on the past. Landscapes in less industrialized regions still contain many features from an earlier time when agriculture was the predominant activity, and villages and towns provided services to sustain agricultural economies. Farms in the modern world, in contrast, are an appendage to cities, which are now the main focus of productivity, whereas agriculture is still the basis for much of the productivity in the traditional world. *Traditional landscapes* are characterized by utilitarian and spartan elements. The landscape elements of the modern world, such as transport systems and skyscrapers, are represented in the traditional world, but they are exceptions rather than the norm. Typical landscape elements of the traditional world include the following:

1. Emphasis on field and farm, with low proportions of urban residents.
2. Structures built with local materials in response to the needs of the cultural ecology of a specific group, creating important regional contrasts in architecture.
3. Economies dominated by the use of animal or human labor.

In both the industrialized and the less industrialized worlds, landscapes can be subdivided into those that have been modified only slightly by humans and those that have undergone eons of intensive human change and manipulation. Urban regions of both the industrial and less industrialized worlds have been highly modified by human activities. Regions with less modification include parts of the tropical rain forest of the Amazon and the Congo and high-latitude or high-altitude locations in the Arctic regions or mountains of Europe and North America. It is important to recognize the division between highly and insignificantly modified landscapes, since it reflects the dichotomy between cultural landscapes dominated by human activities and those dominated by natural landscape features.

A continuum of *environmental modification* exists, ranging from those regions least affected by people (such as Antarctica) to those that are

CONTINUUM OF ENVIRONMENTAL MODIFICATION OF OCCUPANCE TYPE

FIGURE 6–3 Continuum of Environmental Modification of Occupance Type The degree of environmental modification increases from the minimal impact of hunters and gatherers to the widespread impact of modern industrial societies on all aspects of the environment.

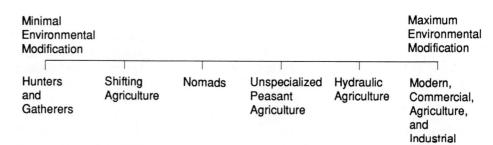

largely the result of human actions (such as New York City) (Figure 6–3). Places falling between these extremes include national parks or wilderness areas with only minor modifications; grazing lands with little evidence of human habitation but major modification of wildlife and vegetation; intensive farming systems where the natural vegetation has been removed and fields and farms have been developed to use the land resource; and villages with more densely developed human habitation.

Whether a landscape is part of the industrialized or less industrialized world, highly modified or largely natural, the cultural geographer is concerned with the landscape, the processes that created it, and the changes now affecting it. Through stimulus or relocation diffusion, landscape elements of the modern industrial world move to the less industrialized world. Consequently, the traditional agrarian world contains nodes of the modern world, particularly in the large cities. The diffusion of such seemingly diverse elements as iron and steel manufacturing technology, television, and McDonald's restaurants to the large cities of the traditional world makes these places more like those of the modern world. At the same time, such elements widen the gulf between the traditional landscapes of the countryside and the urban landscapes in the less industrialized world.

Some human activities in the industrial world cause widespread landscape changes in both the industrial and the less industrialized world. For example, acid rain is spreading across North America and Europe (Figure 6–4); the Amazon rain forest is being destroyed in part to provide pulp for paper production in Japan; and mining is being carried on throughout much of the less industrialized world for the benefit of inhabitants in the modern world. Each of these examples illustrates how the industrialized world can affect the less industrialized world, creating local variations in its broader landscapes.

Landscapes of Low to Moderate Human Modification

Examination of the world's landscapes reveals that the majority of the globe has been subjected to only moderate change by humans. Areas of moderate change include regions that are isolated by rugged landforms, have harsh climates, or are otherwise of marginal utility for the traditional cultivation of crops (Figure 6–5). Consequently, landscapes with low to moderate modification are characterized by *extensive* human activities, such as hunting and gathering and grazing. In the industrial world, landscapes of low to moderate modification are primarily sparsely inhabited regions of low-intensity agriculture such as ranching. In the less industrialized world, areas of low to moderate modification are those that are climatically or topographically difficult to use or where shifting agriculture or nomadic herding are the common forms of economic activity.

Areas of low modification in the industrial world may also result from a decision to maintain certain areas in a *pristine* or semipristine state.

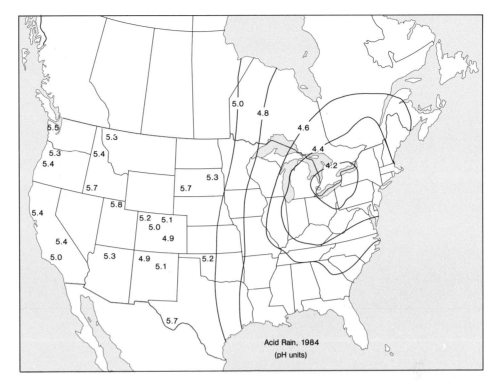

FIGURE 6–4 Acid Rain in Anglo America Levels of acidity in precipitation, 1984, as measured in pH units. The lower numbers are more acidic, and the old industrial core of the continent has the highest acidity in precipitation.

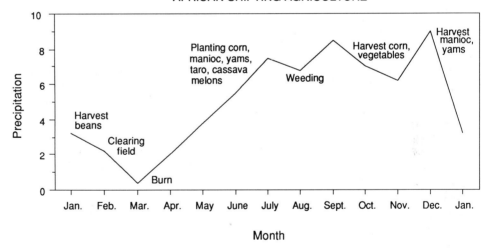

FIGURE 6–5 African Shifting Agriculture The cycle of clearing, burning, planting, weeding and harvesting associated with shifting agriculture primarily modifies the vegetation of an area.

For example, small areas of the modern world that have been little modified are often preserved as national parks or wilderness areas (Figure 6–6). Such lands range from wilderness areas, which by definition in the United States can have no roads, motorized vehicle paths, or permanent settlements and are reserved for low-intensity recreation uses, such as backpacking, to national parks, which provide more *intensive* recreation activities, such as improved hiking trails and camping areas, or fishing, and are accessible by automobile. National parks do try to minimize human modification of the environment outside developed camping and recreation areas.

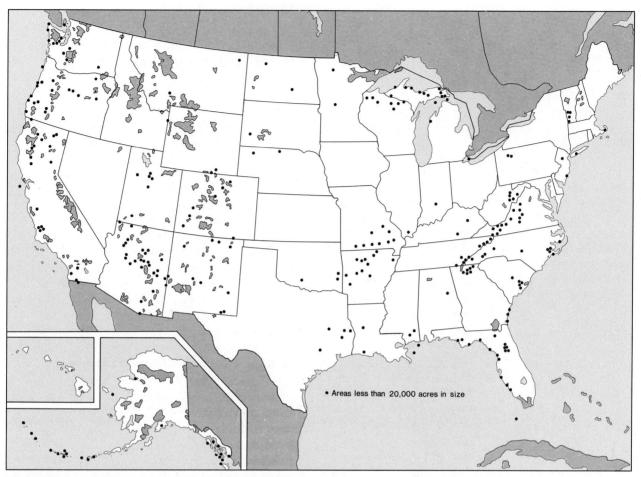

FIGURE 6–6 National Parks and Wilderness Areas in the United States Locations which are unsuitable for permanent human occupance or which have unique natural features are preserved in many countries.

Larger areas of the industrial world can be classified as landscapes of moderate modification; these are used for activities that largely consist of harvesting a natural product, such as ranching or forestry. Modification may take the form of deforestation through lumbering, clear cutting of large areas, replacement of native plants with introduced species, or controlled grazing. The cultural elements in landscapes of moderate modification are characterized by functional and scattered phenomena such as ranch headquarters, reservoirs to provide water for animals, or logging camps. The major long-term change to the natural environment is overgrazing and resultant erosion or plant substitution. Population densities are extremely low, reflecting the marginal utility of lands that are associated with such extensive activities.

Landscapes of low to moderate human modification in the less industrialized world are associated with hunting and gathering or pastoral nomad societies. Hunting and gathering societies have a minimal impact on the landscape. Environmental modification tends to be temporary and consists of foraging for plants and animals. The low density of the human population among hunters and gatherers results in only minor change to the natural environment except where use of fire changes or destroys the natural vegetation or wildlife. Structures are commonly made of naturally occurring vegetative materials, and al-

Arches National Park. National parks are established in many countries to protect unique natural landscape features.

though the posts that support huts made of grass or other vegetation may be used over and over, the vegetation itself needs to be replaced often. Such structures disintegrate quickly when abandoned.

In the modern world the few remaining hunting and gathering peoples (such as the Inuit of North America and aborigines of Australia) create landscapes similar to those of the hunters and gatherers in traditional societies. Before contact with industrial society, the Inuit and aborigines also created a landscape characterized by low-

Members of the Efe tribe in Africa create utilitarian shelters from naturally occurring materials. Although temporary, these structures are highly practical in the environment and lifestyle of this people.

density populations and semipermanent habitations made from local materials. The Inuit winter habitations (igloo) and summer habitations made of seal skins and the homes of the Australian aborigines made of local vegetation were temporary. In each case, however, the encroachment of the modern world has changed much of the extensive landscape of these formerly traditional peoples (See Figure 3–2 on page 79).

Nomads
The landscape of nomads in the traditional world is characterized by change associated with grazing. Most pastoral nomadism is adapted to dry climates where permanent settled agriculture is difficult or impossible. Relatively few people practice nomadism in the world, and their numbers continue to decline, yet they occupy a large territory. With the exception of the reindeer-herding groups of northern Europe, nomads are found primarily in the world's dry lands. Nomads are concentrated in the arid lands of Africa, across the Middle East, and through central Asia. The nomadic population is only a fraction of that of permanent farmers, but they occupy nearly twice as much of the earth's surface (30 million square kilometers or 11.6 million square miles versus 15 million square kilometers or 5.8 million square miles).

Pastoral nomadism creates a landscape modified by livestock and overgrazing. Human habitations tend to be semipermanent and are constructed of materials that can be easily transported to a new location. The key element of the human use of the dry regions of the world associated with nomadism is the movement of both animals and humans, and it occurs on a seasonal basis in definite patterns. The nomad does not wander aimlessly, but usually moves in cycles in order to provide fodder for the animals.

Nomads in the less industrialized world generally rely on sheep and goats since they reproduce rapidly, can tolerate very dry climates, and require less food than larger animals. Camels and cattle are often preferred by nomads because their larger size allows them to be used for transport, and they produce more meat and milk, but the advantages of sheep and goats make them more widespread. Camel nomadism is centered in the dryer portions of the Middle East, in Africa north of the

Nomadic peoples in Eastern Sudan. Landscapes in regions utilized by nomadic peoples are dominated by the natural environment, animals, and the people themselves. Structures are small and portable to allow the people to move with the herds.

Sahara Desert, and in the deserts of Saudi Arabia. The true deserts are rarely capable of supplying even the grazing needs of camels, so nomads tend to occupy the desert's margin (Figure 6–7). Sheep and goat herding are practiced across North Africa and the Middle East as well as in interior China and Soviet Central Asia, although in the latter two areas the groups have become more sedentary due to government planning. Cattle are used by nomads in more humid parts of Africa, often as part of a herd that is predominantly sheep and

goats. In nomadic societies where cattle represent a small minority of the animals, they are normally used to carry tents and other household goods as the nomads move.

The nomadic society and landscape are being transformed in each region of nomadism. Traditionally, the human-built environment of nomadic herders has consisted of simple, temporary mobile structures such as tents and yurts. Now, however, members of nomadic societies are establishing permanent homes and villages, and usually only male family members herd the livestock. This transformation of nomadic societies is changing the landscape as the traditional nomadic structures designed for ease of movement are being replaced. Changing nomadic practices reflect population pressures, competition for land with sedentary farmers, and environmental disasters such as the great droughts of the Sahel region south of the Sahara. Cut off from their former grazing lands, the nomads are adopting sedentary agriculture to ensure feed for their animals. Nomadic peoples often begin practicing permanent agriculture during only part of the season and graze the stalks and fodder from their crops after the harvest. In spite of these changes, the nomads continue to occupy a large area of the earth's surface simply because the land is unsuited for other uses. Furthermore, though nomadic landscapes are changing, they still reflect the prag-

TYPES OF NOMADS

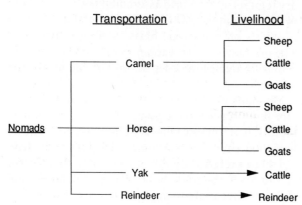

FIGURE 6–7 Types of Nomadism Nomadic peoples rely on a variety of animals for transportation and livelihood.

Bedouin peoples of Israel are increasingly a relic in a society occupied by urban residents. A tent of a Bedouin family is in stark contrast to the urban structures developed in a new Israeli settlement.

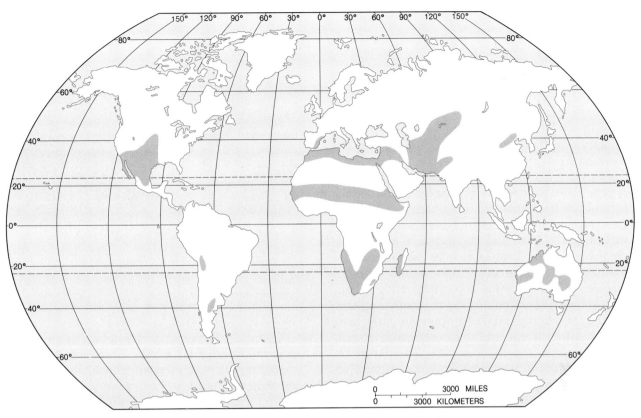

FIGURE 6–8 Areas of Desertification The expansion of deserts often results from overgrazing.

matic adoption of structures and uses that fulfill the traditional needs of these peoples.

Beyond the landscape features for human or animal shelter, the greatest impact of nomads is on vegetation. The environmental impact of nomads is much greater than their small population suggests. Because numbers of livestock often represent wealth in nomadic cultures, the typical nomad is more interested in quantity of livestock than in quality. Animals are sold only to provide for essential needs, and numbers increase until some natural catastrophe reduces herd size. In consequence, large herds cause overgrazing and destruction of wildlife. Vegetation loss allows erosion and makes the land more desertlike in a process called *desertification*. It is difficult for grasses and other plants to revegetate such an area; hence the result of overgrazing and erosion is the expansion of the deserts into areas formerly suitable for grazing (Figure 6–8).

Landscapes of Sedentary Agriculture: The Less Industrialized World

Lands used by hunters and gatherers, nomads, and modern ranchers are associated with low population densities. Although such land uses cover much of the earth's surface, they support only a tiny minority of the world's population. Most of the world's people rely on sedentary agriculture either directly or indirectly. The world's population is dominated by the masses of peasant farmers of India, China, Southeast Asia, Africa, and Latin America whose use of the land is generally intensive. The intensive use of lands by sedentary cultivators over millennia has created distinctive agricultural landscapes that are an integral part of the sense of place of each such locale. Each agricultural landscape changes as crops, technology, population, or political systems change, but it still reflects the underlying cultural characteristics that have created the built environment. Since the agricultural landscape represents the human imprint of the group that occupies a specific location, it may vary markedly within short distances (see the Cultural Dimensions box).

The elements that combine to create a specific agricultural landscape consist of village form and structure, land ownership characteristics and associated field patterns, architectural styles, and intensity of land use. These landscape elements reflect the development of a specific location's agriculture over time and often contain an abundant variety and array of individual features dating from the past and the present. Individual elements in the landscape may be *functional*, indicating a present need or function, or *relic* features surviving from a previous time when they served a function no longer required by the inhabitants. Agricultural communities of the less industrialized world tend to be conservative, and change has historically been slow. The nature of the peasant farming system is such that revolutionary changes have rarely occurred without the impetus of external political, economic, or cultural forces. Changes in the human-land relationship in such areas tend to add newer elements to the landscape without destroying the old. The resultant culture landscape includes numerous features representing the culture history of the area.

Land Tenure

A fundamental element in understanding the agricultural landscape of any region is the laws and customs affecting land ownership or use. Land ownership or rights to its use are referred to as *tenure*, and the rules affecting land tenure are as important today in defining the agricultural landscape as they were in the past. Although the types of land tenure have changed over time, four main types are recognizable:

1. Communal tenure
2. Estate or manor
3. Freehold tenure
4. Dependent tenancy

Agricultural landscapes in regions occupied by sedentary agriculture are highly transformed from the natural environment. Rice paddies in China occupy the level lands, while villages are located on lands less suitable for cultivation.

Communal tenure refers to ownership in common by a particular group or community. Communal tenure was once widespread among agrarian societies, but today exists only in specialized settings. Each member of the community or group who held the land had the right to a parcel of land for his or her own use. Redistribution of individual farm plots after a period of time was not uncommon. Historic examples of the communal tenure system include the European farming village with its communal lands that were plowed and planted as a group effort, the *mir* of czarist Russia in which the village members had scattered parcels of land that they were entitled to farm; the Chinese village in which the plots farmed by individuals were rotated; and a whole variety of communal land holdings held in tribal groups in North and South America, Africa, and Australia (Figure 6–9).

Today communal land tenure survives in several broad types. In Mexico, the *ejido* is a village-based communal land system with two types of tenure. One allows the member farmers to occupy and use a parcel of land during their lifetime, while in the other type land is held collectively by the village. Rules and laws that have been formulated since the Mexican Revolution of the early twentieth century define the specific rights of

CULTURAL DIMENSIONS
A Home Is a Castle

Slipping off his shoes at the bottom step with as much respect as if he were entering a sultan's palace, twenty-nine-year-old Harun Mamat pads barefoot up a set of open wooden steps and plops down on a bench on the tiny porch of his house.* The place is a single-story, unpainted, timber-plank house with a steeply pitched roof, perched chest-high above the ground on posts: not much, perhaps, but to this young father of three, it's a dream house. Having owned little more than the clothes on his back for most of his life, he now owns his own home in Rasau Kerteh, Malaysia. As he relaxes in jeans and T-shirt after a day's work in a nearby oil palm plantation—ten acres of which he also owns and tends communally with twenty-three other families who each own adjoining ten-acre plots—the little porch affords a breezy refuge from the sun.

As Harun squints out across his front yard where flowering bushes cling to the sand, he recalls the harsher life he led just a few years ago as a fisherman in a coastal village twenty-five miles from here on the South China Sea. Fishermen in villages such as his earn the equivalent of

about US$75 per month—well below the poverty line. New-town settlers working in oil palm plantations make $500–$650 a month.

He heard about a series of five new towns being carved out of the jungle in Trengganu (a state on the eastern coast of peninsular Malaysia that ranks as one of the country's least developed states) with the aid of a $16 million loan from the Asian Development Bank (ADB), a Manila-based multilateral lending institution financed by industrial nations.

Forsaking his all-too-often empty fishing nets, Harun moved to Rasau Kerteh and was introduced to the oil palm whose clusters of reddish fruit are processed into cooking oil, soap, and other household products. After three years as a renter, he exercised the option to buy his house (over seventeen years, with his three years of rent considered as down payment). The cost was about $900.

---

*Peter C. Stuart, "Beyond the Projects: People's Profit," *Development Forum* (May 1982): 16.

---

individuals within the *ejido* to use the land and pass it to a descendant as well as their responsibilities toward the rest of the community.

Another variant of the communal system is the *kolkhoz*, or collective farm, of the Soviet Union. Based in part on the earlier Russian village *mir* system, the *kolkhoz* is a communal system in which all of the land is owned by the government and individuals work on the collective and receive a proportion of the profit in return. At the present time the residents of a collective farm are more like workers in an industry than farmers. Agricultural workers are guaranteed a minimum income regardless of the productivity of the farm, with additional income if production targets are exceeded; thus it remains a communal system. In practice, the residents of the *kolkhoz* are assigned specific tasks in which they specialize. Thus

there are work brigades of dairy workers, grain or other crop producers, and so forth. The leadership of the Soviet Union began modifying the *kolkhoz* system in 1988, allowing farm families to lease a part of the *kolkhoz* lands for up to fifty years. The Soviet Union also uses another type of land tenure called the *Sovkoz*. Farmers on the *Sovkoz* do not share in the profit of the farm, but receive a wage just as if they were working in a factory. The *Sovkoz* is technically a form of communal tenure since the lands are owned by the government for the benefit of the entire populace of the country.

A communal form of land tenure known as the *kibbutz* (plural *kibbutzim*) has been established in Israel. Based on the Russian *mir*, the *kibbutzim* were established by Russian immigrants to Israel. The *kibbutzim* lands were originally purchased by the International Jewish Movement as a way

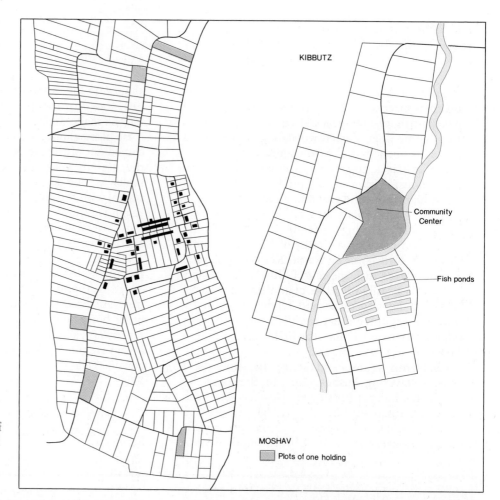

FIGURE 6–9 Diagram of Representative Communal Systems Communal agricultural systems utilized in Israel.

to provide a communitarian life-style for poor migrants from the Soviet Union and Eastern Europe. Lands in the *kibbutzim* are still owned by the movement rather than by individual farmers. The *kibbutzim* are farmed cooperatively and other *kibbutz* operations such as education, housing, and the growing industrial activity that some practice are also cooperative.

A communal system patterned after the *kolkhoz* of the Soviet Union was implemented in China after the Communist Revolution of 1949. The Chinese system is based on a *commune*, in which ownership of the land is retained by the government for the benefit of the people. Since the economic reforms of the 1980s, this communal farm system has developed into a *contract* farm system. Under the contract system a farmer

The sacred sites at the Dome of the Rock and the Wailing Wall in Jerusalem, Israel are symbols of the great civilizations based on early agriculture in the Eastern Mediterranean.

and his family contract to till a parcel of land and return a specified portion of their crop to the commune as rent, but anything over the rent they can keep for their own use or disposal.

Communal tenure is also practiced in African tribal societies. There it generally takes the form of a group cultivating the land around its village and allocating the plots to ensure some relative level of equality (see the Cultural Dimensions box).

Communal systems are under pressure from alternative tenure systems everywhere in the world. Except in Mexico or countries that have based their land tenure on the model of the collective from the Soviet Union, such as China, the nations of Eastern Europe, Cuba, and Ethiopia, there is a constant tendency to move to some form of private ownership (*freehold tenure*). One example of freehold tenure is the manor or estate (Figure 6–10). Estates have ancient origins, having existed for centuries throughout the Mediterranean, Middle East, and South Asia.

The central element of the manorial system is a landed aristocracy that controls the majority of the land in a village or region and allocates individual farming activities to peasant farmers who cultivate the land in return for a share of the product. The manorial system was common in Europe prior to the Industrial Revolution, but today has largely been replaced by other forms of tenure. Before it disappeared in Europe, it diffused to Latin America, where manorial estates are known as haciendas in most countries. Land reform movements have eliminated many of the haciendas in Latin America, and similar efforts have eliminated most estates in the Middle East and much of South Asia.

CULTURAL DIMENSIONS
Communal Agriculture in Zambia

Although her husband works as a security guard in Zambia's capital city, Lusaka, Dorothy Mbozi has never really left her lifelong occupation: farming.*

Dorothy moved to Lusaka five years ago from the Mazibuze communal lands in central Zambia to be with her husband, Timothy. However, his monthly salary of 280 kwacha (US$30) proved inadequate to support her and nine children. To make ends meet, Mbozi sent six of the children back to the rural areas to be cared for by relatives.

"We then requested some land from chief Mungure just outside Lusaka," said the frail, middle-aged woman recently. The chief gave them a one-acre plot about nine kilometers out of town.

The Mbozi home is a two-room mud structure with a grass roof. Timothy leaves home for work in the early evening and walks back home at dawn. Dorothy does most of the plowing by hand. "We grow mostly maize, enough to feed the family," she says. "Last year we used two bags of fertilizer and harvested six bags of maize. This took care of most of our basic needs, though we still had to buy meat and vegetables."

Her main complaint is not too much work, but too little. "The work is not heavy, because there is not enough land. I wish I had a few more acres so that I could grow cotton and groundnuts. I could earn enough money to bring the rest of my children home. The other thing I would like is a plow, and just enough cattle to do the plowing."

Her other main complaint is theft. "Since we live near town, and as people's wages are no longer adequate, we get a lot of thieves. We used to report them, but now the police turn a blind eye."

"Perhaps," she reflects, "we should return to Mazibuza. If the government can help us with loans, we might be able to make a better living there."

"If I can get a bigger plot of land and if I am sure that I will be able to sell my crops at a good price," Timothy adds, "I will give up my job here in town and go back to farming full-time."

———
*African Farmer 1 (1988): 55.

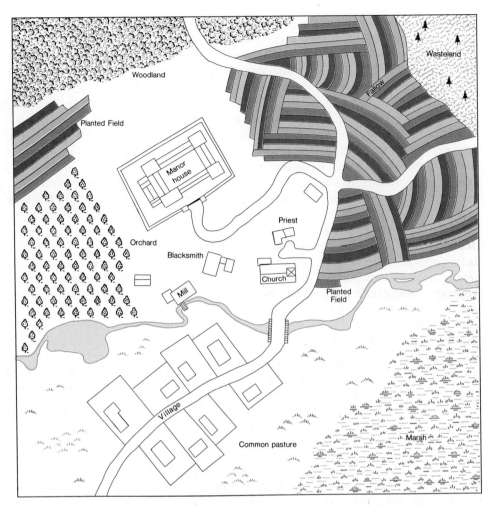

FIGURE 6–10 An Estate (Freehold Tenure) A manorial estate is one form of freehold tenure. The land is owned by the lord who may rent land to dependent farmers.

The other variant of freehold tenure is that of the individual small farmer. Such farms are known as *freeholds* because title is held by one absolute owner or family. Such freeholds are generally small in peasant farming regions, with the average size in Asia being less than 5 hectares (12 acres) and in Africa even smaller. Freehold systems often result from the destruction of a preexisting manorial system. The division of old estates into individual farms often creates unprofitable farms since the estate is rarely large enough to give each farm family sufficient land to care for its needs in a reasonable fashion. Nevertheless, the desire for land ownership is high, and programs to redistribute land have taken place in many regions of the world. Important examples include Mexico after 1917 (although part of the land was converted to the communal *eji-*

dos), Peru and Japan after World War II, large portions of the Middle East, Korea, India, and the Philippines.

A strong government commitment to land redistribution to create a freehold peasantry is necessary for attempts at land distribution to be successful. The increasing population in peasant agricultural systems, combined with no improvement (or an actual decline) in the standard of living, often leads to tremendous pressure for land redistribution. Often these pressures are the basis for revolutionary movements that may bring about change in governments, as occurred in China in 1949. The most successful farming freeholds are found in the modern industrial world, particularly in the United States, Canada, and Australia, which never had a long tradition of estate or manorial tenure (except in the plantations of the southern United States), and individual farms are much larger.

A last form of land tenure is *tenancy*. Tenancy is the legal condition under which a farmer rents land from a landowner and pays for the use of that land. Tenancy of many different kinds is common in both traditional peasant societies and modern industrial societies. Common tenancy arrangements include *sharecropping,* under which the tenant pays the owner with part of the crop. Sharecropping is particularly common in Asia and Latin America and tends to hinder economic development because the tenant does not receive increased income from any improvements he or she might make in the land. The landlord tends to simply collect rents rather than improve the property. Consequently, where sharecropping tenancy is common, the standard of living tends to remain low. Another form of tenancy involves cash rents. Where cash rents are paid, a contract is drawn between the tenant and landlord that allows the tenant to benefit from improvements. Under such conditions, a tenant system can function efficiently and effectively, as is the case in much of the modern world.

Patterns in Agricultural Landscapes

In the majority of the landscapes of the traditional world, tenancy or small freeholds dominate. The

A rural landscape in the traditional world is characterized by small fields and farms associated with the use of human and animal labor for cultivation.

resultant landscapes are dominated by small fields and farms, whose specific pattern and distribution of lands are based on local tenure practices. Such landscape patterns are related to the method of surveying land to create individual parcels. Three types of patterns are evident in agricultural landscapes: survey patterns, cadastral patterns, and field patterns. *Survey patterns* are the large land divisions made by surveyors. The *cadastral pattern* refers to the legal description of property ownership. The *field pattern* refers to the way in which an individual farmer has subdivided his or her land for agricultural use (Figure 6–11).

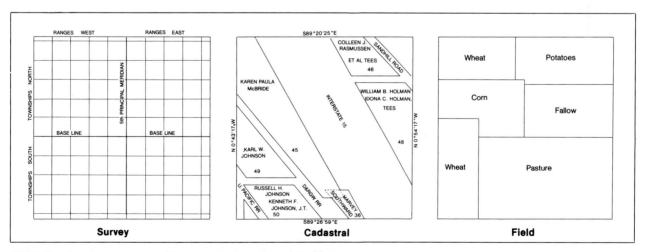

FIGURE 6–11 Examples of Survey, Cadastral, and Field Patterns

Survey patterns divide the land into units that can be easily described and legally recorded so that there is no question about their location and boundaries. The cadastral pattern subdivides the survey pattern into the individual farm holdings. The field pattern is the lowest subdivision; it divides the farm holdings so that the individual farmer can cultivate his or her land effectively.

Two general types of survey, cadastral, and field patterns are recognizable: regular geometric (or rectangular) and irregular. Irregular survey patterns characterize much of the landscape of the less industrialized world. Boundaries of farms, fields, and villages reflect the pragmatic, tradition-rooted decisions of those who use them. Irregular patterns are also the norm in parts of the modern industrial world where lands have been cultivated for centuries. The irregular patterns in these landscapes survive as landscape features from earlier agricultural practices because they still reflect underlying farm ownership patterns.

Regular geometric or rectangular survey and cadastral patterns are associated with standardized spatial arrangements. Such standardization is common in landscapes of the modern industrial world, as in the United States and Australia. Regular survey patterns are also found in the traditional world in lands colonized or cultivated by representatives of the modern world or in new farmlands developed by reclamation or expansion of agriculture. For example, regular survey patterns are found in portions of Latin American countries where agriculture was initiated under Spanish direction, parts of India where the British reclaimed land through irrigation, and some new lands in the western portion of China.

On a world scale, irregular survey patterns are the norm for both survey and cadastral divisions of land. Property ownership lines in most of the less industrialized world tend to reflect natural features rather than geometric survey grids. The surveying system that relies on natural features, such as landforms, streams, trees, or other phenomena to establish property boundaries, is called *metes and bounds*.

Within cadastral patterns, it is possible to distinguish between individual landholdings that are *fragmented* and those that are *compact*. Fragmented holdings are either a reflection of an earlier communal system or the result of inheritance systems in which lands are divided among the heirs. In a fragmented system, an individual freehold owner may have a number of small landholdings scattered in widely separated fields (Figure 6–12).

Fragmented freeholds are the common cadastral pattern over much of the traditional world. Although the property lines appear to have been drawn totally at random, the fragmented cadastral pattern normally reflects attempts of a communal landholding system to provide equity in land ownership. Providing each farm family with parcels of good crop land, grazing land, and woodlands for fuel ensures that each will have the basis for livelihood. Fragmented patterns are found in places as disparate as India, Mexico, Kenya, and China.

Exceptions to the irregular cadastral pattern in traditional rural landscapes are found in areas settled by the Romans or in areas where a system known as *long lot* was used. The Roman *centuriation pattern*, which was based on a regular geometric survey pattern, is still found around

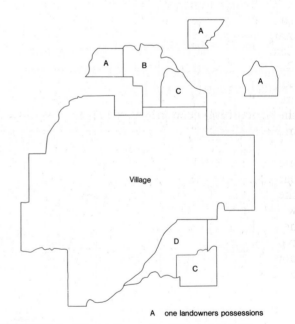

FIGURE 6–12 An Example of Fragmented Property Ownership Fragmented property ownership separates an individual landowner's property into several parcels.

the Mediterranean. Long lot farms consist of a long, narrow property in which the length is at least ten times as long as the width. The long lot system is found in parts of Brazil and Argentina, as well as in parts of Europe, Indonesia, and in areas settled by the French in North America. Long lot division is effective in maximizing a natural transportation system such as a river. Individual farmers had access to the river at the front of their long lot, and then the long lot extended away from the river or transportation artery across a variety of land types to provide the same benefits of variety as the fragmented farm systems.

Field patterns generally reflect the underlying cadastral system. Agricultural landscapes in most of the less industrialized world have field patterns that are even more fragmented than the cadastral pattern of ownership. Except in a few areas (such as India), the agricultural landscape in much of the traditional world has a field pattern characterized by irregular boundaries, which are often marked by fences, ditches, or hedgerows that enclose individual fields and protect the crops from destruction.

Where rice farming is the predominant form of agriculture, terraces and ditches provide the demarcation between individual fields. Regions of Central America and the Yucatán of Mexico and Guatemala have stone fences separating fields. Log or post fences serve to divide fields in other areas. Where animal grazing is an integral part of the agricultural economy, fields tend to be much more open.

The overwhelming dominance of the fragmented pattern in survey, cadastral, and field patterns makes the individual farming regions of the traditional world look remarkably similar when viewed from the air. Closer examination, however, reveals great contrasts in kinds of crops, types of fences, and specific tillage methods that make each individual agricultural landscape unique.

Architecture

The architectural styles of the less industrialized agricultural landscape also contribute to the uniqueness of each place. In traditional agricul-

tural landscapes, the population is concentrated in villages. Villages, like all human settlements, manifest a degree of stability that is in keeping with the resources available to the community that built them. Whereas the settlements of hunters and gatherers, shifting agriculturists, and nomads are characterized by temporary structures, sedentary farmers create a landscape of relative permanence. These rural settlements are fundamentally different from urban settlements because the occupations of their inhabitants are different. For residents of a rural village, their homes are more than just dwellings; they include space and outbuildings central to the farming occupation of those who live there. Thus a rural village landscape is in essence an agricultural workshop that cannot be separated from the cultivated fields surrounding it.

The use, shape, and architecture of the structures of farms and rural villages often reflect the kind of work and the agricultural techniques of the occupants. Each farmer has a home and associated buildings in the village from which he or she goes out to cultivate the land. The resultant landscape pattern is one of a village surrounded by the irregular fields of the traditional agricultural system. The architecture and structures of the village vary from society to society, reflecting the environment, availability of building materials, and the cultural heritage of the residents.

The typical structures in areas of sedentary agriculture, such as China, are low and small, reflecting the importance of the village as a home for farmers rather than as a provider of urban services.

Individual structures are constructed from locally available rock or vegetation. Buildings reflect the environment in the level of protection required against cold, wind, or heavy precipitation. Building materials include wood, rocks, animal products (such as leather), or mud (Figure 6–13). Whatever material is used in construction, the general pattern for villages of the traditional world is broadly similar. Houses tend to be low, rarely exceeding two or three stories, and are generally small. There is a relatively low level of specialization of activity by room, with rooms serving as sleeping or eating rooms depending on the time of day. Within this broad categorization, it is possible to recognize general patterns of house types. The village landscape is one consisting of individual houses and their associated farm buildings in a fenced or walled compound. These compounds may be circular, rectangular, or irregular in shape (Figure 6–14).

Cultural geographers classify houses and other structures in terms of form, function, material, and grouping. The form includes such features as the number of stories in a house, the shape of its roof, and its floor plan. The function of a house may vary; some simply house people, whereas others also provide protection for animals or storage for crops or are used for marketing purposes. The particular materials used in construction reflect the availability of natural materials in the traditional world. Wood houses are dominant in the landscape in forested regions, but because wood is lightweight, wood houses may be common even in nonforested regions. Wooden structures are built of a variety of materials. Posts with bamboo thatching are common in some humid tropical areas. Log houses are relatively uncommon in the traditional world because of the amount of wood they require; more commonly, sawed lumber is used for construction of wooden

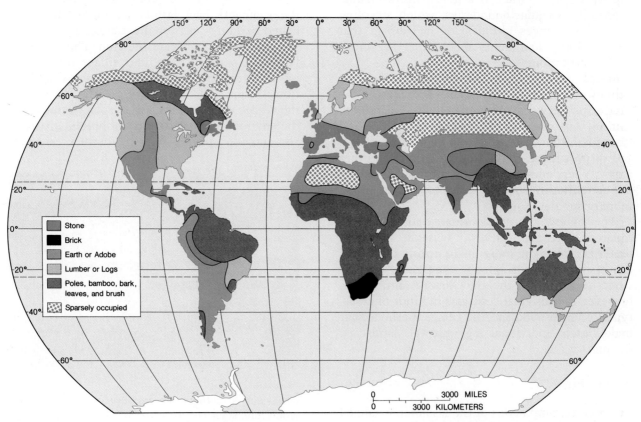

FIGURE 6–13 Building Materials and Structures The distribution of predominant building materials used in construction of small structures such as homes or outbuildings is depicted in this map.

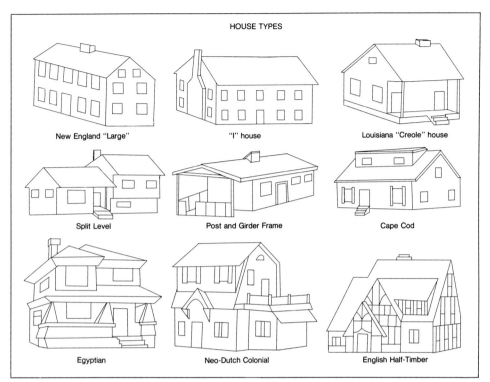

HOUSE TYPES

New England "Large"   "I" house   Louisiana "Creole" house

Split Level   Post and Girder Frame   Cape Cod

Egyptian   Neo-Dutch Colonial   English Half-Timber

FIGURE 6–14 House Types in the United States Each cultural region has its own distinctive house types which are readily recognizable.

houses. The advantage of log or wooden houses is their relative ease of construction. A log house requires only an axe for construction, and a sawed lumber house can be assembled rapidly once a frame is constructed.

Houses of mud or stone are more common in the less industrialized world. Mud houses are constructed in a variety of ways: in some, earth (called rammed earth) is packed into a wooden frame; in others tree branches or willows are plastered with mud (mud and wattle); and others are built of sun-dried bricks known as *adobe*. Fired brick is uncommon in the landscapes of the traditional world because of its cost. Adobe or other rammed-earth houses are unsuitable for damp climates, but are cheap and relatively easy to build in a dry climate. Adobe bricks are made by mixing mud and placing it in forms to dry; then these large adobe blocks are stacked and bound together with either mud or some other mortar. Roofs can be made of vegetation, clay tiles that have been made resistant to rapid erosion by firing in a furnace, or some type of steel, aluminum, or plastic sheets. As long as the mud or

adobe walls are protected from precipitation or from groundwater, they will last indefinitely. Traditional adobe houses are particularly unsafe in areas that are prone to earthquakes. Their weight and the lack of reinforcement in the walls makes them highly susceptible to collapse.

Stone houses are common in areas that have an abundance of either soft stone, which is easily shaped, or large quantities of loose rubble stone that can be used for building. The stone house is more permanent than the adobe house, but unless flat stones or stones that can be easily split along lines of cleavage are available, construction is difficult. Although stone houses tend to be more permanent than houses constructed of other materials, they tend to be less widespread in the traditional world because of their greater difficulty in construction, the limited distribution of suitable local stone, and their tendency to collapse during earthquakes.

Whatever the building materials, the landscape characteristic of the built environment in the traditional farming world is one of simplicity and low elevation. The emphasis on a traditional

Construction materials in rural landscapes often utilize local materials with little transformation, as these adobe structures in Mexico demonstrate.

farming economy prevents the accumulation of the wealth required to build and maintain a large home, and most residents of traditional villages live in small, single-story structures. Within the general pattern of small, low, simple housing, there is great variation as individual cultures make their own imprint upon their built environment. Landscape variations include paint or its absence and landscaping with its distinctive plant varieties. Houses may focus inward to a court or outward to a street. In North Africa, for example, houses are built around an internal court, presenting a blank wall to the street. Houses of the village landscape of China lack an internal court, but often the area around the structure is enclosed with a wall, providing privacy for the occupants.

Farming in the Industrial World

The same processes that interact to create the agricultural landscapes of the less industrialized world are found in the industrial world. Land tenure and the survey, cadastral, and field patterns combine with the built environment to create distinctive agricultural landscapes. In the United States, Europe, Australia, New Zealand, and Japan, land ownership is primarily in freehold tenure form. Exceptions to freehold tenure in the modern world are found in the communal systems of the Soviet Union, Eastern Europe, and Israel and in the use of cash tenancy for a minority of farmers in industrialized countries.

Individual countries in the industrial world use distinctive survey systems. In the United States, two survey systems are used on the majority of the land of the country. The first is a regular geometric survey based on the township and range system (Figure 6–15). One of the first acts of the Congress of the newly independent United States was the passage of the Northwest Ordinance (1789), which provided a plan for bringing new states into existence and for selling lands owned by the government. The township and range system provided a simple survey system that allowed land to be easily located, legally described, and transferred to other owners.

The eastern United States uses a metes and bounds system, reflecting the system in use in Europe at the time of colonization. Most of Canada relies on a geometric survey system similar to that of the township and range, except along the St. Lawrence River. The Ontario peninsula region of Canada uses a metes and bounds system derived from the system in use in Europe when the area was colonized.

The metes and bounds and long lot survey systems of North America were borrowed directly from Europe, where the majority of land surveys are based on an irregular pattern inherited from land division in times past. The patterns of Australia and New Zealand reflect a more regularly geometric pattern as would be expected from their time of settlement. In North America, Australia, and New Zealand, the slow colonization of large tracts of land allowed time for surveying the land before settlers occupied it.

Europe, Japan, and the Soviet Union represent centuries of land use, which led to cadastral systems that are as irregular as their survey systems. The European field pattern is irregular, reflecting its origin. A major exception to this occurs where revolutionary activity has resulted in changes in the cadastral system, as in Eastern European countries and the Soviet Union where a collective farming system replaced the traditional fragmented cadastral and field systems. The field

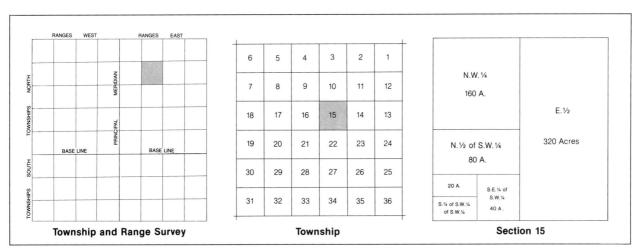

FIGURE 6–15 Anglo America's Survey System The township and range survey system used in most of the United States west of the Appalachian mountains.

pattern tends to be rectangular in such cases, allowing cultivation to the boundaries of the farm, which are also rectangular. The cadastral patterns of the industrial world reflect the underlying survey system. In North America, Australia, and New Zealand, the field patterns are normally rectangular. Even when irrigation systems reliant on a circular pattern are used, they are superimposed over existing rectangular field patterns.

The rural landscapes in areas of commercial agriculture create distinctive patterns, such as these fields irrigated by a center pivot sprinkler system.

One primary difference between the survey, cadastral, and field systems of the industrial world and those of the less industrialized world is that of scale. The individual farm units and fields are generally much larger in the modern world than in the traditional world because of the technology and tools used by the different farming groups. Some form of mechanization is the norm in the industrial world, allowing cultivation of larger areas. Manual labor remains the primary means of cultivation in the less industrialized world, and in consequence fields remain small. Another factor that effectively minimizes large-scale farm activities in the less industrialized world is the high population densities found in rural areas. By comparison, most rural villages and towns in the industrial world have been losing population, and farms have been growing larger as people migrate to the cities to work.

The Built Environment in the Industrial Farming Economy

The built environment of farming in industrial societies is also larger scale than that of the less industrialized world. Structures tend to be larger, with larger buildings to house more livestock, store more crops or machinery, and provide a higher standard of living through a bigger home.

The borough of Sunderland, England is an example of a dispersed settlement pattern.

Farmers of Carcassonne, France lived within the walled city and travelled out to cultivate their fields, an example of a nucleated settlement pattern.

Two types of settlement patterns can be recognized in the built environment in farming in the industrial world. The first is the *grouped* or *nucleated* pattern in which farmers reside in a town or village and travel to their farms to cultivate them. In a nucleated pattern the inhabitants build their houses in a rural community surrounded by cultivated lands (Figure 6–16). The second settlement pattern consists of individual farmsteads with the home and farm buildings located upon them and is referred to as a *dispersed* or *scattered* pattern. The dispersed pattern is common across the Midwest and Great Plains of the United States and Canada, while the nucleated pattern is

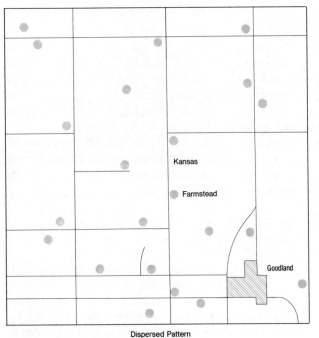

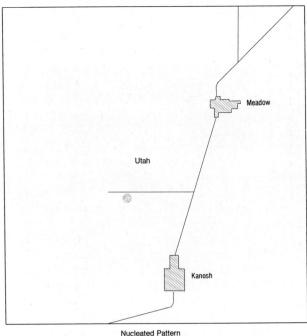

FIGURE 6–16 Nucleated and Dispersed Settlement Patterns in the United States The presence of nucleated versus dispersed farmsteads is an important factor affecting landscapes in rural areas.

more common in Europe, the Soviet Union, and the eastern United States.

Whether dispersed or nucleated, the built environment associated with farming in the industrial world often reflects a functional emphasis. The large steel sheds in the American Midwest, for example, are cheap and easy to build. Functionalism does not prevail everywhere, however, especially in farming landscapes that have many relic features from previous farming periods. The farmstead in northern Germany and Poland, for example, consists of a single structure with the farmer and his or her family residing in one end and the livestock occupying the other end. Switzerland has much more elaborate structures in which livestock occupy the lower floors and the farm family the upper stories. These combinations of barn and house are a relic feature that is being replaced on larger and more economically viable farms.

The towns and villages of the modern farming landscape are more diverse and larger scale than those in the less industrialized world. Specialists provide implements, groceries, tools, clothing, medical care, and other services to the farmers who reside in or around the village. Houses are often two or three stories in height, business buildings may exceed three stories, and the general landscape features are larger.

The farming landscape in the industrial world is similar to the nomadic landscape in one respect—the importance of movement. Farmers rely heavily on modern transportation systems to move raw materials to the farm (seeds, fertilizer, equipment) and to market their products. Whether they live on a dispersed farmstead or reside in a village or town, the farmers of the industrial world must travel to maintain their livelihood, (see the Cultural Dimensions box below.) Movement in the farming sector of the

CULTURAL DIMENSIONS
Sidewalk and Suitcase Farmers

The importance of movement in the modern agricultural economy was emphasized by the geographer Leslie Hewes.* Hewes used terms coined by another geographer, Walter Kollmorgan, to describe two types of farmers who traveled extensively, "sidewalk farmers" and "suitcase farmers." Sidewalk farmers are defined as those who live in town but commute to their farms on a daily or weekly basis. Suitcase farmers are those who travel over a larger area during the course of a season to plant and harvest crops. For example, a suitcase farmer may own land in Texas, Oklahoma, Montana, and Canada and travel to plant grain in the spring or fall and then travel again in the summer to harvest the grain.

The suitcase farmer is a type of "nomad" in that he or she travels in a cyclical manner, but the economy that prompts the movement is far different from that in nomadic societies. Farmers in the modern world are part of the broader industrial society. Farm products reflect market conditions, and the farmer is primarily a businessman

or woman. Capital investment in land, machinery, buildings, and animals rivals that in other businesses or industries. The suitcase farmer, like many other business people, must travel to transact his or her business, which is tilling the land.

The sidewalk farmer also reflects the industrial economy. Modern, efficient transport systems allow sidewalk farmers to commute to their work like most other workers in industrial societies. Instead of commuting to a larger city where offices and factories are concentrated, she or he commutes from town or village to the farm. Such farm commuting reflects agriculture in the industrial world where machines replace manual labor in cultivating the land or caring for animals, and animals are permanently or semipermanently penned until they are shipped to market.

---

*For a complete discussion of sidewalk and suitcase farmers, see Leslie Hewes, *The Suitcase Farming Frontier: A Study in the Historical Geography of the Central Great Plains* (Lincoln: University of Nebraska Press, 1973).

industrial world differs from that in the less industrialized world in both frequency and speed. Railroads, superhighways, paved roads, and airports are a part of the farming landscape of the industrial world, but play only a small role in the agricultural landscapes of the less industrialized world.

Conclusion

The rural regions of the world with low population density have distinctive landscapes that reflect their use. Rural landscapes include those with little apparent human modification and those occupied by farms, fields, and rural settlements. Whatever the specific landscape elements, the landscape itself is a record of the interaction between people and place that reflects the technology, goals, and values of those who occupy it. From the gathering of plants and animals by hunters and gatherers to the corporate farming of the American Midwest, rural landscapes represent the complex interaction of people and place that creates the everchanging cultural mosaic of the geography of these locales.

QUESTIONS

1. Describe the origin and development of the term landscape.
2. Identify and discuss three major types of cultural landscapes.
3. Compare and contrast the terms cultural landscape and landscape.
4. What are the various forms of traditional landscapes? Describe each form.
5. Describe two types of nomadism and indicate their distribution.
6. Describe the four main types of land tenure, giving examples of each.
7. What are some basic forms of land division?
8. What are some problems that can occur from metes and bounds property descriptions?
9. Describe the uniqueness of architectural styles of the traditional agricultural landscapes.
10. Compare and contrast traditional versus modern farm landscapes.

SUGGESTED READINGS

Allen, G. Noble, and Gayle A. Seymour. "Distribution of Barn Types in the Northeastern United States." *Geographical Review* 72 (1982): 155–70.

Amos, Rapoport. *House Form and Culture*. Englewood Cliffs, N.J.: Prentice-Hall, 1969.

Arnold, R. Alanen, and Joseph A. Eden. *Main Street Ready-Made: The New Deal Community of Greendale, Wisconsin*. Madison, Wis.: The State Historical Society of Wisconsin, 1987.

Barker, Randolph, and Robert W. Herdt. *The Rice Economy of Asia*. Washington, D.C.: Resources for the Future, 1985.

Barlett, Peggy F. *Agricultural Choice and Change: Decision Making in a Costa Rican Community*. New Brunswick, N.J.: Rutgers University Press, 1982.

Berry, David. "The Sensitivity of Dairying to Urbanization: A Study of Northeastern Illinois." *The Professional Geographer* 2 (May 1979): 170–76.

Boserup, Esther. *The Conditions of Agricultural Growth*. Chicago: Aldine Publishing Co., 1965.

Bower, Sidney. *Design in Familiar Places: What Makes Home Environments Look Good*. New York: Praeger, 1988.

Bowers, J. K., and Paul Cheshire. *Agriculture, the Countryside and Land Use: An Economic Critique*. London: Academic Press, 1983.

Brewer, Michael. "The Changing U.S. Farmland Scene." *Population Bulletin* 36 (December 1981).

Bunce, Michael. *Rural Settlement in an Urban World*. London: Croom Helm, 1982.

Burbach, Roger, and Patricia Flynn. *Agribusiness in the Americas*. New York: Monthly Review Press, North American Congress in Latin America, 1980.

Chakravarti, A. K. "Regional Preference for Foods: Some Aspects of Food Habit Patterns in India." *Canadian Geographer* 18 (Winter 1974): 395–410.

Chisholm, Michael. *Rural Settlement and Land Use*, 3d ed. London: Hutchinson, 1979.

Christaller, Walter. *The Central Places of Southern Germany*. Englewood Cliffs, N.J.: Prentice-Hall, 1966.

Cochran, Willard W., and Mary E. Ryan. *American Farm Policy 1948–73*. Minneapolis: University of Minnesota Press, 1976.

Cromley, Robert G. "The Von Thünen Model and Environmental Uncertainty." *Annals of the Association of American Geographers* 72 (September 1982): 404–10.

Dahlberg, Kenneth A., ed. *New Directions for Agriculture and Agricultural Research: Neglected Dimensions and Emerg-*

ing Alternatives. Totowa, N.J.: Rowman and Allanheld, 1986.

Datoo, B. A. "Toward a Reformation of Boserup's Theory of Agricultural Change." *Economic Geography* 54 (1978).

DeBlij, Harm J. *A Geography of Viticulture.* Miami: Miami University Geographical Society, 1981.

De Lisle, David deGaris. "Effects of Distance Internal to the Farm: A Challenging Subject for North American Geographers." *The Professional Geographer* 3 (August 1978): 278–87.

Dickinson, J. C., III. "Alternative to Monoculture in the Humid Tropics of Latin America." *The Professional Geographer* 24 (1972): 217–32.

Dickinson, Robert E. "Rural Settlements in the German Lands." *Annals of the Association of American Geographers* 39 (December 1949): 239–63.

Doolittle, William E. (with B. L. Turner, II). "The Concept and Measure of Agricultural Intensity." *The Professional Geographer* 3 (August 1978): 297–301.

Ebeling, Walter. *The Fruited Plain: The Story of American Agriculture.* Berkeley, Calif.: University of California Press, 1979.

Elbow, Gary S. "Determinants of Land Use Change in Guatemalan Secondary Urban Centers." *The Professional Geographer* 1 (February 1983): 57–65.

Fernea, Elizabeth Warnock. *Guests of the Sheik: An Ethnography of an Iraqi Village.* Garden City, N.Y.: Doubleday, 1969.

Furuseth, Owen J., and John T. Pierce. *Agricultural Land in an Urban Society.* Washington, D.C.: Association of American Geographers, 1982.

Geertz, Clifford. *Agricultural Involution: The Process of Ecological Change.* Berkeley, Calif.: University of California Press, 1963.

Getis, Arthur. "Second-Order Analysis of Point Patterns: The Case of Chicago as a Multi-center Urban Region." *The Professional Geographer* 1 (February 1983): 73–80.

·regor, Howard F. *Industrialization of U.S. Agriculture; An Interpretative Atlas.* Boulder Colo.: Westview Press, 1982.

·rigg, David B. *An Introduction to Agricultural Geography.* London: Hutchinson Education, 1984.

———. *The Agricultural Systems of the World: An Evolutionary Approach.* London: Cambridge University Press, 1974.

———. *Population Growth and Agrarian Change: An Historical Perspective.* Cambridge, England: Cambridge University Press, 1980.

Hart, John Fraser. "Change in the Corn Belt." *Geographical Review* 76 (January 1986): 51–73.

———. "The Demise of King Cotton." *Annals of the Association of American Geographers* 3 (September 1977): 307–22.

———. "Land Use Change in Piedmont County." *Annals of the Association of American Geographers* (December 1980): 492–527.

Heathcote, R. L. *The Arid Lands: Their Use and Abuse.* London: Longman, 1983.

Hewes, Leslie, and Christian I. Jung. "Early Fencing on the Middle Western Prairie." *Annals of the Association of American Geographers* 71 (June 1981): 177–201.

Hill, R. D. *Agriculture in the Malaysian Region.* Budapest: Akademiai Kaido, 1982.

Hudson, John C. *Plains Country Towns.* Minneapolis: University of Minnesota Press, 1985.

Hudson, Tim. "A Geography of Cocaine." *Focus* 35 (January 1985): 22–29.

Ilbery, Brian W. *Agricultural Geography: A Social and Economic Analysis.* New York: 1985.

Jackle, John A. "Roadside Restaurants and Place-Product-Packaging." *Journal of Cultural Geography* 3 (1982): 76–93.

Johnson, Hildegard B. *Order upon the Land.* New York: Oxford University Press, 1976.

Johnson, Warren, Victor Stolzfus, and Peter Craumer. "Energy Conservation in Amish Agriculture." *Science* 198 (October 28, 1977): 373–78.

Jordan, Terry G. *Texas Log Buildings: A Folk Architecture.* Austin: University of Texas Press, 1978.

Kniffen, Fred B. "Fold-Housing: Key to Diffusion." *Annals of the Association of American Geographers* 55 (December 1965): 549–77.

Konrad, Victor A., and Michael Chaney. "Madawaska Twin Barn." *Journal of Cultural Geography* 3 (Fall-Winter 1982): 64–75.

Kunze, Donald. *Thought and Place: The Architecture of Eternal Places in the Philosophy of Giambattista Vico.* New York: Peter Lang, 1987.

Lamme, Ary J., III, ed. *North American Culture,* vol. 1. Stillwater, Okla.: Society for the North American Cultural Survey, 1984.

Lawrence, Henry W. "Changes in Agricultural Production in Metropolitan Areas." *The Professional Geographer* 40 (May 1988): 159–74.

Lewis, Pierce F., Yi-Fu Tuan, and David Lowenthal. *Visual Blight in America.* Washington, D.C.: Association of American Geographers, 1973.

Lewis, Thomas R. "To Planters of Moderate Means: The Cottage as a Dominant Folk House in Connecticut before 1900." *Proceedings, New England–St. Lawrence Valley Geographical Society* 10 (1980): 23–27.

Lewthwaite, Gordon R. "New Zealand Milk and the Map." *Annals of the Association of American Geographers* (December 1980): 475–91.

Lonsdale, Richard E., and John H. Holmes, eds. *Settlement Systems in Sparsely Populated Regions.* New York: Pergamon Press, 1981.

Maos, Jacob O. *The Spatial Organization of New Land Settlement in Latin America.* Boulder, Colo.: Westview Press, 1984.

Meining, D. W., ed. *The Interpretation of Ordinary Landscapes.* New York: Oxford University Press, 1979.

Morrison, Peter A., and Judith P. Wheeler. "Rural Renaissance in America?" *Population Reference Bureau* 31 (October 1976).

Nelson, Robert H. *Zoning and Property Rights: An Analysis of the American System of Land-Use Regulations.* Cambridge, Mass.: The MIT Press, 1980.

Norton, William. "Agriculture Evolution on the Frontier." *The Professional Geographer* 1 (February 1982): 64–73.

Penning-Rowsell, Edmund C., and David Lowenthal, eds. *Landscape, Meanings and Values.* London: Allen and Unwin, 1986.

Perelman, Michael, *Farming for Profit in a Hungry World.* Montclair, N.J.: Allanheld, Osmun, 1977.

Platt, Rutherford H. "The Farmland Conversion Debate: NALS and Beyond." *The Professional Geographer* 4 (November 1985): 433–42.

Rapoport, Amos, *House Form and Culture.* Englewood Cliffs, N.J.: Prentice-Hall, 1969.

Rappaport, Roy A. "The Flow of Energy in an Agricultural System." *Scientific American* 224 (September 1971): 117–32.

Reed, Michael, ed. *Discovering Past Landscapes.* London: Croom Helm, 1986.

Roet, Jeffrey B. "Land Quality and Land Alienation on the Dry Farming Frontier." *The Professional Geographer* 2 (May 1985): 173–83.

Rooney, John F., Jr., and Paul L. Butt. "Beer, Bourbon and Boone's Farm: A Geographical Examination of Alcoholic Drink in the United States." *Journal of Popular Culture* 11 (Spring 1978): 832–56.

Rooney, John F., Jr., Wilbur Zelinsky, and Dean R. Louder, eds. *This Remarkable Continent: An Atlas of United States and Canadian Society and Culture.* College Station, Tex.: Texas A and M University Press for the Society for the North American Cultural Survey, 1982.

Rowntree, Lester B., and Conkey, M. W. "Symbolism and the Cultural Landscape." *Annals of the Association of American Geographers* 70 (December 1980): 459–74.

Salih, Tayeb. *Season of Migration to the North.* Translated by Deny Johnson-Davies. London: Heinemann; Washington, D.C.: Three Continents, 1969.

Sauer, Carl O. "Morphology of Landscapes." *University of California Publications in Geography* 2 (1925): 19–54.

Scofield, Edna. "The Origin of Settlement Patterns in Rural New England." *Geographical Review* 28 (October 1938): 652–63.

Shortridge, James R. "The Emergence of 'Middle West' as an American Regional Label." *Annals of the Association of American Geographers* 74 (June 1984): 209–20.

_____. "The Vernacular Middle West." *Annals of the Association of American Geographers* 75 (March 1985): 48–57.

Smil, Vaclav, Paul Nachmand, and Thomas V. Long. *Energy Analysis and Agriculture: An Application to U.S. Corn Production.* Boulder, Colo.: Westview Press, 1983.

Smith, Everett G., Jr. "America's Richest Farms and Ranches." *Annals of the Association of American Geographers* 70 (December 1980): 528–41.

_____. "Anthrosols and Human Carrying Capacity in Amazonia." *Annals of the Association of American Geographers* 70 (December 1980): 528–41.

Spencer, Joseph E., and R. J. Horvath. "How Does an Agricultural Region Originate?" *Annals of the Association of American Geographers* 53 (1963): 74–92.

Symons, Leslie. *Agricultural Geography,* rev. ed. London: G. Bell, 1979.

Tarrant, John R. *Agricultural Geography.* New York: John Wiley & Sons, 1974.

Trewartha, Glen T. "Types of Rural Settlements in North America." *Geographical Review* 36 (October 1946): 568–96.

Turner, B. L., II, Robert Q. Hanham, and Anthony V. Protararo. "Population Pressure and Agricultural Intensity." *Annals of the Association of American Geographers* 67 (September 1977): 384–96.

Uhlig, Harald and Cay Kienau, eds. *Types of Field Patterns, Basic Material for the Terminology of the Agricultural Landscape, vol. 1.* Giessen, West Germany: W. Schmitz, 1967.

Vogeler, Ingolf. *The Myth of the Family Farm: Agribusiness Dominance of United States Agriculture.* Boulder, Colo.: Westview Press, 1981.

Wessel, Thomas R., ed. *Agriculture in the Great Plains, 1876–1936.* Washington, D.C.: The Agricultural History Society, 1977.

Wacker, Peter O. "Folk Architecture as an Indicator of Culture Areas and Culture Diffusion: Dutch Barns and Barracks in New Jersey." *Pioneer America* 5 (July 1973): 37–47.

Yelling, J. A., *Common Field and Enclosure in England 1450–1850.* London: Macmillan, 1977.

Zelinsky, Wilbur. "North America's Vernacular Regions." *Annals of the Association of American Geographers* 70 (March 1980): 1–16.

Cultural Landscapes in Urban Regions

The idea of the town is a familiar one; the world's big towns have a personality that is sometimes embodied in a single monument or in some architectural group which is famous all over the globe. . . .
AIME VINCENT PERPILLON, 1977

THEMES

Landscapes of the industrialized world
Development and growth of cities
City landscapes of the world
The morphology of cities
Urbanization in the traditional less-
    industrialized world.

CONCEPTS

Transportation landscapes
Mining landscapes
Raubwirtschaft
Point settlements
Suburbs
City morphology
Innovation theory
Hydraulic model
Stimulus-response theory
Ziggurat
Greek cities
Roman cities
Cathedral cities
Industrial cities
Hinterland
Central place
Threshold
Christaller model
Urbanization
Concentric zone model
Sector model
Multiple nuclei model
Urban village
Sunbelt
Urban flight
Gentrification
Primate city
Cultural convergence

237

The agricultural landscape of the less industrialized world is characterized by low population density, but the landscape of urban regions (first developed in the modern industrial world) is characterized by high population density. Population in the industrial landscape is concentrated because activities there require constant interaction. Residents of rural landscapes may not need to interact daily with other individuals or groups, but urban residents in both the modern industrial world and the less industrialized, traditional world must do so. They must come into contact to transact business, accumulate raw materials for manufacturing, distribute manufactured products, and provide medical, educational, and other services to large numbers of people. Consequently, the cultural landscape inhabited by the majority of the residents of the modern world and a significant minority of the traditional world is one dominated by large towns and cities (Table 7–1).

The landscape elements of the modern industrial landscape are related to the industrial system. In particular, improved transportation systems developed as humans attempted to increase the speed and frequency of interaction between places. *Transportation landscapes* include a variety of features, beginning with canals developed in the English countryside in the late eighteenth and early nineteenth centuries to allow transportation of large volumes of raw materials and

Paris, France. Elements of the landscape of some places are so distinctive that the city is easily recognizable.

manufactured goods. Railroad lines, stations, and yards eventually supplemented or replaced the canals and their related features in the transportation landscape. The railroads gave way in turn to the automobile with its highway system. The final element of the transportation landscape is the concentration of human activity at points where transportation routes intersect. The ability to concentrate people in one place, to feed and house them, and to provide them with jobs is an outgrowth of the Industrial Revolution. The increasing pace of transportation change associated with the nineteenth and twentieth centuries accelerated the growth of cities, the focus of the transportation landscape elements.

Landscapes of the Industrial World

Mining Landscapes

Examination of the landscape elements of locations and regions with high population densities

TABLE 7–1
Urban Population as a Percentage of Total Population by Major World Region*

REGION	1988	1980	1970	1960	1950
North America	74	74	70	67	64
Europe	75	69	64	58	54
Soviet Union	65	65	57	49	39
Latin America	68	65	57	49	41
Asia	36	29	25	22	17
Africa	30	28	23	18	15
World total	45	41	38	34	29

* Definition of urban varies from region to region, but generally refers to places of 2500 to 25000 inhabitants.

Source: Patterns of Urban and Rural Population Growth (New York: United Nations, Population Studies No. 68, 1980); *1988 World Population Data Sheet* (Population Reference Bureau, Inc., 1988).

indicates the distinctive characteristics of industrial regions' landscapes. The first of the landscape characteristics of the modern industrial world reflects its heavy reliance upon natural resources. Mines, steel mills, factories, and refineries typify the modification of the natural geographic elements for human use. The first of these landscape elements to appear on a large scale in the industrial world was the mine. Mining originated in the Bronze Age, but the scale and nature of mining activities in the industrial world is significantly different. Watt's invention of the steam engine provided the means of bringing larger quantities of ore to the surface, digging deeper by pumping water from underground mines, and moving larger volumes of ore to refineries and steel mills. Watt's engine also made it possible to process large quantities of iron through mechanical rather than manual means. The ability to use larger quantities of iron ore affected coal mines as well. Coal was needed both for the steam engines and for the larger iron and steel processing centers. The demand for iron ore and coal after the Industrial Revolution fostered the emergence of the mining landscape as a distinctive feature of the industrial world.

The Bessemer converter (1856) and the open hearth furnace (1857) greatly increased the ability to produce steel, increasing in turn the demand for coal and iron ore, and causing a rapid expansion of the *mining landscape* of Europe. As the iron industry changed from a series of small, dispersed mills to a few large centers, mining activities transformed the landscape of areas that had coal and iron ore resources. Mine shafts and associated structures, mountains of useless earth or gravel, and the mining communities themselves became important elements in the mining landscape. Because steel production is dependent on coal (approximately two tons of coal to one of iron ore), the iron and steel mills of the United Kingdom also concentrated near the coal mines, creating additional elements in mining landscapes in that country. The steel mills in other industrial countries are in industrial cities farther from the mines because of better transport systems. In either case, the resulting mining landscape is one of the distinctive features of the industrial world.

Resource demands in the industrial world also affect the landscape of the rest of the world. Exhaustion of high-quality ores in the industrial world prompted mining around the world. Expansion of the Industrial Revolution's technology from iron and steel to other sectors of the economy stimulated demands for ever larger quantities of copper, bauxite (the ore for aluminum), manganese (an alloy of steel), and other minerals. Extraction of minerals creates such a distinctive landscape that Germans have coined a term for the economy that produces it, *Raubwirtschaft*, literally robber economy. The mining landscapes of the world can be classified into three types:

1. The old mining landscape of the early industrial countries such as the United Kingdom.
2. The mining landscape associated with more recent industrialization in countries such as the United States and Australia; it is characterized by large-scale strip mining over extensive tracts of land.
3. *Point settlements* related to extracting resources in isolated regions of low population in the industrial world (e.g., Alaska), or point settlements in the less industrial world established by the modern world to extract resources.

Mining landscapes are very visible signs of the technology and associated economy of the indus-

Mining landscapes are often one of the destructive features of the industrial world as in this corn field being strip mined for coal in the American Midwest.

trial world. They are often regarded as the end product of the human capacity to destroy the natural environment. Many mining landscapes show severe environmental degradation, especially where mining activity occurred before the adoption of environmental regulations. Relic landscape features such as ore piles, overburden dumps, or abandoned pits and shafts often make mining landscapes gloomy and ugly. Some mining landscapes associated with point settlements in mountainous areas have attempted to turn these relic landscapes into tourist attractions. Towns like Park City, Utah, no longer actively mine gold and silver, but development of ski resorts on the surrounding slopes has grafted a new element over the old mining landscape. Landscapes of active mining areas continue to exhibit the traditional elements of the mining landscape, including shafts or open pits, ore and overburden piles, and mining-related structures. Modern mining landscapes are widespread; they include the huge strip mining activities for coal in the United States; the open pit copper mines of Utah, Nevada, New Mexico, and Arizona; the diamond mines of South Africa; and the new tin and gold mines of Brazil's tropical rain forest. Each mining landscape includes settlements, mines, and related landscape elements reflecting the destructive aspects of the cultural ecology of mining.

Transportation Landscapes

Transportation is a second element that characterizes the landscape of the modern industrial world. The first major landscape change associated with transportation in the industrial world came in 1759 when Francis Eagerton hired James Brindley to build a canal (completed in 1761) between Manchester and Worsley in the United Kingdom. This canal initiated a canal boom that affected the United States, the United Kingdom, Western Europe, and portions of Russia. Canals and associated river dredging were soon extended inland from the coast, allowing ships to travel far from the sea (Figure 7–1). The necessity of maintaining the canals as level waterways required complex lock structures to raise ships from one elevation to another. The engineering achievements associated with the canal age are reflected

Some mining landscapes have been reclaimed and converted to other uses. Park City, Utah, a silver and gold mining boom town in the Rocky Mountains, has become an international ski resort. Remnants of the old mining landscape are visible in the small homes in the left foreground.

today in relic landscape features such as the canals themselves, the tow paths alongside the canals, and locks or lock remnants. Today the relic landscape features associated with canals are sometimes used for a completely different function; tow paths are used for jogging or bicycle paths, for example. Canals in Western Europe have been enlarged and modernized and remain an important landscape feature. Overall, however, the canal boom was short-lived, as the railroad provided a more efficient means of transportation after the first decades of the nineteenth century.

The railroad combined Watt's steam engine and an iron (later steel) rail to provide an effective and relatively inexpensive means of moving people and goods. Whereas canals could move higher volumes and weights of goods than wagons, the pace was very slow. The railroads provided both speed and bulk transport capability. The first steam locomotive ran on rails in 1812 in England. The first public railway was opened in the north of England between Stockton and Darlington in 1821. The advantages of the railroad were evident after a widely publicized race in 1829 between a horse and a train on the Liverpool and Manchester railway in which the locomotive averaged 24

FIGURE 7–1 Rivers and Canals of Europe Early canals in Europe connected the natural waterways, creating an important landscape element.

miles per hour. The success of railroads in England led to their widespread adoption across the industrial world. The first rail lines were built in the United States in 1830 and the first on the European continent shortly after (Figure 7–2). Between 1830 and 1900, the rail lines spread like a web over the industrial world, playing a central role in the opening and development of the great interior plains regions (Figure 7–3). Although railroads are less important in passenger transport today than fifty years ago, they still move the majority of global land freight. The significance of the railroad as a landscape feature in the industrial world is illustrated by the fact that no part of the densely inhabited industrial southern half of

the island of Great Britain is more than five miles from a railroad.

Specific landscape features associated with railroads are the tracks, railyards, loading and unloading docks, bridges, and trestles. Increased weight combined with topographic requirements (level or nearly level) prompted construction of numerous bridges to meet railroad needs. The diffusion of railroad towns to maintain tracks, provide services for passengers, maintain engines and other equipment, and accumulate freight from the surrounding countryside represents another landscape feature introduced by the railroad. In time many of the early railroad towns (Omaha, Albuquerque, Cheyenne) outgrew their

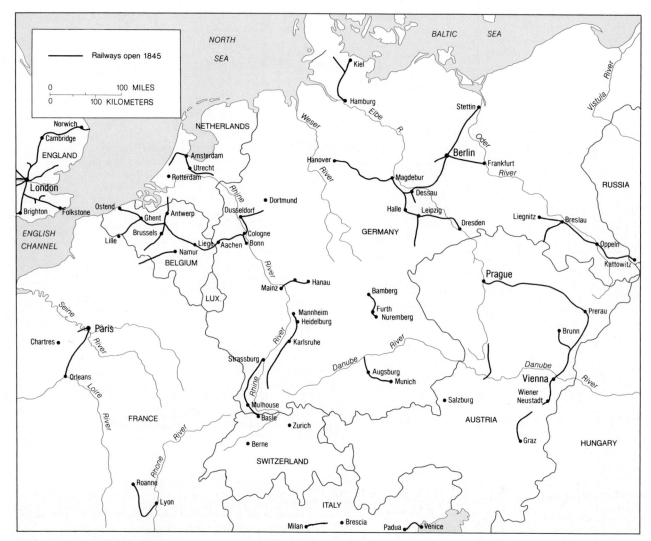

FIGURE 7–2 Early Railroad Network of Central Europe Construction of railroads in Europe in the nineteenth century added a new element to the cultural landscape.

strictly railroad function. Thus railroads fostered urban growth through concentrating goods and peoples, ultimately leading to the paramount landscape feature of densely populated regions, large cities.

The highway systems of the industrial world, another transport element of the landscape, have also contributed to urban growth. The term "highway" comes from the need for early roads to be built above the level of swamps, marshes, or other barriers to travel.

Modern highways, which are designed to move people and goods rapidly, date from the development of *macadam* in England in the late nineteenth century. Macadam was used in building the first level, permanent road surfaces. Macadam roads became more important as experimentation in Europe and the United States led to the internal combustion engine. Combining the internal combustion engine with a buggy created the first motorized vehicle capable of traveling independently of tracks. Advances in automobile technol-

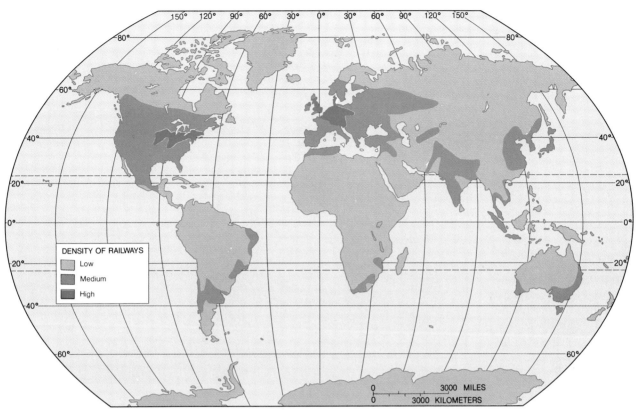

FIGURE 7–3 Density of Railroad Networks For much of the less-industrialized world the presence of a railroad in a landscape is an exception.

ogy in the United States and Western Europe in the 1890s and early 1900s culminated in the mass production of automobiles and trucks.

Travel by automobile in the early decades of the twentieth century tended to be highly problematic, primarily because of the absence of all-weather roads. Not until the 1930s did the United States undertake a construction program to provide permanent, reliable highways across the country, which enabled automobiles and trucks to travel beyond the major thoroughfares within cities. Advances in highway technology in Europe—especially Germany—in the 1930s led to the concept of the Autobahn. The Autobahn is a wide, carefully engineered road that allows for high-speed travel of vehicular traffic. After World War II, the concept of the Autobahn diffused to the United States where it became the basis for the freeway or interstate system. Diffusion of

multilane, high-speed transportation networks to Western Europe, Japan, and Australia reflect an important element in the modern landscape. The present highway system in any large city is a major consumer of space and is one of the most obvious elements of the landscape (Figure 7–4).

The development of transportation systems prompted modifications in other landscape features. The development and diffusion of the railroad and iron and steel industries created the first large cities of the industrial world. The use of railroads to transport city residents (by trolleys or commuter trains) allowed both continued urban growth and suburbanization. Suburban expansion paralleled the railway lines and limited the extent of a city's expansion to those sites easily accessible by rail (Figure 7–5).

The first impact of the highway system was to allow the expansion of cities away from the rail

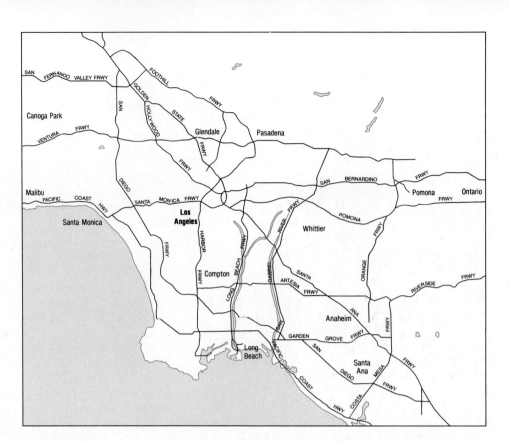

FIGURE 7–4 Freeways in Los Angeles and Its Vicinity The multi-lane, high speed highway is a landscape element of the twentieth century.

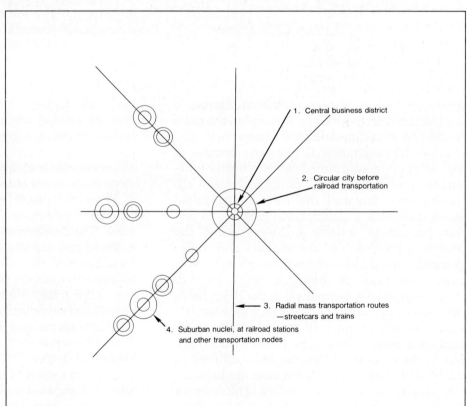

1. Central business district

2. Circular city before railroad transportation

3. Radial mass transportation routes —streetcars and trains

4. Suburban nuclei, at railroad stations and other transportation nodes

FIGURE 7–5 Early Suburbs in the United States Early suburbs developed near rail lines which provided access to the central city.

lines. The adoption of the automobile, trolley, bus, and other motor vehicles hastened the process of suburbanization, causing major changes in the morphology of cities.

Closely related to changes in transportation were changes in the size and use of buildings. Initially, the railroads' ability to move large quantities of people or products concentrated commercial and manufacturing activity at rail junctions, prompting construction of large factories, mercantile centers, and office buildings. The subsequent development of the automobile and highway systems freed commercial and industrial activities from the necessity of being concentrated in the single point in a community where rail systems came together. The resulting diffusion of activities changed the urban landscape of industrial cities as newer commercial and industrial activities expanded away from the focal point at the rail junctions, or older activities relocated, reducing the dominance of the old central core.

The railroad and later the automobile also allowed thousands of people to congregate at a central point for commercial activities in the daytime and return to their homes in the suburbs at the end of the workday. The best example of this in modern times is found in the commercial core of large cities such as New York or London. Nearly two million people in New York come into Manhattan each day for work, and return in the evening to suburban homes. This heavy concentration of people and activities has prompted the vertical landscape of large cities such as New York.

A less visible feature of the transportation landscape is also related to the concentration of people and activities in the large cities of the industrial world. The transmission of words and ideas (first by telegraph and then by telephone) continues to affect population distribution. Rapid developments in the technology of electronic transmittal of messages and the ability to link computer systems are allowing greater diffusion from the center of the city and increasing the importance of suburbs as workplaces at the expense of the central city. Proximity to such activities as the stock market, publishing houses, or other commercial functions was once mandatory,

The urban landscape of New York City epitomizes the high density of large urban centers. The high rise buildings that are the stereotype of New York City result from the concentration of activities and people in the downtown of such centers.

but is now rendered less important by the interconnection of computer systems by telephone lines. This growth of the suburbs is thus another illustration of how the urban regions of the modern industrial world have changed over time in response to improved technology and transportation.

Cities as Landscape Features

The changes in transportation and technology that began with the Industrial Revolution in England have concentrated the population of the industrial world in urban areas. Approximately three-fourths of the inhabitants of the industrialized world live in cities, compared with fewer than one-third of the inhabitants of the less industrialized, traditional world. In both the modern and traditional worlds, however, the proportion of the population living in cities varies widely from country to country.

Despite these variations, cities are increasingly important landscape features. The emergence of urban centers as the dominant landscape feature of the modern industrial world is a phenomenon of the last century. Only 2 percent of the world's population lived in cities of 100,000 or more as late as 1850 (Table 7–2). Today, more than half of the population of the United States and over 20 percent of the global population live in cities of more than 100,000.

The recent emergence of cities as a landscape feature is demonstrated if we condense all human

TABLE 7–2
Cities of Europe, 1850

France		**Holland**	
Paris	1,314,000	Amsterdam	225,000
Lyon	254,000	Rotterdam	111,000
Marseilles	193,000		
Bordeaux	142,000	**Scandinavia**	
Rouen	104,000	Copenhagen	135,000
Toulon	68,000		
		Italian States	
Britain		Naples	416,000
London	2,320,000	Palermo	182,000
Liverpool	422,000	Rome	170,000
Manchester	404,000	Turin	138,000
Glasgow	346,000	Florence	107,000
Birmingham	294,000	Genoa	103,000
Dublin	263,000		
Edinburgh	193,000	**Germany (Prussia)**	
Leeds	184,000	Berlin	446,000
Bristol	150,000	Hamburg	193,000
Sheffield	141,000	Munich	125,000
Wolverhampton	112,000	Breslau	114,000
Newcastle	111,000		
Plymouth	100,000	**Austria**	
Bradford	100,000	Vienna	426,000
		Milan	193,000
Spain		Budapest	156,000
Madrid	263,000	Venice	141,000
Barcelona	167,000	Prague	117,000
Valencia	110,000		
Seville	106,000	**Russia**	
		St. Petersburg	502,000
Portugal		Moscow	373,000
Lisbon	259,000	Warsaw	163,000
Belgium			
Brussels	208,000		
Ghent	108,000		

Source: Studies in Population.

experience into a single 24-hour day. The beginning of large-scale urban life would not have occurred until a few minutes before midnight on such a time scale. If we extend our definition of urban places to include the earliest settlements containing only a few hundred people that developed after the Neolithic Revolution, the total time in which people have lived in such settlements would occupy only the last half hour of our day. The factors that have affected urban growth help to explain the impact of cities on the landscape. The earliest permanent settlements were villages that developed after the Neolithic Revolution. The emergence of the first permanent settlements (such as Jarmo and Jericho in the Fertile Crescent) reflected three changes: food surplus, labor specialization and social stratification. The earliest settlements were small and emphasized agriculture or agriculture-related activities. Transportation was by foot (either human or animal), limiting both the number of people the village could support and its size. Archaeological research indicates Jarmo had only twenty-five permanent dwellings clustered around grain storage facilities.

Organized along the lines of family relationships, early settlements did not necessarily grow into towns or cities. Without social stratification to ensure that agricultural surpluses were taken from producers and redistributed to the elite, they simply remained small farming villages. The emergence of true towns, with populations of several thousand people, occurred only when the three factors of surplus, stratification, and labor specialization were fully developed. The function of towns and cities differs by definition from that of the agricultural village. Towns and cities are primarily inhabited by people whose occupations do not in general include agriculture. Trade, manufacturing, finance, business, and other occupations create a settlement whose scenery, landscape, and life-styles are distinct from the village with its emphasis on agriculture.

Examination of the distribution of the earliest towns reveals their correlation to the hearths where agriculture first developed (Figure 7–6). Agriculture remained the primary economic activity in the regions around these early cities, but cities emerged in response to the need to administer empires, control trade, or protect centralized leadership. The causes of the change from a landscape dominated by agricultural villages to a landscape with cities are complex, but several theories attempt to explain the transformation.

One theory explaining the emergence of cities is the *innovation theory*, which hypothesizes that innovations occur only at locations that have a high potential for interaction with other places. The innovation theory maintains that some sites, in the center of hearths of domestication, were better located to receive ideas diffused from elsewhere and thus able to become centers of innovation. The adoption of an innovation, such as irrigation or a better tool for farming, allowed a particular village to grow as it produced a greater food surplus. Increased surplus and control of the innovation allowed such centers to increase their degree of labor specialization as some individuals benefited from trading or selling new tools or agricultural surplus to others. According to this theory, the resultant trade and related population growth prompted the emergence of early urban centers in Mesopotamia where some of the earliest agriculture occurred in the foothills of the Fertile Crescent. These early cities, according to the innovation theory, were the sites for innovations, such as the development of irrigation. Innovations like irrigation allowed some families or groups to move to the lower river plains where they had a more reliable source of water and more fertile soil; this allowed greater agricultural production, fostering in turn greater surplus and new cities that were even larger.

A variant of the innovation theory of city growth is the *hydraulic model*, which hypothesizes that the growth of cities did not occur until irrigation was invented. This allowed settlements to move to the river plains where water was diverted to increase agricultural production and food surpluses. To maintain the sophisticated irrigation system required to sustain agricultural production, a governing body was necessary, resulting in political organization and stratification of power in irrigation-based societies.

According to the hydraulic theory, the expansion of irrigation systems led to an ever greater administrative role for cities reliant on irrigation, thus creating the first true cities of the world and

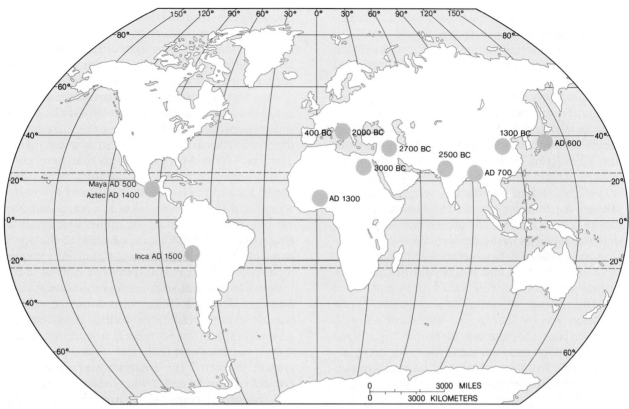

FIGURE 7–6 Areas of Early Urbanization The great civilizations of the past had developed important urban networks in their core areas by the dates indicated.

the emergence of "hydraulic civilizations." However, although it is true that early cities were associated with hydraulic agriculture in the Middle East and China, large cities were also founded in the Americas without the benefit of large-scale irrigation systems.

To explain the emergence of nonhydraulic cities, some observers suggest a *stimulus-response theory*, which states that cities were a response to environmental or human stress. The stimulus-response theory accepts that early domestication occurred in the foothills of the Fertile Crescent and elsewhere during a period of milder and moister climate. Subsequent increases in temperature and aridity as the world entered the interglacial period of the last 10,000 years forced early farmers to migrate to a few sites where they could divert water to ensure reliable crop production.

Environmental change prompted migration, but additional stress occurred as competition

among groups for the limited sites suitable for easy irrigation led to conflict. To protect themselves, farmers were forced into communities for defensive purposes. The organization required for defense marked the beginning of more complex social organizations in cities.

Limited climatological data support the hypothesis that temperatures increased and precipitation decreased in the early hearths for the mid-latitudes during the last 10,000 years. The climatological record is limited, however, and drawing cause and effect conclusions from such data is highly speculative. The stimulus-response theory, postulating cities emerging as defensive centers, is not supported by the evidence (particularly from the Americas), which indicates that early cities played much greater roles as religious or administrative centers than as defensive locations. Cities often included a defensive component in the form of a fortification erected to protect at least some residents in times of conflict, but defense does *not* seem to have been the basis for the establishment of most of the early large towns and cities.

Origin and Diffusion of Cities

The earliest well-documented city is Ur in Mesopotamia (Figure 7–7). Archaeologists have unearthed ruins in Ur dating from 5000 B.C. Approximately one square mile in area, its population never exceeded thirty to forty thousand. The landscapes of the early cities of Mesopotamia were apparently similar in that the major part of the city was enclosed by a wall both for protection and to distinguish it from surrounding rural areas. Within the wall were three distinct sections: the central administrative area, the dwellings of the elite inhabitants, and the other residential sections of the city, which included homes, markets, and workshops of artisans.

The core of the ancient Mesopotamian city focused on the temple and palace. Since religion was the basis for power in the Mesopotamian city, the palace was located next to the temple. The temple in the Mesopotamian city was a pyramid-like structure made of sun-baked bricks (adobes), known as a *ziggurat* (Figure 7–8). Adjacent to the palace and the temple were granaries for storing food. Food was the basis for political power, and the combination of temple, palace, and granaries is often described as the *citadel* of the city. Around the citadel proper were the homes of the wealthy, which were enclosed by a second wall to segregate them from the rest of the city. Paved streets, running water, waste disposal, and other amenities were provided for the residents of this neighborhood.

Most of the people living in early Mesopotamian cities lived around the citadel in lower class neighborhoods. Markets and artisans were found in the squares, surrounded by the two- or three-story, sun-dried brick dwellings of the masses. Population density was high, with each dwelling unit containing at most three to five rooms for all the activities of an extended family. There were no waste disposal systems, and streets were littered with waste materials, which slowly raised the street level as they accumulated. Disease was common, and life expectancy was short. There was no central business district aside from the market squares in the Mesopotamian city, and people lived and worked in the same quarter. Individuals who needed to see one another during the course of the day, for trade or because of their labor specialization, lived in the same quarters, creating distinctive sections of the city. Laborers who worked on farmlands around the city lived closest to the wall in the poorest housing (Figure 7–9).

The cultural landscape of these early Mesopotamian cities was dominated by the temple, palace, and other elements of the citadel. The crowded residential quarters covered the largest area, but the low structures simply seemed to rise

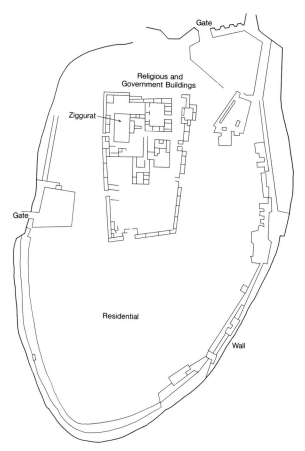

FIGURE 7–7 Plan of Ur This map of the early city of Ur illustrates the general pattern of activity dominated by religion and government activities.

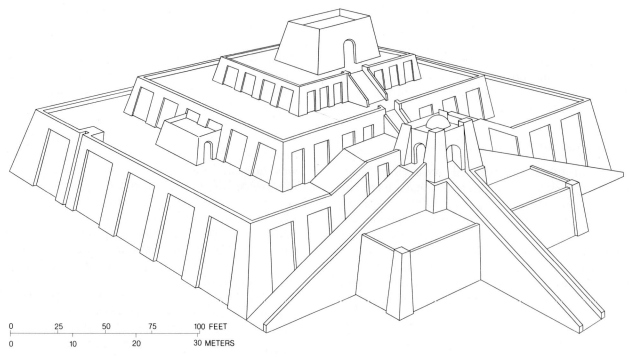

FIGURE 7–8 A Ziggurat The ziggurat was a dominant element in early cities of Mesopotamia.

from the plains on which they were located. The cultural ecology of these early centers revolved around their impact on the agricultural base that supported them. The growing cities prompted expansion of fields and irrigation systems. The accumulation of wastes in the streets and the deterioration of the adobe structures over time eventually caused each city to occupy a mound. These mounds persisted in the landscape long after the cities were gone as silent relics of an earlier culture's use of the earth.

Cities in other locations were similar to those found in the Mesopotamian hearth. Each consisted of a religious administrative center (temple or other structure) and a palace for the ruling class separated from the balance of the residents of the city by some type of wall or major demarcation. Unlike Mesopotamian cities, however, cities in the Nile valley and in the Americas were not walled. The cities of the Americas also had much larger monumental religious centers, such as those in Teotihuacan in Mexico or Tikal, a vast city in the Guatemalan portion of the Yucatán Peninsula of Central America. These cities may have had over 100,000 inhabitants and were characterized by massive temples and official build-

Tikal, Guatemala is the site of remnants of the classic Mayan civilization. The imposing stone structures illustrate the level of urban development in this region nearly two millennia ago.

250

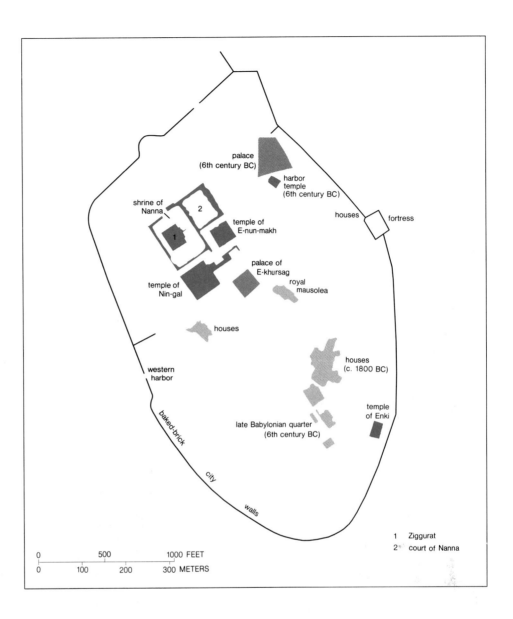

FIGURE 7–9 Districts of a Mesopotamian City

ings. In Teotihuacan, the Temple of the Sun rises to a height of 215 feet (66 meters). The base of the largest structure in Tikal covers 58,500 square feet (18,000 square meters) and rises eighteen stories. These massive structures were built without apparent use of metal tools, an indication of the technological sophistication of the builders who created a distinctive urban landscape (Figure 7–10).

Construction dates of cities in these various locations suggest that there *may* have been diffusion of the idea of a city as a settlement from the

Mesopotamian hearth to other areas. The early American cities date from only a few centuries before Christ, nearly 3,000 years after the earliest cities of Mesopotamia. The evidence is far from conclusive, but the similarity in organization, landscape features, and function of the early cities suggest a possibility of diffusion from a central hearth. Although it is impossible to do more than speculate that the city concept and related technology of the Americas in fact diffused with migrants across the ocean from the Middle East, most observers conclude that the earliest cities of

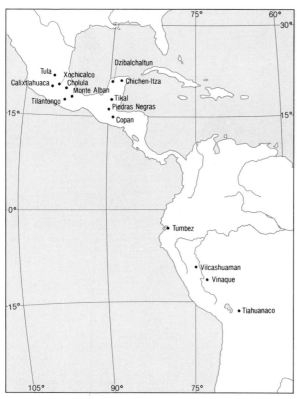

FIGURE 7–10 Early Cities in the Americas Early cities developed in Central America and the Andean highlands.

Egypt, Turkey, Iran, Pakistan, India, and China were based in part on diffusion from the earliest cities in the Fertile Crescent.

The early cities of Mesopotamia (Ur by 3000 B.C., Babylon by 2000 B.C.) were soon followed by other cities (see Figure 7–6 on page 248). These include Knossos on the island of Crete (by 2500 B.C.); Mycenae on mainland Greece (by 1600 B.C.); Troy in what is now Turkey (by 1600 B.C.); Memphis (about 2500 B.C.) and Thebes (by 2000 B.C.) in the Nile valley; Mohenjo-daro in the Indus River valley (2500 B.C.); and those in the Huang-Ho River plain of China (by 1500 B.C.). Eventually, the advances made in urban life in the early cities of the Middle East diffused around the Mediterranean Sea.

Preindustrial Cities of Europe

The first cities to develop in Europe (as opposed to North Africa and the Middle East) date from approximately 1600 to 2500 B.C. in the eastern Mediterranean (Knossos, Troy, and Mycenae). The cities of the Mediterranean were trading centers, reflecting improvements in the technology of shipbuilding that increased trade around the Mediterranean Sea. Early civilizations had founded cities such as Ur and Babylon (the Sumerians), Tyre and Carthage (the Phoenicians), and Memphis and Thebes (the Egyptians) at suitable locations for trade, defense, or political control. These cities remained for centuries, but the rapid increase in the *number* of cities was associated with colonization and trade in the Mediterranean after 2000 B.C. This spread of early towns in the Mediterranean can be divided into two broad periods: the Greek and the Roman.

Greek Cities

Cities spread from Mycenae in Greece in conjunction with growing Mediterranean trade as residents of one town established other trading settlements. New Greek towns were established rapidly, and by 600 B.C. there were over five hundred towns and cities on the Greek mainland and surrounding islands. The expansion of the Greek city throughout the Mediterranean was the basis for later cities of Western Europe where the Industrial Revolution occurred. The *Greek cities* were also hearths of innovation from which came many ideas central to Western civilization that ultimately diffused with European migrants to much of the world.

The ruins of the Agora at Philippi, Greece represent the remnants of the center of an ancient classical city in the eastern Mediterranean.

Athens was the most important of the Greek cities and is estimated to have had 300,000 inhabitants by 500 B.C. The core of a Greek city can be divided into two zones, the *Acropolis* and the *Agora.* The Acropolis was similar to the citadel of earlier Mesopotamian cities in that temples, government buildings, and the treasury for storing valuables were located there. The Agora was originally a place for public meetings, education, and social interaction in the democratic (for men only) society of many cities of Greece, such as Athens. Commercial activities were initially not allowed in the Agora, but in time it became the major marketplace of the city (Figure 7–11).

The landscape of the Greek city is dramatic if only the Acropolis is viewed. The monumental civic and religious structures associated with the Acropolis remain as cultural relics of great interest to historians and tourists today. The bulk of the city of Athens was much different. The streets were narrow, crowded, and unpaved, and the houses were small, crude structures.

Many of Athens' achievements in art, philosophy, technology, education, and politics diffused to the rest of the Mediterranean region and ultimately affected northern and western Europe as cities developed beyond the Mediterranean centuries later.

The colonization of the Mediterranean by Greek city states in the seven centuries before Christ included the establishment of Massilla (Marseilles) in France and settlements on Sicily

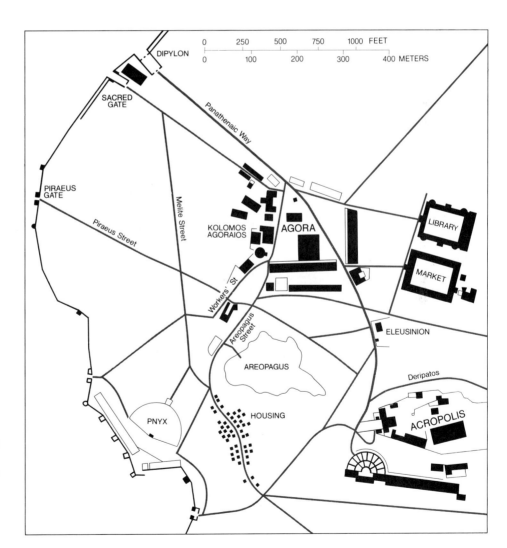

FIGURE 7–11 Diagram of Early Athens The central region beyond the Agora was surrounded by housing for the masses, but only one representative sample is shown.

and the Italian peninsula. Unlike Athens, Greek colonial towns were built in a planned fashion, with a regular grid pattern of streets of uniform width and with relatively uniform city blocks. The theories of city planning associated with these Greek colonial towns were later adopted in the Renaissance period in Europe (between A.D. 1500 and 1700).

Roman Cities

The Roman Empire had replaced the Greek city states as the dominant focus of city life in the Mediterranean and Europe by 200 B.C. Rome maintained its control over much of Europe, North Africa, and the Middle East by establishing settlements to serve as military and administrative centers (Figure 7–12). To encourage trade, maintain political control, and provide security, the Romans constructed roads focusing on Rome. The Romans adopted many Greek ideas, including the regular grid street pattern of the Greek colonial cities.

Rome was the largest city in the world by 150 B.C. with an estimated population of as many as one million inhabitants. Like other Roman cities,

the center of Rome was the focus of activity. The intersection of the two major streets of the Roman grid pattern was called the *Forum*. The Forum was the administrative and religious center of the city as well as the location of the public markets. Around the Forum were the palaces and villas of the elite.

The Forum and wealthy central section of Rome contained aesthetically and architecturally imposing structures, but the majority of the citizens of Rome, like the citizens of other cities up to that time, lived in cramped, dirty, and unsanitary quarters. Although the residences of the elite were served by both underground sewer and water systems, the poor of the city lived in the same squalor and filth that had been associated with Mesopotamian cities nearly 2,500 years earlier. Disposal of animal and human waste and garbage and providing fuel for cooking, food, and housing were problems that plagued Roman cities and remain urban issues today.

The cities of the Roman Empire were an important part of the cultural landscape of the time. The Romans expanded the ideas of the Greeks to add new elements, such as the Forum, to the cultural landscape of cities. Another landscape element of growing significance was the harbor. Increasing trade made its facilities for loading, unloading, and storing of goods a more important landscape feature. The improvement of roads and aqueducts to provide water for the cities created new landscape features beyond the Roman cities themselves.

The cities of Rome flourished for nearly half a millennium, but the fall of the Roman Empire (by A.D. 400) was accompanied by a general decline in urban life in the Mediterranean region. The Romans had provided stability and security for trade, which allowed urban life to flourish, but the political fragmentation of the Roman Empire removed this stability. The banditry and lawlessness of hundreds of different warlords and rulers led to the destruction of large-scale intercity trade and the decline of the cities themselves. By the Middle Ages, the Roman cities were only small towns or villages or were completely abandoned. Even Rome declined to only 30,000 inhabitants. Although a few major cities in highly favorable locations in Asia and the Americas prospered, the

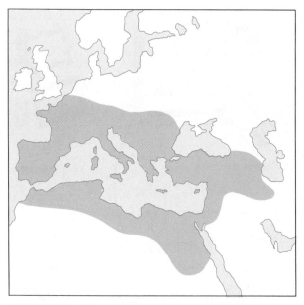

FIGURE 7–12 Expansion of the Roman Empire
Roman expansion included the entire region around the Mediterranean Sea.

decline of the Roman Empire ended an era of urban growth in Europe. The urban landscape of the Roman Empire was replaced by the landscape of the Middle Ages, which was dominated by towns. The great temples and other urban structures erected by the Romans remained only as relics in the landscape. The emergence of cities in Europe in the Middle Ages was in response to a new set of circumstances and reflected a new pattern.

Medieval City Landscapes

Urban life in Europe began to develop again after A.D. 1000. Towns established by feudal lords for military reasons emerged as trade centers and markets for the goods of surrounding farms. The medieval town included four major visual elements in its landscape: the fortress at the center, the church or cathedral, the palace, and the wall around the city (Figure 7–13). The core of the city consisted of the palace and cathedral, with the cathedral as the highest point on the skyline. Adjacent to the cathedral was the palace. Surrounding both was the fortification that protected them in times of conflict. The prominence of the cathedral in the medieval city landscape and the role of the church in politics, society, and the economy has led to the categorizing of major medieval cities as *cathedral cities*. Politically, these cathedral cities were somewhat similar to the Greek city states, since they were independent of government control and responsible for their own defense.

The Industrial City

The medieval cities with the best site-situation relationships, such as London and Paris, emerged as the dominant centers in their respective regions by A.D. 1500 (50,000 and 225,000 inhabitants, respectively). The Industrial Revolution led to even greater growth in these cities as banking, finance, trade, and commerce increased with the rapid growth in industrial production. Prior to the Industrial Revolution European cities were small. London had only 75,000 inhabitants in 1600, and the second largest community in England, Bristol, had only about 20,000. The majority of towns had populations of fewer than 10,000 people.

The Industrial Revolution was associated with rapid growth in communities in England. London by 1800 had 865,000 people, and the population grew to 2,320,000 in 1850. There were 1,314,000 inhabitants in Paris by 1850 (see Table 7–2). London and Liverpool became the trading ports for imports and exports in the British Isles; Amsterdam and Rotterdam developed as the gateways to the Ruhr industrial region of Germany; and Paris grew as a focus of transport routes in France.

The dominance of these large centers of Europe preceded the Industrial Revolution. London, Paris, Berlin, Copenhagen, and other major European cities trace their origins to medieval trade and trade fairs, which were concentrated where natural routes or a source of protection existed. Site characteristics of a good harbor (Rotterdam),

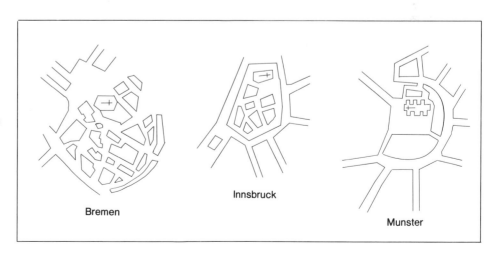

FIGURE 7–13 Examples of Medieval Cities of Europe The circular pattern reflects the city wall that was an integral part of the cityscape, symbolically separating the city from non-urban areas around it.

Bremen

Innsbruck

Munster

Bruge, Belgium is a modern continuation of a medieval city. The medieval city is shown on this map, illustrating the irregular street pattern.

Consequently, the industrial towns that sprang up across Europe after the Industrial Revolution could be located near raw materials (such as coal for steel mills).

The size and relative importance of individual towns and cities of Europe are functions of their *hinterland.* An individual industrial city may have only a limited hinterland, while London's hinterland includes all of the British Isles, and, for selected activities (such as financial services), much of the world (Figure 7–14). The population of cities reflects the demand by their hinterlands for service. The resulting size and pattern of towns create a hierarchy of places providing services, from city to town to village. The smallest villages may contain only a grocery store, pub, post office, and church, and there are thousands of such places in any country. At the other end of the spectrum are cities like London, Rotterdam, Paris, and Copenhagen, which provide all the services of smaller centers plus government, financial, medical, educational, and trade services for a national or international hinterland (See the Models in Geography box on page 258).

The landscape of early industrial cities differed markedly from that of the medieval city. The

an easy river crossing (London), a defensible position as an island in a river (Paris), or a junction of routes (Berlin) provided impetus for growth before the Industrial Revolution. The geographic relationships that affected each European city's initial development and subsequent regional dominance ensured that these places remained important after the Industrial Revolution.

The later development of European cities like Düsseldorf, Manchester, Leeds, and Liège reflected the need for raw materials and illustrates how the location of new industrial cities was related to technology. Since early industrial centers relied on waterpower, they were located where waterfalls or rapids could be harnessed to drive the machinery. Industrial centers were freed from locating at waterfalls or other water power sites with the advent of Watt's steam engine.

Christchurch Cathedral, Dublin, Ireland. Churches of all types often form one of the most imposing features in the cultural landscape in individual communities.

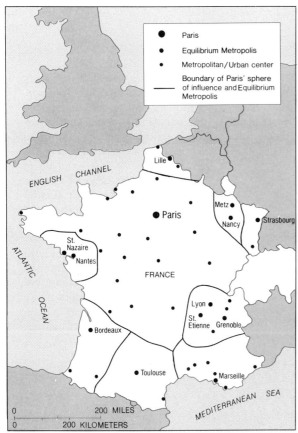

FIGURE 7–14 Hinterlands of Selected Cities in France Paris dominates much of France, but other cities may share a regional hinterland, as shown by equilibrium metropoli.

on either human or animal locomotion. People walked everywhere unless they were among the few who had a horse and buggy. Consequently, the overwhelming characteristic of cityscapes was one of high density and foot traffic. Leaving one's neighborhood was a major adventure, and most people commuted only a few blocks to work. Industrial activities were concentrated near the center of the city where trade and financial activities were located.

The wealthy in the early industrial city occupied mansions with extensive gardens and grounds near the financial district where they worked. Clustered around dirty, noisy manufac-

Row housing, South Boston, Massachusetts. The need to house workers in the growing industrial cities in the United States in the late 19th and early 20th centuries created a new landscape feature.

homes of the wealthy, the halls of the artisans and craftsmen, and the marketplaces in the medieval city surrounded the castle, cathedral, and government buildings. Factories, railroad terminals, port facilities, and offices dominated the city center in the industrial city, with wealthy neighborhoods and mean, crowded tenements for workers occupying specific areas in response to transportation or site amenities. The importance of transportation in the industrial city landscape is exemplified in U.S. cities where different types of transport are associated with specific city landscape patterns.

The first period of transport can be referred to as the era of walking, which lasted until about 1870. During this period, transport in cities relied

MODELS IN GEOGRAPHY
Central Place Theory

An important model related to the hierarchy of towns and cities was developed by the German geographer Walter Christaller. *Christaller's model* defines villages, towns, and cities as *central places*, providing services for a hinterland. Each service requires a minimum market size (number of consumers or volume of purchasing power) to function (known as its *threshold*), if it is to be economically successful.

Some services, such as a grocery store, require a low threshold population. High threshold services are major hospitals, stock markets, expensive clothing stores, and other services that are expensive or in low demand (Table 7–A). Central places offering a similar range of services are smaller in size, with smaller centers providing low threshold services. Christaller's theory concludes that towns of the same size will be spaced equally. Larger towns will be farther apart, thus creating the world's settlement geography of numerous small towns serving local needs and fewer larger centers serving hinterlands that include the smaller centers (Figure 7–A).

TABLE 7–A
Population Threshold for Selected Services

Filling stations	196
Food stores	254
Churches	265
Restaurants and snack bars	276
Taverns	282
Physicians	380
Real estate agencies	384
Appliance stores	385
Barbershops	386
Insurance agencies	409
Fuel oil dealers	419
Dentists	426
Hardware stores	431
Auto repair shops	435
Fuel dealers (coal, etc.)	453
Drugstores	458
Beauticians	480
Auto parts dealers	488
Meeting halls	525
Feed stores	526
Lawyers	528
Furniture stores, etc.	546
Variety stores, "5 & 10"	549
Apparel stores	590
Lumberyards	598
Banks	610
Electric repair shops	693
Florists	729
Dry cleaners	754
Billiard halls and bowling alleys	789
Jewelry stores	827
Shoe repair shops	896
Sporting goods stores	928

Source: William Garrison and Brian J. L. Berry, "A Note on Central Place Theory and the Range of a Good," *Economic Geography*, 34 (October 1958): 304–11

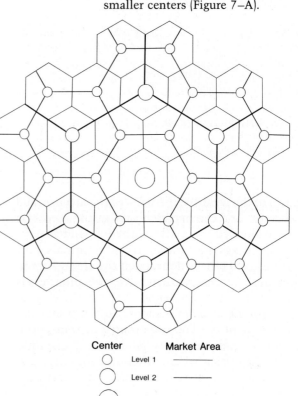

Center		Market Area	
○	Level 1	————	
◯	Level 2	————	
◯	Level 3	————	

FIGURE 7–A The Christaller Model of Hinterlands

turing plants were the workers' homes, generally two- to five-story tenements in which entire families often crowded into a single room without any plumbing or windows. Outside were narrow streets, little or no vacant space, and a streetscape vibrant with carts, buggies, animals, and the ubiquitous pushcarts of vegetables, fruits, fish, and other goods. Public recreational space was limited, but many larger cities, such as New York, subscribed to the belief that large parks were necessary to provide open space and recreational opportunities for the working class. Central Park in New York City is a classic example of this aspect of the industrial city during the walking era.

A major change in industrial cities occurred by 1880 as the electric streetcar and railroads for commuting became more common. Streetcars and trains allowed movement outward from the city center and prompted both the first suburbs and the intensification of commercial and manufacturing activity in the center of the city. People who could afford to live away from the smoke, noise, odor, and crowds in the city center did so, but *economies of scale* continued to concentrate financial and manufacturing in the center of the cities.

The horizontal expansion of the city associated with the rail commuting to suburbs was paralleled by the vertical expansion of the city. Invention of the elevator and its widespread adoption after 1870 and the change from brick or stone construction to steel girders freed buildings from height restrictions of approximately ten stories. Skyscrapers made their appearance in Chicago and New York, and tall buildings became an integral part of the cityscape. The homes of the wealthy were a distinctive landscape element in the suburbs, but the majority of the population of large cities in the industrial world continued to live in relatively low (two- to five-story) apartments. The streetcar allowed large cities to expand, and suburbs of single-family homes for middle-class workers emerged as a distinct landscape element. The homes in these suburbs and the lots on which they were built were generally smaller than in modern suburbs.

Industrial cities during the walking and streetcar eras included small, local service businesses as common landscape elements. Grocers, tailors, doctors, and barbers were scattered throughout residential areas. Church and government buildings were still elements of the industrial cityscape. They preceded industrialization in some cases, but in others (especially in the United States) they reflected a conscious effort to provide a type of monumental architecture as a focal point for the city. Central Park in New York City and state and local government buildings, railroad terminals, and large hotels in cities across the United States represent the spirit of beautification and monumental architecture of the industrial era. Similar features are found in other countries where landscape features such as the Eiffel Tower in Paris and the large hotels and railroad and streetcar terminals of England's industrial cities were built during this period. Increasing city size and the increased role of the railroad in urban transport added monumental steel girder bridges to the industrial city's landscape in the last half of the nineteenth century.

Growing populations in the industrial cities spawned growth in government activities, including police, education, social work, and city planning and land-use regulation. The urban population of the United States, for example, grew from two to twenty-two million between 1840 and 1890. In 1860, New York reached a population of one million, joining London and Paris as the world's largest cities. Rapid growth of industrial cities and increasing world population added seven more cities with over one million inhabitants by 1900. During the twentieth century, the pace of *urbanization* has increased, with thirty additional cities reaching one million between 1900 and 1930, and fifty more by 1955. Over 290 cities had one million or more inhabitants by 1988.

The landscape elements first introduced in cities of the industrial world during the streetcar and railroad era are found in most of these large cities. Tall buildings, monumental architecture, parks and open spaces characterize the central city area of large cities from Canada to Brazil, from Paris to Hong Kong.

The Morphology of Cities

The growth of urban centers in the nineteenth century created not only new landscape elements, but a distinct spatial pattern. The concentration of business and retail activities in the city center, surrounded by manufacturing and residential land uses, created concentric zones of land use (Figure 7-15). E. W. Burgess tried to explain the spatial pattern of land use in cities in his 1923 *concentric zone model.* He hypothesized that cities grow outward from a central area in a series of rings. Although the width of the rings varies from city to city, they appear in the same order in all cities. Zone I is referred to as the Central Business District (CBD). High land values mean that only the most profitable economic activities can afford to locate in Zone I. Residential land uses, manufacturing, or other activities requiring large amounts of land are pushed farther from the center of the city.

Zone II in Burgess's model was a transitional zone of industry and slum areas. Zone II contains the tenements of the workers and large older homes that had been subdivided into small apartments. Immigrants live first in this zone, as do the poor. They inhabit the shabby tenements, rooming houses, apartment blocks, and row houses near manufacturing plants. Zones III and IV were the location of working-class and middle-class housing, respectively. Since most who could afford to live away from the center of the industrial city did so, the wealthy occupied zones farther from the

city center. The landscape in the pre-1920 industrial city was one of densely packed housing. Even middle-class families were apt to live in duplexes. Single-family homes of the middle class occupied small lots with homes nearly touching neighboring residences. Zone V in the concentric zone model is a commuter's zone, beyond the continuous built-up area of the city.

The concentric zone model generalizes land uses in the industrial city, but it illustrates the impact of transportation and other technological changes on the growth of the city. The development of the railroad and streetcars for commuting led to a recognizable change in the land-use pattern as activities moved outward along the transport routes. In 1939 Homer Hoyt developed another model of land use in the city reflecting these changes (Figure 7–15). The *sector model* represents Hoyt's hypothesis that certain areas of a city were initially more attractive for a specific activity, and as the city grew, activities grew out from the center as a series of wedges or sectors. For example, once a district with expensive housing is established, it will attract new expensive housing to its margin. Activities tend to develop outward along transportation routes. The CBD still occupies the center of the city, but manufacturing moves outward along rail lines or other transit routes.

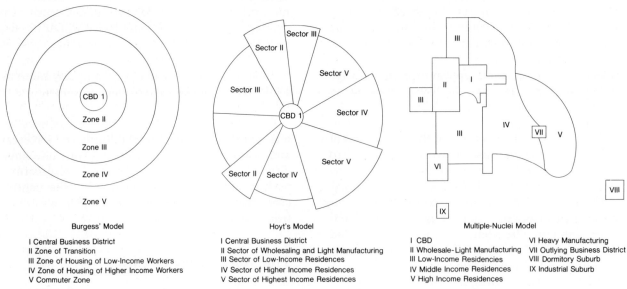

Burgess' Model

I Central Business District
II Zone of Transition
III Zone of Housing of Low-Income Workers
IV Zone of Housing of Higher Income Workers
V Commuter Zone

Hoyt's Model

I Central Business District
II Sector of Wholesaling and Light Manufacturing
III Sector of Low-Income Residences
IV Sector of Higher Income Residences
V Sector of Highest Income Residences

Multiple-Nuclei Model

I CBD
II Wholesale-Light Manufacturing
III Low-Income Residencies
IV Middle Income Residences
V High Income Residences

VI Heavy Manufacturing
VII Outlying Business District
VIII Dormitory Suburb
IX Industrial Suburb

FIGURE 7–15 Traditional Models of Land Use in American Cities

The models of both Hoyt and Burgess were strongly criticized as overly simplistic. In 1945 the American geographers Chauncy Harris and Edward Ullman developed a third model of urban land use, the *multiple nuclei model* (Figure 7–15). This model states that a city has more than one focus, and that activities revolve around these nuclei to create multiple nuclei. Multiple foci in a city could include the CBD, a port, airport, or neighborhood business and shopping center. Land uses cluster around these nuclei in response to the character of the foci. A university will often be surrounded by apartments for students, bookstores, fast-food restaurants and homes of the relatively well educated.

The multiple nuclei model reflects the impact of widespread adoption of automobile transport after 1920. Initially, the impact of the automobile was limited to allowing the areas between rail and streetcar lines to be filled in by middle-class homes. These homes continued to be small, but lot sizes increased as the small detached garage became a part of the urban landscape. A major change in industrial cities dates from the end of World War II, when communication and transportation increased markedly, creating the suburban landscape with single-family detached homes on large lots. Freeways became the symbol of the automobile era as automobile ownership became widespread among Americans, Europeans, Australians, New Zealanders, and Japanese. Freeways and suburban housing sprawled outward from the old central city and affected the entire landscape in response to three processes:

1. Greater mobility of workers
2. Consolidation of production units
3. Expansion of numbers of single-family detached homes.

These changes have decreased the importance of the CBD and encouraged the dispersion of retail, automobile, manufacturing, and residential areas. The dispersion has also created a much lower urban density.

Through automobile commuting, factories could assemble large work forces at one site, freeing them from locating at railroad junctions or the center of the city. The automobile allowed retail businesses to move to suburban areas, cre-

ating a new pattern of land use in cities of the industrial world. The role of the automobile in transportation encouraged expansion of highways, eventually leading to superhighway systems such as the interstate system of the United States, (see the Cultural Dimensions box on page 262.)

A major change in cities (particularly in the United States) since World War II has been in the area that they occupy and in the landscapes that typify them. Whereas an average city in the United States in 1940 was 15 miles or less in diameter, the growth of suburbs in the automobile era has created many metropolitan areas 50 to 75 miles in diameter. The growth of large metropolitan regions including more than a single city reflects the adoption of the single-family house in the suburbs as the residence of choice.

Freedom of access made possible by the automobile allows a lower density of population, and cities and suburbs sprawl across the landscape. In response to urban sprawl, businesses and factories have also moved to the suburbs. The resultant pattern reflects the multiple nuclei model. No longer bound by subway or railway lines that go only to the downtown, human activities have become dispersed throughout the city. Retail trade is dominated by suburban malls where climate is controlled and parking is available. The problems of the old central city district of earlier cities (high land costs, shortage of space, deteriorating physical environment, inadequate highway access) have fostered a rapid movement of activities and people to suburban areas.

Cities whose major growth occurred after World War II have adapted readily to the changes prompted by the automobile, but the old retail centers of the industrial world based on water or railroad transportation have been adversely affected. The movement of population, industry, and trade to the periphery has caused deterioration in the central city. Empty buildings, unprofitable businesses, and a declining tax base combine with the concentration of those least able to afford the cost associated with moving to the periphery (the elderly, the poor, and minorities) to make it difficult to maintain the old city center. The generally poor people who are left behind in old cities of the industrial world inhabit an oft-

Urban Crawl

A single mother of three living in Thousand Oaks, California, spends three hours a day for her 80-mile (round-trip) commute to work in Los Angeles.* She commutes because her salary is one-third higher than what she can earn in her own community. She averages only 26.6 MPH in her commute, and her employer understands that she may be late on any day because of traffic.

Increasing commuting time is prompting a growing chorus of angry commuters. There are 6 million registered vehicles in Los Angeles County alone, according to government estimates, consuming 628,000 hours a year and 72 million gallons of gas over 504 miles of congested highways (compared with 30 miles in 1956). A recent *Los Angeles Times* poll found that 71 percent mentioned some problem in getting around as a major issue. Many note that when they go out for the smallest errand, they have to plan around "peak hours."

Peak hours are already three hours long in the morning and four in the evening. And projections for the future are even more dire. The Los Angeles County Transportation Commission (LACTC) estimated in 1987 that what is now a 90-minute trip down Route 405 would take three hours in the year 2000. And on one particular stretch of the Ventura Highway, average traffic flow would slow to 7 MPH.

Part of what confronts any citizen, public official, or legislator is the tangle of county, city, state, and federal jurisdictions—each with various plans at various levels of funding and completion. Besides the 84 municipal entities that make up Los Angeles County, there are intercounty state and federal regulatory agencies. Many experts have pinpointed southern California's "love affair with the private automobile" as the major obstacle to plans for easing traffic. The answer to that, say recent studies, is "traffic demand management" that focuses on making better use of existing transport systems. At present, transportation and traffic initiatives in southern California are focused on four alternatives.

The first segment of a $6 billion system of mass-transit subways and light-rail lines is under construction with a 4.4-mile segment from downtown's Union Station to Wilshire Boulevard. This $1.2 billion segment is expected to be finished by 1993. Various plans to continue the system to outlying areas—most notably a 17-mile stretch into the San Fernando Valley—are in various stages of proposal; many are stalled in disagreements and funding battles. Controversy rages over whether enough people will choose to ride over Los Angeles's sprawling metropolis to make MetroRail financially viable, and whether use of the rail will remove cars from clogged freeways.

In 1987 the mayor of Los Angeles introduced a major, thirteen-part plan to improve the free flow of traffic and reduce air pollution. Four parts have been implemented by executive order, nine more are up for approval by the city council. The plan includes ride sharing, parking enforcement, gridlock fines, automated traffic signal surveillance, and a provision to remove 70 percent of the large, heavy-duty trucks from streets and roads in the morning and evening rush hours.

The Southern California Association of Governments (SCAG) 1989 proposals advocate flextime strategies, modified workweeks (with such features as work at-home, telecommunication options), parking management, truck rescheduling, and "jobs-housing balance." Counties and jurisdictions need to increase the number of jobs in house-rich areas and increase the number of houses in job-rich areas.

The LACTC plan calls for $4.5 billion worth of short-term freeway construction projects—extra lanes, some new roads, better exits and entrances—but says the most optimistic projection of funds falls $1.5 billion short.

LACTC is also proposing what is known as the "smart street" concept. Electronic sensors in the street would enable traffic controllers at a separate location to assess traffic congestion and manipulate area traffic lights to alleviate congestion. The first such study is already underway in a 12.3-mile section of the Santa Monica Freeway and adjacent streets. The project, which would cost about $50,000 per intersection, would add up to $400 million if all major intersections in Los Angeles were wired.

The keys to all these programs, say officials, are awareness, personal commitment, and—ultimately—money. Some aspects of each plan are already funded, many are not—and shortfalls mean battles for taxpayer dollars with other social programs.

---

*Adapted from *Christian Science Monitor*, "Urban Crawl," May 3, 1989, p. 12.

South Bronx, New York City. The adoption of the automobile and urban sprawl have combined with time to lead to deterioration of old apartment complexes in the inner cities. The resultant landscape of abandoned buildings and wastelands of urban renewal are a unique landscape feature in some large American communities.

times deteriorating landscape of empty storefronts, obsolete housing, and narrow crowded streets with high crime rates.

The suburbs are areas dominated by residential land use. The suburban city of the industrial world is typified by expanses of one- and two-story single-family homes. The absence of an established downtown retail area in the suburbs does not necessitate travel to the central city for shopping or business, however. Instead, the spread of population, the related development of ever larger highway systems, and the demand for services by suburban residents has prompted a fourth urban model, the *urban village.* The urban village model is a variation of the multiple nuclei model, in which multiple nuclei spring up at the intersection of major highway systems. Focal points for automobile traffic, these intersections become the location of suburban shopping malls. Concentration of retail trade in the shopping mall leads to the development of low-rise office buildings to house the professionals who provide services for the retail trade. Land developers, lawyers, insurance agents, and wholesale trade representatives locate near the suburban shopping mall to provide services for mall business owners or suburbanites. The same factors that attracted the suburban mall to junctions of major

highway systems also attract industries that rely on suburban labor. Factories are built near the suburban mall and its associated office complex, creating a miniature city within the larger urban area. These urban villages of commercial and industrial activity are surrounded by a near uniform landscape of single-family suburban homes (see the Cultural Dimensions box on page 264).

The urban village is typical of the rapidly growing suburban regions in the United States, especially in the "sunbelt" (southern California through Virginia), and to a lesser extent in European cities. The impact of urban villages on the old central cities has been devastating. Older cities that benefit from unique site or situation aspects or that have maintained a distinctive role in finance, government, arts, education, or tourism are able to maintain their importance despite the dispersion associated with the urban village.

Populous cities like New York, London, Paris, and Tokyo continue to maintain a viable central city nucleus in part because of the investment they represent and in part because of their unique role. The importance of such cities has prompted the governments of various countries and states to invest heavily to maintain central cities and prevent their relative abandonment. The result-

Housing Project, Chicago, Illinois. Efforts to provide adequate housing for large low-income populations took the form of giant housing projects in American cities in the decades following World War II. The projects represent a distinctive and immediately recognizable landscape feature in several American cities.

CULTURAL DIMENSIONS
Suburban Growth Threatens Walden Pond

A tiny legal notice appeared in the Concord (Mass.) *Journal* on December 25, 1987, informing those who saw it that an obscure realty trust was seeking approval for a plan to develop land "at approximately 641 Walden Street at Route 2."* To the casual observer, there was little to distinguish the project from countless others in Boston's booming high-tech suburbs.

What the notice hadn't mentioned, however, is that the proposed building site is across the highway from the entrance to Walden Pond, where Henry David Thoreau lived in a cabin and wrote his celebrated *Walden;* the pond is now a state park. The building's driveway will cut across a town forest. It will be the only significant development for close to a mile in either direction.

For decades, Walden has been under siege from summertime bathers with plastic coolers, radios, and stereos. Now some see the office building as a new kind of threat, the violation of what remains of the surrounding hills and woods by the building boom that is overtaking much of New England. Not far away, a developer is pressing for approval of a 250-unit housing development.

The threat of suburban development to the nation's historic legacy could not come into sharper focus than at Walden Pond. Walden will be destroyed if the development continues, according to members of the Concord Historical Commission. Nevertheless, the permit was quickly approved by the Board of Appeals. Opponents put most of the blame not on the developers, but on the indifference of Concord's local officialdom and citizenry. Opponents argue that although the notice may have met the minimum requirements of the law, more publicity should have been given to the hearing because something more was at stake than a typical commercial development.

The developers respond that they are doing no more than the local zoning allows. They point out that they are providing extra landscaping, putting the parking lot behind the building, and taking other steps that will make the site "much better" aesthetically than today.

There's hope the office building won't be so bad. The town planner maintains that under the circumstances, the developer did a good job in planning the site. The building will be only two stories, set back from the road, and the company agreed to close its parking lot to weekend bathers. The planner points out that the spreading suburbs have already transformed the rural landscapes of eastern Massachusetts into part of the urban scene.

Adapted from Jonathan Rowe, Christian Science Monitor, "Suburban Sprawl Threatens Walden Pond" August 26, 1987.

---

ant landscape is one of a CBD area that retains its vitality as a business, communication, entertainment or government center (as in London, New York, Paris, or Tokyo), but is surrounded by a zone of derelict housing, obsolescent factories, and undesirable neighborhoods. The problem of obsolescent housing and factories in some of the older cities has led to government programs to provide new housing for low-income groups, resulting in a new landscape feature, the housing project of the inner city. These apartment buildings house thousands of individuals and are communities within themselves. Housing projects represent a landscape feature that is often dysfunctional and is associated with a high incidence of crime, high levels of unemployment, and high poverty rates. The residents of these projects with sufficient income generally flee to the suburbs, creating a problem known as *urban flight*.

Recognition of the changing landscape and activity in industrial cities has led geographers to define three distinct areas within the cities of the industrial world: the central city, the suburb, and the "exurban fringe" or "nonmetropolitan area."

The geographical characteristics of these three areas reflect a gradation from rural to urban. The nonmetropolitan region is an area of small towns surrounding metropolitan areas. The suburbs are the landscape of single-family detached housing with office or retail malls (Table 7–3). The central city reflects the vertical dimension of high-density activity created during earlier periods of slower transport (Figure 7–16).

An indication of the changes affecting cities in the industrial world is represented by the residential relocation of people from the central city. The population growth of central cities continued in the United States until 1970, after which the process of urban concentration was reversed. The central cities of many large American metropolitan areas lost population between 1970 and 1980. Baltimore declined 19 percent, Buffalo 25 percent, Pittsburgh 21 percent, and Washington, D.C., 25 percent.

Two processes now beginning may reverse some of the landscape changes associated with suburbanization: urban to rural migration and *gentrification.* Urban to rural migration is the movement of residents from either the central city or the suburbs to nonmetropolitan areas. Surveys in industrial countries (especially the

TABLE 7–3
Population of Cities and Suburbs in the United States

Residence	Population (MILLIONS)			
	1950	1960	1970	1980
Central city	53.7	59.9	64.3	67.9
Suburbs	40.9	59.6	75.2	101.5

Source: Statistical Abstract of the United States 1975 and *1988* (Washington, D.C., U.S. Government Printing Office, 1975 and 1988).

United States) reveal that individuals prefer to live in non-urban environments. People perceive that smaller communities have lower crime rates, a slower pace of life, and less crowding, yet still provide access to jobs, shopping, and recreation opportunities in the growing urban villages or through commuting. In consequence, in the last decade nonmetropolitan places have experienced rapid growth.

Gentrification is the process by which older, dilapidated portions of the central city are renewed. Gentrification often occurs in old residential districts inhabited by lower income families, who occupy old homes that have been subdivided into apartments. Upper middle class families obtain the homes, remodel them, and create an upper income neighborhood. Gentrification improves the landscape of the city by restoring old neighborhoods but it also displaces the poor who can no longer afford to live there (see the Cultural Dimensions box on page 266).

Not all residents of the central city who could move are attracted by the suburbs or nonmetropolitan areas because their activities still require proximity to downtown. These individuals and some well-to-do suburbanites (the gentry) who are moving back to the city are the chief forces behind gentrification. Gentrification occurs both because of the need or desire of some of these individuals to be near the central city and because of government programs that provide tax incentives for rehabilitating or rebuilding older areas of the central city.

The individuals or families who return to the central city gentrified zone tend to be professional people, have high incomes, and invest heavily in remodeling and upgrading the housing that they

POPULATION DENSITY OF URBAN AREA

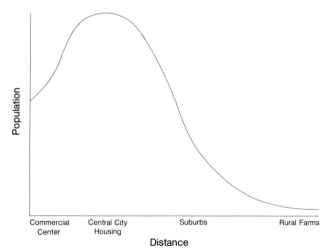

FIGURE 7–16 Population Density of Urban Areas
The population density of an urban area peaks in the high density housing of the central city and decreases in the suburbs and surrounding rural areas.

Central City Residents

The loss of affordable central city housing is one of the by-products of suburbanization, central city decline, and growing gentrification in American cities.* There is adequate housing for the affluent, but for the poor residents of the central city, housing is becoming scarce. Deteriorating, abandoned, and condemned apartment buildings force some central city residents to become homeless street people who sleep in public shelters, subways, cardboard boxes, or other places affording some shelter. Others are forced into "welfare hotels," hotels rented by government agencies to house low-income families crowded into one or two rooms. Neither choice is pleasant, but examples of each illustrate the dilemma facing older central cities.

Under a bundle of blankets, in makeshift beds and "cardboard condos," between twenty and thirty homeless (mainly men) spend the cold winter nights in the 57th Street subway station in New York City. Many collect cans (New York has a five-cent-a-can deposit and return law), panhandle, scrounge for food, or work at temporary jobs during the day. Food cooked over open fires in vacant lots, alcoholism, and drug use are common.

No accurate figures exist for the truly homeless. Estimates range from 100,000 to 500,000 for U.S. cities in 1988–1989. A survey of twenty-six cities by the U.S. Conference of Mayors in 1987 provides some insight into the problem of the homeless. The number one reason for homelessness reported in the cities surveyed was the shortage of housing that low-income people can afford. Mental illness coupled with a lack of services, unemployment, underemployment, and other economic factors (in ranked order) were also cited as causes of homelessness. The survey found that 22 percent of the homeless were employed either full- or part-time, but they earned too little to rent a room or apartment. The plight of the working homeless is often related to older buildings that traditionally rented rooms on a weekly or monthly basis ("bachelor flats") being destroyed due to urban renewal and gentrification projects.

Residents of "welfare hotels" are only marginally better-off, as typified by Manhattan's Martinique Hotel. The Martinique stands across the street from the famed Macy's and Gimbels (now closed). The Martinique was home to over 400 welfare families in 1987, including more than 1,500 children.

The families that live in the Martinique are predominantly single-mother, black, and Hispanic families. They are the victims of batterings, burnouts (both literal and figurative), plain bad luck, and perhaps more than anything, the lack of low-income housing in New York City. A welfare mother of five, receiving a monthly rent allotment of approximately $300, would be hard-pressed today to find an affordable apartment in any of the city's five boroughs. Should she happen to lose her low-income apartment, be displaced from an apartment of a friend or relative, or arrive in Manhattan homeless, she would have little choice but to seek emergency assistance from the city. That could mean being housed in the Martinique or another hotel used to house homeless families temporarily. There are an estimated 4,000 homeless families in New York.

There is a bitter irony at the Martinique that sets it apart from poverty elsewhere. The mothers receive food stamps, but have no stoves on which to cook their food. They must use hotplates, which are forbidden by hotel regulations. The most heinous problem is overcrowding. Young children in cramped space have no place to play. They must wait for an older brother, sister, or their mother to take them to a park or for a walk.

The city pays the owners of the Martinique from $2,000 to $3,000 a month per family. Rents in mid-Manhattan make no distinction between misery and luxury. One often-mentioned solution to the problem of the homelessness is for the city to renovate abandoned buildings and rent the apartments at low cost, but that takes time. Bureaucratic red tape cuts slowly, while the number of homeless families in the city is growing rapidly.

A family's stay at the Martinique may range from a few months to four years. Some residents find apartments that are reasonable and never return. Others find apartments that are terrible and have to return. Still others try their luck at different hotels, but usually find they are equally bad.

---

*Portions adapted from Robert Hirschfield, *Commonwealth*, February 13, 1987, pp. 71–72, and Victoria Irwin, *Christian Science Monitor*, January 19, 1988, pp. 1, 8.

occupy. Thus the gentrification of cities is creating a landscape of affluence in the midst of the decaying central city. At the same time, gentrification is also removing low-cost housing and forcing traditional inner-city residents to move outward to the early suburbs, which are becoming obsolescent, and contain less expensive homes. The process of gentrification can never restore the central city to its pre-auto importance because of space limitations, costs of rehabilitating structures, and limited demand for central city proximity, but it reflects an important geographical change in the modern industrial world.

Distinctions between urban and rural, metropolitan and nonmetropolitan differ from country to country, but reflect the general level of urbanization. The terms urban and rural represent a simple dichotomy based on population rather than function. *Rural* in the United States refers to all settlements with fewer than 2,500 people according to the census, while urban refers to those of 2,500 or more. Other countries use a higher threshold number to divide urban from rural. The distinction between urban and rural is between the concentration of population, industry, and commercial activity of a city and the landscape of the agricultural world with low densities of population and land use. The U.S. Census Bureau recognized a *Standard Metropolitan Statistical Area* (SMSA) until 1983 when the term was condensed to *Metropolitan Statistical Area* (MSA). An MSA is defined as a county with one city of 50,000 or more people or twin cities whose trade and business are closely interlinked and whose borders are adjacent. (Counties without such a single or twin city of 50,000 are by definition nonmetropolitan.) The Census Bureau also distinguishes between a county in a largely rural area with only one city of over 50,000 and larger metropolitan areas like New York or Los Angeles. The Census Bureau defines an urbanized region with one million or more inhabitants as a Consolidated Metropolitan Statistical Area. (CMSA) The CMSAs are the largest urban areas of the country, representing the highest orders in Christaller's hierarchical model of city size and function.

The 1980 U.S. Census indicated that 75 percent of the population lived in counties defined as

TABLE 7–4
Satisfaction with Residential Locations by Size of City

CITY SIZE	PERCENTAGE VERY HAPPY
1,000,000 and over	35
500,000–999,999	38
50,000–499,999	50
2,500–49,999	51
Rural	47

Source: Gallup Report, April 1981, no. 187; June 1981, no. 189.

metropolitan. Within these metropolitan counties, 11 percent of the population resided in places with 2,500 or fewer people. Only 9 percent of the inhabitants in nonmetropolitan counties resided in urban places, 16 percent lived on farms, and 73 percent lived in villages of fewer than 2,500 inhabitants. The importance of the distinction between metropolitan and nonmetropolitan counties in the United States, or similar divisions in other industrial countries, is in the population growth in nonmetropolitan areas. The movement of population, industry, and business to communities of the nonmetropolitan area reflects the beginning of a trend that may create major geographical and landscape changes in the industrial world. The anti-urban mentality of many residents of the cities may prompt further migration to smaller communities (Table 7–4). The change associated with growth in nonmetropolitan counties is from agriculture to urban living, since the residents do not come to farm, but simply bring urban activities from the city or suburb to the nonmetropolitan area. The resultant change in function and form of rural nonmetropolitan landscapes is creating important geographical changes.

Urbanization in the Traditional World

At the same time the processes of suburban sprawl, nonmetropolitan migration, and gentrification are affecting cities that date from the Industrial Revolution, the traditional less industrialized world is experiencing an urban revolution of its own. The highest rates of urban growth and urbanization in the world today are occurring in the less industrialized world. *Urban growth*, the phenomenon of people moving from rural to urban areas, and *urbanization*, the process of transforming a society from a predominantly rural to a predominantly urban population, are occurring at precipitous rates in less industrialized countries.

Although urbanization has been the norm in the industrial world since the early decades of the twentieth century, the majority of the population in the less industrialized world still resides in rural places (Figure 7–17). Urbanization is now occurring in the traditional world, and its urban growth rates and absolute population size of cities are the highest in the world (See Figure 2–13, page 69).

Cities of the less industrialized world are growing at a rate rivaled only by that of industrial cities during the first period of the Industrial Revolution or by the rapid transformation of small suburban villages into the modern urban village. Communities can grow from thousands to hundreds of thousands in decades (Figure 7–18). The change in the traditional, less industrialized world is compounded by the fact that the growth is occurring not in villages, but in large cities that are themselves doubling in size in the course of one decade. Unlike the industrial world, which has a variety of large metropolitan centers, most of the traditional world is characterized by relatively few large cities. The metropolitan scene in many of these countries is dominated by one city, which controls the government, industry, finance, trade, and other urban activities to such an extent that it is referred to as a *primate* city.

A primate city is one that is at least twice as large as the next largest city in a country and is normally (but not always) the capital city and the city that best expresses the national culture. An example of a primate city is Mexico City, which has an estimated 16 to 18 million people, nearly five times as large as the next largest city, and is synonymous with the culture and economy of the country.

Rapid growth of cities in the less industrialized world is changing the location of the most popu-

URBAN POPULATION GROWTH

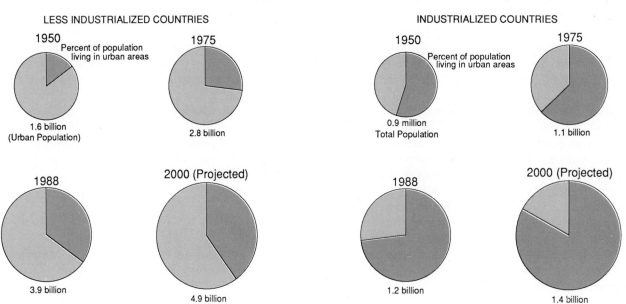

FIGURE 7–17 Urban Population Growth Numbers and percent of population living in urban areas in industrialized and less industrialized countries.

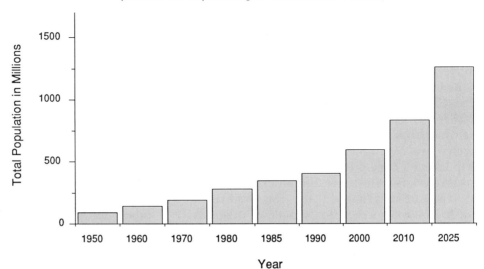

GROWTH OF LARGE CITIES
(Number of People Living in Cities of Over 4 Million)

FIGURE 7–18 World City Growth (Percentage of the Population Living in Urban Areas) Projected population growth in great cities (over 4 million population).

lous cities of the world. The most populous cities were nearly all in the industrial world fifty years ago, but by the year 2000 most will be in the less industrialized world. Urbanization in the traditional world differs from that of the industrial world in two ways—rate and cause. The rate of change is evident in the phenomenal growth of metropolitan centers in the traditional world.

The urbanization that required a century and a half in the industrial world will be realized in a matter of decades in the less industrialized world. Although the majority of the population of the traditional world still lives in rural settings, the growth of their urban populations will give most countries of the world an urban majority in the next few decades (Figure 7–19).

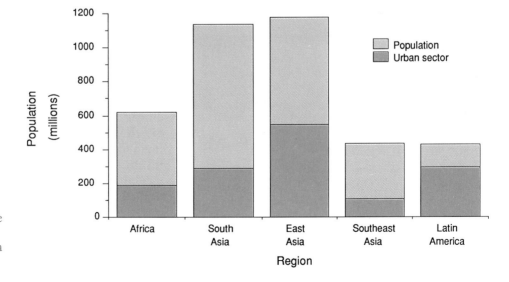

POPULATION AND URBANIZATION IN LESS INDUSTRIALIZED COUNTRIES

FIGURE 7–19 Population and Urbanization in Less Industrialized Countries This graph illustrates the higher degree of urbanization in East Asia and Latin America.

Urbanization in the traditional world is analogous to that in preindustrial cities in that the cities of the traditional world serve as trade and administrative centers. Government, trade, and financial activities provide little in the way of the working-class jobs that characterized urbanization in the industrial world. Migrants to the cities of the less industrialized world often become part of a second economy of self-employed. Artisans, peddlers, errand boys or girls, servants, or small-scale mechanics, they live in the urban realm, but receive few of the benefits normally associated with life in a large urban area.

Mexico City illustrates the problems created by rapid urbanization in the less traditional world. The most dramatic example of urbanization in the traditional world, Mexico City is among the three most populous cities of the world and will clearly be the most populous by the year 2000 with a projected population of over 30 million, if present growth rates continue. Mexico City is growing at a rate of 4.5 percent per year. Each year brings 750,000 new residents. High birth rates in Mexico City contribute an estimated two-thirds of this growth, while the other third is attributable to migrants from the countryside.

Shacks in the Mexico City dump. The metropolitan centers of many less industrialized worlds have landscape elements unknown in the industrial world, as these residences of individuals who collect useable materials from the garbage dump.

The tremendous growth of the city is overtaxing the urban infrastructure, creating slums of the worst magnitude. Air pollution, inadequate water and sewage systems, abject poverty, inadequate housing, few jobs; a litany that would seem to discourage all but the most optimistic from migrating to the city, are present. Nevertheless, rapid population growth and the perception that life will be better in the city continues to attract residents from the country. Life in the city is perceived as better in spite of the problems. As a primate city, Mexico City has more opportunities for government assistance, education, and jobs than exist in the countryside, where there is inadequate land for farming and a relative lack of other jobs. Mexico City and other primate cities of the traditional world are the focus of industrialization, offering further potential for jobs. For migrants, Mexico City presents at least the chance of improving their standard of living.

Life in the Mexico City metropolis is characterized by a *dual society*. Many of the older sections of the city contain the homes of upper income individuals who are long-time residents of the city. This landscape is one of affluence, with all of the benefits associated with urban life. At the opposite extreme, it is estimated that at least 25 percent (4 million) of Mexico City's inhabitants live in squatter housing on the outskirts of the city. These slums provide few or no city services to individual homes. Residents live in shanties made of scrap lumber, cardboard, or other materials. Government services are not provided to such squatter communities until the residents are able to convince the central government that they are permanent, in spite of the seeming impermanence of the landscape. The quality of life in these squatter settlements is similar to that found among migrants to cities during the Industrial Revolution in Europe and North America a century earlier, with overcrowding and inadequate sanitation and medical care resulting in lower life expectancies.

The different nature of urbanization in the traditional less industrialized world should not mask the fact that a *cultural convergence*, the process making the world more alike, is occurring in these cities. The large cities of the traditional world are interconnected with the rest of the world in many ways. Modern airports, modern communications, trade, and government interchange prompt diffusion of many of the characteristics of the industrialized world and its cities to the cities of the less industrialized world. The cities of the less industrialized world contain the universities, government services, business and commercial centers, and emerging industries that capitalize on the abundant and cheap labor of these countries. As the process of urbanization continues in the less industrialized world, it will increasingly make the residents of these cities more like those of the industrial world than like their rural residents.

The squatter settlements of Mexico City and other urbanizing centers in the less industrialized world are zones of *cultural transition*. The residents are in transition from the traditional rural life-style of the less industrialized world to urban life-styles. The urban life-style of the less industrialized world differs in some respects from that of the industrial world, but it is closer to the industrialized world model than to the life-style of the rural majority of the less industrialized world, and continued population growth in these less industrialized cities will continue the process of cultural convergence.

The landscapes of cities in the traditional world share many of the characteristics of cities in the industrial world. The concentration of activities in the central city, the development of middle-class suburbs, and the diffusion of activities across the metropolis of the less industrialized world differs only in degree from that of the industrial world. Each primate city of the traditional world has its high buildings and expensive shops in the central city, surrounded by land-use activities that reflect the level of transportation technology and related suburbanization (Figure 7–20). The cities of the less industrialized world may exhibit different architectural styles or locate specific activities in different areas, but in cultural impact they have the same end result on the landscapes. Thus population movements to urban centers and the resultant transformation of the landscape are among the major changes affecting the cultural geography of the world.

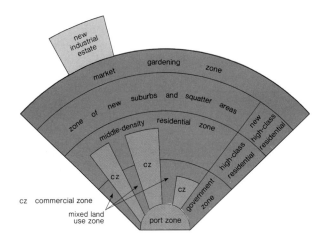

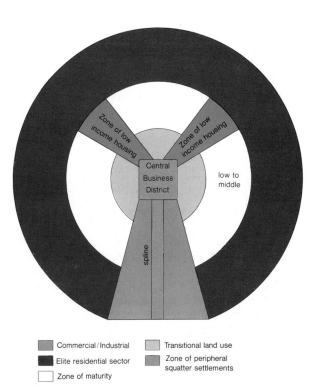

FIGURE 7–20 Models of Land Use in Cities of the Less Industrialized World The top diagram illustrates an Asian city and the lower diagram depicts a Latin American model.

San Salvador, El Salvador. A rapidly growing population in less industrialized countries has created large slum areas in response to their needs for housing.

Conclusion

The urban landscapes of the world have much greater importance than the small area they occupy on a global map indicates. Since the emergence of the first cities, human activity has been focused upon them. Emergence of the great cities of the past as centers of art and learning, trade and commerce marked the change from universal involvement in producing food to labor specialization. The spectacular growth in the cities of the less industrialized world today represents the continuation of forces that concentrate even greater economic, social, and political power in the cities. The diffusion of industrial technology, transportation, and even values in literature and art proceeds in a hierarchical fashion, allowing cities to continue their dominant role in the cultural geography of the region they serve.

QUESTIONS

1. Describe three types of landscapes associated with cities of the industrialized world.
2. In what ways has transportation affected city development in the industrialized world?
3. Which theory on the origin of cities do you think best accounts for the emergence of cities? Why?
4. Compare and contrast the Greek city and the Roman city.
5. Discuss the growth and character of the industrial city.
6. Describe the influence transportation has had on the landscape of the North American city.
7. Which model of the morphology of cities best characterizes cities you know? Explain.
8. What factors could you suggest to explain why the "urban village" is more of an American phenomenon than a European phenomenon?
9. What are the major trends affecting cities in North America? Which is most important? Why?
10. What are the characteristics and problems of rural to urban migration in the less industrialized countries of the world?

SUGGESTED READINGS

Adams, Ian H. *The Making of Urban Scotland.* Handem, Conn.: Archon Book, 1977.

Adams, John S., ed. *Contemporary Metropolitan American.* Cambridge, Mass.: Ballinger, 1976.

Adams, John S. "A Geographical Basis for Urban Public Policy." *The Professional Geographer* 2 (May 1979): 135–45.

_____. "The Geography of Riots and Civil Disorders in the 1960's." *Economic Geography* 67 (January 1972): 24–42.

_____. "The Meaning of Housing in America." *Annals of the Association of American Geographers* 74 (1984): 515–26.

_____. "Residential Structure of Midwestern Cities." *Annals of the Association of American Geographers* 60 (1970): 37–62.

Adams, Robert M. "The Origin of Cities." *Scientific American* 203 (September 1960): 153–68.

Agnew, John, John Mercer, and David Sopher, eds. *The City in Cultural Context.* Boston: Allen & Unwin, 1984.

Alanen, Arnold R. "The Rise and Demise of a Company Town." *The Professional Geographer* 1 (February 1977): 32–39.

Altshuler, Alan, James P. Womack, and John R. Pucher. *The Urban Transportation System: Politics and Policy Innovation.* Cambridge, Mass.: The MIT Press, 1979.

Anderson, Burton L. (with Gary L. Peters). "Industrial Landscapes: Past Views on Stages of Recognition." *The Professional Geographer* 4 (November 1976): 341–348.

Archer, Clark J., and Ellen R. White. "A Service Classification of American Metropolitan Areas." *Urban Geography* 6 (1985): 122–51.

Benevolo, Leonardo. *The History of the City.* Cambridge, Mass.: MIT Press, 1980.

Berry, Brian J. L. *Geography of Market Centers and Retail Distribution.* Englewood Cliffs, N.J.: Prentice-Hall, 1967.

_____. *The Human Consequences of Urbanization.* New York: St. Martin's Press, 1973.

Berry, Brian J. L., and John D. Kasarda. *Contemporary Urban Ecology.* New York: Macmillan, 1977.

Binford, Henry C. *The First Suburbs: Residential Communities on the Boston Periphery 1815–1860.* Chicago: University of Chicago Press, 1985.

Blake, Peter. *God's Own Junkyard: The Planned Deterioration of America's Landscape,* 2d ed. New York: Holt, Rinehart and Winston, 1979.

Bonine, Michael E. "The Morphogenesis of Iranian Cities." *Annals of the Association of American Geographers* 69 (June 1979): 208–24.

Booth, Alan. *Urban Crowding and Its Consequences.* New York: Praeger Publishers, 1976.

Bourne, Larry S., ed. *Internal Structure of the City,* 2d ed. New York: Oxford University Press, 1982.

_____. "Urban Structure and Land Use Decisions." *Annals of the Association of American Geographers* 66 (1976): 531–47.

Bourne, L. S., and J. W. Simmons. *System of Cities.* New York: Oxford University Press, 1978.

Bourne, L. S., R. Sinclair, and K. Dziewonski. *Urbanization and Settlement Systems: International Perspectives.* New York: Oxford University Press, 1984.

Brunn, Stanley D., and Jack L. Williams, eds. *Cities of the World: World Regional Urban Development.* New York: Harper and Row, 1983.

Buswell, R. J., and M. Barke. "200 Years of Change in a 900-year-old City." *Geographical Magazine* 2 (1980): 81–83ff.

Cantor, Leonard, ed. *The English Medieval Landscape.* Philadelphia: University of Pennsylvania Press, 1982.

Carter, Harold. *An Introduction to Urban Historical Geography.* London: Edward Arnold, 1983.

Chang, Sen-dou. "The Changing System of Chinese Cities." *Annals of the Association of American Geographers* 66 (September 1976): 398–415.

_____. "Modernization and China's Urban Development." *Annals of the Association of American Geographers* 71 (December 1981): 572–79.

Chatterjee, Lata, and Peter Nijkamp. *Urban Problems and Economic Development.* The Netherlands: Sijthoff & Noordhoff, 1981.

Christaller, Walter. *The Central Places of Southern Germany,* trans. C. W. Baskin. Englewood Cliffs, N.J.: Prentice-Hall, 1966.

City of Los Angeles. *The Visual Environment of Los Angeles.* Los Angeles: Department of City Planning, 1971.

Clark, W. A. V., and James E. Burt. "The Impact of Workplace on Residential Relocation." *Annals of the Association of American Geographers* 70 (March 1980): 56–67.

Clawson, Marion, and Peter Hall. *Planning and Urban Growth.* Baltimore: The Johns Hopkins University Press, 1973.

Clay, Grady. *Close-Up. How to Read the American City.* New York: Praeger, 1973.

Clements, Donald W. "The Dispersed City: Myth or Reality?" *The Professional Geographer* 1 (February 1977): 26–31.

Conzen, Michael P. "American Cities in Profound Transition: The New City Geography of the 1980s." *Journal of Geography* 82 (1983): 94–102.

_____. "The Maturing Urban System in the United States 1840–1910." *Annals of the Association of American Geographers* 67 (March 1977): 88–108.

_____. *World Patterns of Modern Urban Change: Essays in Honor of Chauncy D. Harris.* Chicago: University of Chicago, Department of Geography, Research Paper No. 5, 217–228, 1986.

Coones, Paul, and John Patten. *The Penguin Guide to the Landscape of England and Wales.* New York: Penguin Books, 1986.

Costa, Frank J., and Allen G. Noble, "Planning Arabic Towns." *Geographical Review* 76 (April 1986): 160.

Creveling, Harold F. "Mapping Cultural Groups in an American Industrial City." *Economic Geography* 31 (1955): 364–71.

Crowley, William K. "Old Order Amish Settlement: Diffusion and Growth." *Annals of the Association of American Geographers* 68 (June 1978): 249–65.

Davis, Kingsley. *Cities: Their Origin, Growth and Human Impact.* San Francisco: Freeman, 1973.

Detwyler, Thomas, and Melvin Marcus, eds. *Urbanization and Environment: The Physical Geography of the City.* Belmont, Calif.: Duxbury Press, 1972.

Dickinson, Robert. *The West European City.* London: Routledge & Kegan Paul, 1961.

Earle, Carville. *The Evolution of a Tidewater Settlement System: All Hallows Parish, Maryland 1650–1783.* Chicago: University of Chicago, Department of Geography, 1975.

Edwards, Arthur M. *The Design of Suburbia: A Critical Study in Environmental History.* London: Pembridge Press, 1981.

Elvin, Mark. *The Pattern of the Chinese Past.* Stanford, Calif.: Stanford University Press, 1973.

Enyedi, György, and Júlia Mészáros, eds. *Development of Settlement Systems.* Budapest: Akademia: Kaidó, 1980.

Erickson, Rodney A., and Marylynn Gentry. "Suburban Nucleations." *Geographical Review* 75 (January 1985): 19.

Estall, R. C., and R. Ogilvie Buchanan. *Industrial Activity and Economic Geography: A Study of the Forces Behind the Geographical Location of Productive Activity in*

Manufacturing Industry, 2d ed. London: Hutchinson, 1966.

Ford, Larry. "Saving the Cities: Urban Preservation in America." *Focus* 30 (1979).

———. "Urban Morphology and Preservation in Spain." *Geographical Review* 75 (July 1985): 265–84.

Fossier, Jean Chapelot Robert. *The Village and House in the Middle Ages.* Berkeley and Los Angeles: University of California Press, 1985.

Fusch, Richard, and Larry Ford. "Architecture and the Geography of the American City." *Geographical Review* 73 (1983): 460–71.

Girouard, Mark. *Cities and People: A Social and Architectural History.* New Haven: Yale University Press, 1985.

Gober, Patricia. "Shrinking Household Size and its Effect on Urban Population Density Patterns: A Case Study of Phoenix, Arizona." *The Professional Geographer* 1 (February 1980): 55–62.

Goldberg, Michael A., and John Mercer. *The Myth of the North American City: Continentalism Challenged.* Vancouver: University of British Colombia Press, 1986.

Gottmann, Jean. *Megalopolis: The Urbanized Northeastern Seaboard of the United States.* New York: Twentieth Century Fund, 1961.

Griffin, Ernst, and Larry Ford. "A Model of Latin American Urban Structure." *Geographical Review* 70 (1980): 397–422.

———. "Tijuana: Landscape of a Culture Hybrid." *Geographical Review* 66 (1976): 435–47.

Griffith, Daniel A. "Evaluating the Transformation from a Monocentric to a Polycentric City." *The Professional Geographer* 2 (May 1981): 189–96.

Grotewold, Andreas. "The Growth of Industrial Core Areas and Patterns of World Trade." *Annals of the Association of American Geographers* 61 (1971): 361–70.

Guest, Avery M. "Population Suburbanization in American Metropolitan Areas, 1940–1970." *Geographical Analysis* 7 (July 1976): 267–83.

Gutkind, Erwin A. *International History of City Development,* 4 vols. New York: Free Press, 1964–1969.

Hall, Peter, and Ann Markusen, eds. *Silicon Landscapes.* Boston, London, and Sydney: Allen & Unwin, 1985.

Hance, William. *Population, Migration, and Urbanization in Africa.* New York: Columbia University Press, 1970.

Hardoy, Jorge, ed. *Urbanization in Latin America: Approaches and Issues.* Garden City, N.Y.: Doubleday (Anchor Books), 1975.

Harris, Chauncey D. "A Functional Classification of Cities in the United States." *Geographical Review* 33 (January 1943): 86–99.

Harris, Chauncey D., and Edward L. Ullman. "The Nature of Cities." *Annals of the American Academy of Political and Social Science* 143 (1945): 7–17.

Hart, John Fraser. "The Bypass Strip as an Ideal Landscape." *Geographical Review* 72 (1982): 218–22.

Herbert, David. *The Geography of Urban Crime.* New York: Longman, 1982.

Herbert, D. T., and R. J. Johnston, eds. *Geography and the Urban Environment: Progress in Research and Applications.* New York: John Wiley & Sons, 1984.

Hoare, Anthony G. "What Do They Make, Where, and Does It Matter Any More? Regional Industrial Structures in Britain Since the Great War." *Geography* 7 (October 1986): 289–304.

Hottes, Ruth. "Walter Christaller." *Annals of the Association of American Geographers* 73 (1983): 51–54.

Hoyt, Homer, ed. *Structure and Growth of Residential Neighborhoods in American Cities.* Washington, D.C.: Federal Housing Administration, 1939.

Hugill, Peter J. "English Landscape Tastes in the United States." *Geographical Review* 76 (October 1986): 408.

———. "Good Roads and the Automobile in the United States, 1880–1929," *Geographical Review,* 72 (1982), 327–49.

Isard, Walter. *Location and Space-Economy: A General Theory Relating to Industrial Location, Market Areas, Land Use, Trade and Urban Structure.* New York: John Wiley & Sons, 1956.

Jakle, John A. "Motel by the Roadside: America's Room for the Night." *Journal of Cultural Geography* 1 (Fall-Winter 1980): 34–49.

Jakle, John A., and Richard L. Mattson. "The Evolution of a Commercial Strip." *Journal of Cultural Geography* 1 (Spring-Summer 1981): 12–25.

Johnson, James H., ed. *Suburban Growth: Geographical Processes at the Edge of the City.* New York: John Wiley & Sons, 1974.

Johnston, R. J. *The American Urban System: A Geographical Perspective.* London: Longman, 1982.

———. *City and Society: An Outline for Urban Geography.* London: Hutchinson Education, 1984.

Jordan, Terry G. "Alpine, Alemannic, and American Log Architecture." *Annals of the Association of American Geographers* 70 (June 1980): 154–80.

Kane, Kevin David, and Thomas L. Bell. "Suburbs for a Labor Elite." *Geographical Review* 75 (July 1985): 319.

Kearns, Kevin C. "Intraurban Squatting in London." *Annals of the Association of American Geographers* 69 (December 1979): 589–621.

Keyfifty, Nathan. "Do Cities Grow by Natural Increase or by Migration?" *Geographical Analysis* 2 (1980): 142–56.

Kiang, Ying-cheng. "Recent Changes in the Distribution of Urban Poverty in Chicago." *The Professional Geographer* 1 (February 1976): 57–60.

King, Leslie J. *Central Place Theory.* Beverly Hills, Calif.: Sage Publications, 1984.

Kipnis, Baruch A., and Eric A. Swyngedouw. "Manufacturing Research and Development in a Peripheral Region: The Case of Limburg, Belgium." *The Professional Geographer* 40 (May 1988): 149–58.

Lenman, Bruce. *An Economic History of Modern Scotland.* Hamden, Conn.: Archon Book, 1977, p. 288.

Ley, David. "Alternative Explanations for Inner-City Gentrification: A Canadian Assessment." *Annals of the Association of American Geographers* 76 (December 1986): 521–35.

_____. "Liberal Ideology and the Postindustrial City." *Annals of the Association of American Geographers* (June 1980: 238–58.

_____. *A Social Geography of the City.* New York: Harper & Row, 1983.

Linsky, Arnold S. "Some Generalizations Concerning Primate Cities." *Annals of the Association of American Geographers* 55 (1965): 506–13.

Lloyd, William J. "Understanding Late Nineteenth-Century Cities." *Geographical Review* 71 (1981): 460–71.

Lo, C. P. "Changes in the Shapes of Chinese Cities." *The Professional Geographer* 2 (May 1980): 173–83.

Lowder, Stella. *The Geography of Third World Cities.* New Jersey: Barnes and Noble Books, 1986.

Lowenthal, David. "The America Scene." *Geographical Review* 58 (1968): 61–88.

_____. "The Bicentennial Landscape: A Mirror Held Up to the Past." *Geographical Review* 67 (1977): 253–67.

Lynch, Kevin. *The Image of the City.* Cambridge, Mass.: MIT Press, 1960.

Manson, Donald M., Marie Howland, and George E. Peterson. "The Effect of Business Cycles on Metropolitan Suburbanization." *Economic Geography* 60 (1984).

Maraffa, Thomas A. "Extended Commuting and the Intermetropolitan Periphery." *Annals of the Association of American Geographers* 70 (September 1980): 313–29.

Marshall, John U. "Christallerian Networks in the Löschian Economic Landscape." *The Professional Geographer* 2 (May 1977): 153–59.

Mayer, Harold M., and Charles R. Hayes. *Land Uses in American Cities.* Champaign, Ill.: Park Press, 1983.

Meinig, D. W., ed. *The Interpretation of Ordinary Landscapes: Geographical Essays.* New York: Oxford University Press, 1979.

Muller, Peter O. *Contemporary Suburban America.* Englewood Cliffs, N.J.: Prentice-Hall, 1981.

Mumford, Lewis. *The City in History.* New York: Harcourt Brace Jovanovich, 1961.

Noble, Allen G. *Wood, Brick, and Stone: The North American Settlement Landscape,* vol. 1, *Houses;* vol. 2, *Barns and Farm Structures.* Amherst, Mass.: University of Massachusetts Press, 1984.

Oakey, Raymond P. *High Technology Small Firms: Innovation and Regional Development in Britain and the United States.*

New York: St. Martin's Press, 1984.

Palen, John. *The Urban World.* New York: McGraw-Hill, 1975.

Pannell, Clifton W. "Cities East and West: Comments on Theory, Form and Methodology." *The Professional Geographer* 3 (August 1976): 233–40.

Park, Robert E., Ernest W. Burgess, and Roderick D. McKenzie, eds. *The City.* Chicago: University of Chicago Press, 1925.

Patten, John, ed. *The Expanding City: Essays in Honor of Professor Jean Gottmann.* London: Academic Press, 1983.

Pirenne, Henri. *Medieval Cities.* Garden City, N.Y.: Doubleday (Anchor Books), 1956.

Procter, Mary, and Bill Matuszeski. *Gritty Cities.* Philadelphia: Temple University Press, 1978.

Pyle, Lizbeth A. "The Land Market beyond the Urban Fringe." *Geographical Review* 75 (January 1985): 32.

Relph, Edward. *The Modern Urban Landscape: From 1880 to the Present.* Baltimore: Johns Hopkins University Press, 1987.

Robson, Brian T. *Urban Growth: An Approach.* Hong Kong: Hong Kong University Press, 1973.

Rogerson, Peter A. "The Demographic Consequences of Metropolitan Population Deconcentration in the U.S." *The Professional Geographer* 3 (August 1982): 307–14.

Rondinelli, Dennis A. *Secondary Cities in Developing Countries: Policies for Diffusing Urbanization.* Beverly Hills, Calif.: Sage Publications.

Rowntree, Lester. "Creating a Sense of Place." *Journal of Urban History* 8 (1981): 61–76.

Rubenstein, James M. "Changing Distribution of the American Automobile Industry." *Geographical Review* 76 (July 1986): 288–300.

Rubin, Barbara. "Aesthetic Ideology and Urban Design." *Annals of the Association of American Geographers* 69 (September 1979): 339–61.

_____. "A Chronology of Architecture in Los Angeles." *Annals of the Association of American Geographers* 67 (December 1977): 521–37.

Saalman, Howard. "The Paradox of the City." *Journal of Cultural Geography* 1 (1981): 98–105.

Scott, Allen J., and Michael Storper, eds. *Production, Work, Territory.* Boston: Allen & Unwin, 1986.

Seargill, D. I. *The Form of Cities.* New York: St. Martin's Press, 1980.

Sit, Victor F. S., ed. *Chinese Cities: The Growth of the Metropolis since 1949.* New York: Oxford University Press, 1985.

Sjoberg, Gideon. *The Preindustrial City.* New York: Free Press, 1960.

Smith, Michael P., and Joe R. Feagin, eds. *The Capitalist City: Global Restructuring and Community Politics.* New York: Basil Blackwell, 1987.

Smith, Neil, and Peter Williams, eds. *Gentrification of the City.* Boston: Allen and Unwin, 1986.

Smith, Wilfred. *Geography and the Location of Industry.* Liverpool: University Press, 1952.

Soja, Edward, Rebecca Morales, and Goetz Wolff. "Urban Restructuring: An Analysis of Social and Spatial Change in Los Angeles." *Economic Geography* 59 (1983).

Stearns, Forest, and Thomas Montag. *The Urban Ecosystem: A Holistic Approach.* Stroudsburg, Pa.: Dowden, Hutchinson and Ross, 1974.

Steele, Fritz. *The Sense of Place.* Boston: CBI Publishing, 1981.

Sternlies, George, and James W. Hughes. "The Changing Demography of the Central City." *Scientific American* 243 (1980): 48–53.

Stilgoe, John. *Common Landscapes of America, 1580 to 1845.* New Haven: Yale University Press, 1982.

Taafe, Edward J., and Howard L. Gauthier, Jr. *Geography of Transportation.* Englewood Cliffs, N.J.: Prentice-Hall, 1973.

Thrift, Nigel, and Peter Williams, eds. *Class and Space: The Making of Urban Society.* London: Routledge & Kegan Paul, 1987.

Todaro, Michael P. "Urbanization in Developing Nations: Trends, Prospects, and Policies." *Journal of Geography* 79 (September-October 1980): 164–73.

Toyne, Brian, Jeffrey S. Arpan, David A. Ricks, Terence A. Shimp, and Andy Barnett. *The Global Textile Industry.* London: George Allen and Unwin, 1984.

Tuan, Yi-fu. *Topophilia: A Study of Environmental Perception, Attitudes, and Values.* Englewood Cliffs, N.J.: Prentice-Hall, 1974.

Vale, Thomas R., and Geraldine R. Vale. *U.S. 40 Today: Thirty Years of Landscape Change in America.* Madison, Wis.: University of Wisconsin Press, 1983.

Vance, James E., Jr. *This Scene of Man: The Role and Structure of the City in the Geography of Western Civilization.* New York: Harper & Row, 1977.

Waller, P. J. *Town, City and Nation: England 1850–1914.* Oxford: Oxford University Press, 1983.

Warner, Sam Bass, Jr. *The Urban Wilderness: A History of the American City.* New York: Harper & Row, 1972.

Wheeler, James O., and Ronald L. Mitchelson. "Atlanta's Role as an Information Center: Intermetropolitan Spatial Links." *The Professional Geographer* 41 (1989): 162–72.

White, Paul. *The West European City, A Social Geography.* London and New York: Longman, 1984.

Whitehand, J. W. R., ed. *The Urban Landscape: Historical Development and Management: Papers by M. R. G. Conzen.* London: Academic Press, 1981.

Winters, Christopher. "Traditional Urbanism in North Central Sudan." *Annals of the Association of American Geographers* 67 (December 1977): 500–20.

Wolch, Jennifer R. "Residential Location of the Service-Dependent Poor." *Annals of the Association of American Geographers* 70 (September 1980): 330–41.

Wonders, William C. "Log Dwellings in Canadian Folk Architecture." *Annals of the Association of American Geographers* 69 (June 1979): 187–207.

Wood, William B. "Intermediate Cities on a Resource Frontier." *Geographical Review* 76 (April 1986): 149.

Yeung, Yue-man. "Controlling Metropolitan Growth in Eastern Asia." *Geographical Review* 76 (April 1986): 125.

The Mosaic of Culture

Some of the most obvious differences in the various regions of the world are associated with specific cultural traits such as ethnic background, religion, language, and political organization. Variations in crops, foods, architecture, dress styles, legal systems, place-names, and organization of space reflect the complex intertwining of these variables in each place. Part 3 explains how these cultural variables combine to create the global mosaic that is evident in the world's cultural geography. Chapter 8 explains how ethnic and racial variations are often confused and discusses the geographic impact of both.

Chapter 9 examines the origin and diffusion of the world's major languages in order to explain the present linguistic map of the world. The impact of religion on land and life is discussed in chapter 10. Taboos on certain foods, restrictions on certain activities, legal systems, and architecture combine to create landscapes and regions that reflect the underlying religions. Chapters 11 and 12 discuss how recognizable regions can be created from the mosaic of culture. The most apparent of these regions are formal political units such as countries, but there are also broad cultural regions that transcend political boundaries, creating recognizable cultural realms.

277

Race and Ethnicity: Conflict and Change

Since ancient times records and opinions have accumulated, and various causes have been invoked as the reasons for the differences between the lives and customs of different groups of mankind. Quite early . . . race was thought to explain many differences. . . . But serious difficulties are encountered in any . . . racial explanation.

It is clear that [it is] the spread of knowledge and customs from one people to another, that is [most] important in [explaining] the final pattern of human life in any one region or among any group.   C. DARYLL FORDE, 1934.

THEMES

The geography of human variation: ethnicity and race.
The geographic distribution of racial groups.
The effects of ethnicity and racial division.
Race and ethnicity in the United States.
Global ethnocentrism.

CONCEPTS

Ethnicity
Ethnic group
Race
Racism
Segregation
Ethnocentrism
Phenotypes
Genotypes
Mutations
Genetic isolation
Genetic drift
Social selection
Homelands
De jure
De facto
Gerrymandering
Tribal groups
Boers
Apartheid
Social Darwinism
Pass laws

This chapter examines two of the most widely misunderstood and abused aspects of culture: ethnic and racial characteristics. The combination of characteristics that make the Japanese distinct from Koreans, or Italians from Spaniards, is called *ethnicity*. Ethnicity refers to an *ethnic group*, a group that has a sense of ancestral identification with a larger segment of the world's population. Ethnicity includes language, religion, and other cultural traits that distinguish an ethnic group, like ethnic food and music. Ethnic differences give character to the world's cultural geography, but sometimes such differences form the basis for persecution or discrimination.

Race is an artificial classification referring to the biological differences (perceived by a society) used to characterize a portion of the human population. Race is culturally defined since biologically there is only one race, the human race. Racial groups are defined as collections of human beings recognized and labeled as alike because of perceived physical similarities in skin color or other attributes. All of us have had the opportunity of indicating "race" on formal documents of one type or another (Figure 8–1). Depending on the country in which you were born, birth certificates may ask for race. When entering another country, visa applications often include a blank for "race." In the name of perceived racial differences individuals and groups have been favored, segregated, persecuted, enslaved, or murdered (Figure 8–2).

The terms race and ethnicity refer to very different concepts, yet are sometimes used inaccurately as if they were interchangeable. Ethnicity refers to the cultural background of an individual or group (Figure 8–3). Race is normally used to refer to minor biological differences such as skin color, hair texture, body structure, eye shape, or blood type. These vary among groups and are biological features that can be measured or observed. *Racism* refers to the theory that the variation in skin color, hair texture, eye shape, or other biological attributes leads to superiority or inferiority on the part of any classified group. Racism is expressed in the generalized, permanent exploitation of real or imaginary biological differences, to the advantage of the accuser and to the detriment of the victim, for the purpose of justifying aggression.

Children of China share physical traits common in the Chinese people.

Proponents of racism have erroneously used such disparate traits as religion, national origin, and geographic location to identify supposed racial groups. Racism helps explain the mass movement of black slaves from Africa to the United States and Latin America, the isolation of American Indians on reservations, and the destruction of six million Jews by the Nazi Germans. Other examples of the geographic impact of racism include the *segregation* of the black American population up to the present time and the segregation and denial of civil rights to the black population of South Africa. The exploitation, murder, or segregation of the individual or group in each example reflects the view that the persecuting group is somehow superior.

Ethnocentrism is the belief that particular ethnic or cultural groups are superior to others. It may not be as overtly discriminatory or damaging as outright racism, but it can be as destructive if it is translated into discriminating action. Ethnocentrism has been used as a basis for the exclusion of groups or individuals from voting, immigrating, or locating in a specific place and for promulgating ideas and views of the superiority of an ethnic or nationality group. Both race and ethnicity as elements of culture have affected the world's cultural geography.

The Geography of Human Variation

Humans belong to a single subspecies, *homo sapiens*. No available scientific knowledge pro-

Here are the QUESTIONS ↓	These are the columns for ANSWERS ← *Please fill one column for each person listed in Question 1.*	**PERSON in column 1**	**PERSON in column 2**	**PERSON in column 3**
		Last name	*Last name*	*Last name*
		First name *Middle initial*	*First name* *Middle initial*	*First name* *Middle initial*
2. How is this person related to the person in column 1? *Fill one circle.* If "Other relative" of person in column 1, give exact relationship, such as mother-in-law, niece, grandson, etc.		*START in this column with the household member (or one of the members) in whose name the home is owned or rented. If there is no such person, start in this column with any adult household member.*	If relative of person in column 1: ○ Husband/wife ○ Father/mother ○ Son/daughter ○ Other relative → ○ Brother/sister — If not related to person in column 1: ○ Roomer, boarder ○ Other nonrelative → ○ Partner, roommate ○ Paid employee	If relative of person in column 1: ○ Husband/wife ○ Father/mother ○ Son/daughter ○ Other relative → ○ Brother/sister — If not related to person in column 1: ○ Roomer, boarder ○ Other nonrelative → ○ Partner, roommate ○ Paid employee
3. Sex *Fill one circle.*		○ Male ■ ○ Female	○ Male ■ ○ Female	○ Male ■ ○ Female
4. Is this person — *Fill one circle.*		○ White ○ Asian Indian ○ Black or Negro ○ Hawaiian ○ Japanese ○ Guamanian ○ Chinese ○ Samoan ○ Filipino ○ Eskimo ○ Korean ○ Aleut ○ Vietnamese ○ Other — *Specify* ○ Indian (Amer.) *Print tribe →*	○ White ○ Asian Indian ○ Black or Negro ○ Hawaiian ○ Japanese ○ Guamanian ○ Chinese ○ Samoan ○ Filipino ○ Eskimo ○ Korean ○ Aleut ○ Vietnamese ○ Other — *Specify* ○ Indian (Amer.) *Print tribe →*	○ White ○ Asian Indian ○ Black or Negro ○ Hawaiian ○ Japanese ○ Guamanian ○ Chinese ○ Samoan ○ Filipino ○ Eskimo ○ Korean ○ Aleut ○ Vietnamese ○ Other — *Specify* ○ Indian (Amer.) *Print tribe →*
5. Age, and month and year of birth *a. Print age at last birthday. b. Print month and fill one circle. c. Print year in the spaces, and fill one circle below each number.*		a. Age at last birthday c. Year of birth /1/ b. Month of birth ○ Jan.–Mar. ○ Apr.–June ○ July–Sept. ○ Oct.–Dec.	a. Age at last birthday c. Year of birth /1/ b. Month of birth ○ Jan.–Mar. ○ Apr.–June ○ July–Sept. ○ Oct.–Dec.	a. Age at last birthday c. Year of birth /1/ b. Month of birth ○ Jan.–Mar. ○ Apr.–June ○ July–Sept. ○ Oct.–Dec.
6. Marital status *Fill one circle.*		○ Now married ○ Separated ○ Widowed ○ Never married ○ Divorced	○ Now married ○ Separated ○ Widowed ○ Never married ○ Divorced	○ Now married ○ Separated ○ Widowed ○ Never married ○ Divorced
7. Is this person of Spanish/Hispanic origin or descent? *Fill one circle.*		○ No (not Spanish/Hispanic) ○ Yes, Mexican, Mexican-Amer., Chicano ○ Yes, Puerto Rican ○ Yes, Cuban ○ Yes, other Spanish/Hispanic	○ No (not Spanish/Hispanic) ○ Yes, Mexican, Mexican-Amer., Chicano ○ Yes, Puerto Rican ○ Yes, Cuban ○ Yes, other Spanish/Hispanic	○ No (not Spanish/Hispanic) ○ Yes, Mexican, Mexican-Amer., Chicano ○ Yes, Puerto Rican ○ Yes, Cuban ○ Yes, other Spanish/Hispanic
8. Since February 1, 1980, has this person attended regular school or college at any time? *Fill one circle. Count nursery school, kindergarten, elementary school, and schooling which leads to a high school diploma or college degree.*		○ No, has not attended since February 1 ○ Yes, public school, public college ○ Yes, private, church-related ○ Yes, private, not church-related	○ No, has not attended since February 1 ○ Yes, public school, public college ○ Yes, private, church-related ○ Yes, private, not church-related	○ No, has not attended since February 1 ○ Yes, public school, public college ○ Yes, private, church-related ○ Yes, private, not church-related
9. What is the highest grade (or year) of regular school this person has ever attended? *Fill one circle. If now attending school, mark grade person is in. If high school was finished by equivalency test (GED), mark "12."*		**Highest grade attended:** ○ Nursery school ○ Kindergarten Elementary through high school *(grade or year)* 1 2 3 4 5 6 7 8 9 10 11 12 ○○○○○○ ○○ ○○○ ○ College *(academic year)* 1 2 3 4 5 6 7 8 or more ○○○○○○○○ ○ Never attended school – *Skip question 10*	**Highest grade attended:** ○ Nursery school ○ Kindergarten Elementary through high school *(grade or year)* 1 2 3 4 5 6 7 8 9 10 11 12 ○○○○○○ ○○ ○○○ ○ College *(academic year)* 1 2 3 4 5 6 7 8 or more ○○○○○○○○ ○ Never attended school – *Skip question 10*	**Highest grade attended:** ○ Nursery school ○ Kindergarten Elementary through high school *(grade or year)* 1 2 3 4 5 6 7 8 9 10 11 12 ○○○○○○ ○○ ○○○ ○ College *(academic year)* 1 2 3 4 5 6 7 8 or more ○○○○○○○○ ○ Never attended school – *Skip question 10*
10. Did this person finish the highest grade (or year) attended? *Fill one circle.*		○ Now attending this grade *(or year)* ○ Finished this grade *(or year)* ○ Did not finish this grade *(or year)*	○ Now attending this grade *(or year)* ○ Finished this grade *(or year)* ○ Did not finish this grade *(or year)*	○ Now attending this grade *(or year)* ○ Finished this grade *(or year)* ○ Did not finish this grade *(or year)*
		CENSUS USE ONLY **A.** ○ I ○ N ○○	CENSUS USE ONLY **A.** ○ I ○ N ○○	CENSUS USE ONLY **A.** ○ I ○ N ○○

FIGURE 8–1 Page from the U.S. Census Asking for "Race" The long form of the 1980 U.S. Census includes questions concerning race.

FIGURE 8–2
Continuum of Possible Types of Cultural Interaction Possible types of interaction range from genocide to peaceful acceptance of another's cultures.

CONTINUUM OF POSSIBLE TYPES OF CULTURAL INTERACTION

Genocide   Enslavement   Segregation   Exploitation   Co-existence   Assimilation   Integration   Cultural Pluralism

281

ETHNICITY IN CALIFORNIA'S PUBLIC SCHOOLS, SELECTED YEARS

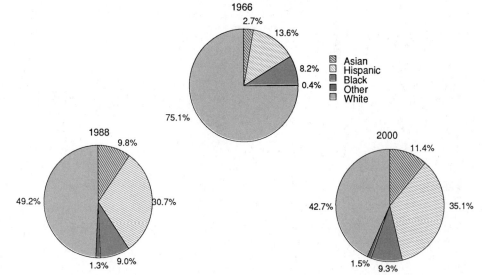

1966
2.7%
13.6%
8.2%
0.4%
75.1%

Asian
Hispanic
Black
Other
White

1988
9.8%
49.2%
30.7%
1.3%   9.0%

2000
11.4%
42.7%
35.1%
1.5%   9.3%

FIGURE 8–3 Ethnic Groups in California: School Enrollment The changing ethnic mix found in California demonstrates the growth of various ethnic groups.

A young woman from Kenya in traditional dress illustrates both physical and cultural characteristics found in Africa.

A common physical characteristic found among residents of Europe is light colored skin and blond hair as in this girl from Switzerland.

Differences in physical appearance between African and Asian peoples are demonstrated by a Nigerian and a Mongolian.

vides a basis for believing that groups of humans differ in their innate capacity for intellectual and emotional development. Biological differences among human beings within a single racial group may have a greater range than differences of the same features among races. Historical and sociological studies support the view that racial differences are of little significance in determining the social and cultural differences between various groups of humans.

Despite the fact that there is but one human race, we can recognize groups having distinctive features. Children inherit from their parents the genetic material that sets a range of blood type, skin color, hair texture, body form, and the assorted characteristics that make each individual unique. An individual's physical characteristics are known as *phenotypes*. A phenotype is the outwardly visible or physically expressed action of the genes. A *genotype* is the actual genetic makeup range that is not visible, but is manifest in the phenotype.

Intermarriage within a group of people over a long period of time tends to make that group and their offspring more alike and emphasizes their differences from those whom they do not marry. Through the effect of dominant and recessive genes, such traits as dark skin, eyes, and hair will prevail over recessive genes for light skin, blue eyes, and blond hair. Dominance of genes holds true for a host of genetically determined characteristics such as left- or right-handedness; hair color; nose, jaw, head, and general body shape; and height. Although differences in nutrition, environment, and child care may affect some of these variables, they are basically selected from the genetic makeup of the parents.

Isolation of a group leads to the convergence of genetic characteristics within the group, creating a fairly distinctive group of people whose physical characteristics can be viewed as racially distinct from others. It is important to reiterate, however, that there is but one race, the human race. What we normally call "races" are simply subgroups artificially classified within the human race.

Variations in physical appearance among human groups may result from *mutations* occurring over long periods of time. Mutations are *permanent* structural changes in genes and in the offspring who receive these genes. Mutations constantly occur in groups or individuals, but few of these last beyond the life of the individual with the mutation. The theory of *natural selection* was proposed by Darwin to explain the success of some mutations. This theory hypothesizes that mutations that help an individual to be more successful in his or her environment or social group will persist in subsequent generations, whereas those that cause a disadvantage will not.

Whether a successful mutation will be added to the gene pool of a group is influenced by several factors. When one genetic group is isolated from others of the same species, the chance of a muta-

tion permeating the entire group is greater (Figure 8–4). *Genetic isolation* of small groups results in their divergence from other groups over time since mutation occurs at random in all groups. Some mutations offer no particular advantage but are simply a variant of existing genetic types that diffuse through a group in a process called *genetic drift*. Such things as eye color or dominant handedness are mutations offering no advantage, and the process of natural selection does not affect them.

Some mutations are not affected by natural selection, but are perceived as valuable by a group and encouraged through *social selection*. Social selection has been used in the past to justify killing of twins or to prevent reproduction by those with limited mental ability. The combination of natural and social selection causes isolated groups to develop distinctive physical characteristics. Examination of specific physical character-

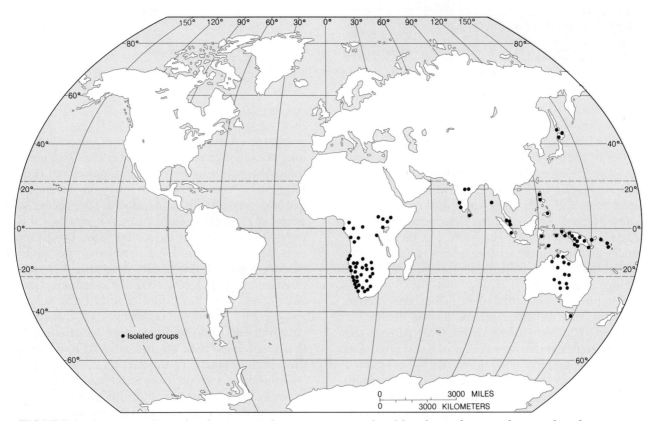

FIGURE 8–4 Genetically Isolated Groups Where groups are isolated by physical geography or cultural mores they slowly develop their own recognizable characteristics.

istics that are commonly assumed to be associated with individual racial groups helps explain the regional distribution of recognizable genetic factors in the world's population

Physical Characteristics

One apparent physical characteristic is skin color. For generations some individuals have attempted to classify human populations into distinct racial groups on the basis of skin color. The apparent justification for this is that skin color is an obvious characteristic. There is a continuum from the complete absence of skin pigmentation found in albinos to the dark skin color in groups in some portions of Africa and Melanesia. Skin color results from the presence of several pigments, such as *melanin*, in the skin. Increased amounts of melanin darken the skin.

Skin color differences persist when genetic groups are relatively isolated or when groups with differing skin color living together are regulated by patterns of social segregation. Efforts to prevent intermarriage between people of different skin pigmentation in some countries or regions are based on a cultural view of what constitutes a race. Skin color itself is not evidence for distinctive classes or races of peoples because a number of genes control skin color, and the variation within any group may be as broad as that between groups.

Less obvious than skin color are physical features such as head shape, hair form, and body structure. Hair form and texture, like skin color, represent visible phenotypes and appear to show distinct variations between racial groups. Characteristics of hair (straight or tightly curled, fine textured or thick, dark or light) have been used at one time or another to classify divisions of people. The straight hair of the residents of North and East Asia marks one end of this continuum, and the tightly curled hair of the West African the other. Like skin color, however, intermediate forms range from straight to wavy to curly across Europe to South Asia. Hair type and color are the results of specific genotypes, but using hair characteristics as a tool for classifying individuals as members of one so-called race or another is very questionable. Like skin color, the variation

within any specified group may be as large or larger than the variation between groups.

Other phenotypes that physical anthropologists have used in trying to classify people into races include such things as the amount of body hair, the shape of the jaw and nose, the presence or absence of the Mongolian (*epicanthic*) eyefold, stature, and head shape or form. Although distinct differences can be recognized in each case, they do not vary in a set pattern with other factors to create distinct groupings of people.

An example is stature. It is possible to recognize in one small area in Africa peoples with tall, medium, and short stature, yet all have relatively dark skins. The Tutsi cattle-herding people in eastern Zaire and the surrounding region are extremely tall, with males averaging over six feet. Yet Pygmies averaging less than four feet six inches and the medium-height Bahutu inhabit the same region. Although differences in stature may in part reflect nurture, the persistence of the tall, medium, and short stature of these groups regardless of the amount of nutrition and cultural selection they receive reflects their distinct genotypes.

Another phenotype that has been studied extensively by anthropologists is head shape. The remains of the earliest ancestors of humans are skeletal, and the proportions of skulls are easy to measure. Anthropologists developed the *cephalic*

The Watusi peoples of Africa are characterized by their height, with many individuals exceeding six feet.

index, the ratio of the breadth to the length of the skull, in the hopes of classifying individuals into distinct groups or divisions. The index has been used to create classes of head shape that essentially divide skulls into long or broad form. Some West Europeans (such as Norwegians) have long heads, while Mongoloids (such as the Chinese) have broad heads. Other West Europeans (such as the Great Russians) have broad heads, however, while other Mongoloid groups (such as the Japanese) have long heads. Thus the use of head shape as a distinguishing trait associated with recognizable racial groups is clearly untenable.

Blood type is another physical characteristic of interest because it might provide clues about the hearths and migrations of early populations. Blood types are classified as A, B, AB, and O and are based on the presence or absence of clotting agents (A or B), which cause red blood cells to clot when mixed with alien blood that does not contain those agents. The four blood types (A, B, AB, and O) denote the specific clotting agents they carry. A and B each carry their respective clotting agent, AB carries both, and type O carries neither. In the genetic system, A and B are dominant over O, but neither A nor B is dominant over the other. The agents that create the distinctive blood types do not mix, hence an AB person has both A and B blood types. Analysis of blood types shows some interesting regional concentrations, but one that is not conclusively correlated with physical features (Figure 8–5).

Examination of the regional distribution of blood types raises interesting questions, but some areas (for example, the modern industrial world) have experienced much greater sampling of blood types than others, and the actual interpretation of the distribution is open to debate. At the present time the incidence of type O is high among the American Indian population (70–90 percent O), in Northeast Asia (over 70 percent), and among Mongoloid peoples of Asia in general (30–40 percent).

High incidence of type A is found in extreme northern North American Indian populations (over 30 percent) and in northern and western Europe. Type A is almost absent in most of the native population of the United States and Latin America (less than 10 percent) and across most of Africa, South Asia, and Northeast Asia (10–20 percent). High incidences of B blood types are found primarily in the highland regions of Asia (more than 25 percent) and in other scattered areas where they are the minority of blood types.

The best conclusion that can be safely drawn from evaluating the phenotype of blood type is that the human population has experienced both wide-ranging movement and migration from a very early date. Other factors are associated with blood type, including the RH factor, enzyme relationships, and other specifics that can be analyzed and mapped, but in each case they fail to co-vary with the other phenotypes.

Thus evaluation of phenotypes demonstrates that humans are one race, related to one another, with physical differences that seem to reflect long periods of adaptation and mutation in specific environmental settings. Nevertheless, it is still useful to recognize the traditional divisions that are accepted among anthropologists and cultural geographers as a means of describing the peoples that are found in a specific area. It must be reiterated that such recognition does not imply any value judgment concerning intellectual ability, physical ability, emotional characteristics, or other value-laden descriptive adjectives. The divisions are intended only to recognize the general physical characteristics of the peoples who live in each broad region of the world, characteristics

Mongolian children share physical characteristics that include a broad head shape.

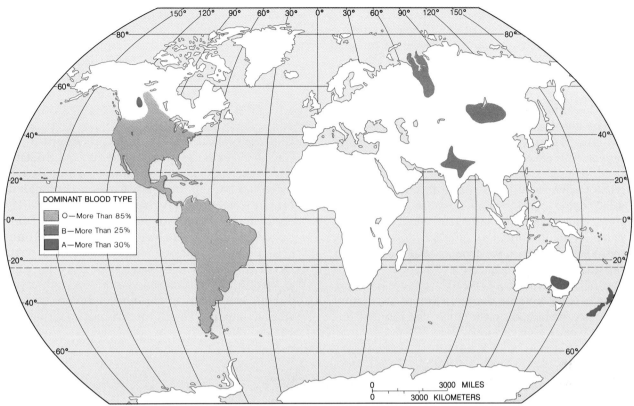

FIGURE 8–5 Distribution of Blood Types This map of world patterns of blood types in original native populations shows the regions with the highest concentration of each type. Most of Europe, Africa, and Asia have complex combinations of blood types.

that contribute to the cultural geography of each place.

Regionalization of the Main Racial Groups

One of the typical divisions of peoples is into Caucasoid, Negroid, and Mongoloid, with their individual subsets. These are sometimes referred to as the main racial groups. The division of peoples into these three broad groups should not mask the fact that they are heterogeneous groups who have a variety of genetic combinations and resultant genotypes and phenotypes. The Negroid (Congoid) phenotypes include dark skin, black curly hair, dark eyes, broad nose, long head, and stocky body build. The hearth region for the Negroid group is Africa.

The Mongoloid classification includes individuals with phenotypes of light yellow to brown skin; brown eyes; flat face and nose; epicanthic eyefolds; straight, coarse black hair; and high cheek bones and short stocky body structure. The hearth region for these people is East Asia, presumably in eastern China and Mongolia.

The Caucasoid genetic group includes phenotypes of fair skin and eyes, prominent and narrow nose, light wavy hair, hirsuteness (abundance of body hair), and thin lips. The original hearth of Caucasians is the region including the tundras and forests of Europe, the Soviet Union, and Turkey.

Within the broad groups of races there are distinct subsets. The Australoid, Bushmanoid, Negritoid, and related subsets are often recognized as distinct genetic groups, isolated until recent times from the three primary racial groups. The uncertain and subjective nature of classifying groups as part of one racial group or another has prompted attempts to classify people into cultural groups based on geographical regions rather than on race. A common division based on geographical regions recognizes European, Asian, Indian, African, and Amerindian groups and subgroups composed of Australians, Melanesians, Micronesians, and Polynesians (Figure 8–6).

The division of groups into Europeans, Asians, and the like is based on cultural history and reflects the effort to separate recognition of cultural groups from racism. Using geographic regions to define cultures is useful, but masks the relationship of individual regional groupings to the broader racial groups. Using the three main racial divisions and four subdivisions provides more insight into how geographical isolation in the past has been associated with the development of distinct phenotypes. The phenotypes that were used in defining the main groups are simply a means of indicating the great variation in hu-

mans found on the face of the earth. Examination of the geographical distribution of each group illustrates the fallacy of assuming that there are so-called pure races, since it is clear that a great deal of intermixing has occurred among human groups over time.

Geographic Distribution of Racial Groups

Caucasoids are the most numerous and the most widespread racial group. Caucasoid populations dominate North and South America, Europe, the Middle East, North Africa, and South Asia. The expansion of Europeans has also spread the Caucasoid group to Australia, New Zealand, and South Africa. Over one-half of the world's population are classified as Caucasoids.

Within the Caucasoid group some observers distinguish subdivisions closely related to geographic regions. The Europeans, for example, can

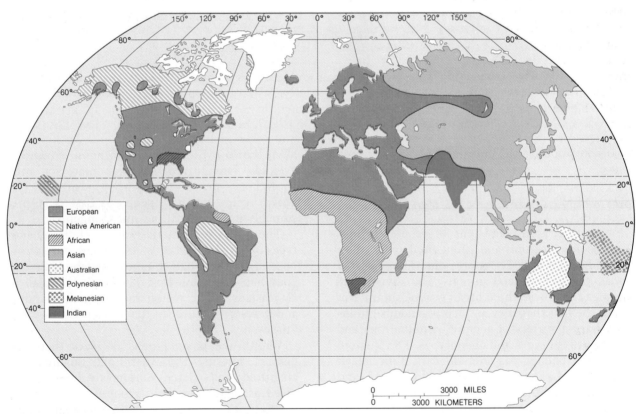

FIGURE 8–6 System of Racial Classification This map shows one system for dividing the world's population into ethnic/racial groups.

be divided into three large groups: Nordic, Alpine, and Mediterranean. Nordic peoples are characterized as having fair skin; blue eyes; wavy, light blonde or red hair; narrow noses; and long heads. Nordics as so defined constitute a bare majority in Scandinavia and represent recognizable population elements in the British Isles, northern Germany, the Netherlands, Belgium, Luxembourg, North America, Australia, and New Zealand.

The Alpine subgroup of the Europeans is characterized as having stocky bodies, light skin color, straight to wavy brown hair, and high bridged noses. They are found in central Europe, the Soviet Union, Czechoslovakia, Romania, Hungary, and France.

The Mediterranean European subgroup occupies the region north of the Mediterranean from Portugal through Italy, Greece, and Turkey; south of the Mediterranean in North Africa; and eastward through India. Mediterranean peoples are defined as having long heads, dark eyes, and hair that ranges from brown to very dark brown and from straight to wavy. Skin color ranges from light olive to dark, and noses are generally prominent.

Representatives of the three European subgroups are found in the Americas, Australia, New Zealand, and South Africa among immigrant populations. The range from Nordic to Mediterranean peoples reflects the great amount of intermarriage and intermixing experienced by the Caucasoid peoples over time. The Mediterranean people in North Africa have characteristics of the Negroid population of Africa, and in India there are elements of the Mongoloid subgroup. Some geographers and anthropologists classify the people of the region from Pakistan and Afghanistan east through India and Bangladesh as an Indian subgroup. As defined, the Indian subgroup differs from other European groups in having darker skin pigmentation and darker, coarser hair.

The difficulty of categorizing an individual as either Alpine, Nordic, or Mediterranean reinforces the fact that it is futile to undertake racial classifications for anything other than the description of the appearance of people in a particular region. Attempts to separate Caucasoid peoples into subgroups largely reflect ethnic rather than racial subdivisions. Where the European peoples have migrated extensively, the individual characteristics of the three European subgroups are mixed, and it is difficult to find individuals who exemplify all of the characteristics supposedly associated with the individual European Caucasoid groups.

The Mongoloid racial subgroups are found in East Asia in China, Korea, parts of the Soviet Union, Japan, and Southeast Asia and are viewed by scientists as including the American Indians. Mongoloids are defined as having phenotypes that include a stocky body, round head, little body and facial hair, flat nose, and epicanthic eyefold. The stocky build of the Mongoloid is replaced by the slender body form in North China, and in Japan the round head is replaced by a slender face with a slight body structure. The Inuit of North America and Siberia in the Soviet Union have many of the phenotypes associated with the Mongoloid group, but have a prominent narrow nose.

American Indians are even more diverse than the Caucasoid group, and the subgroup of Amerindians in the Mongoloid group includes a variety of recognizable phenotypes. Isolated groups of Amerindians in the Amazon, high Andes, and desert Southwest of the United States still exhibit the Mongoloid phenotypes of an epicanthic eyefold, brown to light brown skin color, and straight, dark coarse hair. Most of the native American peoples of North and South America have intermixed to a greater or lesser extent with later immigrants, especially Europeans. Aside from the very early migration of Mongoloid peoples to North and South America (estimated to have occurred some twenty to thirty thousand years ago), immigration from East and Southeast Asia by Mongoloids has been a recent phenomenon. Relatively large numbers are evident in minorities in the Hawaiian Islands, California, France, and in a few places along the Amazon River (Table 8–1). Most of the Mongoloid peoples are still found in East and Southeast Asia, where they number over one and one-half billion.

Negroid (Congoid) population is concentrated in Africa south of the Sahara, the Caribbean Islands and the area around the Gulf of Mexico in North and South America. Negroid subgroups form minority populations in the United States,

TABLE 8–1
Asians in the United States: Immigration by Country
of Last Residence (thousands)

COUNTRY	1820–1986	1961–1970	1971–1980
China	737	34.8	124.3
Hong Kong	252	75.0	113.5
Japan	440	40.0	49.8
Korea	508	34.5	267.6
Philippines	765	98.4	355.0
Vietnam	404	4.3	172.8
Other Asian countries	653	36.5	176.1

Source: U.S. Statistical Abstract (Washington, D.C.: U.S. Government
Printing Office, 1988).

Brazil, Venezuela, and Colombia and the majority
population in Haiti and many Caribbean Islands.

The phenotypes ascribed to the Negroid
group—dark skin, dark tightly curled hair, a flat
nose, prognathism (jutting jaw), everted lips, and
limited body hair—vary greatly from person to
person. Intermixing of Negroids with Caucasoids
in North Africa historically and in North and
South America after migration have diffused Caucasoid characteristics to the Negroid population
and vice versa. The Negroid subgroup of North
America is as apt to have a high-bridged nose and
a light brown skin as to have a dark brown skin
and a low-bridged, flatter nose. The variation
within Africa itself is equally diverse; it reflects
intermixing with colonists and mutation of both
genotypes and phenotypes historically, creating
groups with extreme height such as the Watusi or
extremely short body structure such as the
Pygmy groups.

The great variation found within the main
racial subgroups suggests intermixing among the
earth's population over time. Certain groups have
distinctive characteristics because of isolation,
particularly the Melanesians, Austroloids, and
the Micronesian, Polynesian, and Bushmen
groups. The Melanesian group is found in New
Guinea and the islands to the east. They have
been isolated over several millennia and have the
darkest skin among the world's population
groups. They are generally tall and have high-
bridged broad noses and full but not everted lips.

The Austroloids (or Australnesians) were isolated for thousands of years and are characterized
by dark skin color but with hair that may be red
or even blond in some individuals. Prognathism
of both lower and upper jaw is common, and they
have very flat, broad noses and lips that are not
everted. These Australian aborigines (as the Europeans referred to them) comprised small hunting and gathering populations at the time of the
European occupation of their lands. Numbering
fewer than 300,000 at the time of the European
conquest, they have been reduced to some
100,000 today.

The Micronesian population group is located
east of New Guinea with its Melanesian peoples.
These people were generally isolated on the Pacific islands northeast of Melanesia and are characterized by dark skin, brown eyes, and black,
tightly curled hair. The Polynesian subgroup extends from the Maori of New Zealand to the
Hawaiian Islands. The Polynesians are character-

Polynesians have brown skin, stocky bodies, and dark
curly hair, but intermarriage with Europeans and
others provide a wide range of phenotypes.

ized by strong stocky bodies, brown skin, and curly dark hair.

The Bushmen (!Kung San) are found in the desert areas of Southwest Africa. They are characterized by black hair, brown to yellow-brown skin, flat faces, epicanthic eyefolds, short stature, and *steatopygia* (storage of fat in the buttocks). The Bushmen have been decimated since Europeans occupied the southern portions of Africa.

The broad division of the human race into subgroups based on appearance represents but one cultural variable characterizing the inhabitants of a specific place. Ethnicity is another major variable in the culture of a place. The ethnic background of a group is probably more important to them than any supposed racial differences, and the ethnic map of the world is highly complex because of the great population movements of the past three centuries.

The Effects of Racial Division

The impact of racism, ethnocentrism, and related prejudice toward groups who are in some way different is manifest in a variety of cultural expressions and landscapes (see the Cultural Dimensions boxes on pages 292 and 293). The significance of race for the cultural geographer is found in such diverse examples as segregation of living quarters, differentiation in income and life expectancy by race or ethnic group, distinctive economic and housing patterns, and racial conflicts and racial tension. Consequently, the impact of race or ethnicity on the cultural geography of an area is an important factor in understanding that place. Race and ethnicity have a variety of types of geographical impact including *slavery* (which is still practiced today—in Sudan in 1988 a research report by a professor at the University of Khartoum on the incidence of slavery led to his arrest on charges of anti-state activity); the South African designation of certain areas as **homelands** for the Bantu population; the concentration of black and other minority group members in the central cities of American communities; and ethnocentrism among nationalities or groups from a specific geographic area.

Race and Ethnicity in the United States

The Black Population

The impact of race, ethnicity, racism, and ethnocentricity is apparent in the cultural geography of the United States. As the recipient of a variety of migrant groups over the last four hundred years, the United States provides a broad cross section of both ethnic and so-called racial groups. The impact of these groups on the cultural geography of the country is found in the population geography, housing patterns, economic activity, and community form and shape. Individuals and groups in the United States have been persecuted, enslaved, killed, and discriminated against on the basis of their ethnic or racial background for hundreds of years. Slavery became illegal after 1860, and other types of persecution or discrimination on the basis of race or ethnicity became illegal after 1964, but geographical manifestations of racism and ethnocentrism still occur.

Racism, discrimination, prejudice, and resulting geographical phenomena can be either *de jure* or *de facto*. *De jure* activities are sanctioned by law, while *de facto* activities are illegal but occur nevertheless. It is illegal to enslave people in the United States or to deny access to housing, schools, clubs, employment, or political office on the basis of race and ethnicity. Nevertheless, many instances of persecution or discrimination on the basis of race continue to persist. For example, during the period between the two world wars, in Forsythe County, Georgia (a suburb of Atlanta), all blacks were forced to move after an alleged rape of a white woman by a black male. Blacks have never been allowed to return and are excluded at the present by informal agreements not to sell property to them. In 1987, the white suburb of Howard Beach, New York, was the scene of the death of a black man when a car with three blacks broke down and the individuals entered a bar to phone for help. Patrons of the bar chased the three individuals, causing the death of one who ran into the path of a car. While some of the white individuals involved received jail sentences, the incident itself represents *de facto* racism.

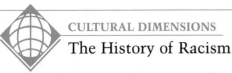

CULTURAL DIMENSIONS
The History of Racism

Racism is an ancient phenomenon. The ancient Greek philosopher Aristotle argued that the social order should be based upon slavery, which he believed was justified by the natural inferiority of the barbarians (non-Greeks) who should be slaves to the Greeks. Widespread adoption of the thesis that some races or cultural groups were inherently inferior appears to date from the beginning of the colonizing efforts of the Europeans. In the sixteenth century, the Spaniards justified their aggression toward and domination of the Indian groups in America by contrasting the civilization of Spain with the supposedly inferior cultures of the Indians.

Almost all groups are ethnocentric to some degree, but the emergence of the idea among the Spanish that the Indians were an inferior race marks the beginning of an aberrant concept that in the nineteenth century came to be known as *scientific racism*. This aberration ultimately led to the Holocaust. Six million Jews died because Hitler and the Nazis accepted scientific racism as a justification for their own bigotry and greed.

Central to the idea of inferior races was the emergence of *Social Darwinism*, which maintained that there were rankings of human groups both within and between societies that reflected varying genetic ability. Invoking the name of Darwin to justify their pseudoscience, European philosophers argued that the gulf between factory owners and factory workers reflected the innate ability of the two groups and concluded that certain groups such as the Jews were inherently inferior.

An influential book by the pseudoscientist Madison Grant (published in 1916) was even more emphatic in proclaiming that Jews, southern and eastern Europeans, blacks, and others were all distinct and inferior races.* Using phenotypes to identify distinct racial groups in Europe, Madison concluded that the Nordic type with blonde hair and long face was superior. In his book *Mein Kampf*, Adolf Hitler quotes Grant almost directly to justify his thesis that there was a pure Teutonic race, which had been defiled by the Jews.

Even today, the ideals of scientific racism persist in the United States, where a racist group has established a "homeland" in Idaho to protect itself from "contamination" by Negroes and Jews (Figure 8–A). Calling itself the Aryan Nation, this group maintains that the so-called white race is being overwhelmed by an influx of genes from Jews and blacks. The Aryan nation has created a tiny region in which racism is an accepted practice.

---

*Madison Grant, *The Passing of the Great Race* (New York: Charles Scribner's Sons, 1916), pp. 184–86.

continued

As an example of *de facto* and *de jure* discrimination, blacks have been entitled to vote and hold office in the United States since the Fourteenth and Fifteenth Amendments to the U.S. Constitution were passed immediately after the Civil War. But more than two-thirds of black adults were effectively prevented from holding office or voting for a hundred years in the American South by such tactics as voter registration fees, literacy tests, or pressure from white employers. Only after passage of the 1964 Civil Rights Act and subsequent civil rights activism in the late 1960s did blacks begin to exercise the right that had been given them by law.

The proportion of southern blacks registered to vote increased from 29 percent to 63 percent between 1965 and 1975. In the last fifteen years, the number of black elected officials has more than tripled, and black mayors have been elected in some of the largest cities in the United States (Los Angeles, Philadelphia, Chicago, and Washington D.C.), and a black presidential candidate gained widespread support in 1984 and 1988.

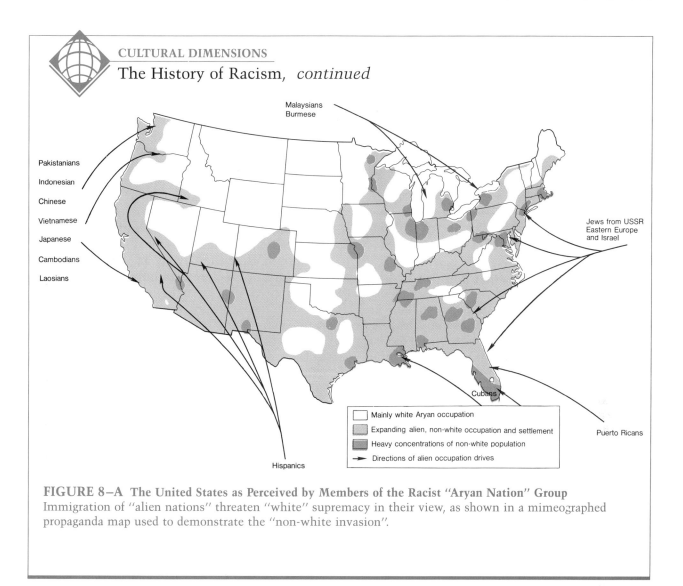

FIGURE 8–A The United States as Perceived by Members of the Racist "Aryan Nation" Group
Immigration of "alien nations" threaten "white" supremacy in their view, as shown in a mimeographed propaganda map used to demonstrate the "non-white invasion".

Examination of the black population of the United States reveals, however, that the geographical impact of race and associated discrimination continues to be a real component in the cultural geography of the nation. Black Americans represent the nation's largest minority group, an estimated 12 percent of the population (Table 8–2). Designation as a black American is based on the U.S. Census. The census designation of race or minority group membership reflects self-identification by respondents. Census enumerators do not attempt to determine membership in

a specific group or race on the basis of ancestry, but merely record self-classification. The census recognizes white, black, American Indian, Eskimo, Aleut, Asian and Pacific Islander, and other. The census categorizes as "white" those who listed themselves as white or as a member of specified ethnic groups (Canadian, German, Polish, or other European). The category of "black" includes persons who indicated their race as black or Negro on the questionnaire, as well as those who did not classify themselves in one of the specific race categories but indicated they were

TABLE 8–2
Black Population in the United States

YEAR	NUMBER (THOUSANDS)	PERCENTAGE OF POPULATION
1800	1,002	18.9
1850	3,639	15.7
1900	8,834	11.6
1950	15,042	10.0
1980	26,683	11.8
1986	29,306	12.2
1988	30,597	12.4

Source: U.S. Statistical Abstract Washington, D.C.: U.S. Government Printing Office, 1988).

Jamaican, West Indian, Haitian, Nigerian, or other African nationals, not all of whom are necessarily black. The categories of American Indian, Eskimo, and Aleut include those who classified themselves as such, as well as those who did not report themselves in one of the specific categories but entered the name of an American Indian nation (See Figure 8–1 on page 281).

The black population of the United States differs from other racial or ethnic groups in the United States in two major ways: its migration and its status in the United States. Until the twentieth century essentially all black migration was involuntary. Over 90 percent of black Americans are descendants of people who came to America before the 1860s as slaves. As involuntary migrants, ancestors of today's black Americans were here before the Swedish, Norwegian, Italian, Irish, Polish, Asian, and other migrant groups who came after the mid-nineteenth century.

The black population of the United States in 1800 was over 90 percent enslaved. Most were agricultural laborers in the rural South. Over time, the freedom granted by the Thirteenth, Fourteenth, and Fifteenth Amendments, which abolished slavery and guaranteed blacks citizenship rights, and other developments since the Civil War have led to important geographical shifts among the black population and changes in their role in the U.S. society and economy. Initially, however, emancipation and the breakup of the slave economy in the South had relatively little effect on the distribution of black Americans.

After the Civil War, black Americans were largely concentrated in the South, particularly the rural South. A few black Americans attempted to establish new black Utopian communities in Kansas and Oklahoma, but these were uneconomical and today have either ceased to exist or are small villages (Figure 8–7). Most black Americans who had fled the South to escape slavery or who moved to the Northeast after the Civil War held menial positions as servants, laborers, or other low-skill occupations.

Segregated schools, legal restrictions on the use of public facilities, and other forms of *de jure* and *de facto* discrimination generally characterized the South after the Civil War. To some degree blacks were also segregated into distinct sections of the cities characterized by low standards of housing, roads, hospitals, and other urban amenities. Such residential segregation by race was not the practice in all southern cities, however. Prior to the twentieth century the social distance that existed between blacks and whites led white employers of blacks to allow them to live in the same neighborhood so they would be readily available for work.

After World War I, blacks began to migrate in increasing numbers to the cities of both the South and the North where jobs were becoming available as foreign emigration to the United States

Brooklyn, New York. The inner city areas of many American cities often have predominantly black populations, in part because of lower average incomes that affect ability to obtain housing.

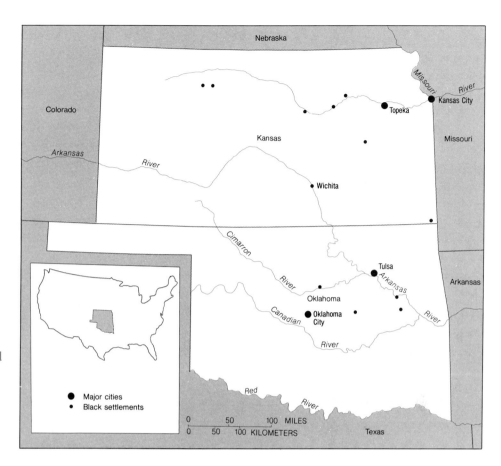

FIGURE 8–7 Black Settlements in Kansas and Oklahoma. These black communities represent efforts to create farming communities for freed slaves. Like others not shown, most failed.

was greatly reduced at the same time that the rate of industrial growth was increasing. The large movement from rural to urban and South to North characterized the distribution of the black population between 1920 and 1970 (Table 8–3). There were 925,000 black farmers in America in 1920, but by 1980 there were fewer than 55,000 black-operated farms. The change in residence from rural to urban as a percentage of the total black population during this half-century was even greater. The 1920 U.S. Census recorded 51 percent of the total population located in towns and cities, compared to only 33 percent of the black population. Nearly 7 million black Americans lived on farms or in rural hamlets in 1920. Only 240,000 of the 27 million black Americans enumerated by the census lived on farms in 1980, and fewer than 4 million resided in places classified as "rural" (Table 8–4).

The Changing Distribution of the Black Population: Urbanization and Migration   The great migration of black Americans to cities has resulted in the present distribution in which the majority of the black population is urban, but an urban population that still leaves the majority of black Americans in the South (Figure 8–8). The reasons for the continued urban migration of black Americans are complex, but include the increasing size and mechanization of American farms (which allowed southern land owners to replace black tenant farmers with machinery); the increasing cost of operating a farm to secure an adequate living, which was biased against blacks with little savings; and the difference between the standard of living in rural and urban areas and in northern and southern cities.

In the 1960s the average rural black family had an income less than half that of all urban blacks.

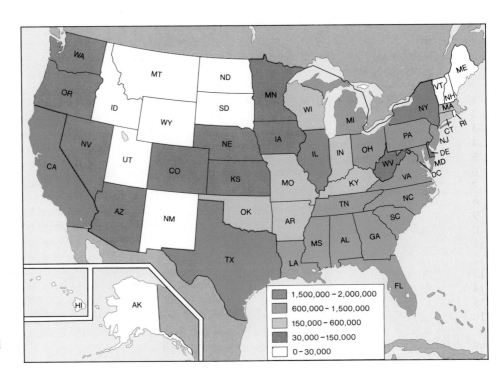

FIGURE 8–8
Distribution of Black Population in the United States

TABLE 8–3
Distribution of Black Population: Regional and Urban–Rural

	1920		1970		1980		1988	
	Number (thousands)	Percentage of the Total	Number (thousands)	Percentage of the Total	Number (thousands)	Percentage of the Total	Number (thousands)	Percentage of the Total
Rural	6,904	65.98	4,213	18.7	3,915	14.8	—	—
Urban	3,559	33.02	18,367	81.3	22,580	85.2	—	—
Northeast	679	6.5	4,344	19.2	4,848	18.3	5,565	18.7
North central	793	7.6	4,572	20.2	5,337	20.1	5,862	19.4
South	8,912	85.0	11,970	53.0	14,048	53.0	15,925	52.8
West	79	0.75	1,695	7.5	2,262	8.5	2,812	9.3
Total	**10,463**		**22,580**		**26,495**		**30,164**	

Source: Historical Statistics of the United States: Colonial Times to 1970 (Washington, D.C.: U.S. Government Printing Office, 1975); *U.S. Statistical Abstract, 1989* (Washington, D.C.: U.S. Government Printing Office, 1989).

TABLE 8–4
1980 Census by Race and Place

PLACE	WHITE POPULATION	PERCENT OF WHITES	BLACK POPULATION	PERCENT OF BLACKS
Central Cities	46,409	24.6	15,144	57.2
Suburbs	63,740	33.9	4,962	18.7
Urban outside Urbanized Area	24,173	12.8	2,488	9.4
Rural	54,050	28.7	3,901	14.7

Source: U.S. Statistical Abstract, 1989

TABLE 8–5
Black Population by Major Cities

RANK	MAJOR CITY	BLACK POPULATION
1	New York	2,363,807
2	Chicago	1,710,173
3	Los Angeles–Long Beach	1,288,468
4	Philadelphia	1,079,183
5	Detroit	833,070
6	Washington, D.C.	749,995
7	Baltimore	597,855
8	Houston	561,704
9	Newark	543,971
10	San Francisco–Oakland	491,212
11	St. Louis	462,571
12	Dallas–Forth Worth	433,094
13	Atlanta	420,388
14	New Orleans	411,084
15	Cleveland	410,157
16	Memphis	324,310

Source: U.S. Statistical Abstract, 1988 (Washington, D.C.: U.S. Government Printing Office, 1988).

The income of a black family in a northern city was higher than that of a black family in a southern city, by sometimes twice as much. The decrease in overt discrimination and segregation in northern cities also made them more attractive, as did the emergence of state welfare programs to assist the poor and jobless. The combination of pull factors causing black population to migrate to the Northern cities created large black populations in cities that previously had only a small resident black population, while still leaving large black populations in southern cities. (Table 8–5).

Washington, D.C., was the only city of at least 100,000 population in which blacks constituted more than half of the population in 1960. There were nine such cities in 1980: Atlanta, Birmingham, Baltimore, Detroit, New Orleans, Newark, Richmond, Jersey City, and Washington. The rapid growth of the black population in the cities of the North and the South has led to ever greater neighborhood concentration. Harlem emerged as the cultural mecca for black artists and intellectuals after 1920 in New York, for example, and continues as a major black neighborhood to the

present time even though it is part of the larger New York urban area.

The segregation of blacks into the central cities in the post–World War II period has created a broad pattern of white suburbs and black inner cities. The present distribution of the population in Atlanta, Georgia, for example, has nearly 90 percent of the white population residing in the suburbs, while 55 percent of all blacks reside in the central city. Of 1.6 million whites in Atlanta in 1980, 1.43 million resided in the suburbs. Only 236,000 of 526,000 blacks lived in Atlanta's suburbs. Thus the whites outnumber the blacks in the suburbs by nearly seven to one, while in the central city, the blacks outnumber whites by a ratio of two to one (Figure 8–9).

The continued persistence of the black inner city is one geographical manifestation of differences in income that are related to race. The concentration of blacks in the central city today

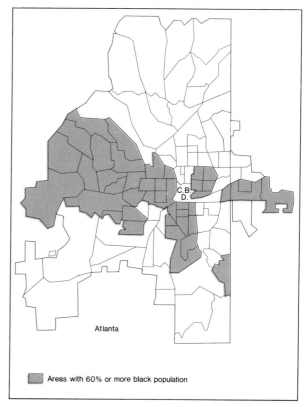

Atlanta

■ Areas with 60% or more black population

FIGURE 8–9 Distribution of Black Population in Atlanta, Georgia

TABLE 8–6
Incomes in Selected Census Tracts (1980)

CITY	LOCATION OF TRACT	FAMILY INCOME	PER CAPITA INCOME
Kansas City	Central city	4,810	2,810
	Central city	8,203	2,585
	Central city	12,439	3,863
	Suburbs	40,626	11,538
	Suburbs	96,119	32,266
	Suburbs	70,484	21,469
Dallas	Central city	7,979	2,908
	Central city	8,646	2,749
	Central city	8,150	3,176
	Suburbs	150,166	47,654
	Suburbs	152,859	49,838
	Suburbs	76,955	29,233
	Suburbs	38,308	13,158

Source: 1980 Census of Housing (Washington, D.C.: U.S. Government Printing Office, 1984).

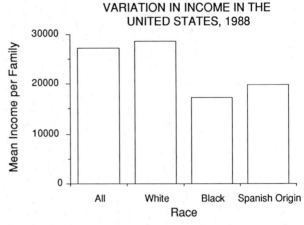

FIGURE 8–10 **Variation in Income in the United States, 1988** The average income in the United States is lower for blacks.

is a function of economics. Black migrants to the cities, like other migrants before them, moved into the lower cost housing near the central city where they could work in the menial and low-paying jobs near the urban center. Persistence of the income gap between races and ethnic groups creates landscapes dominated by blacks, the elderly, and other low-income groups who cannot afford to move to higher cost housing in the suburbs (Table 8–6). With only half the income of the average white family, the average black family is trapped in the central city.

Examination of population statistics for the United States indicates reasons for the segregation of blacks in the central city. Between 1960 and 1988, the number of black families headed by women more than tripled, and the proportion of black families headed by women has doubled from 21 to 44 percent. Fifty-nine percent of all black families with children are headed by women, compared to 20 percent for white families. Overall, the average income among black, female-headed families in 1988 was only 32 percent of the average American family income. The central city, then, is not only increasingly black, but it is increasingly black and characterized by families with a single female parent. The poverty rate among black families headed by a female is 52 percent. Sixty-five percent of black children

living with a single female parent are living below the poverty level, compared to only 21 percent for black children in a two-parent home. Forty-five percent of children in a white, single female parent home are living below the poverty line, and 11 percent of white children in a two-parent home are in poverty (Figure 8–10).

The cultural geography of black Americans shows their concentration in central cities, in poor housing, and in poverty, the impact of which is reflected in overall statistics for black families, (Table 8–7). The unemployment rate among blacks is normally twice that among whites, and examination of birth rates, life expectancy, and other statistics indicates a similar dichotomy.

TABLE 8–7
Income by Race and Place, 1988

PLACE OF RESIDENCE	INCOME PER FAMILY		
Farm	21,884	—	—
In metropolitan areas	31,074	18,203	20,006
In central cities	28,239	16,661	18,030
Outside central cities	32,581	21,942	22,922
Outside metropolitan areas	23,905	14,176	18,068

Source: U.S. Statistical Abstract, 1989 (Washington, D.C.: U.S. Government Printing Office, 1989).

LIFE EXPECTANCY AND INFANT MORTALITY IN THE U.S. (1988)

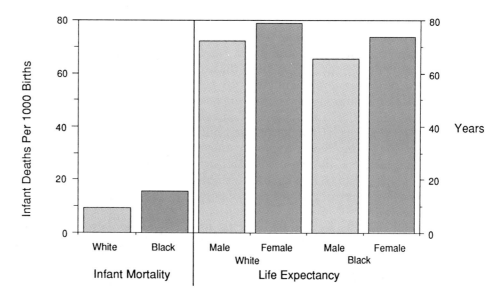

FIGURE 8–11 Life Expectancy and Infant Mortality in the United States (1988) Both life expectancy and infant mortality rates in the United States vary according to race.

The life expectancy for a black infant born in America is lower than that for whites, and the infant mortality rate among blacks is twice that of whites, 18.9 versus 9.3 per thousand, respectively. (Figure 8–11).

The geographic impact of economic differences among racial and ethnic groups goes far beyond the black ghettos of large American cities with their attendant slum housing and myriad social problems. Construction of highway systems frequently disrupts black residential areas rather than white areas; public hospitals for the poor are generally built in low-income white areas rather than black areas; and in spite of efforts to equalize the quality of education, black schools remain significantly behind white schools in terms of facilities, equipment, and quality of education. Although technically illegal, school district and city boundaries have been modified over the years to exclude large black populations. A well-known instance of changing the boundaries to exclude blacks occurred in Tuskegee, Alabama, in 1957. The boundaries of the city were changed from a simple square to a curious 28-sided figure resembling a stylized seahorse to exclude most of the blacks from the city limits and thus prevent their voting (Figure 8–12). Similar boundary changes (known as *gerrymandering*) have occurred in

school districts across the United States. Although court decisions have attempted to force integration through busing, many black inner city residents still attend schools that reflect *de facto* segregation.

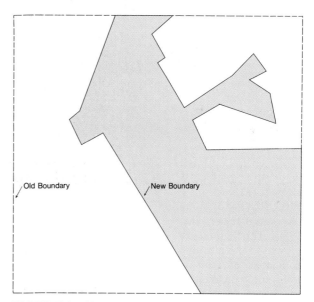

FIGURE 8–12 The Tuskegee Gerrymander Tuskegee, Alabama attempted to redraw its boundaries to limit black participation in voting.

Some significant changes affecting the distribution of black Americans are occurring now, including migration back to the South. Prior to the civil rights movement of the mid-1960s, only one out of every ten blacks who migrated North ever returned home. By the 1980s, more than 50 percent were returning. Approximately 20 percent of the black migrants to the South were born in the North, a rate four times higher than ever before. The return flow of blacks to the South reverses the long-time South-to-North movement. At the same time, the West is exhibiting a net gain in black population due to continued migration and lower return migration from this region (Figure 8–13).

The movement of blacks to the South has not reversed the traditional movement from rural to metropolitan and central city areas. The decrease in the black rural population continues to reflect black movement to and within metropolitan areas. This migration is primarily from smaller to larger metropolitan areas.

These changes in the distribution of blacks have not eliminated *de facto* discrimination against blacks, which continues in a host of subtle ways. One of the more important from a geographical standpoint occurs as the result of zoning decisions in suburban communities. Land in most suburbs is zoned for single-family residences, which prevents the construction of multiple-unit housing for lower income families. Courts ruled that such zoning was discriminatory and hence illegal in two different cases in 1988. The first occurred in the New York suburb of Yonkers, where zoning regulations prevented multiple-unit housing for low-income families from being constructed except in a tiny area close to New York City (Figure 8–14). The court ruled that since most of the residents of the multiple units were black, the zoning ordinance caused segregation and was illegal. In the second case, the U.S. Supreme Court held that zoning ordinances that caused segregation by restricting multiple-unit or low-income housing were illegal whether they represented an unconscious or conscious attempt to prevent blacks from moving into a neighborhood. The effect of these two rulings should create greater movement of black Americans to the suburbs in the next few decades.

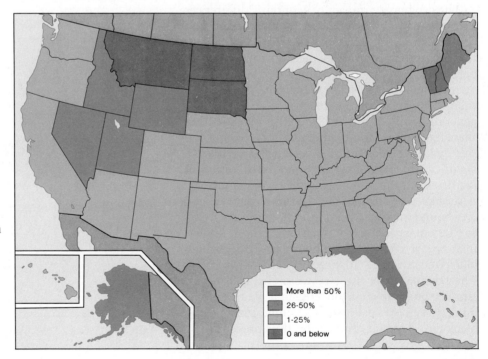

FIGURE 8–13 Migration Flows of Blacks in the 1980s The high rate of growth in the black population of states in the intermountain West and the Dakotas results from their extremely low total black population.

More than 50%
26-50%
1-25%
0 and below

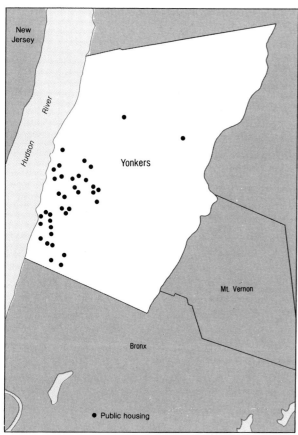

FIGURE 8–14 Multiple-Unit Housing in Yonkers, New York 1988

American Indians

Another group that has been subjected to centuries of persecution and discrimination are the American Indians (or Native Americans). Although not enslaved, they were otherwise subjected to the same sorts of treatment by the white majority population as the blacks. Some Native American nations initially welcomed the European settlers and helped them, but growing numbers of European settlers quickly led to conflict. Native American concepts concerning land provided for communal use rather than ownership in fee simple. The Europeans failed to recognize that land that was not actually occupied or cultivated was still part of the Indians' resource base. The encroachment of settlers on lands used by Native Americans for grazing, hunting, gathering, and

periodic cultivation ensured that there would ultimately be conflict.

When the number of Europeans did not prevent the Indian nations from obtaining a livelihood, the two were able to coexist peacefully. As the Europeans increased in number, however, they attempted to remove the Native Americans from their traditional lands, and major conflicts ensued. European settlers engaged in genocide, segregation, discrimination, and cultural integration with the American Indian groups. Genocide resulted both from the bounties that some colonial governments paid for male Indian scalps and from the Indian wars that killed thousands. The primary means of dealing with the American Indians, however, was to exclude and segregate them from the white settlers. Initially, segregation took the form of forcing the Native Americans to move farther west into land not desired by the European settlers.

The removal of most Native Americans from the eastern portion of the United States left a few behind on small reservations, but the major Indian reservations were established in the western United States (Figure 8–15). Concentration of the Indian nations in the American West on reservations effectively removed them and their plight from the public view. Isolated on large reservations in regions with harsh environments and low population densities, Native Americans are poorly educated and have the lowest income and greatest level of poverty of any ethnic group in the United States.

Historically, blacks were slaves and were viewed as personal property, but they were also citizens of the United States. Native Americans were not granted citizenship until 1924, nearly three-quarters of a century after the Emancipation Proclamation freed the slaves. The efforts of the white population with regard to the Indian population reflect the general problem created when two cultural groups with differing values, perceptions, goals, and priorities compete for the same resources.

The European settlers' goal of maximizing individual profit through occupying and cultivating the land conflicted with the Indians' goal of group well-being and shared wealth. At one time the

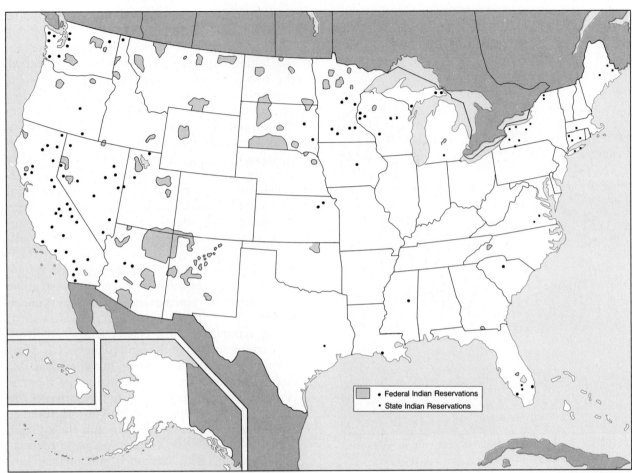

FIGURE 8–15 Indian Reservations in the United States

Europeans responded by attempting to acculturate the Indian. Christian missionaries were encouraged to proselytize among the Indians. Teachers were provided to teach European attitudes and
values. Property was often removed from tribal
ownership and granted to individual Indians.

The failure of acculturation led to the decision
to concentrate Native Americans on reservations.
The estimated 2 to 4 million Native Americans
inhabiting the United States when the first permanent European settlers arrived had declined to
only 350,000, largely resident on reservations, by
the census of 1950. Since 1950, however, the
American Indian population has increased more
rapidly than the general population. In 1980, the
U.S. Census listed 1,361,869 Americans as Indians, 42,149 as Inuits (Eskimos), and 14,177 as
Aleuts. The actual number in each group is prob-

ably higher, since only those who identified themselves in one of these categories or identified
themselves as members of a specific Indian or
Inuit tribe were classified as Indian or Inuit. The
location of the Indian population on often inac-
cessible reservations and the lower levels of literacy among Indians guarantee that a fairly substantial number were not counted by the census,
especially if they had to respond to a mailed
questionnaire.

Native Americans comprised an estimated 0.6
percent of the total U.S. population in July 1987
(1,511,582). They are concentrated in the western
states, with California, Arizona, the Dakotas,
New Mexico, and Utah having large numbers
(Table 8–8 and Figure 8–16). Historically, Indians
have resided in rural areas and been engaged in
farming activities. Today's distribution of Native

Americans reflects their concentration on reservations at the beginning of the twentieth century, plus migration from the reservations in the last two decades. There are more than 280 Indian reservations, the largest of which is the Navajo, which covers 22,000 square miles in Utah, New Mexico, and Arizona. The Navajo reservation is larger in area than the states of Rhode Island, Massachusetts, and Connecticut combined. Nevertheless, this reservation, which is nearly ten times larger than any other, cannot adequately support all of its residents, in part because of its arid to semiarid climate.

Other reservations face similar problems, resulting in low incomes, low life expectancies, and limited economic opportunities. The movement of Indians from the reservations reflects the harsh geographic reality of their land. Over half the individuals who define themselves as "Indians" lived in towns and cities off the reservations in 1989. Many Native American nations now have

Navajo Reservation, Arizona. Isolated on reservations in the west, Native Americans are generally the poorest group in the country.

TABLE 8–8
American Indian Population, 1980

RANK	STATE	POPULATION	PERCENTAGE OF TOTAL INDIAN POPULATION
1	California	198,095	14.5
2	Oklahoma	169,297	12.4
3	Arizona	152,610	11.2
4	New Mexico	104,634	7.7
5	North Carolina	64,519	4.7
6	Washington	58,159	4.3
7	South Dakota	45,081	3.3
8	Michigan	39,702	2.9
9	Texas	39,374	2.9
10	New York	38,117	2.8
11	Montana	37,153	2.7
12	Minnesota	34,841	2.6
13	Wisconsin	29,318	2.2
14	Oregon	26,587	2.0
15	Alaska	21,849	1.6
16	North Dakota	20,119	1.5
17	Utah	19,158	1.4
18	Florida	18,981	1.4
19	Colorado	17,726	1.3
20	Illinois	15,833	1.2
21	Kansas	15,254	1.1
	Total	1,361,869	

Source: The New Book of American Rankings, 1984.

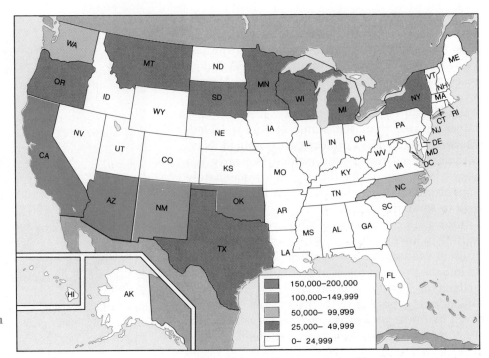

FIGURE 8–16 American Indian Population of the United States

Pueblo Indians in Taos Pueblo, New Mexico, still inhabit traditional structures whose form and function pre-date Anglo-immigration to the continent.

more members living in towns and cities than in rural communities on the reservations. Several American cities have larger Indian populations than any reservation except the Navajo: New York City, Phoenix, Los Angeles, San Francisco, Tulsa, and Oklahoma City. North Carolina, Wisconsin, and the western states of Utah, Alaska, North and South Dakota, Montana, New Mexico, and Arizona are the only states where more Indians are residing on reservations than in urban areas.

Examination of the socioeconomic characteristics of the American Indian reveals marked contrasts with other groups in the United States and between reservation and nonreservation Indians. The Native Americans as a group have the lowest incomes, highest unemployment rates, highest proportion living in poverty, and lowest life expectancy. Those who have left the reservation have an unemployment rate only slightly higher than that of blacks, but all native Americans combined have an unemployment rate nearly four times that of the black population and eight times that of the white. Ten percent of the total American population has an income classified as below the poverty level, but 55 percent of all Indians on reservations are so classified, and 25 percent of Indians living in metropolitan statistical areas live in poverty.

The cultural landscape of Native Americans reflects their history and economic condition. The reservations represent distinctive landscape types, with structures ranging from the pueblos of the Southwest to the hogans of the Navajo, detached single-family homes, and mobile or modular homes. The dominant landscape feature is the large expanse of open space separating the isolated homes or towns. Tuba City, Lame Deer, Moenkopi, Crow Creek, and Uintah are names that reflect the landscapes of isolation. Isolated by space, culture, and economy, native Americans are an ethnic group divided along reservation and urban lines. Reservation residents remain the most isolated, as urban native Americans are increasingly assimilated into the broader American culture.

Hispanic-Americans

The Hispanic population of the United States is second in numbers only to the black population as a minority group and may exceed the black population by the year 2000. The actual number of individuals of Hispanic descent in the United States is difficult to measure. Like Native Amer-

icans, they tend to be underenumerated. The census defines Hispanics as those individuals who list themselves as Hispanic or those from Mexico, Central America, or other Spanish-speaking areas who do not. Many individuals may choose not to indicate their ancestry. An even larger group is resident in the United States illegally and not counted by the census. A 1988 estimate suggests there are approximately 16 million Hispanics in the United States, of whom an estimated 2 million are here illegally. Other estimates of Hispanic population are much higher, with some estimating that as many as 10 million reside in the United States illegally.

Like the Indian population, Hispanics have a regional concentration in the American Southwest (Figure 8–17). New Mexico has the highest percentage of Hispanics, many of whom are descendants of Spanish colonists who arrived in the Southwest before European settlers arrived in the eastern United States. California has the largest total number of Hispanic individuals. Other large Hispanic populations are found in Illinois, New York, and Florida. (Table 8–9).

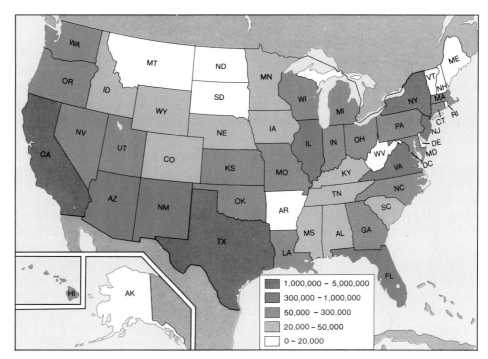

FIGURE 8–17
Distribution of the Hispanic Population of the United States

TABLE 8–9
The Hispanic Population by State

STATE	HISPANIC	PERCENT	STATE	HISPANIC	PERCENT
Alabama	33,923	0.88	Montana	10,103	1.28
Alaska	9,057	2.24	Nebraska	28,262	1.80
Arizona	444,102	16.32	Nevada	54,130	6.76
Arkansas	16,976	0.76	New Hampshire	5,248	0.56
California	4,541,300	19.20	New Jersey	494,096	6.72
Colorado	341,435	11.80	New Mexico	477,051	36.60
Connecticut	125,256	4.04	New York	1,660,901	9.44
Delaware	9,540	1.60	North Carolina	56,039	0.96
District of Columbia	17,777	2.80	North Dakota	3,474	0.52
Florida	858,105	8.80	Ohio	120,002	1.12
Georgia	60,974	1.12	Oklahoma	57,831	1.92
Hawaii	71,399	7.40	Oregon	66,164	2.52
Idaho	36,560	3.88	Pennsylvania	153,579	1.28
Illinois	634,617	5.56	Rhode Island	18,906	2.00
Indiana	86,518	1.56	South Carolina	33,667	1.08
Iowa	26,274	0.92	South Dakota	3,815	0.56
Kansas	62,656	2.64	Tennessee	34,026	0.76
Kentucky	27,094	0.76	Texas	2,982,583	20.96
Louisiana	99,699	2.36	Utah	60,045	4.12
Maine	5,331	0.48	Vermont	3,377	0.68
Maryland	63,196	1.48	Virginia	79,722	1.48
Massachusetts	141,380	2.48	Washington	121,286	2.92
Michigan	157,626	1.72	West Virginia	13,118	0.68
Minnesota	32,115	0.80	Wisconsin	62,782	1.32
Mississippi	24,178	0.96	Wyoming	24,535	5.24
Missouri	51,853	1.04	Total	14,603,683	

Source: U.S. Statistical Abstract, 1989 (Washington, D.C.: U.S. Government Printing Office, 1989).

Individuals classified as Hispanic have various origins. Although a substantial portion are from Mexico or Central America, there are also large numbers of Spanish-speaking individuals from Cuba and Puerto Rico. Although all of these are called Hispanics, they are distinct groups, who view others as different from themselves.

Mexican-Americans, who are the largest Hispanic group in the United States, are of two types; descendants of early settlers in New Mexico and Arizona and migrant laborers. Mexico has served as a source of cheap labor for the United States since the 1920s. The labor system was formalized initially as part of the *Bracero* program, which allowed temporary contract farm workers to come to the United States from Mexico. This program was phased out in 1964, and a substitute program granting green or blue documentation cards for migrants was introduced. Green-card holders are immigrants from Mexico who have acquired an immigrant visa that entitles them to reside and work in the United States as legal resident aliens. Blue-card holders can enter the United States legally, but their residency is limited to not more than seventy-two hours at one time. Many Mexicans simply cross the border illegally along the 1952 miles (3141 kilometers) of border between the two countries (Figure 8–18).

Mexican immigrants to the United States are fleeing the high population growth and low wages in Mexico and are attracted by the demand for low-cost labor for menial occupations in the United States. The actual number of Mexican-Americans in the United States at any specific time varies widely because many are only temporary migrants. Large numbers come seasonally to

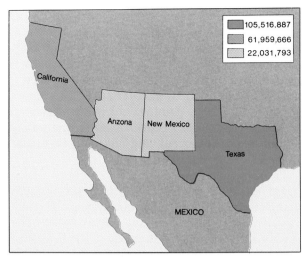

FIGURE 8–18 Border between the United States and Mexico The significance of the border between the United States and Mexico is indicated by the numbers who enter each year. In 1987, 70 percent of all immigrants and visitors to the U.S. crossed the Mexican border.

San Diego, California. Hispanic Americans create neighborhoods with distinctive characteristics.

work on farms in California, Texas, Arizona, Utah, Idaho, and other western states. Others are daily commuters between the twin cities along the Mexican border (El Paso–Ciudad Juarez and San Diego–Tijuana, for example). Many come for the majority of the year but return to be with their families for a few months.

Hispanic-Americans from the Caribbean are quite different. Puerto Ricans are citizens of the United States and can move legally between Puerto Rico and the mainland. Large numbers are found in the New York urban area. Cuban Hispanics, who are concentrated in Florida and the Southeast, are migrants and their descendants of three immigration periods: the first occurred before 1960; the second occurred in the early 1960s after Castro adopted a Marxist government for Cuba; and the third took place in the early 1980s when Castro allowed many individuals to leave the island and rejoin their families in the United States. In 1960 6.8 percent of the residents of the Miami metropolitan area were of Spanish origin, but by 1985 this had increased to 42.3 percent.

The geographical impact of Hispanic-Americans is evident in such things as place-names reflecting their Spanish heritage, such as

San Francisco, San Diego, Los Angeles, and El Paso; the diffusion of Spanish foods into the United States; and the use of Spanish as a second language in portions of California, Arizona, Colorado, New Mexico, New York, New Jersey, Texas, and Florida. Examination of the socioeconomic characteristics of Hispanic-Americans reveals that they have generally higher incomes, higher life expectancy, and higher educational status than the black population. Individual Hispanic subgroups differ substantially from each other, however. Incomes for Hispanics of Cuban ancestry are highest, ranking nearly at the general average. Incomes of those of Mexican origin are higher than for the black population, but incomes for families of Puerto Rican origin are lower than for even the black population.

Race and Ethnocentrism on a World Scale

The geographic manifestation of racial and ethnic groups in the United States exemplified by the three groups just discussed is mirrored by similar patterns in other countries. The majority of black Africans who made up the slave population of the United States came not directly from Africa, but from the islands of the Caribbean where they had replaced indigenous peoples who died largely from diseases carried by the Europeans who colonized the region. The use of slaves in the production of crops for sale in Europe in the seventeenth and eighteenth centuries greatly modified the population map of the Caribbean region. The islands of the Caribbean, the northeastern coast of Brazil, and, to a lesser extent, Venezuela relied on black slave labor, creating a complex cultural geography.

Haiti

One example of the impact of slavery is the island nation of Haiti. Haiti occupies the western third of the island of Hispaniola. Originally conquered by the Spanish, the area that is today Haiti was obtained by the French in 1697. By that time the native Indian inhabitants had died from disease, and black African slaves had been introduced. The French greatly expanded the slave population, and by 1800 there were 700,000 slaves and 28,000 people of mixed racial descent (mulattos) controlled by only 40,000 French.

The slave population was involved in both subsistence agriculture to provide for their own needs and the production of sugar cane, coffee, and other commercial export crops. The slaves revolted in 1791, gaining independence from France in 1804. Today Haiti has an estimated 6.3 million residents, occupying only 10,714 square miles (27,750 square kilometers).

The rapid population growth in Haiti has helped create one of the poorest countries in the Western Hemisphere. Average per capita income in 1987 was only $380 dollars per year. (Figure 8–19). The literacy rate, at 20 percent, is one of the lowest in the world. Life expectancy in Haiti (54 years) is 20

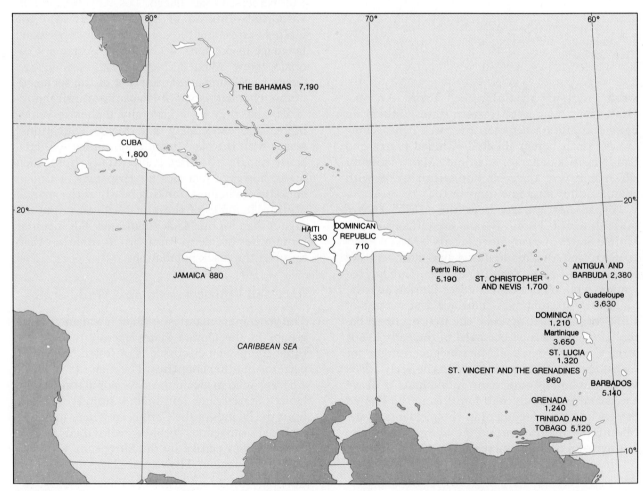

FIGURE 8–19 Per Capita GNPs of Caribbean Countries Variation in the per capita gross national product of selected Caribbean countries.

The Iron Market, Port Au Prince, Haiti. Rapid population growth has helped make Haiti one of the poorest countries in the world.

Brazil's population includes representatives from numerous ethnic backgrounds.

years less than that for the countries of the industrialized Western Hemisphere.

Overpopulation and an economy that relies on subsistence agriculture have destroyed most of the vegetation of Haiti. Eighty percent of the population of Haiti is rural, and nearly everyone cooks with wood. As a result, the country has been deforested to obtain fuel. An estimated 60 percent of Haiti was covered with forest in 1920, but less than 5 percent of the land is forested today. Malnutrition is the norm rather than the exception, and the poverty of life in Haiti is rivaled by only a few African nations.

Brazil

Movement of Africans to Brazil for use in sugar cane production by the Portuguese, Dutch, and other groups brought new population groups to South America. Limited native populations in the region prompted the Europeans to bring Africans as slaves, and the most important geographical relic of this population transfer is visible in the religions, population makeup and distribution, and relative distribution and level of income in Brazil. The black migrants to Brazil and their descendants make up distinct groups within the country today. Brazil has officially adopted a democratic and nonracist society, but the employment statistics and quality of life of the Brazilian blacks and mulattos indicate that there is still a wide gap between the law and the practice of social equality among ethnic groups.

The intermixing of the blacks with the limited Native American population in Brazil and with Europeans led to distinct racial subgroups in Brazil. Mulattos are the descendants of black and European ancestry, and the Zambo are descendants of black and Indian ancestry. Although blacks have migrated to other areas of Brazil in response to job opportunities—in particular to the growing urban areas in the south—the greatest concentration of blacks is still in the Northeast (Figure 8–20).

FIGURE 8–20 Distribution of the Black Population in Brazil

Suriname

Another country in the Western Hemisphere where slavery was part of the cultural ecology is Suriname, which was a Dutch colony from the late 1600s until 1975. The Dutch originally brought black Africans to work their plantations and then, after slavery ended in the nineteenth century, imported workers from other lands, mainly from Java in today's Indonesia. Today the nation of Suriname is an ethnic mix of descendants of slaves, Native Americans, and contract laborers from India and Java (Table 8–10).

Political conditions in Suriname have caused an estimated one-third of the population to emigrate. Former residents of Suriname now form an important enclave in the Netherlands, where 250,000 of the former residents have fled. Conflict continues between East Indians and mulattos, who dominate the government, and a guerilla movement led by black rural residents.

The colonial legacies of population heterogeneity and political unrest in Suriname typify the problems faced by many countries where slavery was once formally practiced. Although laws may change, values and attitudes developed over several centuries are much more persistent. Efforts by Brazil, the United States, Suriname, and other countries to create a democratic social setting for all of their ethnic and racial groups proceed slowly. The persistence of ethnocentrism and racism contribute to both the suffering of individuals and the loss of human potential due to inadequate opportunities for education, training, and employment.

TABLE 8–10
Ethnic Groups in Suriname*

GROUP	PERCENTAGE OF POPULATION	NUMBER
East Indian (from India)	37	148,000
Mulatto	31	124,000
Javanese	15	60,000
Black	10	40,000
European	4	16,000
Native American	3	12,000
Total		400,000

*1988 estimates.

Racism in Africa

Racism, slavery, and ethnocentrism in the Americas have counterparts in Africa. European colonialism came later to Africa, but gave rise to many of the same problems we have seen in the Americas. The great majority of African countries were once European colonies, where the impact of racism and ethnocentrism are a constant reminder of the legacy of European colonialism.

Most African countries gained their independence in the post-1960 era and have attempted to create new nations in the face of economic and social upheaval. Racism has been evident in many of these countries, such as Uganda, where Europeans brought Indians or others as contract laborers because slavery had been made illegal. The rise to power of native black Africans resulted in racial conflict. The Indians in Uganda had become the middle class under the European colonial masters. Owning the shops, banks, and other commercial and trading activities, the Indians had a higher standard of living than the black majority and were visibly different. The emergence of the dictator Idi Amin in the 1970s was followed by overt persecution of Indians. Claiming that the Indians prevented black Africans from having an equitable part in the economy, Amin tortured and ultimately expelled Indians from Uganda. He was equally harsh with Africans who were not members of his tribe, however. More native Ugandans lost their lives than Indians.

Similar examples of racial conflict abound in recent African history as the black population has attempted to gain control of its political and economic destiny. The effects of racism are compounded by the struggles between individual African *tribal groups*; these conflicts are often worse than those between racial groups (Figure 8–21). Estimates place the number of tribal groups at over five hundred. Individual Africans pay greater allegiance to their tribe than to their country, and in most countries one tribe dominates the political-socioeconomic arena at the expense of others. Conflict between African tribal groups is a severe form of ethnocentrism, periodically resulting in armed clashes, death, and human suffering.

Zimbabwe (formerly Rhodesia) is an example of racism and black ethnocentrism in Africa.

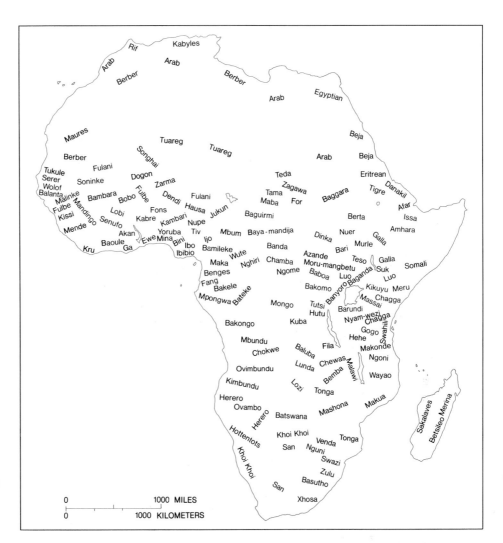

FIGURE 8–21 Selected Examples of Tribal (Ethnic) Groups in Africa These examples of tribal groups in Africa illustrate the complexity of the ethnic map of the continent.

Rhodesia was established as a colony by the British industrialist Cecil Rhodes. The country declared itself independent from Britain in 1965, creating a government in which the minority white population dominated politics and the economy. The inequity in such a situation ultimately led to guerilla conflict and the overthrow of the white government by the black majority in 1980. The racial makeup of Zimbabwe consisted of 8,300,000 black Africans and 200,000 whites of European descent at the end of 1988. The white population controls a disproportionate share of the best agricultural land, which is the basis for Zimbabwe's economy.

The black majority population had originally agreed to a government in which a proportion of the representation was based on race to ensure white representation. The government changed the constitution in 1987, dropping the provision that a certain number of members of Parliament should be of a specific race. Residents of Zimbabwe of European ancestry feel that they are the focus of racism and will be forced from the country. Emigration of whites from Zimbabwe was greatest just prior to and after the revolution, but the white migration rate is again increasing. Fear that their property will be seized by the government is creating an increased air of uncertainty among the white population of Zimbabwe. The dilemma faced by the residents of this country is how to overcome the inequities produced by over a century of overt racism. The whites feel that allowing the majority population to have a fair opportunity negatively impacts those of European ancestry.

South Africa: De Jure Racism   South Africa is synonymous with racism for most observers in the world. The country contained some 35 million people in 1988 divided on the basis of ancestry into four broad groups (Table 8–11). The largest group are native Bantus (blacks), who comprise 70 percent of the total population (24.5 million). The second largest group are the whites (6.25 million or 17.8 percent), who are descended from European migrants. Third are the *coloureds*, who comprise 3.3 million people (9.4 percent). The coloured population is descended from a mixture of the African population with early European conquerors and immigrants to South Africa. The final group are the Asians (sometimes called Indians) who comprise some 3 percent of the population (1 million).

The history of South Africa explains racism in the country today. European influence was introduced by the Dutch East India Company, which brought the first settlers to the Cape of Good Hope in 1652 to establish a way station for its trading ships en route to Asia. By the end of the eighteenth century, the Dutch colony numbered about 15,000 people. Known as *Boers* ("Boer" in Dutch means farmer) or *Afrikaners* (they speak a Dutch dialect known as *Afrikaans*), the settlers

tried to establish an independent republic as early as 1795. Britain occupied the colony in 1795 and took permanent possession in 1814. The British brought 5,000 settlers, a government system based on the British model, and freed the slaves. In the 1830s to 1850s, conflict with the British over the abolition of slavery led about 12,000 Boers to make the "great trek" north and east, where they came into direct conflict with larger groups of Bantus. The Bantu population came from the north in several waves, the first as early as the tenth or eleventh century. Other waves seem to have followed them in the thirteenth, fourteenth, and seventeenth centuries.

The South Asian population is overwhelmingly of Indian descent and resides in the province of Natal, especially in the city of Durban, where 80 percent of all Asians in South Africa are found. Most Asians are descendants of contract laborers brought to Natal during the latter part of the nineteenth century to work on the sugar plantations after the British ended slavery. Approximately 20 percent of the Asians are descendants of people who paid their own passage to South Africa and became traders. The coloureds are concentrated in the Southeast in Cape Province, but are also found in lesser numbers in cities of other provinces.

Racial distinction permeates South African society, government, and economy. It reflects the interaction between the immigrant and native populations and the changing dynamics of their respective populations over time. The original population of Dutch and British settlers in South Africa grew rapidly after the discovery of dia-

TABLE 8–11
Population Characteristics of South Africa 1988 (thousands)

PROVINCE	TOTAL	ASIANS	BLACKS	COLOUREDS	WHITES
Cape	7,139	43	2,195	3,125	1,775
Natal	3,748	925	1,912	127	783
Transvaal	11,718	162	7,921	320	3,314
Orange Free State	2,714	—	2,176	80	457
White areas	25,321*	1,000*	14,206*	3,653	6,331*
Black Homelands	9,679	15	9,609	18	38
Total—South Africa	35,000*	1,000*	24,500*	3,290*	6,230*

Source: South Africa 1986: Official Yearbook of the Republic of South Africa (Pretoria: South African Government Printing Office, 1987).

monds (1867) and gold (1876) brought an influx of immigrants into the country. Conflict between Britain and the Dutch-speaking colonists culminated in the Boer War of 1899, which ended in a British victory in 1902. The Union of South Africa was formed in 1910, with membership in the British Commonwealth of Nations. It became the independent Republic of South Africa in 1961, severing all formal ties to the British Commonwealth. The Afrikaners still constitute the majority among the Europeans and dominate the economy and politics of the country.

The evolution of South Africa from colony to independent state has paralleled an evolution in the treatment of the nonwhite population. The influx of European immigrants after the discovery of diamonds led to increased pressure on the South African landscape and resulted in conflict with the black population. The relationships between the two were formalized by the Native Lands Act of 1913, which preserved approximately 13 percent of the total land for Bantu use only. In essence, reflecting the U.S. policy of segregating the Indians on isolated lands, the South African Native Lands Act attempted to establish reservations for the Bantu population. Unlike the United States, however, the European migrants were never more than a small minority of the total population, and it was difficult to isolate themselves from the majority African population. Industrialization of South Africa's economy in the twentieth century combined with the use of black laborers in mines and on farms to attract Bantus from the native lands to the towns and cities of white South Africa. Rapid increase in the population of nonwhites in South Africa in the twentieth century, especially in and around the white towns and cities, led to the adoption of the Native Consolidation Act of 1945 (also known as the Urban Areas Act). This act declared all white urban areas were off-limits to blacks except at the invitation of the white rulers.

The intent of this act was to allow Bantus to be invited to live in dormitory areas in or around the white communities where they would provide low-cost labor, but not allowed to bring their families or to take up permanent residence. The policy of *apartheid* (literally meaning "separateness" or "apartness") became the official policy in

South Africa in 1948. Apartheid was reinforced in 1950 by another Group Area Act, which established racially segregated neighborhoods in those communities that already had resident Bantu, coloured, and Asian populations. The 1950 Population Registration Act further reinforced efforts to segregate on the basis of race. This act required every individual to be registered according to his or her presumed racial origins as white, coloured, or black. The Registration Act was amended in 1959 to separate Asians as a distinct group as well.

The Bantu Homelands Citizenship Act of 1970 made every black African a citizen of ten proposed homelands (Bantustans), which occupied the 13 percent of the land set aside in the Native Lands Act of 1913 (Figure 8–22). The homelands, as envisioned in the 1970 act, were to become independent nations, with every black African having citizenship in one of them rather than in South Africa itself. To date, four of these homelands have been granted "independence"; Bophuthatswana, Transkei, Venda, and Ciskei (see the Cultural Dimensions box on page 314). Although these four homelands and the other proposed six are viewed as the legal home of the black Africans, they are in fact very similar to the Indian reservations of the United States. They

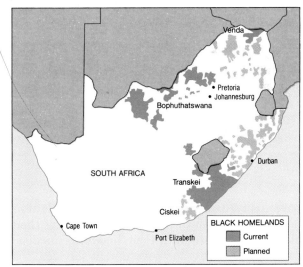

FIGURE 8–22 Black Homelands in South Africa
Black homelands in South Africa represent a modern example of segregation.

South African *Terrae Incognitae*

Terrae incognitae (Latin for "unknown lands") have attracted explorers for centuries. Columbus and subsequent Europeans roamed the globe to illuminate the geography of unknown lands. Today there is a new *terra incognita* for most of the world, an unknown land created by descendants of European colonists in South Africa. Few in the world know where (or what) a Venda, Transkei, Ciskei, or Bophuthatswana are. Fewer still recognize the names of the landscape of this new *terra incognita*.

Kwa Ngema, Driefontein, and Mogopa are towns designated by the white South African government as "black spots" in the 87 percent of South Africa designated as white land. The black residents of numerous small rural settlements as well as specified urban districts are being removed to make the geography of South Africa conform to the government's allocation of territory on the basis of race. For the black residents it means leaving ancestral homes and land that may have been in their families for generations. It means removal to new locations selected by the white South African government, locations that rarely have the geographical amenities that prompted establishment of the black farms and towns originally. All of the new black towns are in the black homelands, where they represent a hidden geography—hidden because they are known only to the government workers involved in the relocation and because they are out of sight and off-limits to even South Africa's white population.

Examine any atlas. See if you can find towns such as Indermah or Onverwacht. These are black "towns" designed for over 100,000 people, yet they have none of the services commonly associated with towns. These towns, like other relocation settlements in the black homelands, have more in common with temporary refugee camps associated with people fleeing drought, famine, and war than they do with the urban geography they replace. Just as the American Indians were forced to relocate in harsh settings by the advancing European settlers, the blacks of South Africa forced to relocate to the homelands face a grim prospect. Uprooted from familiar space, they are expected to recreate their human geography in inferior locations. Like the American Indian, the blacks of South Africa are losers in a confrontation with a technologically more advanced group.

The distribution of black African farms, homes, and towns may someday be replaced as Indermak, Onverwacht, and other black towns are developed, but the disruption of African society and life will reverberate through generations yet unborn. The white South African government may well have only sown the seeds of greater conflict in the future in its efforts to deprive present and future generations of black South Africans of citizenship and human dignity. The American Indians represented a tiny minority, unable ever to regain their lost lands. The black South Africans are a majority, a majority whose growing numerical dominance is reinforced by the rapidly growing population throughout the whole of Africa who view white South Africa as a "white spot." It is doubtful that the white spot of Africa can continue to discriminate indefinitely against the blacks who form the fabric of the continent's geography.

have inadequate resources for their populations and rely upon South Africa for the bulk of their electricity, communications network, medical care, employment, and government services.

The homelands represent the end process of the underlying philosophy of apartheid. Apartheid is a racist thesis that reflects the views of scientific racism with its emphasis on Social Darwinism; the idea diffused to South Africa with early European migrants and was reinforced by the influx of migrants during the nineteenth century. Isolated from social changes in Europe, and perceiving the growing nonwhite populations as a threat, the white minority has made apartheid a significant part of the culture of South Africa. The geographic and social impact of apartheid is wide-ranging and includes identification by race of each group, segregation in communities in areas for whites, coloured, and Asians, exclusion of permanent black residents from white communities, and restrictions on citizenship and employment by racial background (Figure 8–23). Until 1985 under the *pass laws*, the coloured, Asian, and African residents of South Africa were required to carry an identity (pass) card specifying their race and a work pass to allow them to be in nongroup areas. Thus the division of the South African space under the apartheid rules has effec-

FIGURE 8–23 Cape Town and Its Environs The city of Cape Town, South Africa illustrates the spatial impact of creating distinct neighborhoods for each ethnic group.

Segregation in South Africa takes many forms. White women cashiers are paid much more than Black women, who serve as baggers.

tively created a racial geography, evident in the black homelands, the segregated areas for other groups, and daily life.

Legally, the cities of South Africa are the province of whites only, and other ethnic groups may live there in adjacent black townships only by white permission. Black visitors to white cities can stay only seventy-two hours without special permission from the police. Servants or others who work in the cities must have a work pass to prove they are entitled to live in black-only, Asian-only, or coloured-only suburbs (townships) around the white cities. One result of this policy is the huge township of Soweto (Southwest Township) near Johannesburg. It is one of the largest cities in Africa south of the Sahara, and yet legally it is not even a city. Before 1980 black residents could not even own their own homes in Soweto. Other dormitory townships for black workers associated with the great mining areas of South Africa are often male-only preserves. Black workers come to the mining cities and stay for months or years without their families; they live crowded in tiny concrete block structures with space only for a bedroll on cement bunks and a shelf for a few personal possessions. These worker dormitory townships include Bantus from the homelands as well as approximately one million guest workers from other black African countries.

Another landscape feature associated with the racial geography of South Africa is the creation of an entirely new network of black communities as blacks are moved from rural areas they may have occupied for centuries to the new black homelands. Relocation of blacks to new communities creates a population migration not unlike that of refugee movements associated with Afghanistan or other areas of conflict. The resulting geography reflects the creation of new communities that became relatively large in only a few years. The largest city in Bophuthatswana is Mobopane. Mobopane was established in 1981 and had 53,000 people in 1987. The largest city in Ciskei, Mdantsane, was established in 1981 and had 159,000 residents six years later.

At the same time that the South African government has created these new communities for dispossessed blacks, it has destroyed old communities and squatter settlements around the urban areas. The latter have sprung up because rapid population growth among the black population has created a push from rural areas to the urban portions of South Africa. The black population increased 400 percent between 1946 and 1986 while the white population increased only 250 percent. The increase in the black population from only 7.8 million in 1946 to 25 million forty years later has placed greater pressure on the white government to resolve the racial issue in South Africa. The coloured and Asian populations have also grown more rapidly than the white, but not as fast as the black.

The white South African government's attempts to create new homelands for the blacks have been paralleled by efforts to recognize the coloured and Asian populations politically. A new constitution became effective in September 1984. It provides for a parliament divided into three separate chambers made up of whites, coloureds, and Asians. The three divisions are the House of Assembly with 178 members (166 of whom are directly elected and 8 indirectly elected by white voters); the House of Representatives with 85 members (80 of whom are directly elected by coloured voters); and the House of Delegates with 45 members (40 of whom are directly elected by Asian voters). The

Soweto, South Africa is a city which by law includes only Black residents.

houses choose 50 white, 25 coloured, and 13 Asian members of an electoral college that elects an executive president.

Each individual section of the parliament passes legislation relating to its own people, and they meet together to pass legislation affecting national actions or activities. The political representation of coloureds and Asians does not change the overall political control of the whites, since the total representation of Asians and coloureds is only 40 percent of the total vote in the parliament. The white minority population is determined to maintain political control over the country, and the geography of the country reflects this continued self-interest. It appears that conflict will increase over time as the black majority recognizes the inherent inequality of a system that reserves 87 percent of the land and more than three-fourths of all the national resources and gross national product of the nation for the minority white population.

Fear of losing their economic and political dominance prevents South African whites from moving quickly to share power with the other ethnic groups. Recent inclusion of coloureds and Asians in the political process may only create a middle class that will support the whites against the blacks in the event of a power struggle. Nevertheless, the numerical dominance of the black population indicates that over time they may emerge as the political and economic power in South Africa.

Racism in Other Parts of the World

Racism and ethnocentrism are found to some extent in all countries. Racial tension in the United Kingdom exists between the English and people from former colonial possessions who have migrated to the island. The largest groups are from the Caribbean (particularly Jamaica), Pakistan, and India. The immigrants from the Caribbean tend to be black, but because they speak English and practice the Christian religion, they are more readily accepted into Britain than are those from Pakistan or India, who must often learn English and practice Islam or Hinduism. Britain has no formal *de jure* restrictions affecting its minority groups, but *de facto* social customs lead to their segregation in central cities and in the lower working class of society. Periodically conflicts occur between the South Asians and the English or between the West Indians and the English as resentment about poor living standards leads to rioting (Figure 8–24).

Ethnic tensions on the European continent are focused on Moroccans and Algerians in France and on Turkish guest workers in Germany. The immigrants are viewed with displeasure because they do not speak the dominant language, have different religious beliefs, and compete with French and German people for jobs. While they were an important source of low-wage labor during economic boom times, they have become a source of high unemployment with associated high demands for government welfare in the last decade. Resentment against these minorities tends to segregate them in the older, poorer quarters of the cities, and in lower paying jobs.

Ethnic conflict in Asia involves the native inhabitants and Chinese or Indians brought by colonial powers to work in mines or plantations, because the native peoples could not be forced to work there. Ethnocentrism in Southeast Asia reflects the ethnic complexity the Europeans originally found compounded by the mass movement of Indian and Chinese people into the areas under European colonial domination. The resulting ethnic tension takes many forms, from periodic

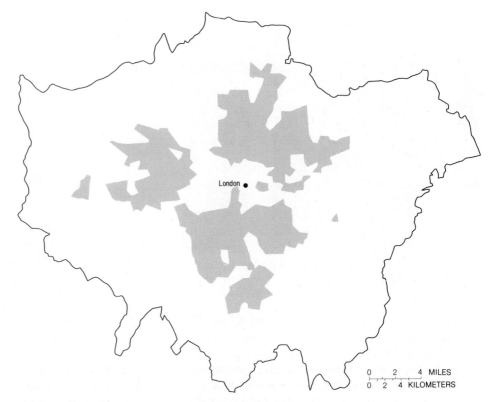

FIGURE 8–24
Distribution of
Pakistanians in London
The distribution of
Pakistanians in London
indicates the areas where
high concentrations of
Asians or West Indians are
found.

armed conflict in Indonesia to the Vietnamese effort to expel the Chinese from the mainly Chinese city of Cholon near Saigon (Ho Chi Minh City) when they gained control of South Vietnam in the late 1970s. More than a million Chinese have felt compelled to leave Vietnam since 1978 as "boat people." In Malaysia the Chinese and Indian minorities are denied equal political rights, and the dominant Malays are given preferential economic treatment by the government, which is controlled by the Malays. Similar patterns of discrimination on the basis of race or culture are found in many areas of the world.

Conclusion

The impact of ethnocentrism and race on the cultural geography of the world cannot be ignored. Although there are no significant differences between specific ethnic or racial groups in terms of intellect, ability, or inherent value, race has been used to justify discrimination against

various peoples throughout history. Racial or ethnic prejudice seems to be the result of five factors:

1. Economics
2. Nationalism
3. Territorial ambition
4. Traditions and stereotypes from history
5. Ideological factors

These factors may interact with one another, but racism and ethnocentricity occur when any or all of these are important in allocating the resources, activities, and rights of a group or region. The cultural geography of South Africa demonstrates the economic and territorial impacts of a racist policy. Malaysia reflects a combination of cultural and historical factors that combine to limit the full participation of the Chinese and Indians in the economic and political life of the country. Slavery in the United States was justified by its supporters because of economic factors, which eventually led to ideological doctrines. Discrimination against other groups on the basis of race or ethnic background is a self-serving phenomenon that is learned rather than biological in origin. The geographical manifestations of racism and ethnocentrism are important in explaining the culture, cultural history, cultural ecology, cultural landscape, and cultural regions of individual places within the world. Understanding the persistence of racism and ethnocentrism provides important tools for predicting and explaining conflicts in the world.

Racial or ethnic prejudice (with its associated bias and discrimination) is a result of learning. If racial intolerance and prejudice can be learned, it would appear that they can be unlearned. The cultural geographer, as well as all rational men and women, hope that understanding the impact of racism and ethnocentrism will foster increased efforts to overcome it. Until and unless the peoples of the world do so, the one race (human) will continue to inflict unneeded pain and suffering upon many of its members.

QUESTIONS

1. Compare and contrast ethnicity and race.
2. Describe the three broad racial groups and their general world distribution.
3. What impact has migration had on the geographic distribution of racial types?
4. Discuss how racism and ethnocentrism are manifest in the cultural geography of the world.
5. What impact has rural-urban migration had on the distribution of the black population in the United States?
6. Discuss the present distribution of the American Indian. What factors explain this pattern?
7. Describe the migration of Spanish-speaking people into the United States, and indicate the regional pattern of their current distribution.
8. Discuss two countries other than the United States where racial or ethnic differences have created distinctive cultural regions.
9. Explain the difference between *de facto* and *de jure* ethnocentrism or racism, giving examples of each.
10. Discuss how the South African policy of apartheid has shaped the cultural geography of that country.

SUGGESTED READINGS

Aiken, Charles S. "Race as a Factor in Municipal Underbounding." *Annals of the Association of American Geographers* 77 (1987): 564.

Anderson, Charles S. "The Idea of Chinatown: The Power of Place and Institutional Practice in the Making of a Racial Category." *Annals of the Association of American Geographers* 77 (1987): 580.

Bowen, William A. "American Ethnic Regions, 1880." *Proceedings of the Association of American Geographers* 8 (1976): 44–46.

Brewer, John D., *After Soweto: An Unfinished Journey*. Oxford: Clarendon Press, 1986.

Brown, Linda K., and Kay Mussel, eds. *Ethnic and Regional Foodways in the United States*. Knoxville: University of Tennessee Press, 1984.

Clarke, Colin G. *East Indians in a West Indian Town: San Fernando, Trinidad, 1930–70*. London: Allen & Unwin, 1986.

Clem, Ralph S. "Russians and Others: Ethnic Tensions in the Soviet Union." *Focus* 31 (September-October 1980).

Cole, John W., and Eric R. Wolf. *The Hidden Frontier: Ecology and Ethnicity in an Alpine Valley*. New York and London: Academic Press, 1974.

Cromley, Robert G. (with Mark Woodall, R. Keith Semple, and Milford B. Green). "The Elimination of Racially Identifiable Schools." *The Professional Geographer* 32 (November 1980): 412–20.

Davis, G. A., and O. F. Donaldson. *Blacks in the United States: A Geographic Perspective*. Boston: Houghton Mifflin, 1975.

Dawson, C. A. *Group Settlement: Ethnic Communities in Western Canada*. Toronto: Macmillan, 1936.

Doran, Michael F. "Negro Slaves of the Five Civilized Tribes." *Annals of the Association of American Geographers*. 68 (1978).

Duckson, Don W., Jr. "Creeks, Runs, and Hollows." *The Professional Geographer* 33 (August 1981): 361–65.

Fellows, Donald K. *A Mosaic of America's Ethnic Minorities*. New York: John Wiley & Son, 1972.

Frenkel, Stephen, and John Western. "Pretext or Prophylaxis? Racial Segregation and Malarial Mosquitos in a British Tropical Colony: Sierra Leone." *Annals of the Association of American Geographers* 78 (1988): 211.

Gann, L. H., and Peter J. Duignan. *The Hispanics in the United States: A History*. Boulder, Colo.: Westview Press, 1986.

Gerlach, Russel L. *Immigrants in the Ozarks: A Study in Ethnic Geography*. Columbia, Miss.: University of Missouri Press, 1976.

Glebe, Gunther, and John O'Loughlin, eds. *Foreign Minorities in Continental European Cities.* Stuttgart: Franz Steiner Verland Wiesbaden GMBH, 1987.

Hall, Bruce F. "Neighborhood Differences in Retail Food Stores: Income versus Race and Age of Population." *Economic Geography* 59 (1983): 282.

Harris, Richard, "A Political Chameleon: Class Segregation in Kingston, Ontario, 1961–1976." *Annals of the Association of American Geographers* 74 (1984): 454–76.

———. "Residential Segregation and Class Formation in the Capitalist City." *Progress in Human Geography* 8 (1984): 26–49.

Hart, T. "Patterns of Black Residence in the White Residential Areas of Johannesburg." *South African Geographical Journal* 58 (1976): 141–50.

Isajiw, Wsevolod W. "Definitions of Ethnicity." *Ethnicity* 1 (1974): 111–24.

Jackle, John A., and James O. Wheeler. "The Changing Residential Structure of the Dutch Population in Kalamazoo, Michigan." *Annals of the Association of American Geographers* 59 (1969): 441–60.

Jett, Stephen C. "The Origins of Navajo Settlement Patterns." *Annals of the Association of American Geographers* 68 (1978).

Jones, Richard C. "Undocumented Migration from Mexico: Some Geographical Questions." *Annals of the Association of American Geographers* 72 (1982): 77–87.

Jordan, Terry G. *German Seed in Texas Soil: Immigrant Farmers in Nineteenth-Century Texas.* Austin: University of Texas Press, 1966.

Louder, Dean R., and Eric Waddell, eds. *Du continent perdu à l'archipel retrouvé: le Québec et l'Amerique française.* Québec: Travaux du Département de Géograhie de l'Université Laval, No. 6, 1983.

Mannion, John J. *Irish Settlements in Eastern Canada: A Study of Cultural Transfer and Adaption.* University of Toronto, Department of Geography, Research Publication No. 12, 1974.

Marantz, Janet K., Karl E. Case II, and Herman B. Leonard. *Discrimination in Rural Housing.* Lexington, Mass.: D.C. Heath, 1976.

Marston, Sallie A. "Neighborhood and Politics: Irish Ethnicity in Nineteenth-Century Lowell, Massachusetts." *Annals of the Association of American Geographers* 78 (1988): 414.

Morrill, Richard L. "The Negro Ghetto: Problems and Alternatives." *Geographical Review* 55 (1965): 339–61.

Morrill, Richard L., and Fred O. Donaldson. "Geographical Perspectives on the History of Black America." *Economic Geography* 48 (1972): 1–23.

Nostrand, Richard L. "The Hispanic-American Borderland: Delimitation of an American Culture Region." *Annals of the Association of American Geographers* 60 (1970): 638–61.

———. "The Hispanic Homeland in 1900." *Annals of the Association of American Geographers* 70 (1980): 382–96.

O'Loughlin, John. "The Election of Black Mayors, 1977." *Annals of the Association of American Geographers* 70 (September 1980): 353–70.

———. "The Identification and Evaluation of Racial Gerrymandering." *Annals of the Association of American Geographers* 72 (June 1982): 165–84.

O'Loughlin, John, and Dale A. Berg. "The Election of Black Mayors, 1969 and 1973." *Annals of the Association of American Geographers* 67 (1977): 223–38.

Peach, Ceri. "Ethnic Segregation and Intermarriage." *Annals of the Association of American Geographers* 70 (September 1980): 371–81.

Raitz, Karl B. "Ethnic Maps of North America." *Geographical Review* 68 (1978): 335–50.

———. "Themes in the Cultural Geography of European Ethnic Groups in the United States." *Geographical Review* 69 (1979): 77–94.

Reid, J. "Black America in the 1980s." *Population Bulletin* 37 (December 1982).

Rogers, Tommy W. "Race and Geographic Differentials in Mississippi Housing Characteristics." *Mississippi Geographer* 6 (1978): 19–31.

Schreuder, Yda. "Labor Segmentation, Ethnic Division of Labor, and Residential Segregation in American Cities in the Early Twentieth Century." *The Professional Geographer* 41 (1989): 131–43.

Thernstom, Stephan, ed. *Harvard Encyclopedia of American Ethnic Groups.* Cambridge, Mass.: Harvard University Press, 1980.

Todd, D., and J. S. Brierley. "Ethnicity and the Rural Economy: Illustrations from Southern Manitoba." *Canadian Geographer* 21 (1977): 237–49.

Vogeler, Ingolf. "Ethnicity, Religion, and Farm Land Transfers in Western Wisconsin." *Ecumene* 7 (1975): 6–13.

Waldinger, Roger D. *Through the Eye of the Needle: Immigrants and Enterprise in New York's Garment Trades.* New York and London: New York University Press, 1986.

Ward, David. "The Emergence of Central Immigrant Ghettos in American Cities: 1840–1920." *Annals of the Association of American Geographers* 58 (1968): 343–59.

Western, John. *Outcast Cape Town.* Minneapolis: University of Minneapolis Press, 1981.

Williams, Colin H. "Ethnic Resurgence in the Periphery." *Area* 11 (1979): 279–83.

Wishart, David J. "The Dispossession of the Pawnee." *Annals of the Association of American Geographers* 69 (September 1977): 382–401.

Wong, Bernard P. *Chinatown: Economic Adaptation and Ethnic Identity of the Chinese.* New York: Holt, Rinehart & Winston, 1982.

The Mosaic of Language

Every language serves as the bearer of a culture. If you speak a language you take part, in some degree, in the way of living represented by that culture. . . . To the extent that you have learned to speak a foreign tongue, to that extent you have learned to respond with a different selection and emphasis to the world around you. . . .   E. BLOOMFIELD

THEMES

The definition of language.
Language families of the world.
Linguistic geography and landscapes.
Linguistic conflict and diffusion.
Language, landform, and refuge.

CONCEPTS

Language
Toponyms
Dialect
Vernacular
Official language
Bilingualism
Creolization
Linguistic culture region
Romance
Linguistic family
Latin Vulgate
Lingua franca
Relic languages
Sino-Tibetan
Ideograms
Malayo-Polynesian
Afro-Asiatic
Dialect regions

Language has been called the cornerstone of culture. Language is essential for transmitting ideas from person to person within a culture and from generation to generation. The distinctiveness of language separates groups from one another; it serves as one of the most easily recognizable boundaries in the cultural geography of the world. Examination of language illustrates the key geographic concepts of hearth, diffusion, culture, region, cultural landscape, and cultural ecology. Language is a culture variable to which people cling tenaciously. For example, the French, who are a minority in Canada, feared losing their cultural identity, an identity tied to the French language (Figure 9–1). To prevent dilution of the "Frenchness" of Quebec, the provincial government passed laws in the 1970s requiring that businesses use only French in their signs and advertisements, that children attend French schools unless one or both of their parents graduated from an English-speaking school, and that business and government transactions be conducted in French (Table 9–1).

The defensiveness of the French of Quebec illustrates the way in which language can be a fundamental expression of cultural identity. Language is also important to the geographer because it provides a record of the movements of people, the diffusion of cultural groups, and the changes in occupance of an area. Landforms or communities in a particular locale are referred to by *toponyms* (place-names). Toponyms reflect the language of the occupants of that landscape. If migration has occurred, the existing toponyms may or may not pre-date the language of the people currently inhabiting a site. By analyzing the distribution of toponyms, it is possible to map the diffusion of certain, specific cultural groups or

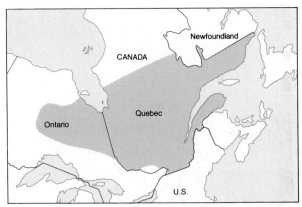

FIGURE 9–1 Areas in Canada Where Most Persons Speak French The French language is dominant in the province of Quebec.

to recreate the historic pattern of occupation of a region (Figure 9–2).

Equally interesting to the cultural geographer is the present distribution of languages. A map showing the distribution of all languages in the world would be highly complex, since there are between 3,000 and 5,000 languages. The difficulty of specifying how many languages exist and the resultant variation among maps showing the languages of the world reflect the difficulty of establishing what constitutes a language. Language has been defined in a variety of ways, but central to all definitions is the concept that language is a medium for communicating ideas.

Language: A Definition

Although it is possible to speak of animal or sign language as a means of communication, most definitions of language state that it involves the

TABLE 9–1
Population Changes of French Canada

	1951	1961	1971	1986
Total	14,009,000	18,238,000	21,568,000	25,625,000
French Parents	4,319,167	5,540,346	6,180,120	6,918,750
French as mother tongue	4,068,805	5,123,151	5,793,650	

Source: Lena Newman, *Historical Almanac of Canada* (Toronto: McClelland and Stewart, 1967), and *The Canada Yearbook, 1987* (Ottawa: Minister of Supply and Services, 1987).

Merchants in Montreal, Quebec utilize bilingual signs even though it is officially illegal.

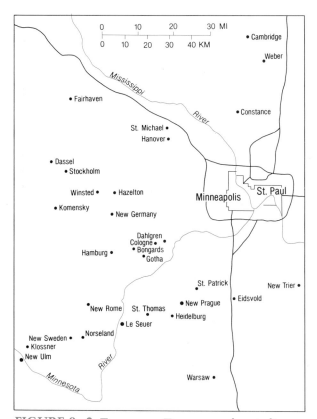

FIGURE 9–2 Toponyms Toponyms often indicate the origin of the people who occupy an area. Near Minneapolis, Minnesota communities bear names indicating German, Swedish, Polish, English and French origins for migrants.

use of words as symbols for communicating ideas among humans. Language is symbolic, and the symbols have conventional meanings and sounds. The meaning and sounds of a language are mutually intelligible to the general members of the group to which they are common.

The uniqueness of the sounds and symbols of specific languages explains why languages form a strong reflective barrier to the diffusion of ideas. The English speaker with no training in Chinese cannot understand the sounds of Chinese. The Navajo who speaks no other language finds English equally unintelligible. Without mutually accepted symbols and sounds for the language components (or a translator), it is impossible to communicate complex verbal ideas.

Language, as defined here, does not require the use of written symbols. Preliterate societies communicated with one another and transmitted ideas. The absence of a written form of a language means only that its role in transmitting culture reflects an *oral tradition.* Although the majority of the largest language groups in the world have a *written tradition,* numerous languages still exist without written symbols or with symbols that have been introduced by outside forces.

The presence or absence of a written literary tradition in a language is important. Languages are constantly changing and evolving. New words are adopted, old words are changed or abandoned. Pronunciation varies over time. Adoption of written symbols often leads to a slower rate of change. Without a formal written tradition, there may be wide disagreement on the actual spelling, pronunciation, or precise meaning of a word.

Each individual has his or her own distinctive speech patterns that are generally acquired during childhood through adoption of a *mother tongue* (first learned language). A *dialect* is the distinctive speech pattern of a region or group that differs from the standard language of a country. Individual dialects share similar pronunciation, grammar, and vocabulary. A language is used by a broader group than a dialect and is intelligible to all its users, including each dialect group. Language allows specific ideas, customs, skills, techniques, and so forth to be explained and made available to anyone who understands the language. Variations in dialects do not prevent the transmission of ideas, but variations in language do. The difficulty of determining whether a specific dialect is a separate language is one reason for disagreement on the total number of world languages. Some dialects may be so extreme as to be almost mutually unintelligible to other members of a broader language group. For example, *Cockney*, (English) a dialect in London's inner city, includes words that are totally unknown to other users of English (Table 9–2). Dialects may result from borrowing terms from languages in an adjacent country or from the retention of words from a language no longer used.

Regional variation in the sound of a spoken language may not be sufficient to classify a distinctive speech pattern as a dialect even though the variation is easily recognized by outsiders (Figure 9–3). Spoken language in the southern United States, for example, is recognizable by people from the North or West, but there is disagreement as to whether this is a dialect or simply an *accent*. An accent differs from a dialect in that it shares the same basic words as the rest of a language, but stresses different elements of a word (Figure 9–4).

Family, community, or regional differences in accent or dialect are learned. Each individual is exposed to a mother tongue by the speech patterns of his or her parents, siblings, playmates, and teachers. Thus initial learning of the language occurs from the random utterances that children first encounter.

Languages are categorized as being either *living* (still spoken) or *dead* (no longer spoken). Latin is an example of a dead language. It is studied today for many cultural reasons, including the insight it provides into languages derived from Latin. The term *vernacular* refers to the common spoken

TABLE 9–2
Examples of English and Cockney Words

ENGLISH	COCKNEY
eat	ett
drought	drawt
tadpoles	polly wags
see-saw	Tippeny-Tawter
hello	ello
fool	faird
money	lolly, gelt
oaf	klutz
silly	schmarel
rubbish	schmutters
nose	schnozzle
mallet	beedle
gossip, chat	clatter
man	bloke
friend	mate or cock
untidy	frutty
cat	moggy

Source: Peter Wright, *Cockney Dialect and Slang* (London: B. T. Batsford, Ltd, 1981).

form of a language as opposed to a literary, cultural, or foreign language. Vernacular also refers to a nonstandard variant of a language used in a region or country. The vernacular is used in everyday speech, while a more formal variant of the language is used in government, education, or other official business.

A *standard* or *official* language is the version of a language recognized for public affairs. For example, a number of English dialects were spoken as the language evolved over the last 1,500 years. The standard English that emerged was that spoken by the royalty and those associated with the English court in London. Over time, standard English was adopted by people in business who

THE CONTINUUM FROM INDIVIDUAL SPEECH TO LANGUAGE

FIGURE 9–3
Continuum from Personal Speech Patterns to Language The continuum is one of increasing formalization and commonality in sound, use, and spelling of words.

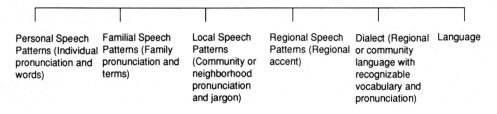

Personal Speech Patterns (Individual pronunciation and words)

Familial Speech Patterns (Family pronunciation and terms)

Local Speech Patterns (Community or neighborhood pronunciation and jargon)

Regional Speech Patterns (Regional accent)

Dialect (Regional or community language with recognizable vocabulary and pronunciation)

Language

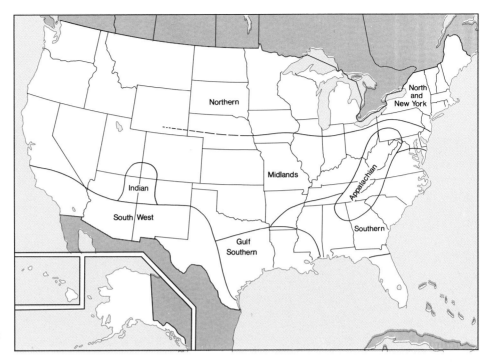

FIGURE 9–4 Dialect Regions of the United States Regions of the east coast and south are most easily recognized. The southwest is recognizable by its use of Spanish words and phrases. Migration makes definition of specific dialect regions in the west and northwest difficult.

wanted to demonstrate their level of knowledge and social position. Classical Arabic is the standard form of Arabic used in religious rituals and government communications across the Middle East. Mandarin Chinese became the standard form of Chinese centuries ago because examinations for government civil service positions reinforced its importance as the standard language. In spite of the existence of a standard language, all users modify language to reflect their own idiosyncrasies. Slang represents one such variation, as do dialects.

Bilingualism refers to the ability to use two languages fluently, while *multilingualism* indicates mastery of more than two. Bilingualism or multilingualism may be acquired by learning two or more languages simultaneously while growing up or by learning a second language after mastering a first (*sequential bilingualism*). In the simultaneous acquisition of bilingualism, a child is exposed to both languages. For example, a migrant to the United States from Vietnam may use only Vietnamese in the home but, through playing in the streets, watching television, and attending school, may learn English simultaneously. Sequential bilingualism

may be associated with changes in political control of an area, changing job requirements, or educational demands. Bilingual or multilingual individuals generally consciously or subconsciously emphasize one language over the other. Bilingualism or multilingualism is a fact of life in some countries (Belgium, Switzerland) and is a major factor contributing to linguistic change and conflict (Figure 9–5).

Linguistic change occurs in all languages over time, as can be seen by examining any language that has a written tradition over several centuries. For example, compare present-day English with Shakespearean English, Chaucer's English, and Old English forms surviving in written texts. Although the language used by Shakespeare is constructed differently from modern English, it is generally clearly intelligible, whereas Chaucer's Middle English is much less so (Table 9–3). For most English speakers, Old English texts are as unintelligible as texts in German.

The evolution of any language over a relatively few centuries involves changes in word meanings, grammar, and pronunciation. Such changes occur through *borrowing, creolization*, and the invention or introduction of *pidgin*. Borrowings

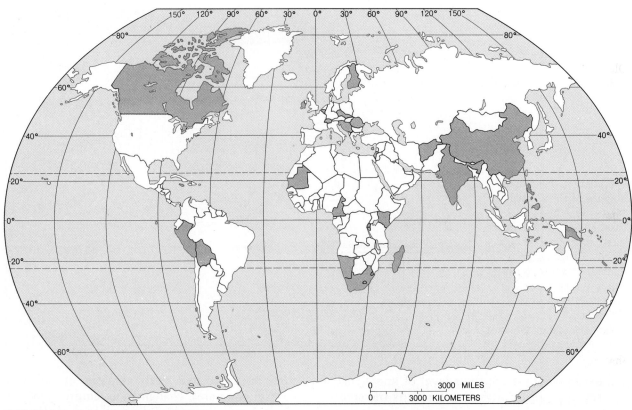

FIGURE 9–5 Countries That Are Formally Bilingual Other countries may have large numbers using a second language even though it has no formal bilingual policy, as the United States.

are the adoption into a language of words from another language. In the last few decades the accession into English of words such as "detente" (borrowed from French), "enchilada" (from Spanish), or "ciao" (from Italian) provides examples of such borrowings. Creolization occurs when two languages merge into a distinctive language understood only with difficulty by speakers of the original languages. The French Creole in Haiti, which evolved from a combination of African languages and French, is an example (Table 9–4). Pidgin is a simplified tongue with only a limited vocabulary selected from an amalgamation of several languages. Pidgin is useful only for trade or simple communication. The classic example of pidgin is the Southeast Asian Malay pidgin, which combines terms from Chinese, English, and Malay. The Swahili language of East Africa was also originally a pidgin language used for trade. Composed of Arabic and Bantu terms, Swa-

hili has developed a literary tradition and become more than a pidgin language (Figure 9–6).

World Linguistic Regions

The division of the world on the basis of language creates *linguistic culture regions* illustrating the present distribution of languages. Linguistic culture regions can be defined in a number of ways. Linguists recognize linguistic regions in terms of word sounds or other commonalities, but for geographers the most useful categorization is a genetic classification based on the relationships of languages as revealed by their origins and development.

Where languages have a long literary tradition, as in Asia and Europe, it is possible to research their origins and development to create such a genetic classification. In regions such as the pre-

TABLE 9–3
Examples from Chaucer and Shakespeare

Olde English: (Beowulf)	Wæs him se man to þon leof þæt he þone breostwylm forberan ne mehte, ac him on hreþre—hygebendum fæst æfter deorum men—dyrne langað beorn wið blode. (That man was so dear to him that he could not hold back the surge in his breast; on the contrary, a secret longing in his bosom for the dear man strained against rational restraints, burned in his blood.)
Chaucer:	From the Prologue of the *Reves Tale*: This White top writeth myne olde yeris; Myn herte is mowled also as myne heris— But if I fare as dooth an open-ers. That ilke fruyt is ever lenger the wers, Til it be roten in mullok or in stree. We olde men, I drede, so fare we: Til we be roten, kan we nat be rype. . . .
Shakespeare:	From *A Midsommer Nights Dreame*: And my gracious Duke, This man hath bewitch'd the bofome of my childe: Thou, thou Lyfander, thou haft giuen her rimes, And interchang'd loue-tokens with my childe: Thou haft by Moone-light at her window fung, With faining voice, verfes of faining loue. . . .

TABLE 9–4
Creole Language

TYPE	LOCATION
French Creoles	Louisiana
	Haiti
	Guadeloupe
	Martinique
	St. Lucia
	Trinidad
	French Guiana
	Mauritius
	Réunion
English Creole	Suriname
	(Among descendants of African slaves)
Portuguese Creole	Guinea-Bissau
	Cape Verde Islands

European Americas or present-day Southeast Asia or Africa, a multiplicity of languages without a literacy tradition make it difficult to classify the relationships among languages. Africa alone contains as many as 600 to 1,000 languages. Analyzing them is difficult since the extreme differences in dialects within the languages make them almost independent languages. In areas where it is impossible to research the relationships between languages to classify them genetically, languages are normally defined in terms of a simple geographic region. *Linguistic families* are the groups of languages related through a common origin and language hearth. There are about twenty major linguistic families, which include most of the three to four thousand languages of the world (Figure 9–7).

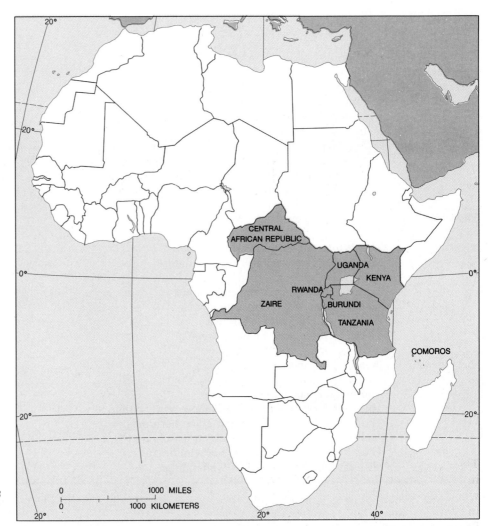

FIGURE 9–6 Countries Where Swahili Is Spoken Some central African countries use Swahili to overcome their linguistic fragmentation.

Source: The World Factbook, 1988 (Washington: Central Intelligence Agency, 1988).

The Indo-European Language

Approximately half the people in the world speak a language in the Indo-European family. Analysis of the origin of the Indo-European language indicates that by about 4000 B.C. a Proto-Indo-European language existed in a hearth area in eastern Europe and the southern plains of the Soviet Union extending around the Black Sea to the Caspian Sea. Archaeological evidence shows that the ancestral Proto-European language diffused across the entirety of Europe in the last 6,000 years, creating the individual distinct languages of modern Europe in the process.

The Indo-European linguistic family includes the primary languages of Europe and areas colonized by Europeans, as well as those found along the north shore of the Mediterranean and into India. The widespread diffusion of individual groups or peoples speaking an Indo-European language creates a broad division into Eastern and Western subfamilies. The Eastern subfamily is found in the areas east of the Mediterranean Sea, the Western in Europe north of the Mediterranean (Figure 9–8).

The Eastern Indo-European Languages

The eastern Indo-European languages seem to have evolved from the language spoken by early cattle-herding migrants (called Indo-Aryans) into South Asia during the second millennium B.C.

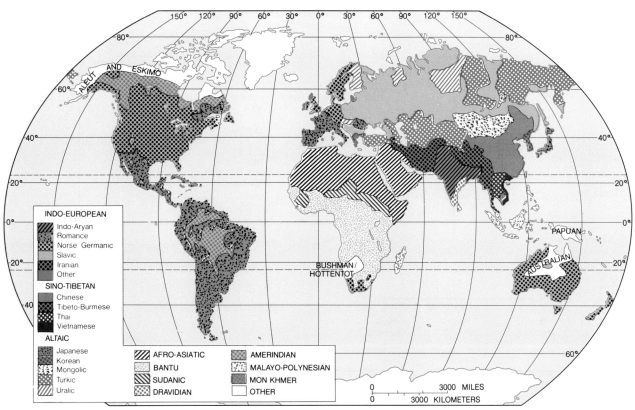

FIGURE 9–7   Linguistic Families of the World

The migrating Indo-Europeans displaced the original inhabitants of the region, the Dravidians, whose languages are still spoken in southern India. The oldest known Indo-Aryan language is Sanskrit, a classical language still used in the Hindu religion. Languages in the area from Bangladesh to Iran are dominated by derivatives of Sanskrit. The main languages recognized in the Indo-Aryan language group include Hindi (one of sixteen official languages of India—see Figure 9–9); Urdu (the official language of Pakistan); Bengali (in the Indian state of West Bengal and in Bangladesh); Punjabi (in northwest India); Oriya (southwest of Calcutta); Bihari (in the central Ganges Basin); and Sinhalese (in Sri Lanka).

The Hindi and Urdu languages typify the problems encountered in trying to create cultural regions based on languages. Although they are both derived from Sanskrit, Hindi is written using the refined characters inherited from Sanskrit while Urdu is written in a Persian form of Arabic. The languages thus are distinct not only in sound,

A large poster map of Bangladesh illustrates the Bengali script.

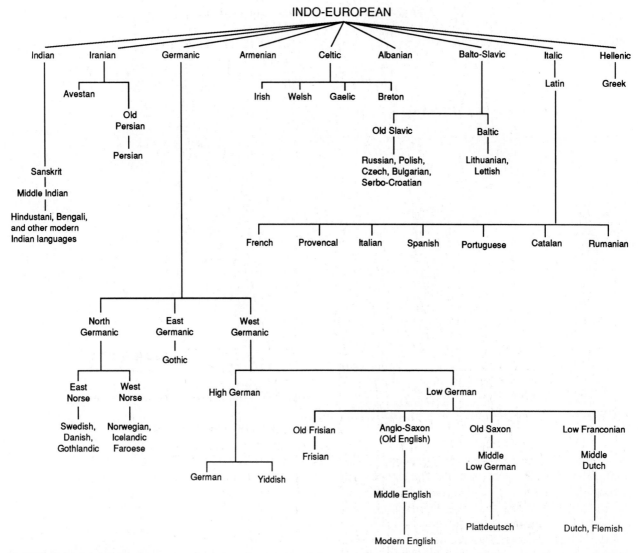

FIGURE 9—8 Indo-European Languages The many subdivisions of the language family illustrate how languages change over time.

but in script. Other Indo-Aryan languages are distinct languages used by significant minorities within the Indian subcontinent (such as Punjabi, Rajasthani, Marathi, Gujariati, Oriyia and Assamese), and each helps define the individual states within India. English is also widely spoken in India, reflecting both the British colonial heritage and its unofficial use as a means of mass communication.

An example of how a language can serve to maintain the identity of a group is Romany. An Indo-Iranian language spoken only by Gypsies (often as a part of a bilingual tradition, and often to make sure they are not understood by non-Gypsies), the Romany language diffused west from India with the Gypsies into eastern Europe and then to France, Britain, and the United States (Figure 9–10). Romany is not shared with non-Gypsies.

Western Indo-European Languages

Western, northern, and southern Europe are occupied by a broad variety of Indo-European-speaking

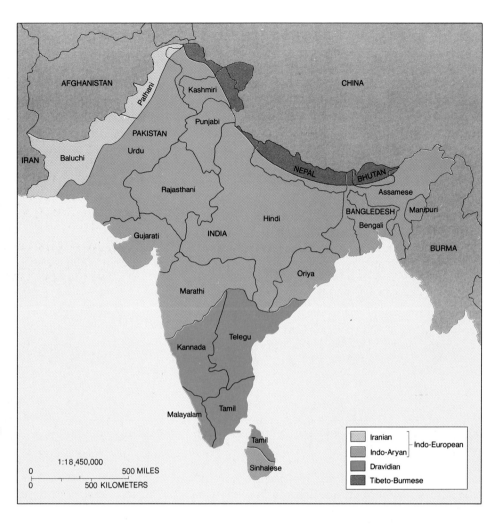

FIGURE 9–9 Languages and States of India Most of the political subdivisions of India are based on language differences.

peoples. Greek, an Indo-European language that is confined to Greece today, was spoken throughout much of Mediterranean Europe at the height of the Greek Empire. Greek shares the distinction with Sanskrit of being the oldest of all Indo-European tongues. As a literary language of great antiquity, classical Greek was a distinctive language by 1500 B.C. Modern Greek is relatively closely related to the forms of three millennia earlier. Today Greek is spoken by only some twelve million people, but many Greek words are familiar to non-Greeks because they have been used in various sciences to describe new developments over the centuries. The terms geologists use to divide geologic history into the Paleozoic, Mesozoic, or Pleistocene periods are from Greek, for example. Greece was also one hearth for much

of Western civilization making it more important than its small population of Greek speakers would indicate.

Romance

Southern Europe is dominated by the Romance language family subgroup. A wide variety of individual languages comprise the *Romance language*, the most important of which are Spanish, Portuguese, French, Italian, and Romanian. All of these languages have descended from Latin (or Italic). The Italic precursor to Latin developed on the Italian peninsula, and with the growth of Rome and the Roman Empire, Latin was spread across much of southern and western Europe. The Latin diffused by the Roman soldiers who occu-

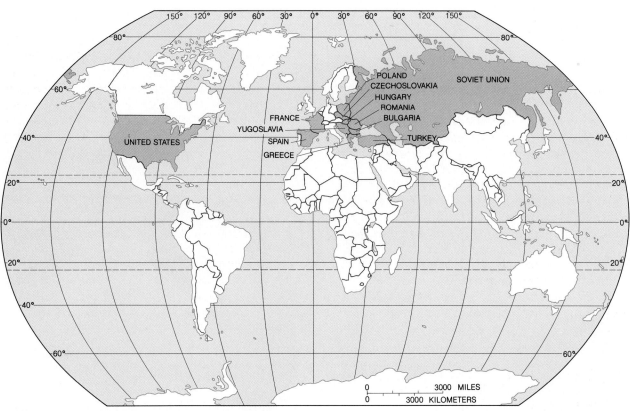

FIGURE 9–10   Countries Where Some Residents Speak Romany

pied various regions from England to Romania and across southern Europe differed from classical Latin in its literary and written form. Modern Romance languages reflect the language spoken by the Roman populace (including soldiers), referred to as *Latin Vulgate*, from the Latin "vulgar," meaning masses. After the fall of the Roman Empire in the fifth century A.D., the breakdown of communication among the former colonies led to ever greater differences in the individual languages.

Italian   Italian is concentrated in the Italian peninsula, but is also spoken by Italians living in France, Argentina, Brazil, Yugoslavia, Switzerland, the United States, and elsewhere. There are many dialects of Italian. Two, Sardinian and Corsican, are sufficiently different to be classed as distinct languages by some observers. Romansch is a dialect that is recognized as a separate language in Switzerland and near the borders of Italy in Austria and Yugoslavia.

Spanish   Spanish is spoken in Spain, eighteen Latin American republics in Central and South America, Cuba, Puerto Rico, the Dominican Republic, the Philippines, former Spanish Morocco in Africa, and the United States. Spanish is now one of the most commonly spoken Indo-European languages because of the population growth in the Latin American countries. Diffusion of Spanish to North America has given the United States the fifth largest Spanish-speaking population in the world (after Mexico, Spain, Argentina, and Colombia).

Portuguese   Portuguese was a language of little importance in the world prior to the voyage of Columbus. With fewer than five million speakers, it was faced with potential absorption into the dialects of the Spanish language. The settlement and growth of Brazil and the division of the Americas between the Spanish and the Portuguese in the Treaty of Tordesillas in 1494 led to the expansion of Portuguese. Today Portuguese is spoken in Portugal and Brazil as well as in former Portuguese colonies in Africa and in Macao in East Asia.

French   French is a major language in Europe; it is spoken in France, the southern half of Belgium, some of the political units of Switzerland, Luxembourg, Monaco, and the island of Jersey in the English Channel as well as in the former colonies of France such as Quebec in Canada (Figure 9–11). Creole languages based on French are spoken in Louisiana and parts of the Caribbean, especially Haiti. French is the official language of nearly forty countries and the co-official language in seven more that were formerly French or Belgian colonies. French is one of the five official languages of the United Nations and is second only to English in the number of students who study it as a second language worldwide.

Celtic

The Celtic languages are descended from an ancient language that once occupied much of western Europe; it has been pushed into the margins of the continent, however, by the expansion of Germanic and Slavic peoples. The principal remnants of Celtic today are Gaelic spoken in Ireland, Welsh in Wales, Breton in the Brittany peninsula of France, and Scottish Gaelic in Scotland. Gaelic is an official language (with English) in Ireland. Scottish and Welsh are vernaculars used by some of the residents of Scotland (100,000 speakers) and Wales (700,000 speakers) to maintain their national identity. In Scotland and Wales, however, the majority of the people are bilingual in English. The Celtic languages continue to be used because of their literary tradition and strong historical ties to the national identities of the individual groups.

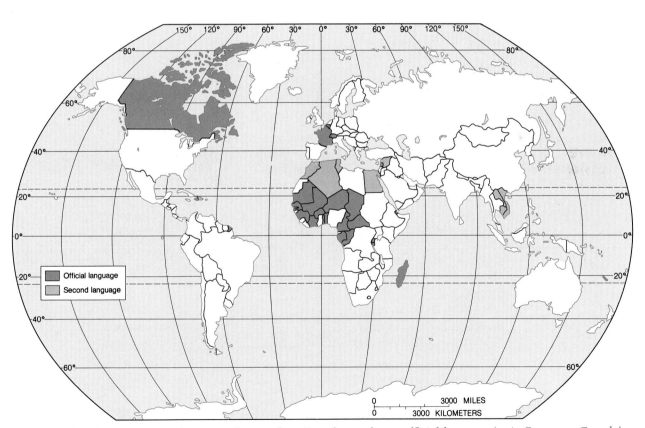

FIGURE 9–11   Countries Where French Is Spoken French may be an official language (as in France or Canada) or an unofficial lingua franca.

Source: The World Factbook, 1988 (Washington: The Central Intelligence Agency, 1988).

Germanic (Teutonic)

Germanic or Teutonic languages dominate most of northern and western Europe. The principal Germanic languages include English, German, Swedish, Norwegian, Danish, and Dutch.

English   English is the most important of the Germanic languages. Including its use as a second language, English is spoken by more people than any other language. English was diffused through the establishment of English colonies in North America, South America, Africa, and Oceania (Figure 9–12). Even in those colonies that have become independent and declared an indigenous language as an official language, English remains as a *lingua franca*. A lingua franca is a language of convenience used for communication over a wide area among peoples of different speech. Countries where English is used as a lingua franca in the modern world include India, Pakistan, Nigeria, Malaysia and the Philippines. English is a co-official language in many former colonies. As late as 1950, one-fourth of the countries of the world used English as their official language.

English is the native tongue of barely a dozen countries in the world today, but it is widely spoken or studied in more than ninety others. Of the more than five billion people in the world, an estimated one billion are familiar with English, making it the most widely spoken language in the world. English is also spreading widely as a lingua franca; in at least fifty-six nations it is required in school or studied widely. In Latin America alone, an estimated fifty million school children study English. Since 1980 in China, those students who go on to secondary school study English. More than half of all secondary school students in the Soviet Union study English. One hundred and fifty professional language schools teach English in France; and secondary students are required to

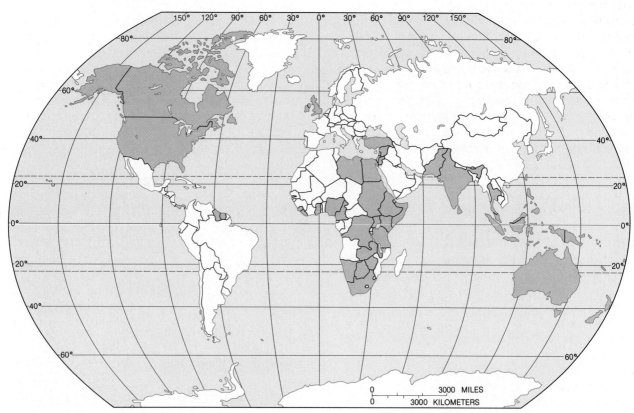

FIGURE 9–12   Diffusion of English From its hearth in Europe, English has diffused to much of the world. In many countries English is primarily important as a lingua franca, especially among the elite of the country.

take four years of English *or* German. It is revealing to note that in the last decade over 80 percent of French secondary students have chosen English.

The distribution of English is an excellent example of the way in which diffusion has changed the cultural map of the world. There were only 2 million English speakers at the time of Chaucer. When Shakespeare wrote, only about 5.5 million people spoke English, and English ranked fifth among languages in Europe in numbers of users. Expansion and colonization from England between the sixteenth and twentieth centuries brought English to preeminence in the world.

There are numerous dialects of English (and even more accents), including at least twenty in the United Kingdom alone. Because of a common school system and mass media, English is more uniform in the United States. Standard English in Britain originated in London, the capital, largest city, and center of trade, and in university centers such as Oxford and Cambridge. Standard English diffused to smaller towns and cities after the mid-nineteenth century with the introduction of the public (actually private) schools. The public schools taught London standard English to those who became church, political, and business leaders.

It is important to note that in the past the use of English as a lingua franca emphasized London standard English. The language used by the British Broadcasting Corporation and in interactions between the United Kingdom and its foreign colonies is based on what is known as British Received Pronounciation (BRP). Used by politicians, actors, and government officials, it is the official English heard in much of the world. The BRP is not used in the United States, and the dialect groups in England do not use it. Nevertheless, in the past it has been the major form of English outside North America.

In the twentieth century the use of standard English by the British Broadcasting Corporation and in schools made it more common in everyday interactions. The increasing popularity of things American, however, makes American English a close rival to BRP. If present tendencies continue, American English will become the dominant form of English in the world.

The variations in spoken English between the United States, England, and former colonies such as Australia or Nigeria have given rise to questions concerning the division of these subsets into distinct dialects or languages. It is clear that the pronunciation of English in Nigeria, for example, is far different from that used in the United States or in England, an illustration of the continual change that occurs in languages (see the Cultural Dimensions box on page 336).

The diffusion of English has caused some countries, notably France and Canada (Quebec), to pass laws to protect their language from words borrowed from English. In France, the High Commission of the French Language supervises media and government to ensure that French terms rather than English ones are used. The commission recommends, for example, that the widely used term "sandwich" should be replaced by *"deux morceaux de pain avec quelque chose au milieu,"* which literally translated means "two pieces of bread with something in the middle."

The French have been unable to prevent the diffusion of English words into their language altogether, in part because things from the American culture are often preferred in France. Typically half of the popular records in France are from England or the United States, and the most popular television programs are American shows such as "Dynasty" and "Dallas." English contin-

Quebec City, Canada. Translation into French does not obscure the origin of well-known trademarks.

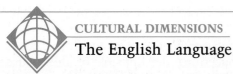

CULTURAL DIMENSIONS
The English Language

English originated in England with the invasion of the Anglo-Saxons (Angles, Saxons, and Jutes), Danes, and Vikings from northern Europe. Each of these groups settled a different part of the island of Great Britain. After the Normans invaded England in 1066, French replaced the English dialects among the nobility. The use of French by the aristocracy until about A.D. 1200 resulted in numerous borrowings of French words for use in the English vernacular dialects.

The diffusion of English to countries colonized by England caused more changes in words and pronunciation. It is estimated that at the time of Chaucer, there were only 50,000 to 60,000 words in English, 70 percent of which are still used in English today. By the time of Shakespeare, another 80,000 to 90,000 words had been added, 75 percent of which are still in use. Today there are at least 750,000 words in the English language, as compiled in the thirteen-volume *Oxford English Dictionary*. The average well-read speaker of English would customarily use about 20,000 words; a lexicographer double that. If we include recognition of words, the number of words a person knows might double, though he or she would use few of these in normal conversation.

An estimated 80 percent of English words in use today are borrowed from other languages. African languages give us such terms as voodoo, zombie, tango, and jazz. From Turkish come words such as coffee, turban, and caviar. Spanish provides a host of words including cannibal, mosquito, tornado, and vanilla. Hindi is the source for bungalow, jungle, shampoo, loot, and pajamas. Words borrowed from Arabic include algebra, almanac, and alcohol. French provides a variety of words including detente, alliance, most words for food (including pork and beef), and numerous others. French has been a particularly important source and has often contributed synonyms for English words. *Begin* (Anglo-Saxon) and *commence* (French) are examples, illustrating how French words persist when an English equivalent is available. The French synonym is often used in more formal settings, perhaps a custom dating from French rule of Britain and the persistence of French terms among English leaders thereafter.

ues to diffuse in part because it is more accommodating to change and accepts new words more readily than most other languages, and in part because of the dominant role played by English-speaking countries in economic and technological development in the world in the last two centuries. The diffusion of English will continue as the use of the language continues to expand.

German   Modern German is a descendant of historic high German, and it is spoken in the two Germanies, Austria, Liechtenstein, and much of Switzerland. There are many German speakers along Germany's borders with Denmark and France as well. Yiddish is a derivative of German, developed by Jewish people in the Jewish quarters of cities in central and eastern Europe and spoken by European and American Jews still. The annihilation of the Jews by Hitler during World War II left few Yiddish-speaking people in Europe. Some still live in the Soviet Union, and some are found in and around New York City. German is not spoken widely beyond the confines of Europe since Germany had extensive colonial holdings for only a few decades (1880s–1918). The colonists from Germany who came to the United States generally adopted English, but migrants from Germany after World War II in Paraguay, Brazil, and other parts of Latin America represent outlying groups of German speakers.

A few German-speaking groups who emigrated to Latin America or North America in the nineteenth century have retained their German language. These include the Mennonite, Amish, and Hutterite religious groups of the United States, Canada, Brazil, and Mexico; these are small groups who have isolated themselves from the broader culture around them (Figure 9–13).

Dutch   Dutch is a derivative of low German and is intermediate between modern German and modern English. It is spoken in the Netherlands and some of its former colonies, such as Indonesia. Derivatives of Dutch include Flemish, spoken in northern Belgium; Afrikaans in South Africa and Botswana; and a creolized dialect in Sri Lanka and a few small islands in the Caribbean.

North European Germanic Tongues   The languages of northern Europe are also Germanic and are descended from one common precursor, Old Norse. The tongue most similar to Old Norse is Icelandic, preserved in its purer form because of Iceland's isolation. The other three (Norwegian, Swedish, and Danish) have been influenced both by other European languages and by one another. Icelandic has changed so little over time that an

Icelander of today can read the great Norse sagas easily without special training. Norwegian, by comparison, has changed markedly and now consists of two competing varieties, sometimes called *Bokmal* ("book language") and *Landsmal* ("rural language"). *Bokmal* reflects the Danish literary tradition since Norway was united with Denmark in one political unit from 1380 to 1814. Danish became the written language in the sixteenth century, excluding Norwegian because the Danes controlled the printing press and the University of Copenhagen, which dominated intellectual life. Rural Norwegian (*Landsmal*) is the spoken Norwegian vernacular that persisted during the entire period of union with Denmark.

Danish is spoken by only five million people, primarily in Denmark, although a few Danish speakers are found in border areas of adjacent Scandinavian countries. Particularly in its structure, Danish has been strongly influenced by German.

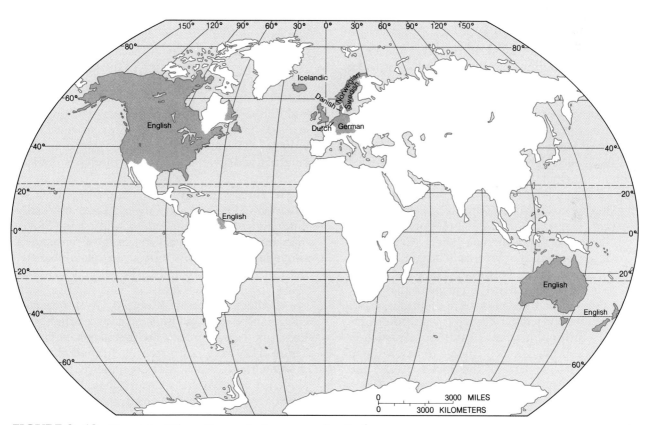

FIGURE 9–13   Countries Where Germanic Languages Are Spoken

Swedish Swedish is the most widespread north Germanic tongue, and is spoken by some nine million individuals today. Swedish is spoken extensively in southern Finland where Swedes have migrated over time (Finland was long part of a Swedish empire), as well as in some regions in the United States. The Swedish language has retained its integrity because the Swedish Academy, founded in 1786, has taken steps to formalize the language.

The Baltic-Slavic Groups

The remaining Indo-European languages are those that were associated with the areas around the Baltic Sea and the continental areas of Europe. The Baltic languages consist of two small groups: Latvian (Lettian) and Lithuanian, each spoken primarily in the Soviet Socialist Republic of the same name. The Lithuanian language demonstrates that isolation can lead to continuity and lack of linguistic change. The Lithuanian peoples were isolated along the Baltic Coast by forests that separated them from other peoples of Imperial Russia. The result is a language that is said to be nearer to Indo-European even than Sanskrit. However, the small size of its population (3 million) and the increasing emigration of Russians into Lithuania threaten the existence of Lithuanian. The Lettian language is spoken in an even smaller area and has been influenced by the Estonian language, which is closely related to Finnish.

The Slavic (Slavonic) languages are comprised of nine principal distinct languages, the main one being Russian (Figure 9–14). Russian is part of a division of north Slavic languages that includes Russian, Ukrainian, and Byelorussian. The central Slavic languages include Polish, Czech, and Slovak, while in the south are found Slovene, Serbo, Croatian, and Bulgarian. The Russian language is by far the largest language of the group, even though it ranked only sixth in Europe in the sixteenth century (after French [14 million], German [12 million], Italian [10 million], Spanish [9 million], and English [5 million]). Today Russian is the official language of the Soviet Union, and its linguistic preeminence is unquestioned as a result of Russian migrations in the sixteenth through nineteenth centuries. Byelorussian and

Movie advertisement, Krakow, Poland. Polish uses the Latin alphabet.

Ukrainian are closely related to Russian, with many of the words being mutually intelligible.

Czech, Slovak, and Polish are distinct from the Russian or Byelorussian languages. Polish and Czech-Slovak have not diffused except through word borrowing by migrants to the United States. Words have been borrowed in communities such as Milwaukee, and these have added to the variety and richness of the English language of America, but the languages are important in few areas outside the individual countries since neither country was ever involved in foreign colonization.

The three southern Slavic languages of Serbo-Croatian, Slovene, and Bulgarian are markedly different from Russian. Serbo-Croatian and Slovenian are spoken in Yugoslavia. Serbo-Croatian is a single language that is written in two alphabets; the Cyrillic by Serbs and Latin by the Croatians. Those peoples who received Christianity from the Orthodox church (Russians, Byelorussians, Ukranians, Serbs, and Bulgars) adopted the Cyrillic script based broadly on the Greek alphabet. Those who were converted by Roman Catholic missionaries (Czechs, Slovaks, Poles, Croats, and Slovenes) adopted the Latin alphabet. The specific script adopted represents diffusion and borrowing of the alphabet. Thus in Yugoslavia's capital city

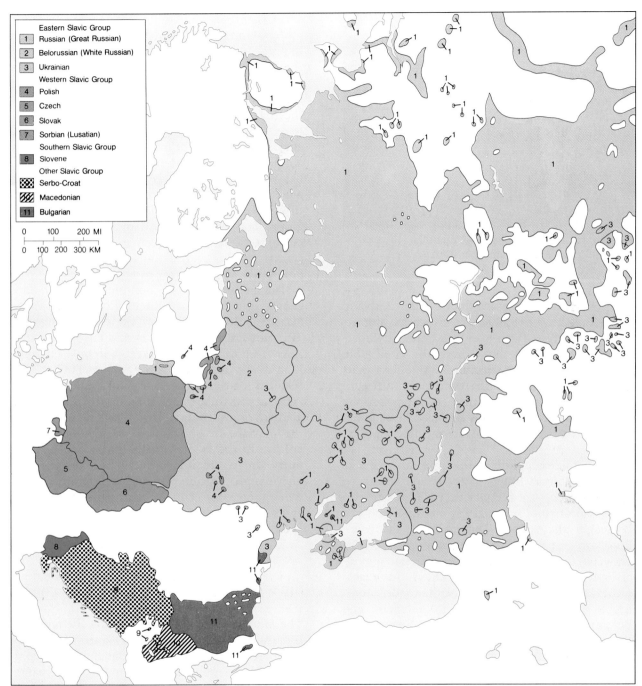

FIGURE 9–14   Countries Where Slavic Languages Are Spoken The Soviet Union and Eastern Europe constitute the major region of Slavic languages.

This church sign in Dubrovnik, Yugoslavia illustrates Serb use of the Cyrillic alphabet (top).

of Belgrade, street signs are duplicated in Serb and Croatian, the written forms of which are extremely different. In spite of this, Serb and Croatian are one language that is mutually intelligible in the spoken form. Bulgarian is the oldest written Slavic language, dating from the ninth century A.D. Outside of Bulgaria it is spoken by inhabitants of communities in Greek Macedonia, in southern Romania, and areas bordering Bulgaria.

The complexity of the Indo-European language and the resultant fragmented cultural landscape should not obscure the fact that it is possible to recognize distinct cultural regions based on the languages themselves. For many of the Indo-European languages, there is a reasonably good fit between the language and the political boundaries of an individual country. Dutch, for example, is largely confined to the Netherlands; Swedish is largely concentrated in Sweden; and Bulgaria, Polish, and Czech-Slovak all exemplify the way in which language serves as the primary unifier of a cultural group.

Relic Languages of Europe

Within Europe there are also some non-Indo-European languages, *relic languages* that have persisted due to isolation, such as Uralic, Basque,

and Caucasian. The Uralic languages include Estonian, Finnish, and Hungarian. Basque is spoken in northern Spain and southern France in and near the Pyrennes Mountains. Caucasian languages are found between the Black and Caspian seas in the Caucasus Mountains. Finnish, Estonian, and the Magyar language spoken in Hungary are related. Like the other non-Indo-European languages of Europe, these languages are relics that are not spoken beyond the boundaries of individual regions except by isolated individuals who have migrated. They persist as the primary identifying feature of the culture of the individual group. The importance of language in defining a cultural group can be seen in the name of the country, as Finland, Estonia, or the Magyar Republic (the official name of Hungary).

The presence of the non-Indo-European languages illustrates the difficulty of attempting to classify culture simply by the use of language. The movement of peoples across Europe and into Asia over millennia has resulted in a complex linguistic map. In countries such as Finland, the majority of the people learn to speak a second, third, or fourth language. The dominance of bilingualism or multilingualism in this region reflects the persistence of language as an identifier of culture in the face of forces that would tend to unify the world linguistically. As the primary purveyor of culture, language is retained even when there is no economic rationale for doing so, and when persistence in using the language may even be an economic handicap.

Sino-Tibetan

The *Sino-Tibetan* languages occupy the region of East Asia centered on China (Figure 9–15). The Sino-Tibetan language family has the second largest number or speakers, with approximately one-fourth of the world's population speaking one of its languages. Mandarin Chinese is the largest distinct individual language in the Sino-Tibetan family; it is spoken by three-fourths of China's population, making it the largest single first-learned language in the world.

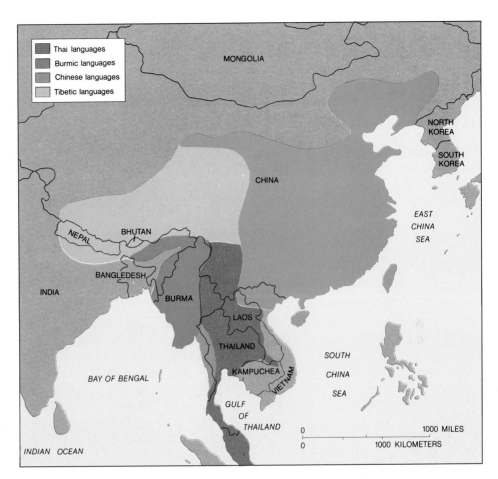

FIGURE 9–15 Areas Where Sino-Tibetan Languages Are Spoken Chinese dominates in terms of numbers of users of the Sino-Tibetan languages.

Chinese

The Chinese family contains a number of mutually unintelligible languages, each with numerous dialects. Mandarin and Cantonese Chinese have the largest number of speakers. Each has its own distinct regional distribution. Mandarin, which had its hearth in northeast China along the Huang He (or Yellow River) is the official language of China today (see Figure 9–16). The written language of Chinese is based on *ideograms* (stylized pictures representing a concept or idea), rather than an alphabet (Figure 9–17). Evolving at the same time as classical Greek, Mandarin Chinese has a rich literary tradition that has served as a basis for the culture of the Chinese, Koreans, Tibetans, and others just as Greek and Latin did for the Indo-European lan-

guages. Chinese is unique because all the Chinese languages use the same ideograms, making written Chinese similar, but since ideograms present only an idea or concept there is no agreed-upon pronunciation.

Mandarin, the northern language, is spoken in China as far south as the Yangtze (Chang Liang) River. Since the Communist Revolution in 1949, the Mandarin language has been taught in schools throughout China. The Wu dialects are found in eastern China in the lower delta area of the Yangtze River. Cantonese is the dominant language in the South of China and in the American Chinese community since the great majority of the Chinese migrants to the United States came from the South of China.

Cantonese is spoken in Hong Kong and in adjacent parts of Thailand, Burma, and Cambodia.

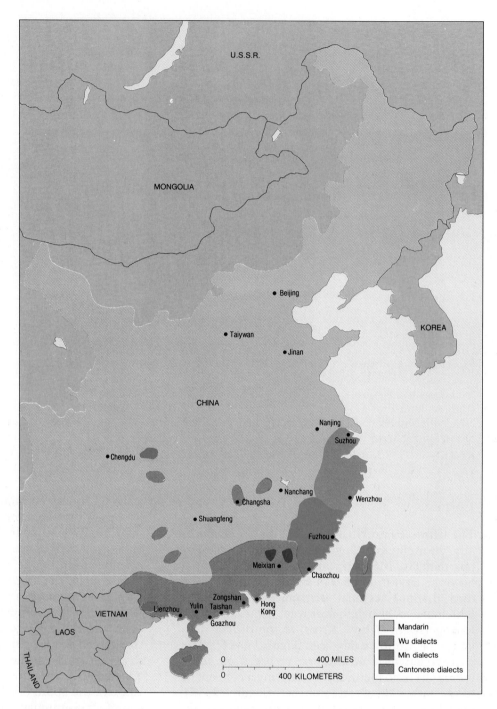

FIGURE 9–16
Languages of China
Chinese is fragmented into a variety of specific languages, but Mandarin dominates.

FIGURE 9–17 Chinese Ideograms The characters for the cardinal directions illustrate how each word is represented by an ideogram.

Min dialects are spoken in Fukien in eastern China, on Taiwan among the Chinese, and in Chinese communities in Indonesia, Malaysia, and the Philippines.

Tibeto-Burmese

The Tibeto-Burmese family is spoken by groups in southern Tibet and Nepal, as well as parts of Assam and northern India, Burma, and parts of northern Pakistan and Thailand. Thai (historically called Siamese) is the largest of this group and is spoken by some 45 million people. Tibetan is the standard literary language of Tibet and is fragmented into a multitude of dialects. Burmese is the language of administration in Burma and is spoken by nearly 30 million people. These languages are used to reinforce the nationality of the individual group. A conscious attempt by the Chinese government to replace Tibetan with Mandarin during the period from 1950 to 1988 has resulted in conflict with the Tibetan people. This conflict is another example of the importance of language to culture as the Tibet-speaking minority of Tibet today see the migration of Mandarin-speaking Chinese and the use of Mandarin in the educational system as threats to their culture, religion, and identity.

Japanese and Korean

Japanese and Korean are sometimes erroneously perceived to be part of the Chinese language family because their literary tradition is based on the borrowed Chinese ideogram system. The Japanese use of Chinese ideograms dates from the fourth century A.D., but Japanese uses additional characters that are more like alphabet symbols in Western languages since each symbol has a specific pronunciation rather than representing an idea. The Korean language also borrowed Chinese ideograms. The Korean and Japanese languages each include dialects, but only in Japanese do we find a subgroup, like Ryukyu spoken on the Ryukyu Islands and on Taiwan.

Uralic-Altaic

The Altaic family of the Soviet Union and Central Asia is part of the same family as the Uralic tongues in the West. The languages of the Altaic group can be divided into three main subfamilies: Turkish, Mongol, and Manchu or Manchurian (Figure 9–18).

Dravidian

The ancient languages of South India are classified as Dravidian. They developed from a language spoken in the Indian peninsula before the invasion of the Indo-Europeans in the second millennium B.C. The major languages are Telugu, Tamil, Kannada, and Malayalam. Nearly 200 million Indians speak these four languages, and each is the basis for one of the Indian states. A minor Dravidian language is Gond spoken in areas of central India (see Figure 9–9 on page 331).

LANGUAGE GROUPS OF ALTAIC

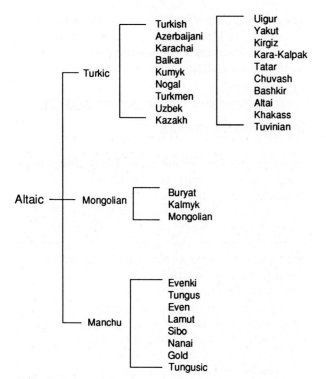

FIGURE 9–18 Languages of the Altaic Group

Malayo-Polynesian (Austronesian)

South and east of Asia are a group of islands that stretch across the Pacific from Taiwan to New Zealand. Languages on the archipelagos of this region are among the most complex in the world, probably because each developed in extreme isolation. Estimates indicate that at least 500 specific languages and even more dialects are spoken across the region. Some 175 million people speak these Malayo-Polynesian languages (see Figure 9–7 on page 329).

The Malay language of southeast Asia has long been used as a pidgin lingua franca in the region from Indonesia to the Philippines. When Indonesia gained its independence, it adopted one version of Malay as the national and official language of Indonesia. Today Malay Indonesian is the largest language in the Malayo-Polynesian group. The Philippines adopted another of the Malayo-Polynesian languages, Tagalog (Filipino) as an official language to replace Spanish and English because of their colonial connotation.

The adoption of Malay-Indonesian and Tagalog as official national languages in regions that had developed a lingua franca based on Indo-European languages illustrates the factors that affect the diffusion of languages. English and Spanish in the Philippines and Dutch in Indonesia were widely used by the ruling classes. Once independence was won, however, opposition to colonial domination by foreign countries made adoption of a native language politically desirable. Consequently, languages native to these countries have become the officially recognized languages. The educated elite continue to use a European language in their communication in spite of the official status of a native language. Such use of a lingua franca is important because it both allows continued interaction in the interconnected world of the twentieth century, and facilitates the diffusion of cultural aspects from the industrial world into the less industrialized world.

Afro-Asiatic

The Afro-Asiatic linguistic family dominates North Africa and the Middle East. The primary

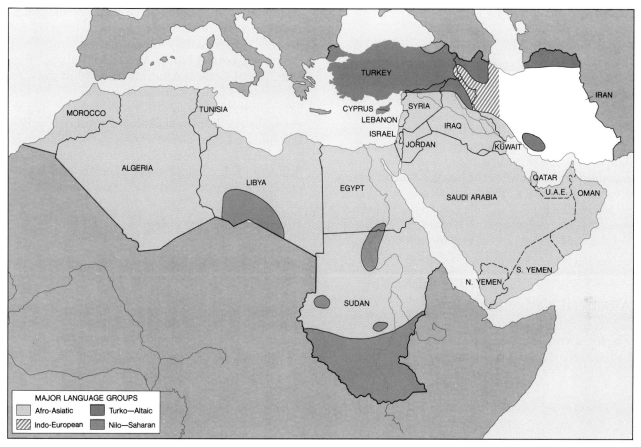

FIGURE 9–19 **Languages of North Africa and the Middle East**

languages of the Asiatic group are Arabic and Hebrew (Figure 9–19).

Arabic is the most widely spoken of the Afro-Asiatic groups. Today it is spoken across North Africa from Morocco to Cairo, south into the Sahara, and in Saudi Arabia and the Middle East. Its widespread use can be traced to its rapid diffusion with the Islamic religion after 630 A.D. Although a standard spoken Arabic (Fusha) has been adopted for mass communications and media use, there are a variety of dialects in Arabic. Literary Arabic is also standardized throughout the entire Arabic world.

Arabic was the basis for the great scientific advances in the Islamic world from A.D. 700 to 1600 and consequently has been a rich source of words for many languages. Borrowings from Arabic include Arabic numerals (even though the numbers themselves were not invented by the

Arabs but diffused from their origin in India to Baghdad in the eighth century and to Venice by the eleventh or twelfth century), the knowledge of the world provided by Arab geographers, and associated scientific and technical advances including geometry and algebra.

Unlike Arabic, which was spread by the expansion of Islam, Hebrew was not associated with a proselytizing religion, and consequently its number of speakers is small, totaling only some four million people at the end of 1988. Classical Hebrew disappeared from use as a common vernacular after the fourth century B.C., but was retained in Jewish religious texts and services. When Israel declared itself a state in 1948, Hebrew was selected as the national language. The Hebrew of today is spoken primarily in Israel and among Jewish congregations in other parts of the world.

Niger-Congo-Bantu

The continent of Africa south of the Sahara is a complex mosaic of languages, primarily because of isolation and the distinct languages within each tribal group. Over 500 languages have been recognized, generally grouped as the Niger-Congo-Bantu family. Over 300 million people in Africa speak the individual languages of Niger-Congo-Bantu. The complexity of the group makes it impossible to describe all of the languages, as even individual countries may have several dozen languages (Figure 9–20). The Congo-Bantu family includes languages such as Fulani (in Guinea), Yoruba and Haunsa (in Nigeria), Mongo and Baluba (in Zaire), and the languages of the Bantu population of South Africa (Zahe, Xhosa, and so forth).

American Native Languages

The linguistic fragmentation in Africa is exceeded by the native languages of America. It is estimated that at one time some 1,250 tongues were spoken in America by the native Indian population. These include approximately 24 among the Inuit-Aleut group, 351 among North American Indians, 96 in Mexico and Central America, and 783 in the Caribbean and South American continent. With a few exceptions, few people still speak these languages. For some the living speakers may number only a few hundred. The classification of these language groups has largely been ignored, in part because the diffusion of English, French, Spanish, and Portuguese has overwhelmed the native languages.

Linguistic Geography

Linguistic geographers study the character and extent of dialects and languages. The occurrence of individual words, specific terms, or geographic features can be used to map dialect and language regions, but the distribution of languages is only one of the factors of interest to geographers. Study of language helps to explain the past or existing

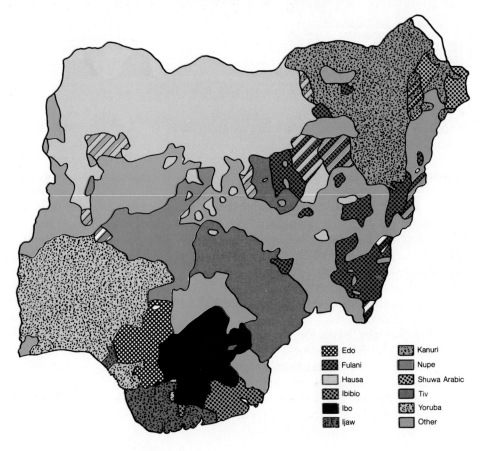

FIGURE 9–20
Language Regions of Nigeria The multitude of languages in Nigeria is typical of many African countries.

Edo Kanuri
Fulani Nupe
Hausa Shuwa Arabic
Ibibio Tiv
Ibo Yoruba
Ijaw Other

cultural regions, culture history, and cultural ecology of an individual area. Geographers are interested in all of the aspects of language that affect the geography of a place. One of the most important of the obvious effects of geography concerns linguistic cultural regions. Some language cultural regions cover large areas with a single language, creating *monolinguistic* regions. In other areas two languages may coexist in a region, creating bilingual linguistic regions. In some instances, multiple languages are spoken in the same area, creating multiple linguistic regions.

Monolinguistic Regions

A single language can become dominant over a large area when any of the following conditions occur:

1. The technological, administrative, or military dominance of one group extends over a region.

2. Large-scale social, cultural, economic, scientific, or other types of interaction develop over an entire region.

3. The lack of physical barriers allows frequent contact among occupants of the region.

Areas where a single language dominates large regions provide numerous examples of these characteristics (Figure 9–21). In China, Mandarin displaced other dialects and languages as the Mandarin-speaking peoples from North China became politically dominant. The expansion of English to North America, Australia, and New Zealand and its adoption as a worldwide lingua franca were associated with British imperialism, technology, and trade. The diffusion of Spanish and Portuguese to large areas accompanied Portugal and Spain's colonization of large areas of the Americas.

The presence of large monolinguistic cultural regions sometimes obscures the fact that there

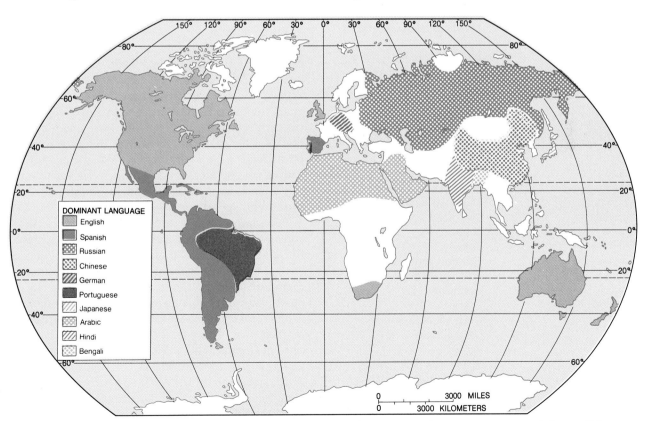

FIGURE 9–21 Major Linguistic Regions By looking at only the dominant language in an area, it is possible to create a simplified linguistic map of the world.

may be important differences within these regions. English spoken in the United States is different from English spoken in England, even though the two are mutually intelligible. The adoption of English as an official language in a variety of countries has led to markedly different dialects (Table 9–5). Nevertheless, the presence of a large monolinguistic region facilitates communication and interaction within that group.

TABLE 9–5
Countries with English as an Official or Dominant Language

Anguilla	Isle of Man
Antigua and	Marshall Islands
Barbuda	Mauritius
Australia	Federated States
Bahamas	of Micronesia
Barbados	Midway Islands
Belize	Montserrat
Bermuda	Namibia
Botswana	New Zealand
Cameroon	Nigeria
Canada	Northern Mariana Islands
Cayman Islands	Philippines
Christmas	Pitcairn Island
(Keeliy) Island	St.Christopher
Cook Islands	and Nevis
Cyprus	St. Helena
Dominica	St. Lucia
Falkland	St. Vincent
(Malvinas)	and the Grenadines
Fiji	Seychelles
Gambia	Sierra Leone
Ghana	Singapore
Gibraltar	South Africa
Grenada	Swaziland
Guam	Tanzania
Guernsey	Tonga
Guyana	Trinidad and Tobago
Hong Kong	Turks and Caicas
India	Tuvalu
Jamaica	Uganda
Kenya	United Kingdom
Kiribati	United States
Lesotho	Vanuatu
Liberia	Virgin Islands, U.S.
Malawi	Virgin Islands, British
Malta	Zambia

Source: The World Fact Book, 1988 (Washington: The Central Intelligence Agency, 1988).

The importance of monolingualism lies in the relationship of language to culture and society. To say that a person is English rather than Russian provides some understanding of the culture of that person. To label large regions of the world as English speaking indicates some general characteristics about the residents of that area in addition to their language. Individual languages have distinctive ways of dealing with space and time. English, like other Germanic tongues, is very time oriented (many verb tenses), while many Asiatic languages tend to overlook the role of time (few verb tenses). English, for example, defines actions according to their occurrence in time, while Chinese is less apt to do so. Chinese has few terms that express opposites or extremes (rich-poor, good-bad) while English and other European languages have many such pairs. Some observers maintain that the language reveals something about the thought patterns of a group and the way in which it structures its time and spatial relationships. The distinctiveness of language also helps to reinforce feelings of patriotism or ethnocentrism.

Language differences are often the basis for cultural conflicts that plague various countries or regions of the world. The inability to communicate easily with peoples having a distinctive language or dialect often leads to value judgments concerning the character of those peoples. In France, the dialects of various regions were a traditional means of recognizing social status. Those who spoke Parisian French viewed themselves as better educated and more refined than those who spoke a rural dialect.

Economic, social, and political organizations and structures often parallel linguistic usage, and the distribution of languages tends to coincide with distinct political units. German in Germany, French in France, Italian in Italy, and Spanish in Spain exemplify this correlation. The correlation is not perfect, however, since outlying areas of language users may be located beyond the boundaries of the individual country. For instance, Austria and Switzerland rely on German; a total of eighteen countries in South America and Central America rely on Spanish; and English is spoken in a number of countries far beyond the boundaries of the United Kingdom. Nevertheless, there is a strong relationship between language and habits, society, and political processes in an individual area.

Dialect Regions

Examination of the linguistic geography of any area indicates that each region will likely contain within it a number of subregions. The United States is generally classified as an English-speaking country, but there are important subgroups speaking Spanish in the Southwest and Vietnamese, Chinese, Philippine, Tagalog, Yiddish, and other languages in certain areas.

Variation in specific terms within a language can also be mapped. The cultural geographer Wilbur Zelinsky, for example, examined names that are applied to small streams. "Creek" is fairly common in the northeast, "brook" is highly concentrated in the New England states, "run" is concentrated in Pennsylvania, Ohio, and West Virginia, and "branch" is localized to West Virginia. Other terms used to refer to a stream include "fork" in the Appalachians of West Virginia and "kill", which is found only in Pennsylvania.

Similar analyses can be made using other words. The boundary between such regions based on word usage or pronunciation is called an *isogloss*. The pronunciation and use of specific words or terms can be used to develop a group of isoglosses that tend to cluster together, creating a dialect region. In the United States, these dialect regions are broadly divisible into northern, midland, and southern regions. Within each of these regions subdivisions are recognizable (see Figure 9–4 on page 325). For example, the dialect of eastern New England is recognizably distinct from that of the inland northern dialect spoken across the region of upstate New York, Illinois, Wisconsin, and Michigan. Lifetime residents of New York City have their own distinct dialect, based on pronunciation. The western portion of the United States is more difficult to regionalize because it has been settled by migrants from all areas of the country. The persistence of recognizable dialects in the United States is uncertain. Some observers believe that the increasing use of mass media will ultimately destroy regional dialects while others maintain that regional identification will cause people to cling tenaciously to their dialects as evidence of their distinctiveness in the melting pot of the American culture.

Languages and Dialects in France

The development of French in France illustrates the way in which dialects and distinctive language regions emerge in a country sharing a common language and culture. France (originally a part of Gaul, the ancient name for Europe south and west of the Rhine River, north of the Pyrenees Mountains, and west of the Alps) was conquered by Julius Caesar between 58 and 51 B.C., becoming part of the Roman Empire. Latin had displaced the Celtic languages by the fourth century A.D. except in refuge areas such as the Breton Peninsula. When the Roman Empire collapsed, the vulgar Latin evolved into a variety of dialects only distantly related to classical Latin, and with the passage of time, the language spoken in northern France became different from that of the south. Germanic invaders from the north (the Franks) occupied what is today northern France and developed another variant of Latin.

Charlemagne attempted to introduce classical Latin as an official language in the ninth century, but the attempt failed. By the twelfth century, Paris had emerged as the political center of France. Parisian French then became the court language and eventually the official written French standard language. In southern France, *Provençal* emerged as a type of literary language used by itinerant songsters (the troubadours), but the written version of Parisian French continued to dominate. By the seventeenth century the standard written French of Paris had been adopted nationwide. By the end of the nineteenth century, a number of events including universal schooling, institution of compulsory military service, and the proliferation of national newspapers strengthened the dominance of Parisian French over other dialects. Today, Parisian French has replaced all other dialects except in a few local areas such as along the border or in refuge locations where Breton, Basque, Catalan (a Spanish dialect spoken in the extreme southeast), and Flemish (in the north) are spoken. In the east, Germanic vernaculars remain in Alsace and Lorraine, areas annexed from Germany in the seventeenth and eighteenth centuries (Figure 9–22). The spread of the official standard language from Paris illustrates the way in which the *language* of France became a symbol of the *country* of France as well.

The Challenge of Bilingualism

Many countries of the world have not been able to establish a uniform standard language and are officially bilingual, with many of the inhabitants fluent in two or more languages. Countries that have officially adopted two languages include Canada, Belgium, the United Kingdom, and South Africa. Bilingualism poses a general problem of communication and can be divisive, making it difficult for the government to maintain unity. The challenge of bilingualism is compounded in countries that have multiple official or vernacular languages, such as India and many African and Latin American countries. Bilingualism can be regional, where the different language family is spoken in a specific area as in Canada or Belgium, or it may be a *polyglot*, in which the languages are

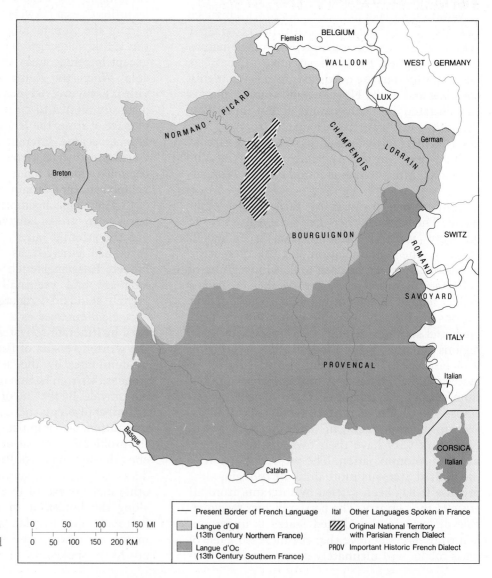

FIGURE 9–22 French Dialects, Past and Present The historic French dialects, such as Provençal, were the original basis for the broad dialect regions of France.

— Present Border of French Language

Langue d'Oil (13th Century Northern France)

Langue d'Oc (13th Century Southern France)

Ital Other Languages Spoken in France

/// Original National Territory with Parisian French Dialect

PROV Important Historic French Dialect

0   50   100   150 MI
0   50   100   150   200 KM

Bilingualism in Canada and other countries adds to the costs of government activities, such as signs.

not confined to a specific locale. The nature of the problem is different in each case, as an examination of selected bilingual countries illustrates.

Canada

French is not equal in importance to English in Canada, but it dominates the province of Quebec (Figure 9–23). The French speakers comprise slightly more than one-quarter of Canada's population today in spite of the limited French immigration since the eighteenth century. After independence from Britain and creation of the Canadian confederation in 1867, Quebec maintained its political identity, and both French and English were adopted as official languages.

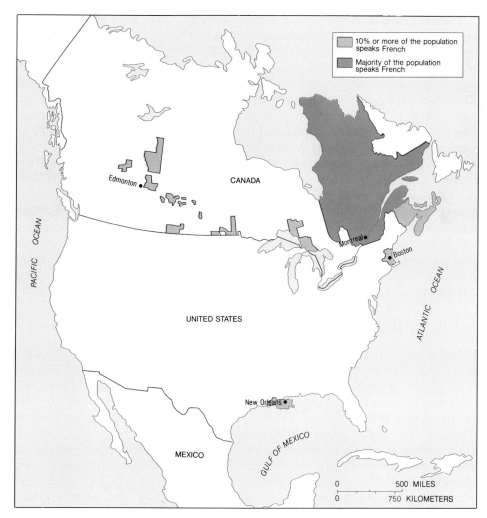

FIGURE 9–23 The French Archipelago in North America Although Quebec is best known as a location of French speakers in North America, there are other locations where a large minority speak French.

In 1977 the provincial government of Quebec approved legislation mandating French as the only official language for government activity. The French-speaking people coined the term "Quebecois" to identify themselves and in 1980 elected a prime minister for the province on a platform of separation from the rest of Canada. The political issues involved in this decision will be discussed in a subsequent chapter, but it is important to note that the distinctive language and French culture were the basis for political action designed to divide one part of the country from the rest. By a bare majority, the citizens of Quebec voted in 1986 not to separate from the rest of Canada. Since that time a new provincial premier has been elected who is less ethnocentric. Nevertheless, the language issue remains a major problem in Canada.

Quebec still has laws requiring that all children attend French-speaking schools unless one or both of their parents graduated from English-speaking schools; the province also requires that all advertisements be in French under penalty of a $760 fine. Between 1980 and 1986, all businesses were required to change their advertising to French, and all official government activities were conducted in French. At the peak of the separatist movement, businesses were even told to stop using English date stamps on incoming mail. Measures to enforce the use of French have caused companies to move from Quebec into the neighboring province of Ontario. Quebec has recently softened its policy on language in response. For example, since 1987, cities in Quebec that have an English majority may use English in day-to-day government communications.

Belgium: The Challenge of Unity

Belgium has a problem similar to that of Canada. Since the fifth century A.D., Belgium has been divided into a northern part where Flemish is spoken (the ancestor of modern Dutch) and a southern half where Walloon (French) is spoken (Figure 9–24). The language difference has caused acrimonious problems between the two regions, and the country has faced a constant challenge to maintain unity. In the 1980s the government granted greater autonomy to each of the two

Street sign in Brussels, Belgium with French (top) and Dutch dialects.

halves of the country, but the question of the language of the capital, Brussels, handicapped efforts to resolve the language issue. The final plan allows Brussels to be independent of the two language regions and to utilize both languages. Estimates in the mid-1980s indicate that 53 percent of Belgium's population is in the Flemish zone, 33 percent in the Walloon area, and 11 percent in the Brussels bilingual area.

Multilingualism: Intensifying the Language Problem

The bilingual nations of the world are exceeded by those in which a variety of languages are spoken as official or unofficial tongues. Maintaining unity, cooperation, and communication is even more difficult for these multilingual societies than for bilingual countries.

The Soviet Union

In the Soviet Union the Great Russians are an ethnic group making up slightly more than half of the total population. Each of the other fourteen republics has its own distinctive language, because the republics were created on the basis of language after the Communist Revolution of 1917. There are also a variety of languages spoken by small groups across the country.

The Russian language has diffused across most of the country, and Russians are a significant

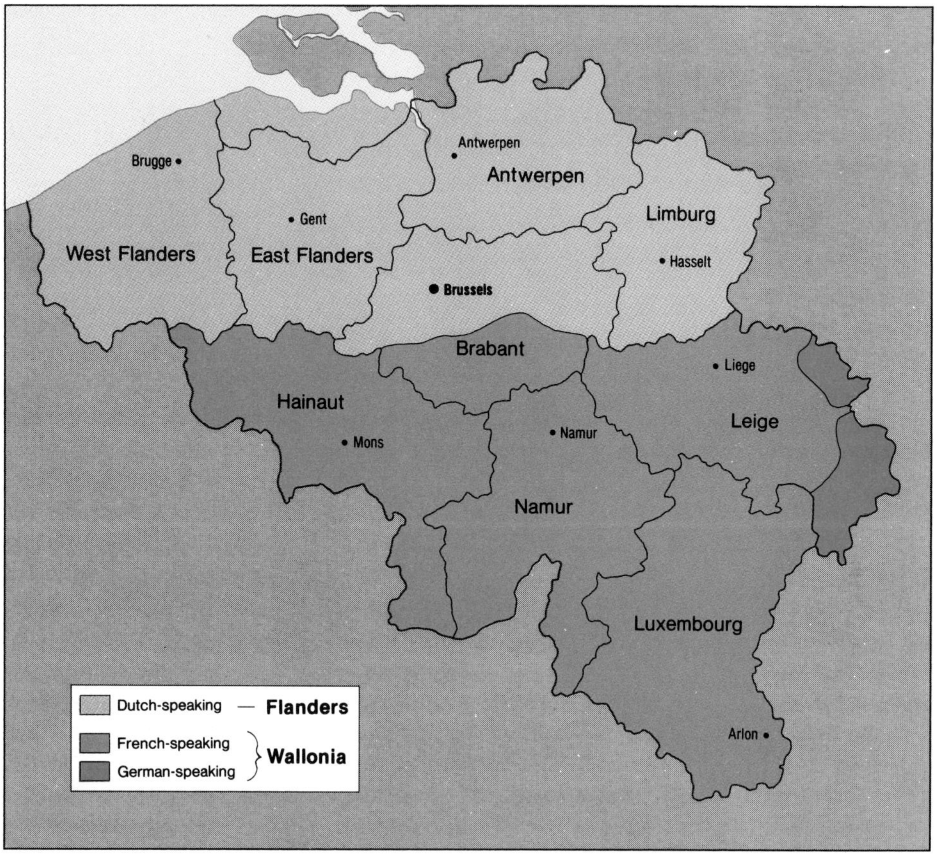

FIGURE 9-24
Language Regions in Belgium The historic division of Belgium into the Dutch north and French south is the basis for a division into the two broad regions of Flanders and Wallonia.

minority population in each republic. "Russification" is the process by which the Russian language and culture are replacing the historically dominant languages of other republics through the migration of Russians and the adoption of the Russian language and culture by native speakers. Russians continue to migrate to all regions of the country because they play a leading role in government, industrial management, higher education, and the military.

The educational system of the Soviet Union uses Russian plus the local language. In order to progress to the top in the Soviet system, however, a knowledge of Russian is essential. All individuals in the Soviet Union carry an internal passport for their individual republic, and at age sixteen they declare their nationality for passport purposes. Many at this time declare themselves Russian, particularly if they are children of mixed marriages (Figure 9-25).

Thus a combination of factors has helped spread Russian at the expense of local languages; nevertheless, the persistence of Estonian, Latvian, Armenian, and other minorities' ethnic identities reinforces the effect of multiple languages in the country. Periodic conflict occurs when an individual ethnic group sees its identity being threatened by Russification. The diffusion of Russian and the process of Russification are unlikely to completely destroy the individual regions of linguistic distinctiveness in the Soviet Union in the foreseeable future.

India

Perhaps more than any other country, India typifies the multilingual or polyglot state. The Indian subcontinent is home to fifteen major languages and some nine hundred minor languages and dialects. The linguistic map can be divided into

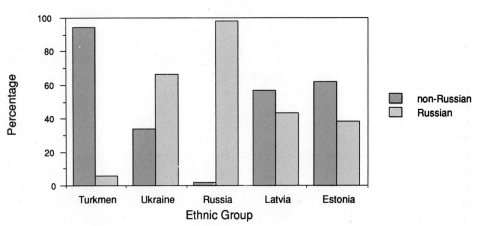

NATIONALITY CHOICE AMONG 16-YEAR OLDS

FIGURE 9–25
Language Selection in the Soviet Union The selection of nationality for internal passports is generally based on language, with most ethnic groups choosing their own group and language rather than Russian.

four language families, two of which, the Indo-European and the Dravidian, dominate. The Tibetan language is found in the northwest of India as part of the Sino-Tibetan family, and Austro-Asiatic representatives are found in the far east of India (see Figure 9–9 on page 331).

The Dravidian language represents the old language of the subcontinent that was pushed to the southern peninsula as the Indo-European Aryans invaded the subcontinent in the second millennium B.C. Today Dravidian languages are spoken by some 25 percent of India's population, and four distinct Dravidian languages are included among the sixteen official languages recognized by the Indian constitution.

The importance of linguistic geography was recognized at the time of India's independence. Hindi, spoken by 30 percent of the population, was adopted as the official language, but English was maintained as a second language to be used for official purposes. English was adopted as the language of government activity in independent India because of its previous colonial status as part of the British Empire. At the time of independence, thirteen additional languages were designated as "languages of India," and a state was created to give each representation. Subsequent demands by distinct linguistic groups have resulted in the creation of a total of twenty-two different states.

The persistence of multiple languages in India exemplifies the problems of attempting to unify a country politically when linguistic fragmentation exists. Problems created by multilingualism abound in the world, and reflect the importance of language as a symbolic representation of the entire culture of a group. The spread of English or other languages should not obscure the fact that the divisiveness of language will continue to create cultural conflicts in many regions of the world.

Diffusion: Basis of Linguistic Conflict

The present distribution of Indo-European languages illustrates the effects of both contiguous and relocation diffusion through time and space as well as the effects of reflective and nonreflective barriers. The patterns of the linguistic map of the world reflect the economic, political, and religious expansion of specific cultural groups over time.

The diffusion of the Indo-European linguistic family from its hearth in the southern steppes of Eurasia was initially for economic reasons, apparently in response to population growth. In more recent times, the spread of English, Spanish, Portuguese, and French represents an example of politically based expansionist diffusion. British political control of North America, Australia, New Zealand, South Africa, India, and other regions of the world was accompanied by the diffusion of the language. In some cases this expansionist political diffusion involved relocation, as when English settlers came to the United States. In other cases, hierarchical diffusion occurred, as in the adoption of English by the civil servant class in India, but not the majority of the population.

Just as English has spread to much of the world in modern times, Latin diffused across the Roman Empire in ancient times, explaining the distribution of Romance languages in Europe today. The expansion of Romance languages from Europe to overseas locations reflects subsequent political diffusion as the individual countries of Portugal, Spain, and France pursued their political interests. The French expanded into North America, but conflict with the British ultimately forced them into only a small area. In Latin America, the two competing empires of Spain and Portugal brought their respective languages to the areas they controlled.

The retention of English, Spanish, Portuguese, French, or any other language in an area that has been colonized through political diffusion, often indicates the nature of the language in the society. In countries such as India, where the language was primarily used by a minority group of administrators, it never became the dominant language. In cases such as the United States where English became a *de facto* official language, other migrants were forced to learn the language. Designation of a language as an official state language usually leads to its dominance over other languages. In the United States, although English has not been formally adopted as a national official language, its informal use in education and the media has caused most migrants to quickly abandon their own tongue in favor of English. Migrants to the United States, both historically and at present, are forced to learn English to become fully involved in American society (see the Cultural Dimensions boxes on pages 356–357).

The continued use of English as a worldwide lingua franca represents an anomaly in a world in which political and economic colonialism is viewed negatively. The great revolutionary movements that led to the independence of African, Asian, and Latin American countries were often accompanied by the desire to eliminate all aspects of the former colonial power's culture. This

has resulted in the banning of the use of the colonial language and the adoption of a native tongue in many countries. Dutch was replaced by Malay-Tagalog in Indonesia after independence at the end of World War II, for example. French has been displaced by Vietnamese in Vietnam, but a strong affinity to the former colonial language remains in both Indonesia and Vietnam because of the cultural ties to the old colonial power.

Even in countries where a language is made illegal, such languages persist. Polish was officially replaced by German or Russian after the partition of Poland in 1772, 1792, and 1795. The Poles, however, maintained their language since they were primarily a rural people and it was impossible to enforce the use of another tongue in everyday communication. The resulting pattern in which a nonnative language was used in government, business, or the arts while a native language persisted simultaneously is similar to what occurred in many areas colonized by foreign powers.

Religion and Language Diffusion

Some languages have been diffused as the result of religion. In Europe, the expansion of the Roman Catholic church was accompanied by the hierarchical diffusion of Latin, which was used in religious services in the Roman Catholic regions of Europe until after World War II. The language map in the Soviet Union and eastern Europe reflects the diffusion of the Orthodox Catholic church from Constantinople (Istanbul).

An even more obvious example of the spread of a language by a religious group is Arabic. The holy book of Islam, the Koran (or Quran) was written in Arabic, and the expansion of Islam after the seventh century A.D. spread Arabic across North Africa into Spain, across the Middle East, and into the southern part of the Soviet Union. Arabic became a formal religious language even where Arabic was not adopted as the speech of the people, as in Iran.

The existence of Hebrew today shows the relationship of religion to cultural history. The origin of the Jews in the Middle East, their subsequent dispersal to Europe, North Africa, and Asia (the Diaspora), and their more recent re-

CULTURAL DIMENSIONS
English: America's Official Language?

The United States does not have an official language, a lack that some people see as a threat. Several organizations are attempting to overcome what they perceive as an oversight by the founders of the country. The most active group supporting the adoption of English as a national language is U.S. English. They are quick to point out that they do not want to make English the *only* language in the United States, but the "only language of official use."

Arguments for and against English as an official language range from accusations that proponents are at best bigots, and at worst racists, whose arguments mask anti-Asian or anti-Spanish attitudes to claims that without an official language we will become a bilingual country. Proponents maintain that bilingualism costs the Canadian people between $500 million and $1 billion per year for duplicate forms, education, courts, and training. Every working day a force of 100 Canadian translators convert about a million words in English documents to French and French to English. Proponents also point to Europe where a large percent of the budget of the European Economic Community is spent on translations. Bilingualism costs more than money and can destroy the unity of a country, even a sophisticated

country with a long and proud history such as Belgium.

Adopting English as an official language would not mean that English would be the only language spoken in the United States. It would mean that ballots would be printed in English, courts would be conducted in English, schools would move as quickly as possible to get children of immigrants to learn in English, and other government services (drivers' license examinations, for example), would be in English.

Opponents of making English an official language maintain that most immigrants already want to learn English and adopting an official language would not speed up the process. Adoption of an official language act might even slow down the learning of English by immigrants because states might choose not to have bilingual classes for non-English-speaking students. Opponents use data from California to show that when children are taught English in one class for the first years of school and learn other subjects in their own tongue, they learn both English and the other subjects better. (Proponents use the same argument to show that students immersed in

continued

grouping in Israel are an example of how religion can even revive a language.

Economics and Language Diffusion

In some instances language diffuses because of economic interests. The most obvious example of this is the widespread use of English in international trade and communication. In air transport, all international flights communicate with air controllers in English. Thus an Iranian plane flying into Canada must have an English-speaking pilot, likewise a Turkic plane flying into the Soviet Union. The Japanese have emphasized learning English because of the importance of the

American market and its technology for international trade.

Pidgin languages are an example of the creation and diffusion of language for economic reasons. Pidgin languages for trade are created when a group borrows words from the languages of the traders to attain some level of communication with them. Many pidgin languages are based on a simplified form of English because of its role in the trade of the colonial era. Thus various forms of English are found in China, Southeast Asia, and the West Coast of Africa. A pidgin language based on Spanish is found in the border areas of the southwestern United States and northern Mexico. Sometimes called "Border Spanish," "Tex-

English: America's Official Language? *continued*

English after only one year in school do better than those learning in their native tongue.) Opponents also point out that we are a nation of immigrants, and tolerance of other groups and their languages should be encouraged to prevent the recurrence of discrimination that has been common in the past.

Across the United States groups are collecting signatures to get proposals on the ballot and encouraging their legislators to pass or defeat bills making a state a part of official English America. The 1988 elections in the United States added three more states (Florida, Arizona, and Colorado), bringing to seventeen the number that have officially adopted English as their language.

Opponents decry the victories, claiming they have racist overtones (all three states have large and growing Spanish minorities), while proponents celebrate the reinforcement of "an American heritage." Residents who voted for the English legislation did so because they felt it was not onerous to urge immigrants to the United States to learn the basic English of their new land, be able to read a ballot in English, and participate in court cases. Less philosophical voters simply voted their instincts; as a Californian noted, "Yesterday I called a store. Three people at-

tempted to help, but none spoke English. I asked for the manager, who told me he understood my feelings but he was obligated by law to hire those people." Or as a Floridian noted, "I believe that immigration is the source of America's economic strength, and more immigrants should be admitted. When more people come to the United States, however, it takes a bigger effort to ensure that they are able to survive and to thrive, to live and to work here; those efforts must include both teaching immigrants English and letting them know that English is the language of the United States."

Whatever your own feelings, a groundswell of support for making English the official U.S. language appears to be growing. Ironically, the increased enthusiasm for such an act is a response to the recent influx of Spanish and Asian immigrants and efforts to help them acculturate more easily. Reacting against bilingual education will certainly not make it easier for immigrants to learn English. Official English laws may make it more difficult for first-generation immigrants to vote or get a driver's license, but it is debatable whether they will have a major impact on the acquisition of English by second-generation immigrants who are already learning it.

mex," or "Spanglish," this pidgin language uses terms borrowed from each language to allow communication between the two groups. It has been diffused away from the border area as Spanish immigration to the United States has increased in the last two decades. Hence, such "border" Spanish is now found in places such as Chicago or Seattle. (See the Cultural Dimension box on page 358).

Language, Landforms, and Refuge

Languages diffuse most readily where there is greatest accessibility and greatest interaction. By contrast, certain areas have become refuge areas

for languages that have been pushed out by a diffusing expansionist language. These refuge areas are often associated with rugged topography, harsh environments, or remote locations. American Indian languages are best preserved among isolated groups such as members of the Navajo tribe in the arid margins of Arizona, New Mexico, and southern Utah. The Basque tongue is preserved in the mountains of the Pyrennes. The Celtic tongues that once were spoken across all of western Europe now exist only in small pockets at the extremities of the continent.

In Latin America, the Native American languages are best preserved in the isolation of the Amazon Basin, the rugged Andean Mountains, or

CULTURAL DIMENSIONS
Esperanto: An International Language

Some 6,000 people from around the world gathered in Warsaw in 1987 for the World Congress of Esperanto.* Although the congress is an annual event for Esperanto enthusiasts, 1987 was special: it was the hundredth anniversary of the international language created by Ludwik Zamenhof.

Dr. Zamenhof, a Russian, was not the first to attempt a world language, but his is the only one that has survived and grown. Although difficult to pinpoint, the number of Esperanto-speakers worldwide is estimated to number in the hundreds of thousands. A more precise indicator of the language's scope is the number of members in the Universal Esperanto Association, (UEA), which in 1986 reached a record 40,589 people representing 104 countries.

Esperantists come from all walks of life, as evidenced by the numerous Esperanto clubs for teachers, doctors, lawyers, scientists, and musicians, to name just a few. Although Esperanto is regarded by some as a Western European fad, UEA membership data indicate that the language is more popular in Poland and Japan than in England or France. Esperanto has also caught on quickly in Brazil, a Portuguese-speaking country in the midst of a Spanish-speaking continent, while

*Adapted from the Christian Science Monitor, July 17, 1987.

growth in the United States and Canada has been slow.

The "building block" structure of the language makes it easy to acquire and expand vocabulary. From a base of 15,000 roots taken from many languages, some 150,000 words can be formed. Sixteen basic grammar rules are without exceptions, and each of the 28 letters of the alphabet has only one sound.

If it so easy, why hasn't Esperanto enjoyed greater success? The lack of native speakers has been one impediment to Esperanto's growth. Another is the fact that several "natural" languages such as English, French, and Spanish are already widely spoken. More people speak English as a second language than any other language in the world, and English is the accepted language in international business, science, and commerce.

A century after Esperanto's debut, many people remain unaware of its existence. Yet over a hundred magazines, journals, and newspapers are published in Esperanto, and several radio stations broadcast regularly in the language. Since 1957, the UEA has averaged more that a thousand new members per year. And while international arenas such as the Common Market and the United Nations still use the cumbersome process of translation to deal with language barriers, scores of international Esperanto conferences are held each year without the need for interpreters.

---

the extreme climates at the southern tip of the continent. The environmental characteristics associated with rugged lands, harsh climates, or forests, marshes, and swamps often serve to protect a minority language. These refuge areas are numerous, but the degree of their isolation is rapidly declining. The isolation that led to the creation of the distinctive languages of the Caucasus Mountains, for example, is giving way in the face of modern transportation and communication.

The distinctive Creole Cajun language of southern Louisiana is spoken by descendants of French settlers from Acadia (an old name for Canada's Nova Scotia and New Brunswick provinces). Forced to move by the British who conquered eastern Canada, the Acadians found security in the swamps of the Mississippi Delta. Today Cajun children attend schools where English is the primary tongue, television brings English-speaking programs into their homes, and their interactions in business and social activities

Basque sign in northern Spain. The sign is in opposition to nuclear weapons.

are with English speakers. Advances in transportation and communication technology make it ever more difficult for a linguistic group to find such refuges.

The present pattern of languages represents the growth and diffusion of individual languages during an age when barriers were more impermeable because of limited transportation and communication technology. The division between French, German, and Italian in Switzerland correlates with mountain barriers. The division between Russian and the various Caucasian languages results from the Caucasus Mountains.

The correspondence of language boundaries and physical barriers, such as that between France and Spain, indicates how the barriers affected a culture and its development over time. Spanish was spoken in the lowlands south of the Pyrennes, French in the lowlands to the north, with the mountains forming a frontier region between the two. When France and Spain emerged as political units, the actual boundary between the two was negotiated on the basis of culture and physical geography. Formal establishment of the boundary between the two countries and subsequent political and economic involvement with France by the French and with Spain by the Spanish solidified the linguistic divide. The creation of the boundary was simplified by

the low population densities associated with mountainous areas. In the United States, the Rio Grande is a political division between two countries, but the language boundary does not coincide with the river because Spanish settlements existed north of the Rio Grande before its occupation by English-speaking Americans and because Spanish speakers continue to immigrate from Mexico (see Figure 8–17 on page 305). The Rio Grande or other rivers are less suitable as linguistic divides because they are more permeable barriers and are in areas usually more densely settled than mountains.

Linguistic Landscapes

Language also contributes to the creation of cultural landscapes. The toponyms (place-names) of a particular region are one indication of the language spoken by users of that landscape, both at the present time and historically. The terms for streams (kill, branch, run) or the distribution of Indian-based place-names in the United States indicates the language used by the inhabitants of that land.

Cultural geographers and linguists use toponyms to trace the diffusion of peoples and languages and to recreate the changing occupance of an area. Such phenomena as the suffixes or pre-

Toponyms are sometimes used as a tourist attraction, as this sign in Wales.

fixes applied to toponyms tell something about the language of the people who named the land. The Slavic suffix *sk* (Novosibersk, Omsk, Tomsk) means "town" in the Soviet Union. The suffix *gorod* and its derivative *grad* are also Slavic terms used to identify a city, such as Leningrad or Volgograd. Other place-names in the Soviet Union reveal former Persian influence; for instance, the prefix *kara* as in Karakum means black, referring to the black sand in the Karakum Desert.

The expansion of the Islamic religion in Spain led to the conquest of the Iberian Peninsula by the Moors. When they were eventually driven out by the Spanish in 1492, many of the names they had given to places were retained, and these are a reminder of the Moorish influence.

Equally a part of the landscape are signs associated with advertisements, streets, or individual personal names. Their presence forms a rich part of the visual cultural landscape that makes each place distinct.

Conclusion

It is clear that language is one of the most important elements of culture. It plays a role in transmitting ideas, shaping views of self, others, and the environment, and in expressing our ideals and values. The visible manifestation of language in culture regions, culture landscapes, cultural diffusion, and as a distinctive cultural element makes it one of the most important aspects of the geography of the world.

QUESTIONS

1. Define and describe "language."
2. Compare and contrast the two broad divisions of Indo-European languages.
3. Discuss the how and where of the diffusion of English throughout the world.
4. Explain the statement "English is the lingua franca of the world."
5. Identify and describe the location of the relic languages in Europe.
6. Discuss the main Sino-Tibetan languages and indicate their locations.
7. What is a lingua franca? Give some examples.
8. What is linguistic geography?
9. What are the problems associated with bilingualism? Give examples.
10. What problems does linguistic geography present to India? Explain.

SUGGESTED READINGS

Aitchison, J. W., and H. Carter. "The Welsh Language in Cardiff: A Quiet Revolution." *Transactions of the Institute of British Geographers* n.s. 12 (1987): 482–92.

Allen, Harold B. *The Linguistic Atlas of the Upper Midwest* 3 vols. Minneapolis: University of Minnesota Press, 1973–1976.

Barrett, F. A. "The Relative Decline of the French Language in Canada: A Preliminary Report." *Geography* 60 (1975): 125–59.

Baugh, Albert C., and Thomas Cable. *A History of the English Language,* 3d ed. Englewood Cliffs, N.J.: Prentice-Hall, 1978.

Bennett, Charles J. "The Morphology of Language Boundaries: Indo-Aryan and Dravidian in Peninsular India." in David E. Sopher, ed., *An Exploration of India: Geographical Perspectives on Society and Culture,* pp. 234–51. Ithaca, N.Y.: Cornell University Press, 1980.

Bickerton, Derek. "Creole Languages." *Scientific American* 249 (July 1983): 116–22.

Bolinger, Dwight, and Donald A. Sears. *Aspects of Language,* 3d ed. New York: Holt, Rinehart and Winston, 1981.

Buchanan, Keith. "Economic Growth and Cultural Liquidation: The Case of the Celtic Nations," in Richard Peet, ed., *Radical Geography: Alternative Viewpoints on Contemporary Social Issues,* pp. 125–43. Chicago: Maaroufa Press, 1977.

Cartwright, Don. "Language Policy and Political Organization of Territory: A Canadian Dilemma." *Canadian Geographer* 25 3 (1981): 205–24.

Cassidy, Frederic G. *Dictionary of American Regional English,* vol. 1. Cambridge and London: Belknap Press of Harvard University Press, 1985.

Comrie, Bernard. *The Languages of the Soviet Union.* Cambridge, England: Cambridge University Press, 1981.

Cooper, Robert L., ed. *Language Spread: Studies in Diffusion and Social Change.* Bloomington, Ind.: Indiana University Press, 1982.

Delgado de Carvalho, C. M. "The Geography of Languages," in Philip L. Wagner and Marvin W. Mikesell, eds., *Read-*

ings in Cultural Geography, pp. 75–93. Chicago: University of Chicago Press, 1962.

Dugdale, J. S. *The Linguistic Map of Europe.* London: Hutchinson University Library, 1969.

Fromkin, Victoria, and Robert Rodman. *An Introduction to Language,* 3d ed. New York: Holt, Rinehart and Winston, 1983.

Gade, Daniel W. "Foreign Languages and American Geography." *The Professional Geographer* 35 (August 1983): 261–65.

Gelling, Margaret, *Place-Names in the Landscape.* London: Dent, 1984.

Greenberg, Joseph H. *Studies in African Language Classification.* Bloomington, Ind.: Indiana University Press, 1963.

Herzog, Marvin I. *The Yiddish Language in Northern Poland: Its Geography and History.* The Hague, Netherlands: Mouton, 1965.

Hughes, Arthur, and Peter Trudgill. *English Accents and Dialects.* Birkenhead: Edward Arnold, 1979.

Katzner, Kenneth. *The Language of the World.* New York: Funk & Wagnalls, 1975.

Kaups, Matti. "Finnish Place Names in Minnesota: A Study in Cultural Transfer." *Geographical Review* 56 (1966): 377–97.

Kearns, Kevin C. "Resuscitation of the Irish Gaeltacht." *Geographical Review* (1974): 83–110.

Keller, R. E. *The German Language.* London: Faber and Faber, 1978.

Kirk, John M., Stewart Sanderson, J. D. A. Widdowson, eds. *Studies in Linguistic Geography: The Dialects of English in Britain and Ireland.* London: Croom-Helm, 1985.

Klima, Edward S., and Ursula Beliugi. *The Signs of Language.* Cambridge, Mass.: Harvard University Press, 1979.

Kurath, Hans. "Dialect Areas, Settlement Areas, and Culture Areas in the United States," in Caroline F. Ware, ed., *The Cultural Approach to History,* pp. 331–45, New York: Columbia University Press, 1940.

_____. *A Word Geography of the Eastern United States.* Ann Arbor, Mich.: University of Michigan Press, 1949.

Kurath, Hans, and Raven I. McDavid, Jr. *The Pronunciation of English in the Atlantic States.* Ann Arbor, Mich.: University of Michigan Press, 1961.

Labov, William. *Language in the Inner City.* Philadelphia: University of Pennsylvania Press, 1972.

Laird, Charlton. *Language in America.* New York and Cleveland: World Publishing, 1970.

Leighly, John. "Town Names of Colonial New England in the West." *Annals of the Association of American Geographers* 68 (June) 1978.

Lind, Ivan. "Geography and Place Names," in Philip L. Wagner and Marvin W. Mikesell, eds., *Readings in Cultural Geography,* pp. 118–28. Chicago: University of Chicago Press, 1962.

Luckmann, Thomas. "Language in Society." *International Social Science Journal* 36 (1984): 5–20.

MacAodha, B. S., ed. *Topothesia.* Galway, Eire: Department of Geography, University College Galway, 1982.

McCrum, R., William Cran, and Robert Macneil. *The Story of English.* New York: Penguin Books, 1987.

McDavid, Raven I. "Linguistic Geography and Toponymic Research." *Names* 6 (1958): 65–73.

Meeker, Josephine. "Canada: Path to Constitution." *Focus* 33 (November-December 1982).

Muller, Siegfried H. *The World's Living Languages.* New York: Frederick Ungar, 1964.

Noble, Allen G., and Ramesh C. Dhussa. "The Linguistic Geography of Dumka, Bihar, India." *Journal of Cultural Geography,* 3 (Spring/Summer 1983): 73–81.

Pool, Jonathan. "National Development and Language Diversity," in Joshua Fishman, ed., *Advances in the Sociology of Languages,* pp. 213–30. The Hague, Netherlands: Mouton, 1972.

Pryce, W. T. R. "Migration and the Evolution of Culture Areas: Cultural and Linguistic Frontiers in North-East Wales, 1750 and 1851." *Transactions of the Institute of British Geographers* 65 (1975): 79–108.

Ramanujan, A. K., and Colin Masica. "A Phonological Typology of the Indian Linguistic Area," in Thomas A. Sebeok, ed., *Current Trends in Linguistics,* vol.5. The Hague, Netherlands: Mouton, 1969.

Renfrew, Colin. *Archaeology and Language.* London: Cambridge University Press, 1988.

Shortridge, James R. "The Dictionary of American Regional English." *Geographical Review* 76 (January 1986).

Sopher, David E., ed. *An Exploration of India: Geographical Perspectives on Society and Culture.* Ithaca, N.Y.: Cornell University Press, 1980.

Sopher, David E. "The Structuring of Space in Place Names and Words for Place," in David Ley and Marwyn Samuels, eds., *Humanistic Geography: Prospects and Problems,* pp. 251–68. Chicago: Maaroufa Press, 1978.

Stewart, George R. *Names on the Land: A Historical Account of Place-Naming in the United States.* Boston: Houghton Mifflin, 1958.

Symanski, Richard. "The Manipulation of Ordinary Language." *Annals of the Association of American Geographers* 66 (1976).

Trudgill, Peter. "Linguistic Geography and Geographical Linguistics." *Progress in Geography: International Reviews of Current Research* 7 (1975): 227–52.

_____.*On Dialect, Social and Geographical Perspectives.* Oxford, England: Blackwell, 1983.

Weiner, Edmund, and John Simpson, eds. *The Oxford English Dictionary,* 2d ed. Oxford, England: Clarendon Press, 1989.

Williams, Colin H., and C. J. Thomas. "Linguistic Decline and Nationalist Resurgence in Wales," in Glyn Williams, ed., *Social and Cultural Change in Contemporary Wales*, Chapter 12. London: Routledge & Kegan Paul, 1978.

Withers, Charles W. J. "The Geographical Extent of Gaelic in Scotland, 1698–1806." *Scottish Geographical Magazine* 97 (1981): 130–39.

————."A Geography of Language: Gaelic-Speaking in Perthshire, 1698–1879." *Transactions of the Institute of British Geographers* n.s. 8 (1983): 125–42.

Wixman, Ronald. *Language Aspects of Ethnic Patterns and Processes in the North Caucasus*. University of Chicago, Department of Geography Research Paper No. 191, 1980.

Wood, Gordon R. *Vocabulary Change: A Study of Variation in Regional Words in Eight of the Southern States*. Carbondale and Edwardsville, Ill.: Southern Illinois University Press, 1971.

Zelinski, Wilbur. "Classical Town Names in the United States: The Historical Geography of An American Idea." *Geographical Review* 57 (1967): 463–95.

————.*The Cultural Geography of the United States*. Englewood Cliffs, N.J.: Prentice-Hall, 1973.

————."Cultural Variation in Personal Name Patterns in the Eastern United States." *Annals of the Association of American Geographers* 60 (1970): 743–69.

————."Generic Terms in the Place Names of the Northeastern United States." *Annals of the Association of American Geographers* 45 (1955): 319–49.

————."Generic Terms in the Place Names of the Northeastern United States," in Philip L. Wagner and Marvin W. Mikesell, eds., *Readings in Cultural Geography*. Chicago: University of Chicago Press, 1963.

————."North America's Vernacular Regions." *Annals of the Association of American Geographers* 70 (March 1980).

Religion, Life and Landscape: An Ideological Triad

Wherever people are found, there too religion resides. Occasionally religion is hard to pin down, but in the great metropolitan capitals and in the most primitive areas of the world, there are physical and cultural temples, pyramids, megaliths, and monuments that societies have raised at tremendous expense as an expression of their religion. . . . Indeed, there is no other phenomena so pervasive, so constant from society to society, as the search for gods.   L. M. HOPFE

THEMES

Religion as a cultural element.
Ethnic religions versus universalizing religions.
The distribution of the world's universalizing religions.
Conflicts and religion.
The geographic imprint of religion.

CONCEPTS

Ethnic religions
Universalizing religions
Buddhism
Exclusivist religions
Monotheistic
Polytheistic
Sacred space
Profane space
Ghetto
Eastern Orthodox
Pilgrimage
Theocracy
Taboos
Jainism
Sikhs
Granth
Gurus
Theravada
Mahayana

Religion is a primary component in the belief system (ideology) of a people; it affects both life and landscape. The effects of religion on life and landscape range from prohibitions on behavior that influence dress, foods, architecture, and individual actions to legal systems that guide the activities of entire civilizations. The legal systems of the Christian world are based upon the Ten Commandments given to Moses. The social and moral framework of India is based on the caste system of the Hindu religion. Islam provides both a legal code and individual guidelines affecting dress and food. Islam and Judaism's prohibitions against pork affect the foods and crops produced in areas where they are the dominant religion. Islam prohibits wine consumption, but vineyards in Catholic countries reflect the use of wine as part of the sacrament.

Religious feelings have contributed to the displacement of entire communities, as in the near annihilation of the Jews of Europe during the Nazi Holocaust between 1937 and 1945, and to mass movements of peoples, as in the journeys of the Crusaders to the Holy Land in the Middle Ages. Millions of Muslims travel each year to Mecca in Saudi Arabia in response to their belief in the doctrines of Islam.

The holy books of the great religions, the Koran (Quran) of Islam, the Bible of Christianity, the Talmud of Judaism, and the Bhagavad-gita of Hinduism, each provide a framework for their respective cultures. Neither the behavior of these peoples nor their relationship with their environment can be understood without an understanding of the impact that their religion plays upon the lives of the groups of believers. The Koran, for example, is not only a holy book upon which religious life is based, but also a code for material practices that may constrain everyday activities.

Each religion relies upon an oral or written tradition to shape the culture of the people of an area of the world. In the name of religion, humans have blessed God and cursed their neighbors; built churches, cathedrals, and temples; founded schools, universities, towns, and countries; and created major literary works. The complex combination of religion with life and land makes it difficult to reach a clear definition of what reli-

Sultan Ahmed Mosque, Istanbul, Turkey. Religious structures are often dominant features in the cultural landscape.

gion is, or to specify what aspects of religion are of interest to geographers.

Religion: Defining the Unknown

Religion has been defined in a number of ways, most of which include some reference to supernatural or divine actions. Webster's Dictionary defines religion as "a system of faith and worship; . . . a body of institutionalized sacred beliefs, observances, and social practices." Another common definition is "the system of shared beliefs and practices by which people invoke supernatural forces to explain their earthly struggles and activities." This definition includes belief systems that attribute personality and will to nature as well as those that encompass one or more gods.

Religions, as a cultural phenomenon, can be classified in a variety of ways. One common division is between *ethnic* and *universalizing* religions. Ethnic religions are associated with a specific group of people, and are usually tied to a specific place. Membership in ethnic religions is received through birth into the cultural group or occasionally, after a long period of indoctrination. Ethnic religions may be associated with a small group of people (referred to as *tribal ethnic reli-*

gions) or with a nation or civilization *(compound ethnic religious systems)*. Tribal ethnic religions are found among the Native Americans, tribal groups of Africa, and Pacific Islanders. Compound ethnic religious systems are generally associated with societies that have written legal and religious codes, economic specialization, some degree of urbanization, and an organized political system for governing the civilization. The first compound ethnic systems were established millennia ago in the Fertile Crescent among early civilizations. Existing examples of compound ethnic systems include Judaism in Israel, Hinduism in India, and Shintoism in Japan (Figure 10–1).

Unlike ethnic religions, universalizing religions believe that their message and doctrine deal with life and relationships with god (or gods) in a way suitable for all people. Universal religions have broken their ties to a specific place and spread their message by proselytizing. Membership in universal religions generally can be obtained after only a short period of contact. The main universalizing religions of the world are Christianity, Islam, and Buddhism (Table 10–1). Each has diffused widely from its original hearth area and attracted large numbers of adherents. Conversion to each can be accomplished after very minimal contact, which may either be formal as with a *proselytizing* missionary force (Christian) or informal as in *contact* conversion (for example, a group having economic dealings with Islamic traders might convert to Islam). Contact conversion occurs as individuals accept the religious beliefs of friends or acquaintances even though there is no formal proselytizing involved.

Some universalizing religions have evolved from ethnic religions, such as Christianity and Islam from Judaism and Buddhism from Hinduism. These religions changed from ethnic to universalizing after an influential and dynamic individual emerged whose teachings transformed the religious beliefs. Such dynamic leaders can attract believers from outside the ethnic group, changing the religion into a universalizing belief.

The universalizing religions include both *exclusivist* and *nonexclusivist* religions. Exclusivist religions believe that their doctrine is not only universally appropriate, but that it is the *only* correct philosophy of life. Christianity and Islam are exclusivist religions. Faithful adherents cannot belong to another religion. Buddhism is a nonexclusivist religion: its believers may accept its doctrine and other religious doctrine as well.

Religions may be either *monotheistic* (one god) or *polytheistic* (multiple gods) regardless of their other characteristics. Judaism, Christianity, and Islam are monotheistic while Buddhism, Hinduism, and many tribal religions are polytheistic.

The geographer is interested in examining the impact of religion on culture and the landscape, the process by which religion is diffused from one area to another, and the resulting cultural regions. Examination of the major religious groups helps to explain their origins, diffusion, and the belief systems that affect the human interaction with the environment.

Ethnic Religions

Tribal Ethnic Religions

Simple or tribal ethnic religions are the most numerous and are normally associated with a small group of people (Figure 10–2 on page 370). Specific details of ethnic religions vary, but each commonly has a formal set of behaviors and rituals designed to placate the gods of an individual locale. One of the most interesting aspects of tribal ethnic religions is their environmental orientation. Tribal ethnic religious groups use religion to provide order in uncertain environments. Environmental uncertainty is dealt with by ritualistic behavior designed to ensure good weather, abundant game for hunting, fertile soils and productive crops, and a suitable human-land relationship as well as to prevent droughts, earthquakes, hurricanes, or other hazard.

Closely associated with placating the gods is reverence given to specific features in the environment. *Sacred space* is regarded as categorically and qualitatively different from the *profane space* that comprises most of the world. Tribal ethnic religions recognize a variety of physical features

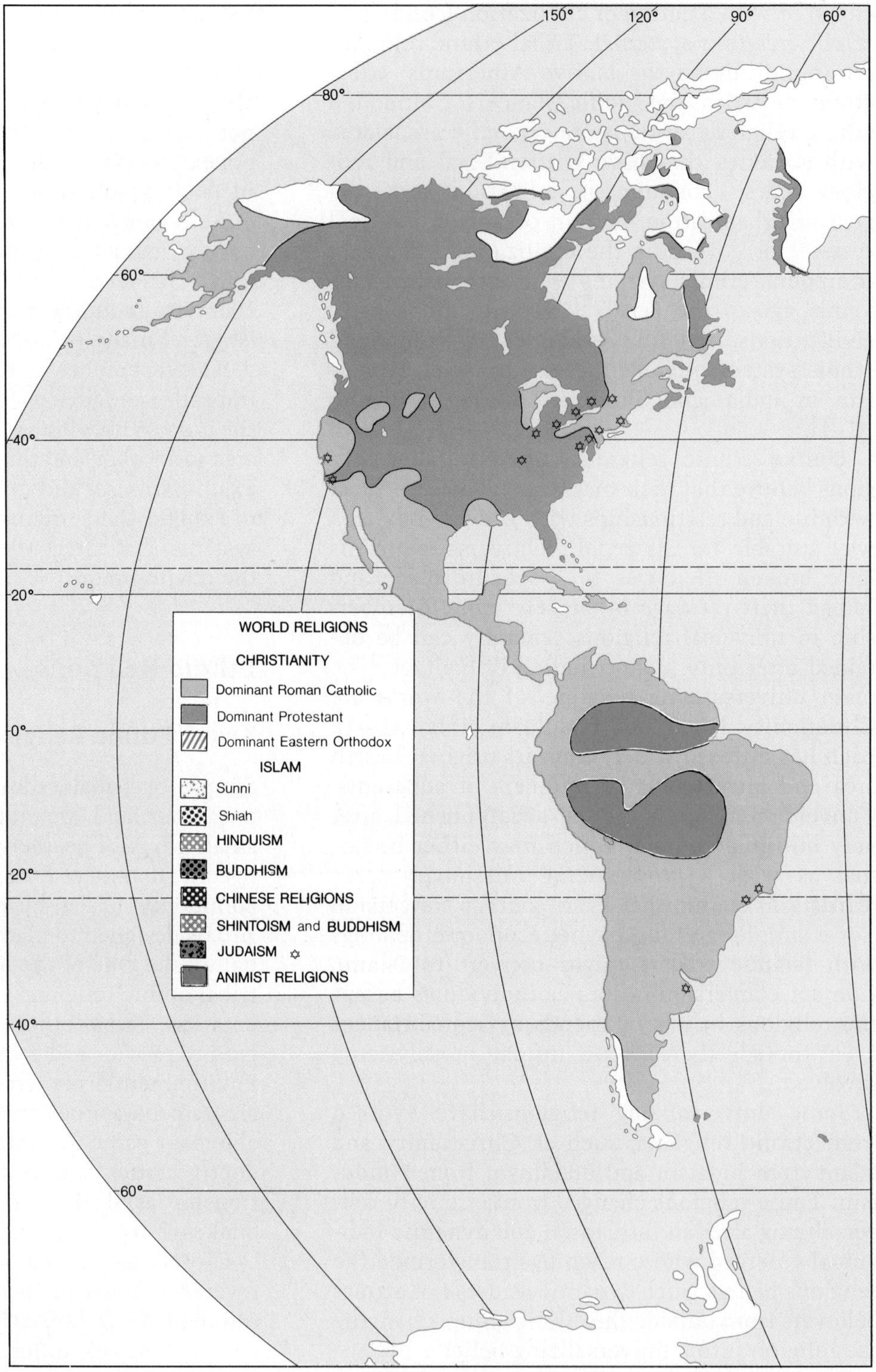

WORLD RELIGIONS

CHRISTIANITY
Dominant Roman Catholic
Dominant Protestant
Dominant Eastern Orthodox
ISLAM
Sunni
Shiah
HINDUISM
BUDDHISM
CHINESE RELIGIONS
SHINTOISM and BUDDHISM
JUDAISM ✡
ANIMIST RELIGIONS

FIGURE 10–1 Types of Religions

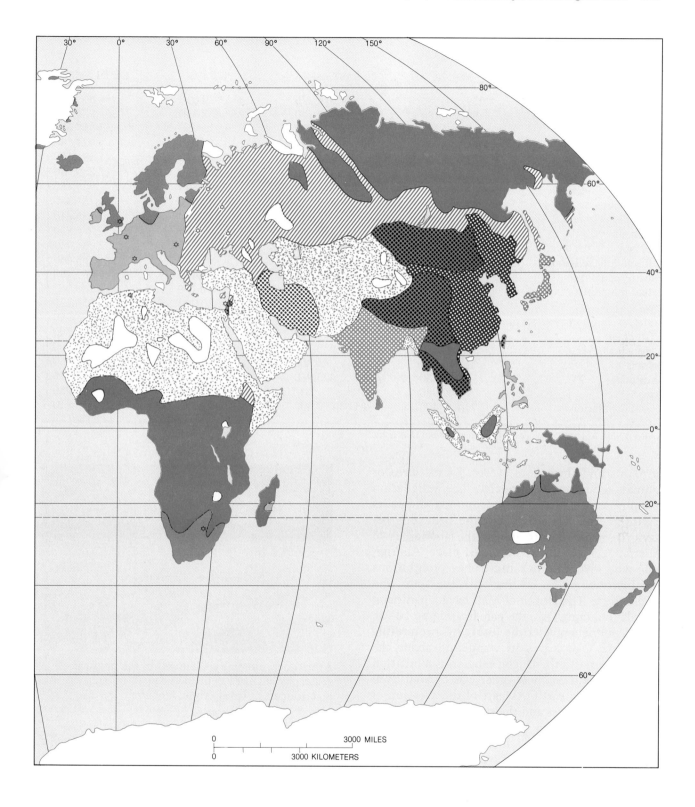

TABLE 10–1
Religious Membership of the World

RELIGIONISTS	AFRICA	EAST ASIA	EUROPE	LATIN AMERICA
Christians	259,544,680	74,614,270	415,529,010	388,863,450
Roman Catholic	98,557,180	8,904,140	253,852,110	365,973,680
Protestants	69,087,120	28,871,510	79,973,650	13,417,650
Orthodox	24,129,950	74,190	35,351,060	421,240
Anglicans	21,496,950	326,770	33,076,370	1,191,100
Other	46,273,480	36,437,660	13,275,820	7,859,780
Muslims	237,067,660	24,143,740	9,042,340	625,180
Nonreligious	1,443,850	619,384,830	48,788,900	12,751,380
Hindus	1,395,390	9,740	585,540	636,340
Buddhists	13,850	155,159,650	209,600	496,980
Atheists	234,470	133,582,030	17,153,270	2,388,410
Chinese folk religionists	10,220	194,049,230	50,350	68,460
New religionists	12,080	41,289,300	33,470	357,830
Tribal religionists	70,170,270	772,770	50	1,168,160
Jews	276,390	1,950	1,520,060	976,230
Sikhs	29,110	1,080	212,240	5,950
Shamanists	1,100	12,828,530	300	500
Confucians	550	5,635,590	990	500
Baha'is	1,319,350	44,200	67,090	543,460
Shintoists	50	3,423,960	390	990
Jains	49,980	430	9,870	1,980
World population	571,622,400	1,264,999,800	493,499,610	415,610,810

Source:World Christian Encyclopedia, (New York: Oxford University Press, 1988).

or plants and animals as sacred. Sacred trees or animals are normally protected. The designation of a physical feature as sacred not only affects its environmental condition, but the believer's behavior associated with it. For the aborigines of Australia, Ayers Rock is a sacred place. Aboriginals who visit the rock must obey prohibitions and instructions for dealing with the gods who reside there. The Dogon of Mali have a pantheon of gods including Ogo, the pale fox (Figure 10–3). Dogon diviners predict the future by interpreting the tracks Ogo leaves in wandering about the earth. Central to the Dogon religion is a creation myth that explains how order was created from disorder on the earth. As part of the creation of order, communities and their religious structures represent the Dogon view of the environment. Villages are established in pairs with an altar located between them. The symbolic pairing with the intervening altar emphasizes the mediating power of the altar between the profane and sacred realms.

Ayers Rock, Australia. Elements of the physical geography are often defined as sacred among practitioners of tribal ethnic religions.

Every tribal ethnic religion has developed a system for dealing with the uncertainty of the world in which its members live. The ritual actions and proscriptions of the religion are designed to ensure that the well-being of the group

NORTH AMERICA	OCEANIA	SOUTH ASIA	USSR	WORLD	PERCENT
231,539,750	21,143,000	125,954,640	102,083,790	1,619,272,560	32.9
88,144,650	7,310,180	72,863,900	4,940,000	900,545,840	18.3
96,293,800	7,449,790	23,049,300	8,409,000	326,551,820	6.6
5,948,530	506,130	3,201,260	88,720,290	158,352,650	3.2
7,760,000	5,463,380	286,990	500	69,602,060	1.4
33,392,740	413,520	26,553,190	14,000	164,220,190	3.3
2,675,720	93,520	535,079,210	31,494,020	840,221,390	17.1
19,310,020	2,859,170	18,386,060	82,981,670	805,895,880	16.4
764,200	284,080	644,218,360	1,300	647,894,950	13.2
193,440	16,880	150,975,920	349,710	307,416,030	6.2
1,029,120	507,920	5,084,340	60,562,030	220,541,590	4.5
116,160	19,170	8,442,280	100	202,755,970	4.1
1,025,050	4,910	66,777,800	200	109,500,640	2.2
65,210	84,730	25,215,740	0	97,476,930	2.0
8,050,100	85,540	3,893,810	3,177,380	17,981,460	0.4
8,660	6,530	15,897,290	50	16,160,910	0.3
290	290	12,000	299,750	13,142,760	0.3
1,020	290	1,500	200	5,640,640	0.1
300,110	56,540	2,228,260	4,900	4,563,910	0.1
710	590	200	100	3,426,990	0.1
2,040	980	3,304,650	20	3,369,950	0.1
265,766,620	25,187,410	1,605,687,810	280,960,220	4,923,334,680	100.0

is maintained. Most tribal religions traditionally had a harmonious relationship with the environment. Since most of these tribal religions relied on hunting and gathering, nomadism, or simple subsistence farming, their major environmental impact involved the protection of selected trees, animals, landforms, or other physical features. Among those who relied on a specific animal for livelihood, as did the Plains Indian with the bison (buffalo) or the Inuit with the seal and polar bear, conservation of limited resources was a necessity and was included in the formal belief system of the religion to ensure that there would be adequate food supplies for their future. Many of the groups with a tribal religion have a much less harmonious relationship with the environment today as modern technology has led to larger populations. Nomadic groups' livestock cause serious overgrazing, the adoption of the rifle allows the Eskimo to harvest more seals or bears, and growing populations in other locations disrupt the human-land relationship in tribal religions.

The religious beliefs of the tribal ethnic groups still provide a spiritual geography that guides their actions with respect to the environment and to one another. The religion specifies the proper procedures for selecting a site for a home, a grave, or a village or interacting with the environment. The resulting cultural ecology of the tribal ethnic religions tends to be markedly distinct from that of other groups. The central role of the environment in the religion and beliefs and behavior designed to prevent humans from offending or destroying the environment and its gods typify the cultural ecology of these groups.

Compound Ethnic Systems

Compound ethnic religious systems are larger and normally have a higher level of political organization and influence than do tribal ethnic religions. The location of the hearths of the largest religions is very similar to the location of the original center of plant and animal domestication

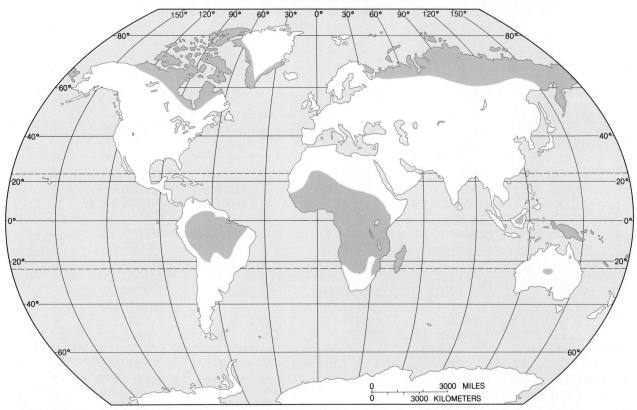

FIGURE 10–2 Areas of Tribal Religions Most tribal religions are located in areas which have been characterized by isolation.

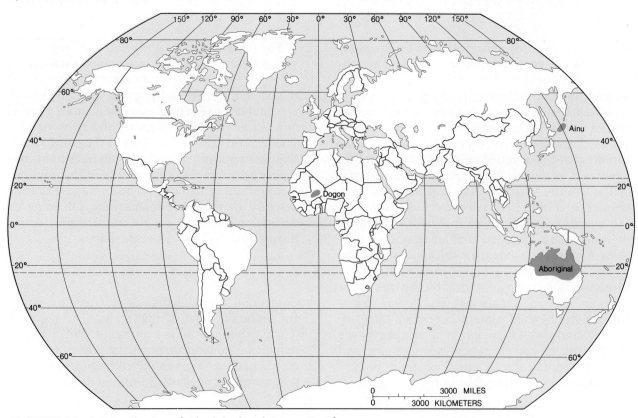

FIGURE 10–3 Distribution of Aboriginal and Dogon Peoples

(Figure 10–4). One of these religions, Judaism, is also the source for the ideas and teachings of two major universalizing religions, Islam and Christianity.

Judaism

Judaism formally began with the flight of Abraham from Ur in the Tigris-Euphrates Delta, a journey that illustrates the diffusion of religion and ideas (Figure 10–5). About 2,000 B.C. Abraham moved his family from among the Semitic tribes who inhabited the uncultivated areas around the Babylonian kingdom, which occupied the Tigris-Euphrates Delta. The Babylonian kingdom was monotheistic and had developed a formal, systematized code of laws (the Code of Hammurabi). Abraham apparently carried with him many of the ideas common to the Babylonian civilization as he traveled north and west to the area along the Jordan River in Palestine (now occupied by Israel and Jordan).

Numerical growth among the descendants of Abraham, their interaction with neighboring tribes, and their political growth and fortunes are related in the Old Testament. Monotheistic and nationalistic, the kingdom of Judah became a powerful state, but it was conquered by the Babylonians in 586 B.C. Over the next five centuries, the Jews had periods of greater or lesser political independence, but all independence ended when Roman rule was imposed at the time of Christ. Revolt against the Romans led to the dispersal of the Jews in A.D. 70 (Figure 10–6).

The dispersal of the Jews led to the development of two groups based on their location, the *Sephardim* and the *Ashkenazim*. The Sephardim were those who lived in the Mediterranean lands of southern Europe, northern Africa and western Asia, while the Ashkenazim were those who resided in central and western Europe.

Jewish Culture History   The dispersal of the Jews began an important odyssey in their culture history, known as the *diaspora*. Over the course of the next 1,900 years, they were persecuted because of their belief system, separated from their homeland, and destroyed on a scale unknown by any other major religious group. After Christianity was adopted as a state religion of the Roman Empire (A.D. 325), the Jews faced increas-

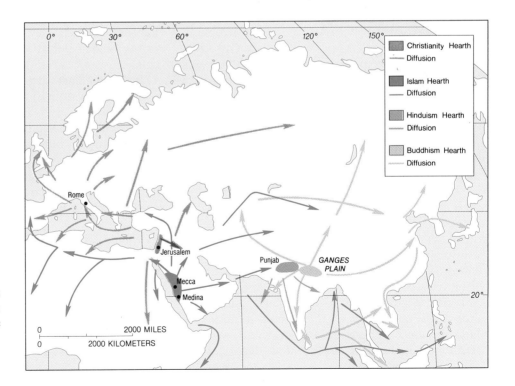

FIGURE 10–4 Hearths of Major Religions From their original hearths the major world religions diffused to much of the world.

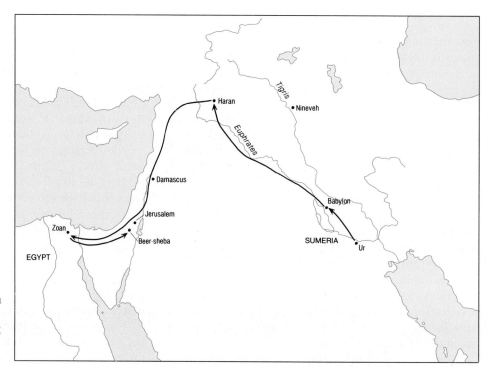

FIGURE 10–5 Migration of Abraham Abraham moved from the area along the Babylonian kingdom to the Jordan River.

ing discrimination. In the thirteenth century, they were expelled from England and not readmitted until after the seventeenth century. Persecution and Judaism's belief that the Jewish people were a chosen people combined to concentrate Jews into specific quarters in cities. The area occupied by Jews in Venice in the Middle Ages was called the *Ghetto*, a term later applied to any Jewish quarter of a city.

Persecution in Germany in the fourteenth century led the Jews to flee to Poland where the king offered them refuge. At that time the kingdom of Poland was one of the most powerful in Europe and included much of present-day Ukraine in the Soviet Union. The migration of the Jews to Poland led to the creation of the unique language known as Yiddish (from the German Jüdisch, meaning Jewish), based on a German dialect with additional Hebrew and Slavic words. Many countries in western and central Europe prevented the Ashkenazim Jews from owning land and restricted the occupations they could practice. Many became traders or artisans, and some loaned money because of the New Testament

injunctions against usury (charging interest on loans), which prevented Christians from providing this vital service. Concentrated in central Europe, Jews became the middlemen between the ruling landed nobility and the majority population.

In Middle Eastern and North African countries conquered by Muslim groups, the Jews enjoyed considerable liberty. Particularly in Spain under Muslim rule, the Sephardim experienced a revival of their culture, but by 1492 the Iberian Peninsula had been freed from Arab rule, and Jews were expelled unless they accepted Christianity. Many of the Spanish Sephardim fled to North Africa or to the Turkish Empire in the eastern Mediterranean.

France was the first European country to give Jews equal rights, as did most other western and central European countries in the nineteenth century. The Jewish population of the Ukraine, Imperial Russia, and Poland, however, experienced continued persecution.

Two events related to the history of persecution of the Jews in Europe play a major role in the

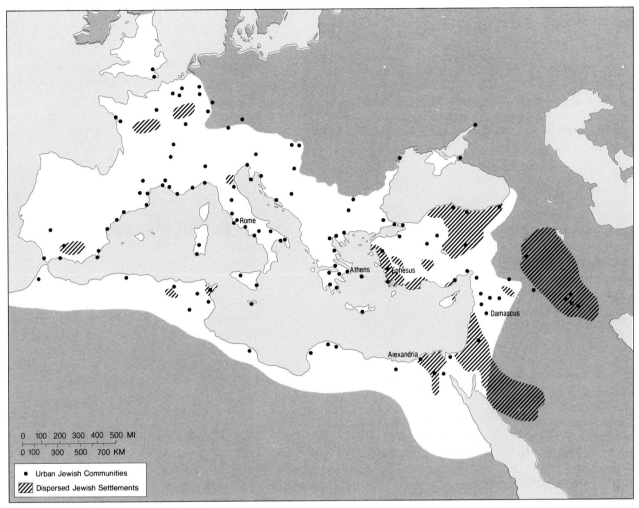

FIGURE 10-6 Jewish Diaspora, A.D. 70-300 The Jews were scattered to all of the region around the Mediterranean Sea after A.D. 70, creating both dispersed rural settlements and communities within urban centers.

present geographical distribution of the Jews. The first was the migration of Jews from Europe in the late nineteenth and early twentieth centuries. Before 1800 only a few thousand Sephardim Jews lived in the United States, but during the great European migrations to the United States, Jews arrived in large numbers. Between 1820 and 1870, nearly 250,000 Jewish immigrants (Ashkenazim) came from Germany alone. As many as one million Jews entered the United States from 1870 to 1915, the majority fleeing Russia, the Ukraine, and Poland. Another 250,000 Jews successfully fled the Nazi terror or survived the concentration

camps to enter the United States at the end of World War II. There are nearly eight million Jews in the United States today, giving it the largest Jewish population in the world (Table 10-2).

The other event that affected the culture history, distribution, and resultant landscapes of Jews in the world was the Nazi German Holocaust of 1937 to 1945. Hitler's Nazi government killed six million Jews in its concentration camps, primarily those in Germany and Poland. Only 10 percent of the Jews massacred were from Germany; the rest represented the bulk of the Jewish population of Poland, the Ukraine, and

TABLE 10–2
World Jewish Population

COUNTRY	NUMBER OF JEWS
United States	7,920,890
Israel	3,254,000
Soviet Union	2,630,000
France	650,000
United Kingdom	410,000
Canada	305,000
Argentina	300,000
Brazil	150,000
South Africa	118,000
Hungary	80,000
Iran	70,000
Australia	67,000
Uruguay	50,000
Rumania	45,000
Belgium	41,000
Italy	41,000
Germany	38,000
Mexico	37,500
Chile	30,000
Netherlands	30,000
Turkey	24,000
Ethiopia	22,000
Morocco	22,000
Switzerland	21,000
Sweden	17,000
Venezuela	15,000
Austria	13,000
Colombia	12,000
Czechoslovakia	12,000
Spain	12,000

Source: World Zionist Handbook (1986).

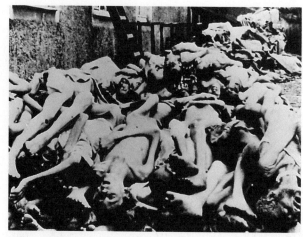

During World War II millions of Jews were massacred by the Nazis.

other areas recently occupied by the Germans. It is estimated that at least one-third of the world's Jews, and a much higher percentage of the Ashkenazim of Europe, were killed in the Holocaust.

The Nazi Holocaust also affected the emergence of the state of Israel in 1948. Jews had migrated to Palestine after World War I in response to their interpretation of the British Balfour Declaration of 1917, which declared Palestine a ''homeland for Jews and Palestinians,'' and the Holocaust increased their determination to create an independent Jewish state. The events surrounding the creation of the state of Israel will be discussed in the next chapter, but its present existence and conflict with its neighbors represent the most recent phase in the Jewish odyssey. Scattered from their Biblical homeland, Jews have once again restored the tie between people and place that is characteristic of an ethnic religion. Although only some 20 percent of the world's Jewish population lives in Israel, its existence is a symbol of the Jewish nation to Jews everywhere, and it has attracted migrants from throughout the world.

Universalizing Religions

The universalizing religions differ in their geographic distribution because they are no longer tied to a specific place, nation, or group of people. The largest of the universalizing religions is Christianity, which began as a reform movement within Judaism.

Christianity

Jesus Christ, a Jew born in Galilee, taught many elements of Judaic thought, combined with Greek thought emphasizing the ideals of love of God and other humans. The origin and diffusion of Christianity is well documented. The message of Christ was apparently heeded by many Jews, but its greatest diffusion occurred as a result of the apostle Paul. A Roman with a classical Greek education, Paul established an organization that allowed Christ's message to be diffused through the Roman Empire. Spreading along the trade routes of the Roman world, Christianity was adopted as a state religion following the conversion of the Roman Emperor Constantine in A.D. 325. (Figure 10–7). The diffusion of Christianity was aided by the Roman Empire, which provided political stability, a common language (Greek for the people in the eastern portion of the Empire and some of the educated elite in the western part of the Empire, and Latin everywhere else), and protected trade routes and roads. Educated Christians could travel anywhere in the Roman Empire and be understood, and they could write letters that would be safely delivered and understood.

The diffusion of Christianity through the Roman Empire is an example of hierarchical diffusion. The movement of early Christian missionaries to the large cities and centers of the Roman Empire was followed by conversions and diffusion to surrounding areas. By the fourth century the Christian church had three patriarchs, in Rome, Antioch, and Alexandria. They were soon joined by the patriarchs of Jerusalem and Constantinople (Figure 10–8). The movement of the seat of the Roman Empire to Constantinople from Rome (A.D. 330) left Rome an outpost of Christianity in the West, while the bulk of the church membership and culture was concentrated in the eastern end of the Mediterranean.

Eastern Orthodox

Although each of the *Eastern Orthodox* patriarchs directed the affairs of the church in his region, the patriarch of Rome ultimately claimed supremacy over the others. Conflict led to the great schism of 1054, which divided Christianity into the western or Roman Catholic church and the Eastern or Orthodox churches. The two divi-

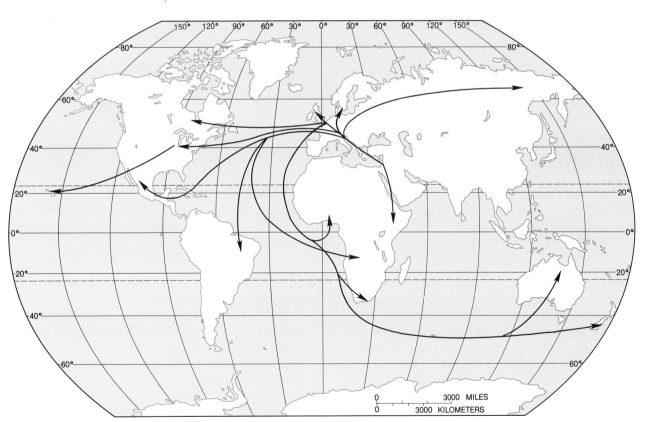

FIGURE 10–7 Diffusion of Christianity From its hearth in Jerusalem, Christianity spread west along the Mediterranean. Ultimately dominating Europe, it spread with colonialism to other areas.

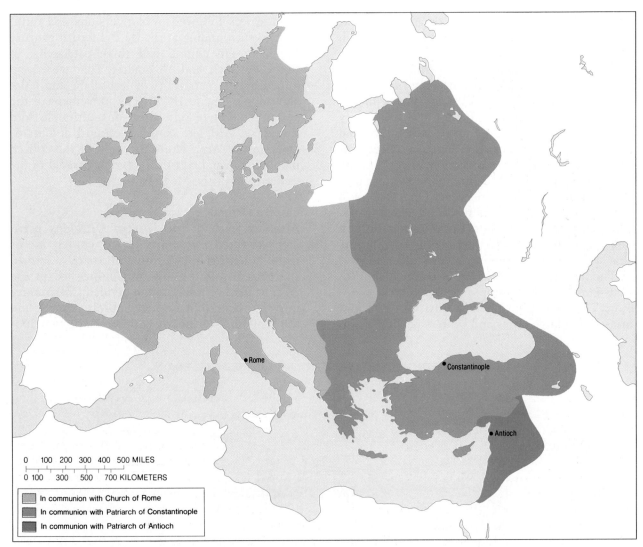

0   100   200   300   400   500 MILES

0 100   300   500   700 KILOMETERS

In communion with Church of Rome

In communion with Patriarch of Constantinople

In communion with Patriarch of Antioch

FIGURE 10–8 Patriarchs of the Christian Church in the Fourth Century Comparison of the regions in communion with three of the five patriarchs at this time (Jerusalem and Alexandria, Egypt are excluded because only small areas were in communion with them) indicates the importance of Rome and Constantinople.

sions of the Catholic church are different, with the Roman church becoming a supranational organization directing a church that spread to all ares of the world. The Orthodox churches (formally the Holy Orthodox Catholic Apostolic Eastern church) followed the lead of the patriarch of Constaninople who became heavily involved in the political affairs of the Roman Empire. Today the Orthodox churches are more nationally oriented and associated with individual countries, such as the Bulgarian, Albanian, Greek, Russian, and Armenian Orthodox churches.

By the eleventh century, the Eastern Orthodox faith had spread across the Black Sea and up the Dnieper River as the missionaries Cyril and Methodus carried the Christian religion to Moscow. The expansion of Islam after the seventh century A.D. caused the decline of Constantinople as the principal center of the Eastern Orthodox church, and the patriarch of the now larger Rus-

The historic Russian Orthodox churches of today's Soviet Union illustrate the characteristic style of the Eastern Orthodox church.

sian Orthodox church became the dominant patriarch of the eastern church. The Russian patriarch maintained his position of dominance until after the Communist Revolution of 1917.

The Eastern Orthodox church has not diffused extensively outside Europe. Members of the individual Orthodox churches have migrated to the United States creating a distinctive element in the architecture of American cities. Each church is associated with the individual national group that it represents, reflecting the historic nationalistic orientation of the orthodox churches.

Several small splinter groups of the Eastern Orthodox church have created unique cultural areas in the area around the Mediterranean. These include the *Monophysites*, who were so named because they believe that the divine and human in Christ were of one (monos) nature (physis). The Monophysites split with the main Eastern Orthodox church in A.D. 451 over the issue of the nature of Christ. Today's representatives of the Monophysites are the Coptic churches. The Coptic church is found in Egypt and in a related form in the Ethiopian Coptic church.

The Maronite Christians of Lebanon form another division of the Eastern church. The Maronite Christians, who have considered themselves an autonomous nation, represent a transition of Christianity from universalizing to ethnic orientation. The Maronites maintain contact with the Roman church as a result of the Crusades, but they are Eastern Orthodox. The political organization of Lebanon initially reflected the division of the country between Muslims and Maronite Christians, but in the last decade civil war has led to anarchy. The resulting religious map of Lebanon leaves part of the territory controlled by the Maronite Christians, part by Lebanese Muslims, and part by invading forces from Syria and Israel (Figure 10–9).

Western Christianity

The church centered in Rome became the most effective of the Christian groups in proselytizing, but during the Reformation various Protestant groups split away from the Roman Catholic church. Today the total Christian world consists of approximately 58 percent Roman Catholic, 35 percent Protestant, and 6 percent Eastern Orthodox. The cultural history of the western Christian branch includes three major stages: first, the diffusion of Catholicism across Europe, second, the Protestant Reformation, and third, the diffusion of both Protestantism and Catholicism during and after the European voyages of discovery in the fifteenth through seventeenth centuries. The cultural ecology and cultural landscapes in the Christian culture regions in Europe and the areas colonized by Europeans differ, exhibiting some of the sharpest cultural boundaries in the world (Figure 10–10).

Roman Catholicism and Protestantism   By the end of the fifteenth century all of northern, western, and southern Europe were part of the Roman Catholic culture region. At about the same time, the Protestant Reformation began a process of cultural fragmentation within the region. The resulting breaks with Rome led eventually to a distinctive division of Europe into a southern Roman Catholic region and a northern and western Protestant region. Some observers believe the

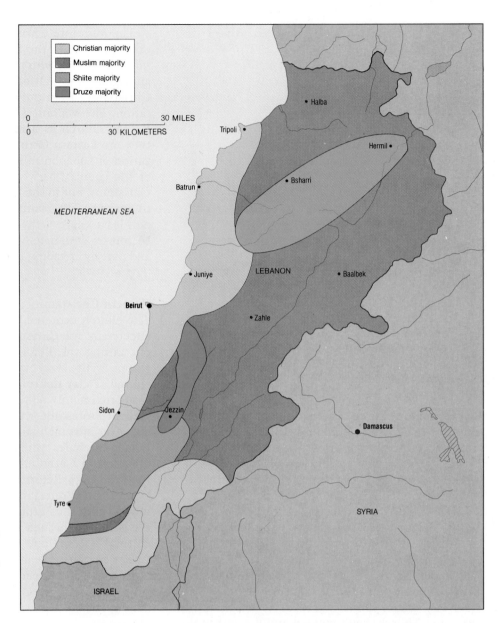

Christian majority
Muslim majority
Shiite majority
Druze majority

30 MILES
30 KILOMETERS

• Halba

Tripoli •

Hermil •

Batrun •

• Bsharri

MEDITERRANEAN SEA

LEBANON

• Juniye

• Baalbek

Beirut •

• Zahle

Sidon •

Jezzin •

• Damascus

Tyre •

SYRIA

ISRAEL

FIGURE 10–9 Map of Religions in Lebanon The control of territory by individual religions in the country is a major factor in the ongoing conflict there.

division between Protestant and Catholic Europe parallels the development of the Democratic and Industrial Revolutions of Western Europe.

The third event associated with western Christianity is the diffusion of the two groups from western Protestant Europe and southern Catholic Europe. Following the discoveries of the Americas by Columbus, the Protestant diffusion from England to Anglo America and the Catholic diffusion from Spain and Portugal to Latin America created the cultural dichotomy between these two regions. The dominance of Protestantism in Anglo America is somewhat less complete than the dominance of Catholicism in Latin America because of the presence of the French Catholics in Quebec and the subsequent diffusion of Catholic peoples from the mid-nineteenth century to the present (Table 10–3).

The map of Christian cultural regions shows the correlation of Protestant or Catholic Christi-

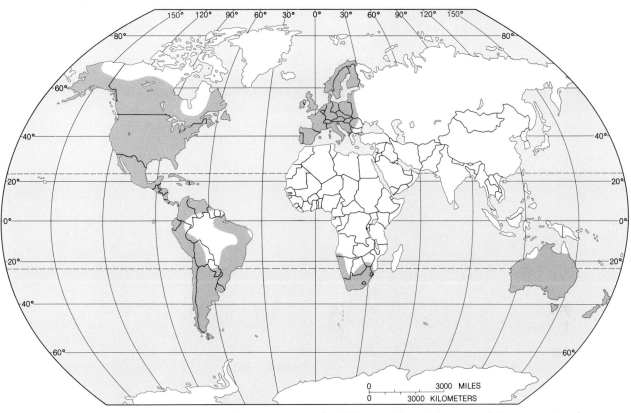

FIGURE 10–10 Western Christianity Culture Regions The diffusion of Catholicism and Protestantism from Europe has created distinct cultural regions in other areas of the world.

TABLE 10–3
Church Membership in the United States

GROUP	MEMBERS	GROUP	MEMBERS
Adventist churches	681,330	Friends	113,138
Baha'i Faith	100,000	Jehovah's Witnesses	730,441
Baptist churches	25,841,292	Jewish organizations	3,550,000
Brethren	181,989	Latter-day Saints (Mormons)	4,054,736
Buddhist Churches of America	100,000	Lutheran churches	8,467,879
Christian and Missionary Alliance	227,846	Mennonite churches	182,445
Christian Church (Disciples of Christ)	1,116,326	Methodist churches	12,782,054
Christian Churches and Churches of Christ	1,051,469	Moslems	2,000,000+
		Old Catholic churches	483,956
Christian Methodist Episcopal Church	718,922	Pentecostal churches	3,551,748
Churches of Christ	1,604,000	Plymouth Brethren	98,000
Churches of God	226,293	Polish National Catholic Church of America	282,411
Church of the Nazarene	522,082		
Eastern Orthodox churches	4,295,898	Presbyterian churches	3,455,296
The Episcopal Church in the U.S.A.	2,739,422	Reformed churches	581,585
Evangelical churches	241,232	Roman Catholic Church	52,654,908

Source: 1988 World Alamanc

anity with individual countries (Figure 10–11). Switzerland is nearly equally divided between a southern Catholic region and a northern Protestant region, as are the Netherlands and Germany. In Canada, Catholics are concentrated in Quebec. In the United States, Catholics are spread widely, but have the greatest numbers in the Northeast, the Southwest, and around the Great Lakes (Figure 10–12).

Religion in the United States   Christianity in the United States is greatly fragmented. Numerous churches were organized on the American frontier in the nineteenth century, reflecting the egalitarian, isolated, and utopian views of many of the immigrants to the United States. Groups such as the Church of Christ, the Pentecostal churches, the Seventh-Day Adventists, the Mormons, Christian Scientists, and Jehovah's Wit-

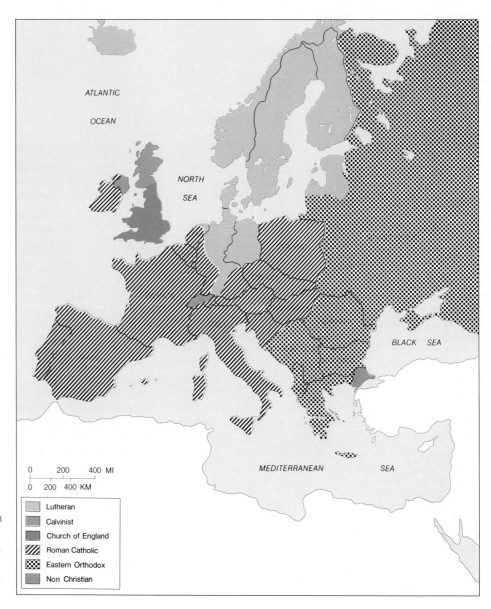

FIGURE 10–11
Protestant and Catholic Churches The distribution of Catholicism (Roman and Eastern Orthodox) and Protestantism creates distinct cultural regions in other areas of the world.

Legend:
- Lutheran
- Calvinist
- Church of England
- Roman Catholic
- Eastern Orthodox
- Non Christian

0   200   400 MI
0   200   400 KM

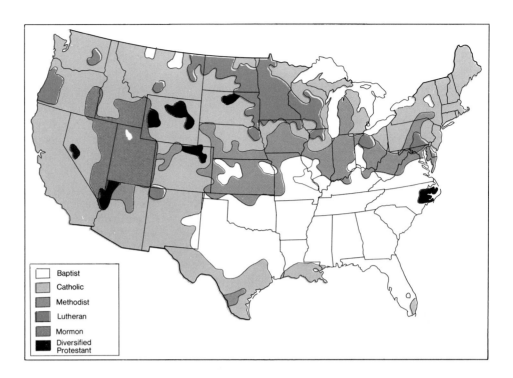

FIGURE 10–12
Religions in the United States

Baptist
Catholic
Methodist
Lutheran
Mormon
Diversified Protestant

nesses began in the United States. Their existence reflects differences in interpretation of scriptures and resulting doctrines, but their distribution creates distinct cultural regions in the United States.

Jews and Catholics in the United States tend to be concentrated in urban areas, since many of their adherents migrated to the United States at the time of the Industrial Revolution and great urban growth in the United States. Baptists, by contrast, are concentrated in the rural areas of the American South. Mormons dominate the intermountain area of Utah and adjacent regions of Wyoming, Nevada, and Idaho. The Lutheran denomination is concentrated regionally in the northern Plains States where Europeans from the Scandinavian countries and Germany migrated following the Civil War. The passage of the Homestead Act in 1862, which allowed every adult to qualify for 120 acres of free land for only a five-dollar filing fee, combined with the removal of the Indian tribes to attract the rural Scandinavian migrants to these states.

Migration of peoples from many lands and the emphasis on individual freedom in the United

States created a mosaic of religious groups. These include Amish settlements in Pennsylvania and adjacent states, Mennonites and Hutterites of the Midwest and Canada, and cults such as the Hare Krishna.

The Old Order Amish originated as a Protestant reform movement in 1520 in Switzerland. Migrating to the United States, they arrived in Pennsylvania in 1727. Their geographic impact is distinctive and impressive. They dress differently from others, with all women wearing dark-colored skirts, blouses, and matching aprons made only from home-dyed cloth. Unmarried girls wear white aprons, but all women wear homemade bonnets in accordance with their belief in the Biblical injunction that women must cover their heads when they pray. Men dress alike in jackets that have no lapels, outside pockets, buttons, or zippers except those used with work clothes. The Old Order Amish reject almost all modern conveniences, including electricity and telephones, and all modern appliances except sewing machines. Forbidden to use automobiles or tractors, they rely on horse-drawn farm equipment, buggies, and wagons. The Amish landscape

An Amish farmer in Lancaster County, Pennsylvania has adapted a tractor drawn hayrake for use with horses. Religious values shape the landscape and economy of each region.

of farms and their reliance on horses or mules create a distinctive cultural region in southeastern Pennsylvania and parts of Iowa, Ohio, and Indiana. The latter settlements have developed because population growth among the Amish (the average family has over six children), has forced them to move outside Pennsylvania to find land at affordable prices.

Most American religions do not create such a unique cultural region as the Old Order Amish, but each has affected the religious landscape of the country. From Catholic cathedrals to Jewish synagogues, from Mormon temples to Muslim mosques, to the "peculiar" landscapes of smaller groups such as the Mennonites or Old Order Amish, each has played a role in the evolution of the landscape of the United States.

Islam: A Judaic Neighbor

Islam is second only to Christianity in number of adherents. Islam is the name given to the religion preached by the prophet Muhammad (or Mohammad), and a Muslim is one who accepts the message of the prophet. Muslim is an Arabic word that means "one who submits" (to God, Allah in Arabic). Islam had its origins in the words re-

vealed to the prophet Muhammad, who was born on August 20 A.D. 570, according to the Gregorian calendar. Living in Mecca, a major trading center of the Middle East, he received his first revelation in 610. Muhammad's strict teaching that "there is no god but Allah" led to conflict with the Meccans whose livelihood depended on visits by pilgrims to shrines important to the various tribal ethnic religions in the area (including the Black Stone of Mecca, a rough oval stone, 18 centimeters across set in one corner of the cube shaped building known as the Ka'bah). As a result, in 622 Muhammad fled to the city of Medina (then called Yathrib). This flight is called the *Hegira*, and the Islamic calendar is dated from this year. Finding success in Medina, Muhammad and his followers returned to Mecca in 630 and occupied the city. Now accepted by the people of Mecca as a prophet, Muhammad made Mecca and Medina the two most sacred cities of Islam.

The revelations of Allah to Muhammad are recorded in the Koran. Muslims accept Christ and the prophets of the Old Testament as prophets, but they do not view them as divine. They trace their lineage to Abraham and believe that his two sons Ishmael and Isaac were the progenitors of the Islamic and Judaic peoples, respectively. The Muslims believe that Ishmael was unfairly banished by Abraham and was the ancestor of Muhammad.

The Koran contains religious doctrine and rules of worship, as well as pronouncements on worldly matters. Muslims regard the Koran as divinely inspired from beginning to end and use it as the basis of all activities, including religion, politics, dress, and foods. Islam is based on the five pillars of the faith: repeated saying of the basic creed, prayer, almsgiving, fasting and *pilgrimage*.

Devout Muslims pray five times daily: at dawn, at noon, in the afternoon, in the evening, and at nightfall. Only on Friday are they expected to attend noon prayers at a mosque. When praying, the believer faces Mecca and recites passages from the Koran and other verses of praise to God, including the basic creed "There is no god but Allah, and Muhammad is His prophet." Almsgiving or charity consists of two types: mandatory

(Zakat) and freewill (Sadaqah). Zakat requires individuals to give 2.5 percent of their wealth each year to a trust fund for the needy.

Fasting is required during Ramadan, the ninth month of the Muslim year. (Table 10–4). Muslims may not eat or drink from dawn to sunset during Ramadan. The fifth pillar is pilgrimage, most importantly, pilgrimage to Mecca that each individual should make at least once during his or her lifetime (see the Cultural Dimensions boxes on pages 384–385). This pilgrimage, called the *Hajj*, includes several ceremonies, the most important of which requires the pilgrim to walk seven times around the Ka'bah and kiss the sacred Black Stone in its wall. According to Islamic legend, the Black Stone was delivered to Abraham by Gabriel, the angel of revelations. (Figure 10–13).

The Diffusion of Islam

Unlike other universalizing religions, Islam diffused rapidly, providing a good example of expansion diffusion (Figure 10–14 on page 386). Its rapid expansion from area to area was facilitated by a number of factors, not the least of which is that it is a simple religion. Unlike other religions that require meditation or great sacrifice, a person who

TABLE 10–4
Islamic Calendar*

MONTH		SPECIAL DAYS
Muharram	1	New Year
	10	Ashura
Safar		
Rabi al Awwal	12	Birthday of Muhammad
Rabbi al Thani		
Jumada al Awwal		
Jumada al Thani		
Rajab	27	The Night of Mi'raj
Shaban		
Ramadan		The Month of Fasting
Shawwal	1	Id al Fitr
Dhu-l-Qaddah		
Dhu-l-Hijjah	10	Id al Adha

*The Islamic year is a lunar year with twelve months, each calculated from new moon to new moon. Thus it has no fixed relation to other calendars.

repeats the creed is a Muslim, and a person who keeps the five pillars of Islam is a good Muslim.

Before the death of Muhammad, followers of Islam began to conquer and unite the Arabian peninsula. After the prophet's death, Islam expanded outside Arabia, conquering Damascus in

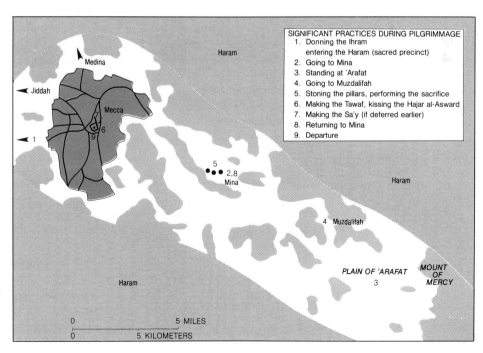

FIGURE 10–13
Pilgrimage Route at Mecca
The specific elements of the pilgrimage are specified by tradition.

The Hajj: An Introduction

The Hajj—the pilgrimage to Mecca—is essentially a series of rites performed in and near Mecca, the holiest of the three holy cities of Islam: Mecca, Medina, and Jerusalem. Because the pilgrimage is one of the five pillars of Islam—that is, one of the five basic requirements to be a Muslim—all believers, provided they can afford it and are healthy enough, must undertake the Hajj at least once in their lives.

The Hajj must be made between the eighth and the thirteenth days of the twelfth month (called Dhu al-Hijjah) of the Muslim lunar year.

Donning the Ihram

In a general sense, the pilgrimage begins with the donning of the Ihram, a white seamless garment, which is a symbol of the pilgrims' search for purity and their renunciation of mundane pleasures.

At the moment they don the Ihram, the pilgrims enter a state of grace and purity in which they may not wear jewelry or other personal adornment, engage in any disputes, commit any violent acts, or indulge in sexual relations.

Uttering the Talbiyah

In donning the Ihram the pilgrims also make a formal declaration of pilgrimage and pronounce a devotional utterance called the Talbiyah—"Doubly at Thy service, O God"—a phrase that they will repeat frequently as an indication that they have responded to God's call to make the pilgrimage.

Entering the Haram

Only after they have donned the Ihram, may the pilgrims enter the Haram, or sanctuary. In a sense, the Haram is merely a geographical area that surrounds Mecca. But because its frontiers were established by Abraham and confirmed by Muhammad, the Haram is considered a sacred precinct within which humans, undomesticated plants, birds, and beasts need fear no molestation, as all violence, even the plucking of a wild flower, is forbidden.

For the duration of the Hajj, Mecca and the sanctuary that surrounds it have a special status. To cross the frontiers of the Haram—which lie outside Mecca between three and eighteen miles from the Ka'bah—pilgrims from outside Saudi Arabia must now have a special Hajj visa in their passport, which entitles them to travel only within the Haram and to certain other places that pilgrims must, or customarily do, visit. Non-Muslims are strictly forbidden to enter the Haram under any circumstances.

Going to Mina

On the eighth day of Dhu al-Hijjah the assembled pilgrims begin the Hajj by going—some on foot, most by bus, trucks, and car—to Mina, a small uninhabited village five miles east of Mecca. There they spend the night—as the Prophet himself did on his farewell pilgrimage—meditating and praying in preparation for "the Standing" (Wuquf), the central rite of the Hajj, which will occur the next day.

Standing at 'Arafat

On the morning of the ninth day, the pilgrims move *en masse* from Mina to the Plain of 'Arafat for "the Standing," the culmination (but not the end) of the pilgrimage. The pilgrims gather on the plain and, facing Mecca, meditate and pray in a simple ceremony. Some pilgrims literally stand the entire time—from shortly before noon to just before sunset—but, despite the name of the ceremony, are not required to do so.

Going to Muzdalifah

Just after sunset, which is signaled by cannon fire, the pilgrims gathered at 'Arafat immediately proceed to a place called Muzdalifah, a few miles back toward Mina. There, traditionally, they worship and sleep under the stars after gathering a number of pebbles for use during the rites on the following days.

continued

The Hajj: An Introduction *continued*

Stoning the Pillars

Before daybreak on the tenth, again roused by cannon, the pilgrims continue their return to Mina. There they throw seven of the stones that they have collected at Muzdalifah at one of the three white-washed, rectangular masonry pillars. The particular pillar that they stone on this occasion is generally thought to represent "the Great Devil"—that is, Satan, who three times tried to persuade Abraham to disobey God's command to sacrifice his son. The throwing of the pebbles symbolizes the pilgrims' repudiation of evil.

Performing the Sacrifice

Now begins the greatest feast of Islam: the 'Id aladha, or feast of sacrifice.

After the throwing of the seven stones, the pilgrims who can afford it buy all or a share of a sheep or some other sacrificial animal, sacrifice it, and give away a portion of the meat to the poor. The sacrifice has several meanings: it commemorates Abraham's willingness to sacrifice his son; it symbolizes the believers' readiness to give up what is dearest to them; it marks the Muslim renunciation of idolatrous sacrifice; it offers thanksgiving to God; and it reminds the pilgrims to share their blessings with those less fortunate.

Doffing the Ihram

Now having completed a major part of the Hajj, men shave their heads or clip their hair and women cut off a symbolic lock to mark partial deconsecration. At this point the pilgrims may remove the Ihram, bathe, and put on clean clothes.

Making the Tawaf

The pilgrims now proceed directly to Mecca and the Sacred Mosque, which encloses the Ka'bah, and, on a huge marble-floored oval, perform "the Circling," or Tawaf. The Tawaf consists essentially of circling the Ka'bah on foot seven times, reciting a prayer during each circuit.

Kissing the Hajar al-Aswad

While circling the Ka'bah the pilgrims should, if they can, kiss or touch the Black Stone (the Hajar al-Aswad), which is embedded in the southeastern corner of the Ka'bah and which is the precise starting point of the seven circuits. Failing this, they salute it. Kissing the Stone is a ritual that is performed only because the Prophet did it and not because any powers or symbolism are attached to the Stone per se.

Adapted from *Aramco World Magazine*, Vol. 25, No. 6, Nov.-Dec. 1974, pp. 2–6 by permission.

634; in turn, Persia (Iran) fell by 636, Jerusalem in 638, and Egypt by 640. During the following decade Islam consolidated its victories until most of North Africa had become Muslim by the end of the seventh century. In 711 the Muslims entered Spain where they were dominant until the late 1400s. Not until the eleventh century did Islam expand into India and China, and Constantinople was not taken by the Muslims until 1453. In the fourteenth century the people of Indonesia were converted to Islam. Since that time Islam has not spread widely except by relocation diffusion through migration of Muslims to other areas.

Some observers, noting the present distribution of Islam and its concentration in the desert lands of North Africa and the Middle East, have concluded that it spread only to areas where the climate was similar to that in Mecca. Yet the largest number of Muslims are found in Indonesia and Bangladesh, in tropical environments. The distribution of Islam more closely reflects the pattern of trade in the world at the time of its origin and diffusion. The trade routes led from the Middle East through central Asia to northern China, where some thirty million Muslims live today. Caravan routes crossed the Sahara to the

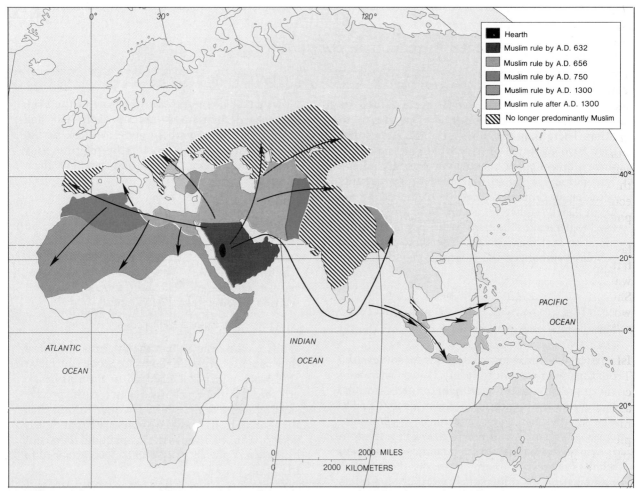

FIGURE 10–14 Diffusion of Islam From its hearth in the Arabian Peninsula Islam spread widely.

Sudan, and overseas routes led from Arabia southward along the east coast of Africa, eastward to India, and across the Straits of Malacca to Indonesia. The traders were missionaries for Islam and spread the religion to the east through hierarchial diffusion.

The spread of Islam had an important impact on language. The Koran prohibits translation from Arabic, so Muslims use Arabic in religious rituals wherever they live. Outside the Arabic-speaking world, youths memorize the ritual Arabic passages required for worship.

Cultural History in Islam

The spread of Islam was not accomplished without conflict. Muhammad did not clearly indicate

a successor or plan of succession, so the period following his death was a time of confusion among the Muslims. Muhammad was the *caliph* (or *iman*), who ruled the Muslims in temporal matters, and after his death the religion split into two groups: the *Sunnis*, who believed that the caliph should be elected; and the *Shi'ites* who believed that the caliph should be a lineal descendant of the prophet Mohammed.

Eighty-five percent of all Muslims are regarded as Sunnis, and are divided into four major branches concentrated in different geographical regions: the *Hanifite* in western Asia, India, and northern Egypt; the *Malakites* in North and West Africa and southern Egypt; the *Shafi'ite* group in northern Egypt, Syria, and Indonesia; and the

Hambelites found today in Saudi Arabia. The Hambelites are the most conservative of the four groups.

The Shi'ites—concentrated in Iran, Iraq, and Syria, but forming an important minority in other Islamic countries—believe that Muhammad should have been succeeded by his son-in-law, Ali (who was also his cousin). Ali was named the caliph in 656 but gradually lost control of the Muslim world. After his death in 661, the Umayad dynasty seized the caliphate. Ali's son, Husaim, challenged the Umayad caliph in 680 but was defeated and executed at Karbala, Iraq, which then became an important pilgrimage site for Shi'ites (Figure 10–15).

Shia Islam is suspicious of the world at large, favoring the true believer who is devoted to the faith, conservative, orthodox, and obedient in all ways to the teachings of Islam. The presence of Shi'ite majorities and minorities in the Islamic world is the basis for important political and religious conflicts.

Islamic Cultural Ecology and Landscapes

The all-encompassing role of Islam in the Muslim countries of the world has had a political as well as religious impact. Although pan-Islamic movements have attempted to unite the Islamic peoples, bitter divisions between individual countries and the major sects of the religion make this difficult. The destructive conflict between Iran and Iraq from 1980 to 1988, for example, included religious hostility, compounded by linguistic and historical differences. The majority population in both Iran and Iraq are Shi'ite, but the government of Iraq is ruled by the minority Sunnites. Iran is viewed by the Iraqis as distinct, because culturally it is Persian rather than Arabic.

The government of Pakistan is based on the principles taught in the Koran, in effect creating a *theocracy*—a political system operated on the basis of religious beliefs and practices. Charging of interest (usury) on loans, for example, is forbidden, and other restrictions taught in the Koran are also observed. These include fasting during the month of Ramadan, closing of shops, ministries, and other activities on Friday, the generally subservient role of women in public activities, and similar prohibitions. In spite of its Islamic heritage, the Pakistan people elected a woman, Benazir Bhuttu, president in 1988. Her election was based in large part on her ability to capitalize on her father's name. He was viewed by many Pakistanis as a martyr after he was overthrown by the military and executed.

In Saudi Arabia, an ultraconservative Sunnite group known as the Wahabbites emerged in the eighteenth century. They expanded Muhammad's injunctions against alcohol to include tobacco

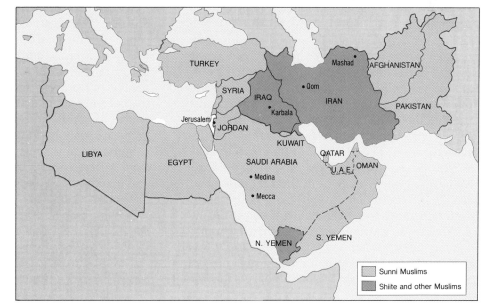

FIGURE 10–15 Present Distribution of Islam Although all of the countries of North Africa are over ninety percent Islamic, the countries from Saudi Arabia to Iran contain the most important sacred sites to both the Sunni and Shi'ite groups. (See also Figure 12–1 on page 459.)

and coffee, and today's Saudi Arabian society is one of strict observance of the Islamic code. Air conditioning, automobiles, and television are not *taboo,* however, since the Koran does not mention them.

The role of Islam in politics differs from country to country, as does the degree to which the governments follow the Koran. Nevertheless, from Indonesia to Algeria and from Turkey to Sudan, the pervasiveness of Islam affects the political process. The role of the Saudi Arabian government as the caretakers of Mecca, for example, brings them into conflict with other groups as believers make their yearly pilgrimage to Mecca (see the Cultural Dimensions box). The arrival of more than two million visitors in a one-month period each year forced Saudi Arabia to build a new airport for the arriving pilgrims. The ongoing conflict between Sunnites and Shi'ites periodically results in attempts by Shi'ites to wrest control of Mecca and its holy sites from the Sunnite majority.

Although Saudi Arabia's Shi'ite element constitutes less than 8 percent of its total population, in 1986 a Shi'ite group seized the Ka'bah, and only after killing several Shi'ite Muslims did Sunni Saudi forces regain it. In 1987, the Saudis accused the Iranians of coming to Mecca with the goal of creating confusion and turmoil amongst the pilgrims.

Iran provides another example of the impact of Islam on the daily activities of Muslims. When the late shah came to power in 1953, he immediately set out to remove what he perceived as the negative influence of Islam on his people's political and economic development. One of his first acts was to strip the Shi'ite clergy of their vast land holdings; he also granted equal rights to women and encouraged westernization and industrialization. The shah was overthrown in 1980 by Shi'ites who believed that the shah's efforts were leading them away from their religious heritage. The Ayatollah (literally a "holy man") Khomeini came to power, ended modernization, and created an austere and strict Islamic state.

At the other extreme, the state of Turkey was secularized by the dictator, Kemal Ataturk, in the 1920s. He adopted the Latin alphabet for the Turkish language and forced women to remove their veils. Turkey has become the most modern and westernized of the Islamic countries, but even here the Shi'ite minority and Sunnite fundamentalists are attempting to increase the role of religion in the government, economy, and society.

Across the Islamic world, cries for return to the strict interpretation of the Koran have led to growing fundamentalist movements. The emotionalism of such groups was highlighted by the assassination of the Egyptian leader Anwar Sadat for his perceived abandonment of Islam when he signed a treaty with Israel. Underground movements in universities in the more westernized Islamic countries have forced female students to wear veils and cover their hair, and active proselytizing of individuals is underway to expand support for a return to the rules of the Koran.

Muslim Landscapes   The rising tide of interest in the Islamic heritage parallels the dominance of the religion in the landscape of the region. Although many westerners assume that Islam is primarily associated with the nomadic peoples of the desert regions of the Middle East, Islam is overwhelmingly a village and urban religion. One of the mandates of the Koran is that the believers gather on Friday at a mosque to pray, forcing true believers in Islam to be sedentary.

At the center of the classic Islamic city is the mosque. The heart of the whole community, it was built on the agora or forum in towns predating Islam. Surrounding the mosque is a concentric arrangement of activities (Figure 10–16). Immediately adjacent to the mosque is the bazaar, the commercial district with its booths and rows of shops, hotels, and warehouses. The individual crafts are separated from one another, and there is a hierarchical arrangement descending outward from sellers of religious products' to booksellers and bookbinders, to clothing and fabric shops, to leather workers, rug and tapestry makers, and jewelers, carpenters, locksmiths, and coppersmiths. Farthest out are merchants selling vegetables, fruits, and meat.

Surrounding the bazaar are residential districts segregated by ethnic and religious groups. The

CULTURAL DIMENSIONS
Pilgrimage: Devotion or Demonstration?

The continued hostility between the Shi'ites of Iran and the Sunnis of Saudi Arabia illustrates the intertwining of religion and politics in the Middle East. Part of the official title of King Fahd of Saudi Arabia is "Custodian of the Holy Islamic Mosques of Mecca and Medina." Iran, in part because of ideological differences, in part because Saudi Arabia supported Iraq in the Iranian-Iraqi conflict, has used the great pilgrimage (Hajj) to demonstrate against the involvement of Saudi Arabia with the United States and other Western influences. Iran views the pilgrimage as an opportunity to spread what they maintain is the "real Islam" among the broad spectrum of Muslims from around the world.

Among other things, Iran's "real Islam" calls for Islamic leaders to end alliances with the United States. Iran and Ayatollah Khomeini, before his death in 1989, have maintained that the Saudis practice "American Islam" because of their close ties to the United States. Demonstrations among the 150,000 pilgrims from Iran at the Hajj in 1987 escalated into violence that took the lives of more than 400 Muslims, mostly Iranian.

As a result, the Saudi government announced a new quota system that began with the 1988 pilgrimage. Entry visas are granted to only 1,000 Hajjis (pilgrims) per 1,000,000 Muslims in a country's population. Consequently, Iran is limited to only one-third the number of pilgrims that attended in 1987. Iran views the quotas as a means of limiting Iranian visits to the holy sites claimed by both Shi'ite and Sunnite Muslims. The Ayatollah Khomeini responded by stating that it is impossible for Iranian pilgrims to visit Mecca and not demonstrate against "world Arabics," (Khomeini's term for Arab Muslims who do not support his brand of revolutionary Islam).

Although the majority of Muslims, particularly those farther from Saudi Arabia, have never and will never make the Hajj, if Saudi Arabia maintains a strict 1,000 per 1,000,000 population quota, most Muslims can never make the Hajj. The significance of the Hajj in the life of a Muslim goes beyond the simple obedience to the commandment of the faith. One who makes the pilgrimage is entitled to use the title Hajj in his or her name thereafter, assuring the pilgrim of a position of honor and importance in his or her community. The Iranians argue that the fundamental right of every Muslim to visit Mecca and Medina transcends the political rights of the Saudi Arabians. They maintain that the pilgrimage should be open to any financially able Muslim regardless of age or national quota.

The Iranians have consistently called for an international committee of Muslims to control Mecca and Medina. Reflecting their militant monotheism, the Iranian leaders maintain that Mecca is the place for the fight against all other gods, which they interpret to mean idols or non-Islamic influences. The Iranians maintain that the political or military alliances of some Islamic countries with the United States and the Soviet Union are tantamount to worshipping false gods. The demonstration of 1987 began with the Iranian pilgrims chanting "Down with America. Down with Israel. Down with Russia. God is great. There is no God but God." (Since Khomeini's death the Iranians seem to be reconsidering their opposition to the Soviet Union).

The growing fundamentalist movement in all Islamic countries makes the issue of whether Mecca and Medina are solely for devotion by the faithful or places of demonstration for the renewal and commitment of Islam's fundamental beliefs a critical one. The annual pilgrimage may become a potential flash point in the political and social life of the Islamic countries.

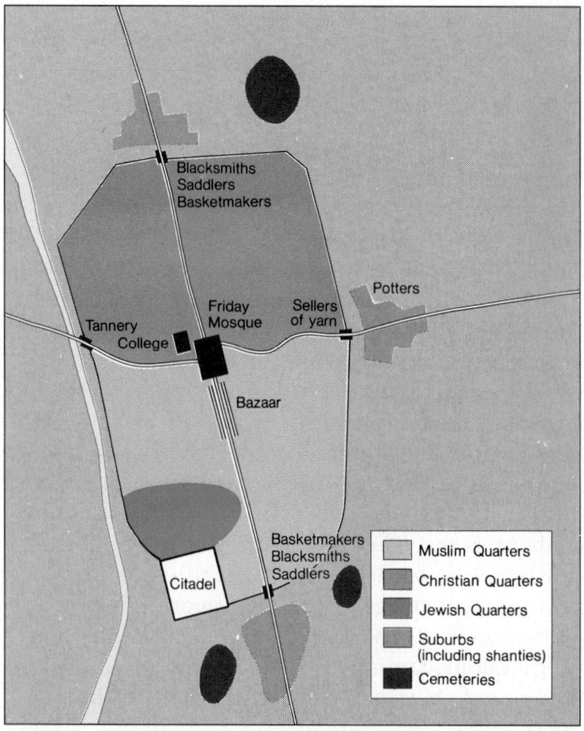

FIGURE 10–16 An Islamic Town The Friday Mosque is at the center of the town with the bazaar adjacent.

residential districts are divided into separate units consisting of courtyards and alleys off the main street, these units can be closed at either end by great gates. The seat of government in the Islamic city is at the edge of the city rather than in the center. This location apparently allowed its rulers to defend themselves against popular uprisings. Historically, Jewish urbanites located their quarter (ghetto) in the vicinity of the palace where they could be protected from mob violence.

This pattern of spatial organization in the Islamic city has changed with modernization and greater urbanization. Nevertheless, the basic pattern is still recognizable. Islamic cities do not always appear orderly because the Koran makes no provision for streets or public spaces other than the requirement that a laden donkey be able to pass. In consequence, Islamic cities initially developed in a random and seemingly capricious fashion.

Residential construction reflects the Koran's injunction against ostentatious houses of multi-

ple stories, the requirement that men's and women's quarters be separate, and that the family life be private. Most Muslims have adopted the Greek house plan with a central court separating the male and female quarters. Islamic houses thus look inward to this court and are characterized by blank walls or walls with barred windows facing the streets.

The Koran also affects rural landscapes. Proscriptions against alcohol in the Koran have affected the number of vineyards in the Mediterranean climates of the Middle East; even Christians were discouraged from growing wine grapes by the imposition of excessive taxes by their Muslim conquerors: a tax on the grape harvest, a tax on the amount of wine produced, a surtax on transporting wine, and a 10 percent tax on the value of the total wine yield in a particular district. Nevertheless, some Sunnite Islamic groups have accepted brandy since it is not mentioned specifically in the Koran, while on the peripheries of the Islamic world in Africa, palm wine and millet beer are widely used.

The Koran's prohibition against pork has also affected agriculture. Sheep and goats are universally raised, and overgrazing has occurred on many of the arid and semiarid lands of the Middle East and North Africa. Migratory Muslim nomads to this day graze their herds over the vast stretches of the region that cannot be cultivated. Mutton and goat meat are the primary sources of animal protein across the region from Pakistan to Algeria.

Hinduism

Perhaps the oldest and most complex of all the religions of the world is Hinduism. Most of today's active religions began sometime after the sixth century B.C., but Hinduism traces some of its themes to the third millennium B.C. Hinduism is the most tolerant of all religions and includes practices and beliefs ranging from simple worship of animate or inanimate objects to sophisticated and elaborate philosophical systems for personal and societal organization. Hinduism is polytheistic, allowing numerous major and minor gods with their temples and priests.

Hinduism has been the basis of three other religions. Two reform movements, Buddhism and *Jainism*, which began in the sixth century B.C., attempted to reform Hinduism and were reabsorbed into Indian Hinduism within a few centuries. Today Jainism is a minority religion in India, whereas Buddhism, though having great influence in other nations, is almost nonexistent in India. The third religion, *Sikhism*, which began in the fifteenth century A.D. after the Muslim invasions of India, is a more recent offshoot of Hinduism. Sikhism combines the monotheism of Islam with Hinduism, and it remains a minority religion in India today.

Hinduism has no one identifiable founder, nor any defined dogma. The word Hindu comes from the Sanskrit for the River Indus, Sindhu. Seminomadic Aryan cattle herders and plow cultivators, speaking an Indo-European language, arrived in lands along the Indus River around 2000 B.C. Over the next thousand years, repeated waves of Aryans came into the Indus valley where they found a region with a highly developed civilization, towns, and cities. Once there the invaders adopted aspects of the local religions to form the modern Hindu belief system.

The principal books associated with Hinduism can be divided into three broad groups. First are the four Vedas, of which the Regveda is the oldest and most important. The Regveda consists of ten divisions or books. One book, the Bhagavad-gita, is the most widely used of all Hindu scriptures and conveys a tolerant attitude towards other teachings and practices. The second group is composed of the Sutras that appeared between the sixth and seventh centuries B.C. Each Sutra contains a comprehensive discussion of a single subject. The third group consists of the Dharamasutras and Dharamasastras, which establish a set of concepts meaning "the pattern of right living." Unlike the Vedas and Sutras, these books go beyond discussion of ritual duties to emphasize moral behavior as a part of the Hindu's total religion obligation. Most influential of all the Dharmasastras was the Manaadharmasatra (Laws of Manu). Compiled between 200 B.C. and A.D. 200, it is the guide for the Hindu society.

The laws of Manu and the Dharmasastras explain the principle of *Dharma*, which implies a world that is rigidly structured. Out of this grew the caste system that to most observers is synonymous with Hinduism.

Castes and Outcastes

The caste system is the basis for social organization in Hinduism. A Hindu is born into a caste, marries within his or her caste, and worships and works according to the rules of his or her caste. The caste system is anchored in the concept of order in the universe taught by the laws of Manu and the other Dharmasastras. This all-embracing order is hierarchical and gives all creatures rank. At the top of the rank stand human beings, with the castes indicating the order of all humans. Above humans in the universal order is the final release from earthly existence. Living things are reincarnated into this universal order until such time as they are able to reach unity with the "universal soul," and a condition known as *Nirvana*. Until reaching Nirvana, each living thing continues to be reborn at a level determined by his or her conduct during the previous existence. Thus *Karma*, or "law of the deed," is based on a strict relationship between past behavior and present form of life.

The law of Manu provides four broad groups of castes. The *Brahmans*, the highest caste, are the religious leaders, scholars, and teachers. The *Kshatriya* are warriors and protectors of society

An untouchables village in India. Members of individual castes occupy distinct areas of a village based on their occupations.

TABLE 10–5
Major Castes and Representative Occupations

CASTE	SELECTED OCCUPATIONS
Brahmans	Scholars
	Priests
Kshatriyas	Landowners
	Warriors
Vaisyas	Trader
Sudras	Carpenter
	Blacksmith
	Barber
	Potter
	Begger
	Basket-maker
	Tanner
	Sweeper
Untouchables (Harijans)	Butcher
	Launderers
	Fisherman

(political leaders); the *Vaisyas* are the traders, herders, and farmers according to Manu, but trading is their distinctive occupation. The *Sudras*, the lowest caste, provide services for society. The highest possible work of the Sudras is to engage in handicrafts and manual occupations. Sudras include florists, brickmakers, carpenters, and blacksmiths (Table 10–5). The four groups are normally subdivided into mutually recognized classes within each caste.

Below the four broad castes are the "outcastes" or untouchables. The untouchables are individuals whose jobs are regarded as sinful or grossly unclean. They include launderers, sellers of liquor, fishers, leather workers, and handlers of dead bodies. The untouchables were believed to cause ritual impurity for members of castes who came into personal contact with them. Historically, untouchables lived in separate communities, had their own separate wells for water, and in the nineteenth century were prohibited from traveling on roads used by the castes. Although some of the barriers between castes, such as the prohibition against intercaste marriage, are breaking down in rural India caste consciousness is common (see the Cultural Dimensions box).

Concern for the lot of the untouchables led Gandhi to champion their cause in pre-independence India. He coined the term *harijan*, literally meaning "child of God," as an adjective to describe the untouchables. Nevertheless, the untouchables are still discriminated against in India and in rural areas are still required to live in separate villages, use their own wells, and otherwise maintain a degree of separateness.

The Cultural Landscape of Hinduism

The landscape of India reflects the Hindu belief system. First-time visitors remark on the special treatment accorded to cattle, even on city streets. Belief that cattle are sacred reflects the Hindu metaphysical world in which all life-forms have their rank. India today has more cattle than any other country, and Hinduism proscribes the consumption of beef by any Hindu. The numerous

The reverence afforded cattle in Hinduism is illustrated in this photo from Calcutta.

A Question of Caste

Chhawla is a village only twenty-five miles from New Delhi. Chhawla's proximity to the capital has ensured attention to its water supply and the construction of two primary schools, two secondary schools, and one nursery school.* One woman from Chhawla (Shashi) has traveled a remarkable and sometime agonizing journey from the village, surrounded by its invisible wall designed to keep its females inside, to a career as a nurse in India, Libya, and Saudi Arabia and finally to the United States where her husband is now doing postdoctoral work at Yale.

The alleys of Chhawla can be negotiated only on foot. The village is a drab collection of windowless brick houses more or less whitewashed, attached to one another along a haphazard arrangement of winding lanes with open drains. The village includes a big pond that dries up during the hot weather, a deserted Hindu temple, and a few tiny shops selling flowers, matches, tobacco, spices, and cooking oil. There is a water line to the Yamuna River, and the village has had electricity for twenty years.

The home where Shashi grew up has six small rooms that open off the courtyard, with another room on the roof. It is now occupied by four adults and five children, all related to Shashi. It is the only one in the village with a latrine, flushed with pots of water. By village standards, it is a fine house, and by Indian standards, the girl's family is well above the poverty line, below which 48 percent of the population of India lives.

The invisible wall around the village was erected by its dominant population, the Jats. Historically a caste of farmers living only in northern India, many Jats in Chhawla today lease their land and go to Delhi for salaried jobs. A small colony of Harijans (untouchables) cleans the alleyways, makes sandals, barbers, repairs farm implements, and molds pots. They live separately, but they are free to use the services of the community health center, and for the last ten years they have been able to send their children to school with the Jats. Jats are stereotyped in India as strong, hard-working, and thrifty people.

After completing school in the village, Shashi enrolled at the Delhi nursing college. It was unthinkable for a Jat girl, not only to go out of her house unveiled, but out of the village, unveiled and *unmarried.* Training for nursing was also unthinkable since nurses dealt with unclean things, and the village members of the Jat caste believe that serving others makes you inferior. To deal with criticism from the villagers, Shashi agreed to accept a husband. Her brothers and father knew that Shashi would never consent to be put on display and sit demurely in a corner with her face veiled, so they conducted negotiations using her photograph, except on two occasions when Shashi chose to be present so she could explain courteously that the suitors' insufficient education would make them poor life companions. Floods in 1977 interrupted the search for a husband.

Although Shashi served as a volunteer nurse for four months during and after the floods, her failure to marry continued to draw criticism from the villagers. Shashi therefore accepted the first overseas offer that came along, in Libya. After a year she returned home and began teaching at the nursing college again, but the innuendos about her failure to marry began all over again. Eventually Shashi met her husband, Ravi, who is from Kerala, in southern India. They met through a friend while he was studying for his doctorate at Jawarharlal Nehru University in New Delhi. Since Ravi had five unmarried sisters at home, and since he was head of the family (his father had died), he felt he should not marry until at least two of his sisters were married.

Three years after they met, they discussed marriage, but since he was a different caste (Vaisya) and came from southern India, Shashi felt her parents would be unable to consent. Shashi accepted a job in Saudi Arabia, and Ravi went to the village to find out what had become of her. Ravi discovered that Shashi's respectable home compared favorably with his own. Ravi could not see many differences between the Vaisya and Jat castes.

Shashi's brother, a sergeant in the Indian Air Force, was able to get permission from the family for Shashi to marry in secret. The wedding took place in a Hindu temple in New Delhi and the two left for the United States. They hope that in the two years he is attending Yale the villagers will accept the marriage.

---

*Adapted from "A Modern Woman out of Old India," by Chris Prouty *New York Times Magazine,* September 14, 1986, pp. 122–28.

A Hindu shrine in Bombay displays elements important in the Hindu belief system.

temples and shrines at which the inhabitants of the estimated 500,000 villages in India worship are a second landscape element associated with Hinduism. The shrines and temples reflect the multiplicity of gods in the Hindu pantheon, Shiva and Vishnu being the most important.

The impact of Hinduism on society is often perceived negatively because of the caste system, but in a rural subsistence economy the caste system provided a secure social arrangement in which each village was a nearly self-sufficient unit in which all members had a specific role and their basic needs were met. The rigid organization, with its emphasis on duty to one's caste, helped maintain the orderly functioning of the society.

Jainism

Jainism developed as a reform movement within Hinduism in the sixth century B.C. Jainism incorporates the Hindu belief in reincarnation and maintains that the ideal way to advance in the next life is to practice asceticism and withdrawal. Jains are vegetarians and extend vegetarianism to not using leather, since this necessitates taking animal life. Jain monks sweep the paths before them when they walk to avoid treading on insects, and they strain the water they drink in order to protect whatever life may be in it. The concepts of aestheticism and vegetarianism have been accepted by many Hindus, but Jainism itself remains a minor sect in the region. There are no more than 3.5 million Jains in India, primarily in and around Bombay.

Sikhs

One of the world's newest religions, Sikhism, originated in the sixteenth century A.D. The Sikhs maintain that they are a new and independent religion, but Sikhism is essentially a reform movement that takes its basic theology and world view from Hinduism. Unlike other reforms of Hinduism, however, Sikhism incorporates a distinct view of Islam.

Sikhism originated in northwestern India in the path of invading Islamic groups that eventually dominated India. Hostility between the Indian converts to Islam and the Hindu majority was often violent. Sikhism grew out of the teaching of some holy men who believed that the Hindus and the Muslims could live in peace. The actual founder is believed to have been Nanak (1469–1538), who received a vision telling him that he was a prophet of the true religion and that there was no Muslim and no Hindu. Nanak's role and that of his disciples (called Sikhs) was to teach this new religion to others. Nanak made the Hajj to Mecca and accepted the Islamic ideas that there is but one god and that, as the supreme creation, humans are entitled to kill animals for their use. Consequently, the Sikhs are among the few Indians who are allowed by their religion to eat meat. Nanak also adopted the principles of reincarnation and passivism. The unwillingness of the Sikhs to accept Islam as the only true religion led to their persecution by later Muslim leaders, causing violent hostility between the two groups.

Over time the principles taught by Nanak were incorporated into the official scriptures of Sikhism, the *Granth,* which includes the writings of nine subsequent *gurus* who taught after his death. (The term "guru" in Hinduism connotes "teacher," but to Sikhs it means "leader.") The tenth Sikh guru, Gobind Singh (1675–1708) was made guru when his father, the previous guru, was imprisoned and killed by Muslims. Singh introduced the worship of the Hindu goddess of death into Sikhism, as well as a ritual baptism in which the initiate was sprinkled with holy water

into which a dagger had been dipped. He also developed an elite warrior class of Sikhs known as Singhs (lions). The Singhs class was open to all adult male Sikhs. To distinguish them from other men, they had long hair on their heads and faces, wore turbans, and carried steel daggers. The Singhs were not allowed to use wine, tobacco, or any other form of stimulant but were encouraged to eat meat. This warrior class among the Sikhs is the basis for much of Sikhism today.

The British used the warrior Sikhs as soldiers and policemen when they governed India. With the independence of India in 1947, the Sikhs lost their privileged status. They remain highly regionalized today, with over 80 percent of the 16 million Sikhs concentrated in the Punjab, and have diffused but little beyond the boundaries of India (Figure 10–17). There were important Sikh communities in Pakistan before 1947, when the Indian subcontinent was divided into Pakistan and India on the basis of religion. Pakistan is primarily Islamic while India is Hindu, and the Sikhs' long tradition of hostility to Islam caused them to migrate to India.

Sikh Ecology and Landscape

The most sacred site in the Sikh world is the Golden Temple of Amritsar where the sacred book, the Granth, is kept. Faithful Sikhs attempt to visit the golden temple of Amritsar at least once during their lifetime, even though the founder Nanak specifically prohibited pilgrimage. Sikh militancy leads to constant demands for separatism from India and the creation of an independent state. Conflict with the Indian authorities peaked in the 1980s, when the Indian leader, Indira Gandhi, was assassinated by a supporter of the Sikhs' separatist movement. Militant Sikhs used the golden temple at Amritsar as a site for storing weapons, and in 1986 the Indian military raided the temple and killed hundreds of Sikhs. Although the death of the Sikhs was tantamount to martyrdom for them, the violation of the sacred space of the holy Amritsar temple left a burning resentment among the Sikhs of the Punjab. Today the Sikhs continue to demand independence (see Figure 10–17)

Boston, Massachusetts. Sikhs are recognizable by their distinctive dress styles, including their turbans, even when they migrate to new areas.

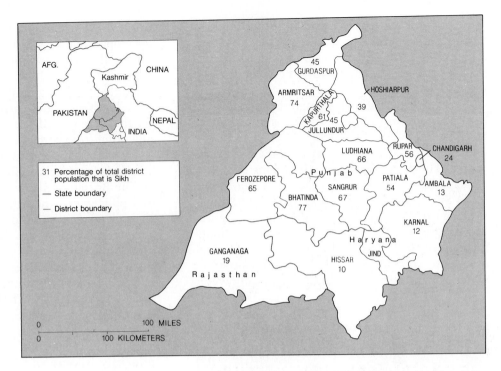

FIGURE 10–17
Distribution of Sikhs in India The Sikhs form the majority in most of the Punjab, leading Sikhs to seek political recognition for it.

The landscape features associated with the Sikhs reflect their willingness to adopt Western economic practices and their construction of temples known as Gurdwas. These temples are the focus of the Sikh community, whether in India or in major cities of Canada, the United States, or England where they have migrated. The adoption of Western economic practices has allowed the Punjab region of the Sikhs to become one of the most economically advanced regions of India, and overseas Sikh communities are generally prosperous as well.

The Amritsar temple in India is the most holy site for Sikhs.

Buddhism

A fourth branch of Hinduism is Buddhism. Buddhism's founder Guatama, is usually listed as having lived between 560 and 480 B.C., but these dates are open to question. Guatama was born a Hindu prince and raised in wealth. He became Buddha, literally the "enlightened one," when he perceived that the path of salvation did not come from asceticism or other extremes, but the middle way, which is based on the *Four Noble Truths:*

1. Life is full of suffering.
2. The cause of suffering is desire.
3. Pain and suffering end with the cessation of desire.
4. Right views and right conduct (honesty, noninjury of any creature, and forgiveness of any enemy) lead to an end of suffering and pain.

Buddha's new teachings did not include reverence for gods, worship as such, sacrifices, or division into castes.

After the death of Buddha disputes arose concerning his actual teachings and the leadership of the disciples. Over time this has caused the religion to divide into numerous branches, of which two—*Theravada* and *Mahayana*—are the most important. Theravada is found primarily in Sri Lanka and Southeast Asia and is known as the "lesser vehicle" (or *Hinayana* by its opponents) because it requires redemption by personal actions and the difficulty of this redemption dictates that few will gain salvation.

Mahayana is primarily found in East Asia, in Japan and China. It is based on teachings by subsequent followers of Buddha who maintained that he was but one of many Buddhas who brought teachings to the earth. These Buddhas were divine beings who became men because of their pity for human suffering and their desire to teach humans how to overcome the sufferings of

this life. Theravada introduced the worship of deities and polytheism as well as their idea that some followers of Buddha lived such holy lives that their good deeds could be used by others to gain salvation. The resulting belief system is known as the "greater vehicle" because more will gain salvation.

The Expansion of Buddhism

Buddhism was basically an Indian religion until the Indian emperor Asoka (third century B.C.) adopted it as a state religion and sent missionaries to surrounding countries, including Thailand, Sri Lanka, and Kampuchea (Figure 10–18). Interactions with other people, particularly Alexander the Great in India, brought Buddhism into contact with Greek art, which has greatly influenced the Buddha figure as can be seen in sculptures in Buddhist countries. Buddhism expanded into China along trade routes in the first century A.D., and from China it diffused into Korea and Japan and southward into Vietnam.

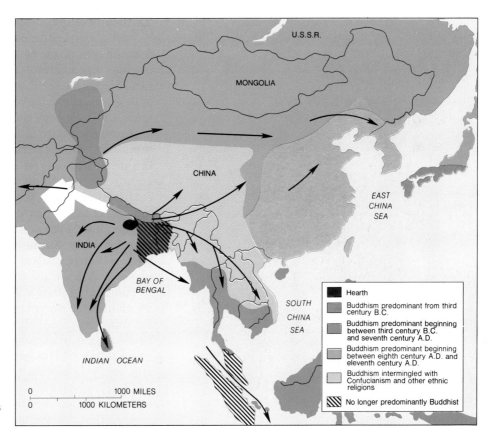

FIGURE 10–18
Diffusion of Buddhism
Buddhism diffused to southeast and East Asia and much of India from its hearth in East India.

Chinese missionaries carried Mahayana Buddhism to Tibet and Mongolia, where major modifications occurred, including reliance on magical phrases to deal with the unknown, the use of a prayer wheel, and the emergence of a priestly class known as lama ("the superior one"). The head of the lamas is the Dalai Lama, the ruler of Tibet. The Dalai Lama is viewed as a divine being who is reborn after his death. When one Dalai Lama dies, an extensive search for his replacement begins. A group of monks scours Tibet for a child who seems to have the qualities and characteristics of the Dalai Lama, and when the group finds such a child, the boy begins a period of training to prepare him to lead the nation. The dominance of the Dalai Lama ended in 1950 when the Chinese government invaded Tibet and the Dalai Lama fled. Riots in 1988 in Tibet by Buddhists who demanded the return of the Dalai Lama led to the death of several lamas and reinforced the continuing association between religion and politics.

Buddha statue from Taiwan.

Today's distribution of Buddhism (see Figure 10–18) reflects the expansion, diffusion, and evolution of the religion over the last 2,500 years. It has essentially disappeared in India, its hearth, while in China and parts of Southeast Asia communist governments have either downplayed its importance, persecuted its leaders and practitioners, or harassed its followers. Buddhism is still estimated to be the fourth largest religion in the world, with over 300 million adherents.

Shintoism

Shintoism, the loosely organized native religion of the Japanese people, embraces a wide variety of beliefs and practices. Shintoism is essentially an ethnic religion similar to Hinduism. In its most basic sense, Shintoism is a fervent religious form of Japanese patriotism. A geographically based religion, its mythology describes the formation of Japan as a land superior to all other lands, and its shrines commemorate great heros and events in the history of the country. Historically, Shintoism has taught that the emperors were literally the descendants of the sun goddess. Shintoism also invokes a worshipful attitude towards the beauties of Japan, particularly its mountains; in a form of pilgrimage, the Japanese travel to sacred mountains and trek between them. The most famous Japanese mountain is Fuji, but a number of other mountains are also recognized as sacred. One of the pilgrimages even requires a 1,000-day walk among the mountains; in practice, this pilgrimage is divided into ten 100-day pilgrimages over ten years. Only a few Japanese, all of them monks, have ever completed these extended pilgrimages. The Shinto religion is also associated with a warrior class known as Samurai that played an important role in Japanese history.

Fundamentalists still practice the fervent patriotism of pre-twentieth century Shintoism, but for most Japanese, Shintoism is a broadly based belief system that supports Buddhism as the primary religion. The geographic impact of Shintoism today is manifest in the ethnocentricity of the Japanese people, and their reverence for their island and its mountains, and in the Shinto shrines that dot the landscape.

Shinto shrine in Japan. For a small donation an individual receives a printed prayer to leave at the shrine.

Eastern Folk Religions

In addition to the three major ideologies just discussed, there also developed in East Asia an amalgam of beliefs sometimes referred to as *Chinese folk religion*, the fifth largest recognized belief system in the world. Based on a combination of Buddhism and the teachings of Confucius and Tao, it was the dominant belief system in China prior to the Communist Revolution of 1949.

Chinese folk religion could exist because Buddhism, unlike other universalizing religions, allowed its converts to continue to practice their native religions. A Buddhist can also be a Christian, a Jew, or member of any other religion. Consequently, the ideas and teachings of such philosophers as Confucius and Tao have merged with Buddhism to create the distinctive Chinese folk religion.

Both Confucianism and Taoism grew out of the long history of Chinese culture and civilization.

Teachings of the two philosophers reflect the ideals of the Chinese people: polytheism, *Yin* and *Yang*, divination, and filial piety and ancestor worship. Polytheism in China began as a form of animism associated with beneficial and negative spirits. Yin and Yang are ancient Chinese concepts that explain the nature of the universe. Yin is the negative force; in nature, it is found in darkness, coolness, femaleness, the moon, shadows, and the earth itself. Yang is the positive force; in nature it is found in lightness, brightness, warmth, maleness, dryness, and the sun. Neither is considered better than the other, and with the exception of a few things (such as the sun and the moon, which are specifically Yang and Yin, respectively), all elements influencing humans incorporate both Yin and Yang forces. Harmony between these two forces leads to a peaceful life.

Filial piety and ancestor worship have been characteristics of the Chinese people throughout their history. Unlike Western civilizations, the Chinese, it appears, have always venerated old age. Historically, the aged father, mother, grandfather, or grandmother dominated the Chinese home. Children were obliged to support the elderly, obey them, and give them a proper burial after their death. Descendants visited the grave site to maintain it and to offer sacrifices, leading Westerners to conclude that the Chinese practiced ancestor worship.

Symbol for Yin and Yang.

The Ming Tombs near Beijing demonstrate how geomancy was used. The tombs are built at the base of south facing mountains with a stream and road flowing south.

Divination, the prediction of the future through magic, is another long-standing characteristic of the Chinese people. Divination, polytheism, Yang and Yin, and filial piety can all be traced to Taoism. Taoism attempts to find harmony with nature, balancing the Yang and Yin forces. Through geomancy (a form of divination) the Taoist finds the proper site and situation relationships for human activities. A Tao priest (geomancer) could be called upon to define the best location for a grave site or tomb, a home, a business, or other activity. Over time the geomancers' prescriptions for locating activities became formalized: mountains to the north, water to the south, and a southern exposure to receive the sun. Elements of this aspect of Taoism are obvious in the location of the tombs for the emperors of China from the Ming dynasty. Even in communist China today, rural Chinese use geomancers to select locations.

The teachings of Confucius provide a system of ethics, a theory of government, and a set of personal and social goals that have influenced the Chinese for almost twenty-five centuries. Confucius was born in 1551 B.C. and died in 1479 B.C. He taught that the basis for life was a series of proper relationships between leader and follower, master and servant, father and son, and that the emperor was responsible for the well-being of the Chinese people. The emperor was superior to all other monarchs, both within and outside China, and his success was measured by absence of war, drought, famine, and general disorder. Peace and prosperity indicated that a particular emperor had been approved by heaven. Ruling officials other than the emperor were drawn from a scholarly elite who exercised all power and controlled the disposition of political favors, jobs, and potential wealth. This scholarly class, known as *Mandarins*, consisted of men who had studied the writings of Confucius for as long as twenty years. Following their studies they took a series of examinations, and the top 2 or 3 percent were admitted into the scholarly administrative order of the Mandarins.

The teachings of Confucius concerning interpersonal relationships and those of Tao concerning relationships to the environment were combined with many elements of Buddhism after it diffused to China. Communism officially attempted to destroy many of the aspects of this Chinese folk religion, especially the rule of the Mandarin class and the class structure that had developed upon the idea of interpersonal relationships. At the present time, many Chinese still practice elements of the Chinese folk religion, especially respect for the elderly and concern for proper relationships between individuals.

Secularization and Religion

The continued practice of many of the elements of the Chinese folk religion in China is an example of a *secularization* of religion. Communism disavows religion, yet in countries dominated by Communist leadership, the life of the people still reflects many religious beliefs. The political, economic, and social systems also perpetuate many elements of the religions on which they were based. Although the Russian Orthodox church in the Soviet Union has lost its position, influence, and control over its great cathedrals and lands, many residents of the areas dominated by the church still believe in the religion. Even those who do not, still observe important religious holidays such as Easter.

A similar pattern has emerged in China where official declarations that religion is a sign of superstition and illiteracy have not destroyed the traditional belief systems and their related activities. In the last decade most Communist countries have allowed greater freedom for religious worship, but the official role of the church is still downplayed. Nevertheless, the values and traditions of religion are often recognizable in the culture of the Communist world.

The Geographical Impact of Religion

Discussion of the major religions and their landscapes and regions illustrates the complex intertwining of religion with the culture of people. Religion affects the environment in several important ways including the following:

1. Designation of certain places as sacred.
2. Circulation and movement of population.
3. Perception of the environment.
4. Taboos that affect the crops or animals produced in an area.
5. Creation of specific landscapes and landscape features.
6. Political boundaries, as between India and Pakistan.

Sacred Space

The sacred space associated with each religion is important for a variety of reasons. Sacred spaces attract pilgrims, protect specific areas, and serve as symbols of the religion. Examples of sacred spaces in the Islamic world include Mecca and Medina, but Muslims recognize other sacred spaces including Jerusalem's Dome of the Rock (where they believe the prophet Abraham ascended into heaven) and the holy city of Qom in Iran. Hindus also recognize a variety of sacred sites. As in Islam, these are arranged hierarchically, from local shrines to temples, to regional or national shrines. Sacred landscape features such as the Ganges River, cities such as Varanisi, places in which miracles have occurred, or locations where an individual is venerated attract people from large areas. The Taj Mahal in Agra, India, built by one of the Islamic emperors of India, has become a place of worship for Hindus and is also a major tourist attraction. Sacred sites among

Christians include the regions in and around Israel (such as Bethlehem and Jerusalem), sites where miracles are reputed to have occurred (such as Lourdes in France), and famous churches (such as the Sistine Chapel or the Cathedral of Notre Dame).

Circulation (Pilgrimage)

The existence of sacred space has a major impact on the circulation of the members of a specific religion. The two million individuals who annually make the pilgrimage to Mecca and the millions of Christians who travel to Jerusalem, Lourdes or Rome are evidence of the impact of religion on circulation even when visitors do not come for reasons of religion. Some religions restrict pilgrimage to official believers; for example, non-Muslims are not allowed to enter Mecca and Medina. In other religions, notably Christianity, pilgrimage to sacred sites may involve an element of tourism. Visitors to Jerusalem may include devout Christians as well as those interested only in their Judeo-Christian hearth. The Mormon religion in the United States does not include any doctrines concerning pilgrimage, but over three million Mormons and non-Mormons alike visit Temple Square in Salt Lake City yearly because of its symbolic importance as the center of Mormondom. Many Catholics visit pilgrimage sites in Europe, the U.S. and elsewhere (Figure 10–19).

Perception of the Environment

Religion's effect on perception of the environment is intertwined with pilgrimages and sacred sites. The definition of a place as sacred reflects the perception of the group. At a more pervasive level, attitudes of a group toward the environment, animals, or plants affect the way in which they deal with the world. Cultural geographers are interested in the way people interact with the environment and deal with such natural hazards as earthquakes, hurricanes, and drought. Tribal residents of Africa or Latin America, for example, tend to believe that their gods are intermediaries between them and the environment, whereas in Protestant western Europe and North America

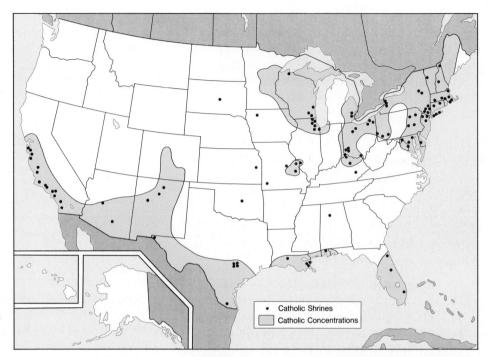

FIGURE 10–19 Catholic Pilgrimage Sites in the United States

there is widespread belief that human action can mitigate or minimize the impacts of hazards.

Taboos

Just as religion affects the way people view the environment, religions influence the crops or animals raised in a region. The Judaic and Islamic prohibition on pork is a clear example of this, as is the Islamic taboo on wine with its impact on vineyards. The use of wine for sacramental purposes in the classical world before Christ and the need of Christians for wine for the sacrament led to the spread of viticulture (grape production) in southern Europe, Latin America, and elsewhere. The unique involvement of religion in wine production is evident throughout western Europe, southern California and Latin America, where the introduction of vineyards was associated with missions and monasteries.

The Catholic emphasis on a day of fasting (Friday) in which meat was at one time taboo has led to an emphasis on fishing in Catholic countries. Hundreds of vessels from Spain, Portugal, and France fished yearly on the northeastern

Vineyards along the south facing bank of the Rhine River in Germany.

shores of the United States and Canada long before any permanent European settlements were established on the continent.

The Cultural Landscape

The pervasive influence of religion on the geography of the world is especially evident in the landscape. It is possible to categorize the landscape impacts of religion into the following:

1. Structures
2. Categorization and organization of space
3. Distinctive dress types
4. Toponyms
5. Distinctive crop or livestock combinations

Structures associated with religion are numerous, ranging from small wayside shrines to imposing cathedrals. They include relic features such as the famous St. Basil's Cathedral in Moscow, which is now a museum, and functional features such as the Cathedral of Notre Dame in Paris, which is also a famous tourist attraction. In the United States, the congregational chapels of New England are architecturally simple, while Catholic churches tend to be more ornate.

The mosques of the Islamic religion found in the Islamic core from Egypt through Iran have a distinctive design. Variants of the mosque that evolved in periphal areas where Islam diffused have domes and minarets whose stylized design is quite different from those in the traditional mosques found in the Middle East. Temples associated with the Hindu religion are often ornate, filled with statues of various gods. These temples have diffused with the religion and are found in such disparate places as the Caribbean Islands, Canada, the United States, and the United Kingdom. The presence of a Hindu temple in a generally Protestant or Catholic region of the United States emphasizes the increasing interconnection of the entire world and the importance of freedom of religious expression in the United States and Canada. (It should be noted that religious tolerance is a relatively recent phenomenon in the United States, however. The first Catholic church built in Worcester, Massachusetts, in the late 1700s, was burned by the Protestants of the community.)

It is possible to identify Catholic landscapes and Protestant landscapes in the United States on the basis of their architecture. Protestant

A Catholic church in the United States.

A Protestant church in New England.

Omayad Mosque, Damascus, Syria. Mosques differ from region to region but share enough similarities to be recognizable.

A Hindu temple in the United States illustrates the styles of southern India (on the left) versus those of the north (on the right) in one structure.

churches tend to be functional places for worship, while Catholic churches—viewed as the earthly abode of God—are much more elaborate and imposing. Signs advertising churches or calling sinners to repent are also a part of the landscape features derived from religion.

Typical Buddhist architecture is concentrated in East and Southeast Asia, but has diffused to North and South America and isolated locations in western Europe. The Buddhist temple provides devotees a place to worship statues of Buddha; in some cases, a Buddhist temple may be simply a shrine with a Buddha and places for leaving written prayers. Wayside shrines where flowers or other offerings can be left or prayers made are common in Buddhism, Hinduism, and Catholicism. They vary in size and importance, ranging from a small crucifix in southern Germany to elaborate pilgrimage shrines in such Catholic countries as France or Yugoslavia.

Categorization and Organization of Space

The presence of cathedrals, temples, or pilgrimage sites reflects the role of religion in categorizing and organizing space. Designation of a space as sacred, holy, mystical, or miraculous affects the role that it plays in the geography of a place or region. The pilgrimage destinations of Mecca, Jerusalem, or Lourdes all involve the organization

Signs related to religion are an important part of the cultural landscape.

of space in a highly formal fashion. This organization begins with the sacred sites themselves and the placement of specific features within them, extends to ancillary activities associated with the pilgrims around the site, and includes such functions as rooming places for tourists, purveyors of religious paraphernalia, and all the service-related activities associated with a major movement of population (Figure 10–20).

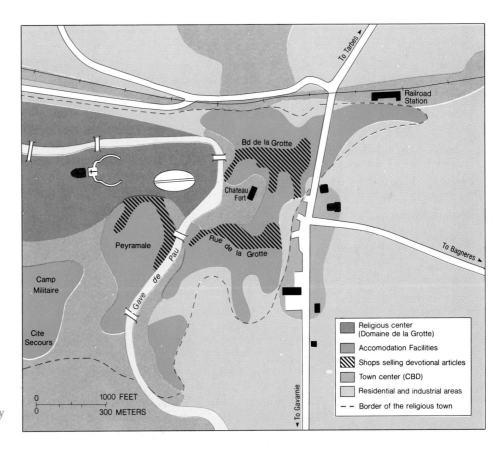

FIGURE 10–20 Activity Space at Lourdes, France

The layout of cities provides another example of how religion affects the organization of space. The Islamic city has been previously mentioned, and the Mormon cities of the Intermountain West, patterned after the City of Zion plan devised by the founder of Mormonism, Joseph Smith, are another example. Primary characteristics of the City of Zion plan are a rigid geometric format with orientation to the cardinal directions. Applied uniformly by the Mormon settlers of the Intermountain West, the resultant pattern emphasizes wide streets, large blocks, and open space (Figure 10–21). This plan has allowed the automobile to be adopted without a great deal of destruction of buildings in Salt Lake City, but in small cities it results in an extensive amount of wasted space since the streets are as wide in the village of 500 people as in a city of 500,000.

Religion also affects the organization of space in the American Southwest. The diffusion of Catholic missionaries to California was associated with a distinctive settlement type called the

mission. Organization of mission space centered on the church, with all of the basic elements specified (Figure 10–22). The New England village also illustrates the combination of religion and civic activity: the dominance of the church during the colonial period led to a spatial organization focused on the church and a common grazing area in the center of the community. As the community grew, satellite communities were established to ensure that children of the original settlers had both access to land and proximity to their church for religious services. Place names such as North Chatham, South Chatham, East Chatham, and so forth indicate the location of these satellite communities. In French Canada (today's Quebec) land division partially reflected religion, as the settlements established along the rivers for transportation and communication were characterized by individual plots of land given out as long narrow strips. This system allowed the settlers to be close to the church for attendance at Mass. The centrality of the church

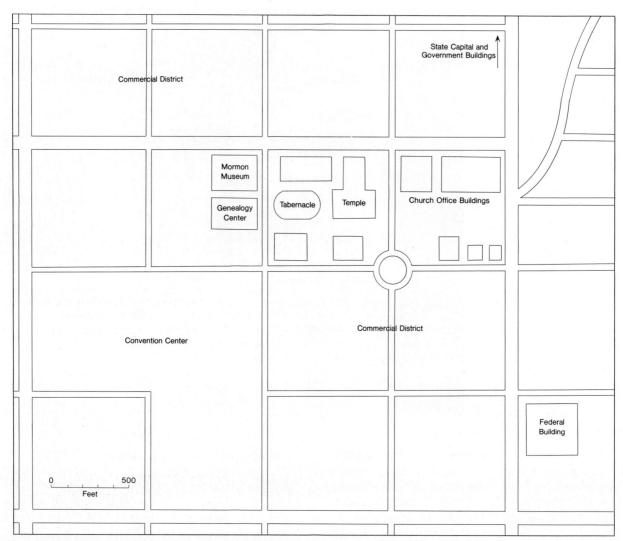

FIGURE 10–21 Salt Lake City: A Mormon City Mormon towns are laid out in a rectangular grid with uniform blocks and lots. Construction of a convention center in Salt Lake City required combining three blocks into one.

in the resulting landscape is obvious to even casual observers of maps of the St. Lawrence River area, even though religion was only one reason for the land division system.

Another example of the impact of religion on spatial organization is found in cemeteries—the landscapes of the dead. Cemeteries reflect the specific views of an individual's religion regarding death. Structures such as the pyramids of Egypt or the Taj Mahal of India, for example, are landscape elements that reflect the veneration of the deceased. For some, it is important to reconstruct a miniature landscape for the deceased. For others,

the passage from this life leaves little evidence as the dead are cremated. Eastern religions (Hinduism, Buddhism, Shintoism) generally practice cremation, after which the ashes are either scattered or kept in an urn in the family home. Christians, Muslims, and adherents of the Chinese folk religion all practice burial of their dead. In so doing they create, literally, cities of the dead that may be as imposing as the homes of the inhabitants of an area. Cemeteries are an important element of open space in large American cities and often protect vegetation that would otherwise be destroyed. Christian cemetery landscapes range

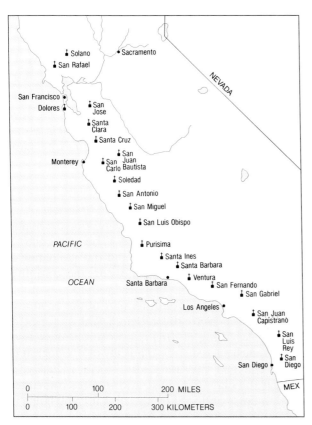

FIGURE 10–22 The California Mission System
Many of the sites initially selected for missions by the Spanish are major cities in California today.

from mausoleums to simple headstones, but each types of structure is symbolic of the religion.

Religion and Dress

In some areas of the world, religion has a major impact on the way people dress. The most noted example is in strict Islamic countries, which require women to wear a veil and a hair cover when they are in public. Where Theravada Buddhism is practiced, the mandate that each individual obtain his or her own salvation through righteousness results in a high proportion of the young men spending at least part of their late teen years as monks. These monks, who dress in saffron-colored robes, travel through the community obtaining food from the Buddhist believers.

Cemetery in San Juan, Puerto Rico. Cemeteries are miniatures of the cities occupied by the living.

The Taj Mahal in Agra, India is one of the best known examples of how religion affects architecture. The use of the dome and minarets (towers) reveals that it was built by a Muslim ruler. Such famous shrines often become destinations for pilgrims and tourists.

Teheran, Iran. Belief about wearing the veil and head covering vary among various Muslim groups.

The distinctive turbans and daggers of the Sikhs of India or the dark conservative dresses and suits of the Old Order Amish or Mennonite communities of America are other examples of the continued role of religion in dress style.

Dress styles do not always originate with a religion, but devotees may still wear specific garments reflecting their adherence to the faith. Flowing robes associated with men in the Middle East, for example, are not mandated by religion, but many Muslims view the traditional dress as a symbol of their opposition to Western influences in the world of Islam.

Conflicts over clothing style occur frequently when tourists travel to a region with different religious values or interpretations. Topless beaches in the strongly Catholic countries around the Mediterranean have resulted in the uneasy juxtaposition of two contrasting groups: although tourists are allowed to wear the skimpiest of clothing while at the beach, visiting the city center and mingling with the local residents in the same clothing can result in arrest. Similar conflicts occur in other areas when tourists from one region fail to recognize the role dress plays in a culture.

Toponyms

Place-names are another example of the impact of religion on landscape. Toponyms reflecting religious influence are widespread. Examples from the Western world include the vast array of towns or cities with the prefix Saint in their name: St. Louis, St. George, and the Spanish variants such as San Francisco or San Juan are only a few examples. Catholic-settled areas of North and South America are particularly likely to have religious toponyms, often reflecting their establishment on a particular saint's day. In Protestant-dominated areas, the place-names are apt to commemorate historic figures.

Agricultural and Other Economic Practices

The impact of religion on the economy of a region has already been mentioned. The presence of vineyards, the eating of fish, and the avoidance of pork all result in distinctive landscape combinations. The production of sacred religious symbols such as crucifixes is another part of the economy related directly to the religion. The practice of *geophagy* (eating of dirt) is also sometimes associated with sacred sites. The clay of one region in Guatemala is the basis for a local industry in which the clay is molded into squares imprinted with a cross and sold by the tens of thousands to pilgrims in the sacred sites around the mine. Eating of these clay squares is largely confined to pregnant women.

The emphasis on sheep in the Middle East and the requirement that sheep be slaughtered in a ritual manner have led to a major industry involving trade between Australia and New Zealand and Middle Eastern countries. Animals that were traditionally shipped live for ritual slaughter by the appropriate authorities are now slaughtered by their representatives before they are shipped. Similar examples of the impact of religion on an economy can be found in kosher food stores and vegetarian markets in Hindu neighborhoods.

Religion and Politics

Finally, religion has an impact on landscape in the area of politics. The impact of religion in the political sphere is felt from Ireland to Israel, and from Italy to Iran. The next chapter discusses the role of religion in politics in detail, so only a few examples will be given here. The city of Belfast, Ireland, is divided between Catholics and Protes-

For the practicing Jew it is essential that meat be purchased from a Kosher butcher. In Dijon, France this butcher indicates in both French and Hebrew that his shop is Kosher.

tants. The continued existence of Palestinian refugees, the occupation of the West Bank of the Jordan by the Israelis, and the issue of the holy sites of Jerusalem are the basis of spatial organization and landscape features. The Vatican's status as an independent city state within Rome is a classic example of religion in politics. The dominance of fundamentalist Shi'ites in Iran in recent years is producing a major change in the landscape. Their reemphasis on traditional values (including dress) and opposition to anything of Western origin affect the landscape. The impact of religion on politics in the United States is widespread, even at the national level, where discussion of a candidate's religion is unofficially frowned upon, but remains a major factor in elections.

Conclusion

The complex role religion plays in culture is apparent to even the casual observer. The impact of religion on its hearth area, the subsequent migration of adherents or diffusion of beliefs, and the associated culture history and cultural ecology combine to create the landscapes that typify much of the world. The presence of religion in functional or relic landscape features and toponyms is a major theme in the story that cultural geographers read in the face of the earth. The continued importance of sacred sites, pilgrimages, and religious-based conflict makes religion one of the most significant elements in explaining human activity on the earth.

The relation of religion to landscape, economy, political activity, dress, and life reflects the pervasive role it plays in human culture. Even in those societies that view themselves as secular or atheistic, religion plays a major role. Relic features such as cathedrals, toponyms, interpersonal relationships and language remain as remnants of the religion that was associated with the culture before the adoption of modern life-styles or secular or atheistic beliefs. The continued importance of religion in the world makes it an area of interest for research and study not only by cultural geographers, but other social scientists as well.

QUESTIONS

1. Compare and contrast ethnic and universalizing religions
2. What is the difference between tribal ethnic religions and compound ethnic systems?
3. Describe the history of Judaism, including its diffusion.
4. Identify and describe the major universalizing religions of the world.
5. Describe the diffusion of western Christianity.
6. Discuss the hearth, diffusion routes, and resulting regions dominated by Islam.
7. Compare and contrast the beliefs and distribution of Shi'ites and Sunnites.
8. Describe the Muslim landscape.
9. What impact has religion had on the Indian subcontinent?
10. Identify and discuss four major geographical impacts of religion.

SUGGESTED READINGS

Al Faruqi, Isma'il R., and Lois Lamaya' Al Faruqi. *The Cultural Atlas of Islam.* New York: Macmillan, 1986.

Al Faruqi, Isma'il R., and David E. Sopher. *Historical Atlas of the Religions of the World.* New York: Macmillan, 1974.

Bapat, P. V., ed. *2500 Years of Buddhism.* Delhi: Government of India Ministry on Information and Broadcasting, 1959.

Barraclough, Geoffrey, ed. *The Times Concise Atlas of World History.* Maplewood, N.J.: Hammond, Inc., 1982.

Bennett, Merrill K. *The World's Foods.* New York: Harper and Bros., 1954.

Bhardwaj, Surinder M. *Hindu Places of Pilgrimage in India.* Berkeley: University of California Press, 1973.

Bigsby, C. W. E., ed. *Superculture: American Popular Culture and Europe.* Bowling Green, Ohio: Bowling Green Popular Press, 1975.

Blouet, Brian W. "Sir Halford Macinder as British High Commissioner to South Russia, 1919–1920." *Geographical Journal* 142, (July 1976): 228–36.

Crowley, William K. "Old Order Amish Settlement: Diffusion and Growth." *Annals of the Association of American Geographers* 68 (June 1978).

Doeppers, Daniel. "The Evolution of the Geography of Religious Adherence in the Philippines before 1898." *Journal of Historical Geography* 2 (1976): 95–110.

Donkin, R. A. "The Cistercian Order and the Settlement of Northern England." *Geographical Review* 59 (1969): 403–16.

Doughty, Robin W. "Environmental Theology: Trends and Prospects in Christian Thought." *Progress in Human Geography* 5 (1981): 234–48.

Eliade, Mircea. *Patterns in Comparative Religion.* Cleveland: World Publishing Co., 1968.

Fickeler, Paul. "Fundamental Questions in the Geography of Religions." *Readings in Cultural Geography,* Philip L. Wagner and Marvin W. Mikesell, eds. Chicago: University of Chicago Press, 1962.

Francaviglia, Richard V. *The Mormon Landscape.* New York: AMS Press, 1978.

Gaustad, E. S. *Historical Atlas of Religion in America.* New York: Harper & Row, 1962.

Gay, John D. *The Geography of Religion in England.* London: Gerald Duckworth, 1971.

Glacken, Clarence J. *Traces on the Rhodian Shore.* Berkeley: University of California Press, 1967.

Halvorson, Peter L., and William M. Newman. *Atlas of Religious Change in America, 1952–1971.* Washington, D.C.: Glenmary Research Center, 1978.

———. *Patterns in Pluralism: A Portrait of American Religion.* Washington, D.C.: Glenmary Research Center, 1980.

Hardon, John A. *Religions of the World,* 2 vols. Garden City, N.Y.: Image Books, 1968.

Heatwole, Charles A. "Exploring the Geography of America's Religious Denominations: A Presbyterian Example." *Journal of Geography* 76 (1977): 99–104.

———. "Sectarian Ideology and Church Architecture." *Geographical Review* 79 (January 1989): 63.

Hiller, Carl E. *Caves to Cathedrals: Architecture of the World's Great Religions.* Boston: Little, Brown, 1974.

Hostetler, John A., *Amish Society.* Baltimore: John Hopkins University Press, 1980.

Ismael, Tareq Y. *Iraq and Iran: Roots of Conflict.* Syracuse, N.Y.: Syracuse University Press, 1982.

Jackson, Richard H. "Mormon Perception and Settlement." *Annals of the Association of American Geographers.* 68 (September 1978): 317–34.

Jackson, Richard H. and Robert L. Layton. "The Mormon Village: Analysis of a Settlement Type." *The Professional Geographer* 28 (May 1976): 136–41.

Jackson, Richard H., and Roger Henrie. "Perception of Sacred Space." *Journal of Cultural Geography* 3 (Spring/Summer 1983): 94–107.

Jordan, Terry G. "Forest Folk, Prairie Folk: Rural Religious Cultures in North Texas." *Southwestern Historical Quarterly* 80 (1976): 135–62.

———. *Texas Graveyards: A Cultural Legacy.* Austin: University of Texas Press, 1982.

Lehr, John C. "The Landscape of Ukrainian Settlement in the Canadian West." *Great Plains Quarterly* 2 (1 Spring 1982): 94–105.

Levine, Gregory J. "On the Geography of Religion." *Transactions of the Institute of British Geographers* N.S. 11 (1987): 428–40.

Ling, Trevor. *A History of Religion East and West.* London: Macmillan, 1968.

Livingstone, David N. "Environmental Theology: Prospect in Retrospect." *Progress in Human Geography* 7 (1983): 133–40.

Lodrick, Deryck O. *Sacred Cows, Sacred Places: Origins and Survivals of Animal Homes in India.* Berkeley and Los Angeles: University of California Press, 1981.

Manyo, Joseph T. "Italian-American Yard Shrines." *Journal of Cultural Geography* 4 (1983): 119–25.

Meinig, Donald W. "The Mormon Culture Region: Strategies and Patterns in the Geography of the American West, 1847–1964." *Annals of the Association of American Geographers.* 55 (1965): 191–220.

Newman, William M., and Peter L. Havorson. "American Jews: Patterns of Geographic Distribution and Change, 1952–1971." *Journal for the Scientific Study of Religion* 18 (1979): 183–93.

Nolan, Mary Lee. "Irish Pilgrimage: The Different Tradition." *Annuals of the Association of American Geographers* 73 (1983): 421–38.

_____. "Types of Contemporary Western European Pilgrimage Places." Pittsburgh, Pa.: Conference on Pilgrimages: The Human Quest, 1981.

Quinn, Bernard, Herman Anderson, Martin Bradley, Paul Geotting, and Peggy Shriver. *Churches and Church Membership in the U.S. 1980.* Atlanta: Glenmary Research Center, 1982.

Rinschede, Gisbert. "The Pilgrimage Town of Lourdes." *Journal of Cultural Geography* (Fall/Winter 1986): 21–34.

Shilhav, Yosseph. "Principles for the Location of Synagogues: Symbolism and Functionalism in a Spatial Context." *Professional Geographer* 35 (1983): 324–29.

Shortridge, James R. "The Pattern of American Catholicism." *Journal of Geography* 77 (1978): 56–60.

_____. "Patterns of Religion in the United States." *Geographical Review* 66 (1976): 420–34.

Simoons, Frederick J. *Eat Not This Flesh: Food Avoidances in the Old World.* Madison, Wis.: University of Wisconsin Press, 1961.

_____. *Eat Not This Flesh.* Madison, Wis.: University of Wisconsin Press, 1967.

Simpson-Housely, Paul. "Hutterian Religious Ideology, Environmental Perception, and Attitudes toward Agriculture." *Journal of Geography* 77 (1978): 145–48.

Singh, Rana P. B. "Distribution of Castes and Search for a New Theory of Caste Ranking: Case of the Saran Plain." *National Geographical Journal of India* 21 (March 1975): 20–46.

Sopher, David E. "Geography and Religions." *Progress in Human Geography* 5 (1981): 510–24.

_____. *The Geography of Religions.* Englewood Cliffs, N.J.: Prentice-Hall, 1967.

Stanislawski, Dan. "Dionysus Westward: Early Religion and the Economic Geography of Wine." *Geographical Review* 65 (1975): 427–44

Tanaka, Hiroshi. "Geographical Expression of Buddhist Pilgrim Places on Shikoku Island, Japan." *Canadian Geographer* 21 (Summer 1977): 116–24.

Thompson, Jan, and Mel Thompson. *The R.E. Atlas: World Religions in Maps and Notes.* London: Edward W. Arnold, 1986.

Topping, Gary. "Religion in the West." *Journal of American Culture* 3 (Summer 1980): 330–50.

Tuan, Yi-Fu. "Discrepancies between Environmental Attitude and Behavior: Examples from Europe and China." *Canadian Geographer* 12 (1968): 176–91.

Tuan, Yi-Fu. "Sacred Space: Explorations of an Idea," in Karl W. Butzer pp. 84–99. ed., *Dimensions of Human Geography: Essays on Some Familiar and Neglected Themes,* Chicago: University of Chicago, Department of Geography, Research Paper No. 186, 1978.

Tweedie, Stephen W. "Viewing the Bible Belt." *Journal of Popular Culture* 11 (1978): 865–76.

Vogeler, Ingolf. "The Roman Catholic Culture Region of Central Minnesota." *Pioneer America* 8 (1976): 71–83.

White, Lynn, Jr. "The Historical Roots of Our Ecologic Crisis." *Science* 155 (March 10, 1967): 1203–7.

The Human World

The final section of this book examines the issues created by the complex interplay of culture and environment. Two major categories of issues are currently changing the world's cultural geography. First are issues related to cultural variables. They include world urbanization with its related problems and opportunities, the diffusion of technology from the more industrialized countries to those that are less so, the diffusion of specific cultural traits such as religion and language, and the resulting cultural convergence that is making the world both more alike and more different.

The second category includes issues related to the ongoing human modification of the earth's physical environment. Environmental degradation is recognizable in the loss of agricultural land, the destruction of the tropical rain forest, the spread of deserts, and changes in the global atmosphere that could lead to climatic change.

The combination of physical and cultural issues directly threatens the continued existence of some groups of people and their life-styles and indirectly threatens the entire human population. Understanding the issues that affect the world's cultural geography may allow us to prevent the destructive effects of our own actions.

413

Culture, Politics, and the Organization of Space

"The patterns of a country's [cultural] geography are shaped not only by the physical factors that govern land use or routeways, but also by the political conditions under which settlement and development take place."
J. H. PATERSON

THEMES

The organization of personal and local space.
The state as a formal organization
Types of boundaries and the shape of states.
Types and characteristics of states.
Conflict and fragmentation.

CONCEPTS

Territoriality
State
Nation
Nation-state
Ecumene
Centripetal forces
Centrifugal forces
Sectionalism
Polyethnicism
Exclave
Irredentism
Nationalism
Frontiers
Demarcation
Antecedent
Superimposed
International rivers
Stateless nation
Insurgent state

Introduction

One of the most familiar examples of culture regions is a map showing the political units of the world. Most people recognize that each political region is related to the culture of its inhabitants; that Japan, for example, is culturally distinct from China or Poland.

Examination of the relationship between culture and political divisions helps explain the complex variation in the world's cultural geography. Recognition of individual countries indicates something of the culture, economy, technological level, language, religion, landscapes, or other cultural variables found there. The boundary between the United States and Mexico illustrates the cultural variation implied by political boundaries. This boundary divides a region whose physical characteristics are identical, yet the culture, life-styles, economies, and cultural landscapes of the two countries are quite different. The United States represents the industrial world with its distinctive landscapes, while Mexico is a part of the traditional world. Thus the boundary is more than a simple political line, for it marks the change from one cultural region to another.

Political boundaries, which represent formal agreements about the organization and division of the earth's surface, may begin as only attempts to define the space claimed by a country, but over time they may become cultural boundaries. This is not always the case, however. Political units often reflect the underlying culture, but in many instances their boundaries cut across cultural boundaries of language, religion, or other cultural variables.

The Organization of Space

Space—its use, organization, and meaning—reflects the culture of the occupants of a particular place, making the study of spatial relationships central to cultural geography. Organization of space is manifest in any cultural group from the personal to the international. Human concern with space is manifest in the concept of *territoriality*, which refers to the need of animals for a specific amount of territory. Some observers believe the behavior of humans with respect to territory is analogous to animal territorial behavior. Animals mark the boundaries of their territory in a variety of ways; for example, bears leave claw marks on trees and elephants break down bamboo or other small trees. Humans also establish boundaries of many types to define their territory, but boundaries are not the only evidence of people's concern with space.

Personal and Local Spatial Organization

The importance of space in cultural geography is easily demonstrated. Consider the extent to which space and spatial relationships affect your activities on a typical day. Each individual, for example, has a zone of *personal space* in which he or she is uncomfortable if others intrude (Figure 11–1). This personal space varies from culture to culture and situation to situation. Personal space in the American culture is between two and three

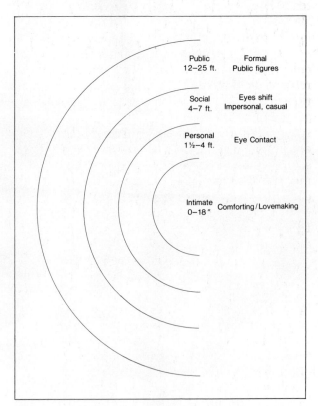

FIGURE 11–1 Diagram of Personal Space
Hypothesized model of individual spatial zones.

feet when standing talking face to face. A subset of that personal space is *intimate space,* a relaxation of the boundaries of the personal territory when dealing with a close friend or loved one.

Beyond the zone of personal space is a zone known as *impersonal space.* Impersonal space extends from some three feet to ten feet and is associated with impersonal relationships. Note the distance your professor stands from the class as he or she is lecturing. Normally it will be from three to ten feet. Classroom organization illustrates the importance of personal space: chairs and desks are arranged so that individuals are not touching one another and have one to three feet of space between them, while the instructor is three to ten feet from the students.

Individuals act in a territorial way concerning their personal space and will mark even temporary space as their own. For example, they leave books on a chair at the library or a coat over the back of a desk in a classroom. Children learn the territoriality implicit in personal space through observation of others in their culture or through formal instruction on acceptable behavior.

The use of personal space is specific to each culture and represents underlying cultural values relating to individualism, privacy, social interaction, and public display of affection. The personal space of an American is similar to the personal space requirements of persons in the United Kingdom, Germany, and northern Europe. Moving south and east from France, however, the demands for personal space are much less. The crowding that Americans complain about in the Mediterranean region of Europe, North Africa, or Latin America reflects cultural differences in the definition of personal space.

Extensions of the concepts of personal space and territoriality help to explain the organization of space. Spatial organization is found at every level and may be *formal* or *informal.* Formal organization of space includes the division of a suburb into specific plots of land, the organization of a house into specific rooms, and the division of countries by political boundaries.

Informal spatial organization is so called because it is not formalized within the entire society. Examples of informal space include recognition by family members that a certain chair is

The barrier between Catholic and Protestant dominated regions in Belfast.

claimed by one individual and others use it at their own risk, the territories of rival gangs, areas within a city that are unsafe for strangers, or informal subdivisions of cities, states, or neighborhoods (Figure 11–2). Sometimes such informal space becomes formal, as has occurred with the boundary between Catholic- and Protestant-occupied areas in Belfast, Northern Ireland. The original territory reflected an informal recognition of which areas were predominantly Catholic and which were predominantly Protestant. Ultimately, a wall was constructed to demarcate the territory of each side.

Formal spatial organization affects much of human activity. A typical middle-class home in the American culture is organized into activity spaces that appear to be mutually exclusive: the kitchen is for preparing food, the dining room is for consuming food, the living room is for entertaining, the family room is for entertainment of the family, the bathroom is for bathing, the bed-

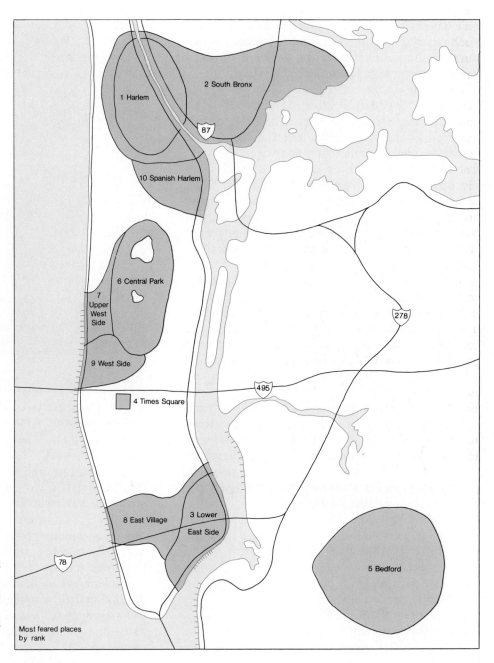

FIGURE 11–2 The Most Feared (Unsafe) Areas of New York City Perception that certain areas of a city are unsafe creates informal regions.

room is for sleeping, and the study or library is for studying. This formal organization breaks down upon closer inspection as certain types of activities cease to be confined to specific rooms. Many people read in bathrooms or study in their bedrooms. However, few Americans eat in bathrooms or sleep in kitchens unless they occupy small homes or apartments or must crowd several families into one apartment or home because they are poor.

The homes of other societies, such as the Japanese, exhibit a much different organization of space. Traditional Japanese homes are smaller, and rooms serve multiple purposes. The room in which sleeping mats are laid down at night will be used for a variety of other purposes during the daytime. Such differences represent cultural characteristics ranging from technology to attitudes about compartmentalization of activities and behavior.

Organization of space in communities also varies greatly. The internal spatial organization of communities in the Western world tends to be of three general types: a radial pattern focusing on a central point, a regular geometric grid, or an irregular pattern with little homogeneity in the division of blocks and lots (Figure 11–3). The radial pattern is found in Paris, where streets converge at the center. Space is ordered outward from this point, with the numbers increasing with distance from the center. This particular pattern of spatial organization is often confusing to Americans since names of streets that extend outward from the center may change at given points. For Americans used to organizing space in a rectangular grid, it takes some experience to recognize that the changes along the arterials extending outward from downtown Paris change with landmarks. Once it is recognized that there is an order to the spatial organization, learning that order is relatively simple.

The grid pattern arranges and assigns numbers to places with respect to a coordinate system, as in Mormon towns where all coordinates are given in the cardinal directions (e.g., 200 North and 300 West), or streets may have numerical designations along one axis and named along the other (e.g., 2000 East Fifth and 5000 North Cedar). In either case, this type of spatial organization is familiar to all who use it, is logical, and allows orientation and circulation within the space.

Cities outside the Western world do not necessarily follow this pattern of organizing space. Japanese cities are ordered with respect to when a structure was built rather than by its location. Thus to find a house, a visitor must first locate the district and then ask a resident for a specific house number since house number 1 may be adjacent to house 40 on one side and to house 22 on the other. Organization of space by time of construction is logical to the Japanese, but to Americans familiar with a rectangular grid pattern, it seems arbitrary and capricious. Whatever the system of spatial organization used in a city, occupants become familiar with it and develop a mental map that allows them to navigate within the city.

Territory is organized in similar fashion at every other level, from town to city to state to nation. Recognition of the organization of space and its importance to the individual and group helps to explain the actions of individuals and groups that affect spatial relationships. Actions

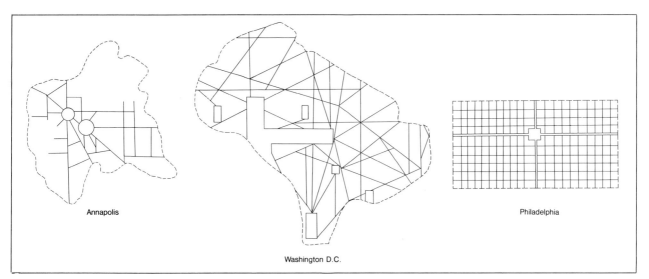

Annapolis

Washington D.C.

Philadelphia

FIGURE 11–3 Internal Spatial Organization of Communities Black lines show major streets; open spaces and squares are shown by circles and squares.

affecting or reflecting the organization of space are constantly changing and are associated with a variety of human activities including conflicts that reflect territoriality, migrations that reflect perception of the value of space, or designation of some space as more important than others, such as sacred space. Examination of the organization of space from the largest unit to the smallest will indicate the ways in which spatial organization affects the geography of the world.

Independent Countries: A Formal Organization

One of the most widely recognized forms of spatial organization is the designation of independent *countries*, which are formally known as states. A number of terms are used in referring to the countries of the world. To prevent confusion, it is important to distinguish the terms state, nation, and nation-state. All refer to territorial divisions of the world based on political recognition and are often used as if they were synonymous when referring to countries.

States

The term *state* refers both to a place and to a concept represented by symbols. Seven qualities must exist before a country technically qualifies as a state:

1. Possession of a specific territory whose boundaries are more or less generally recognized.
2. A permanent resident population.
3. A government that performs the functions needed or desired by the residents of the state, such as defense.
4. An organized economy supervised by the government. This may range from issuing money and supervising foreign trade to direct involvement in specific economic functions.
5. A formal, organized circulation system to move people, goods, and ideas within the territory of the state.

6. *Sovereignty*, which refers to the quality of being completely independent of direct external control.
7. Recognition by a significant portion of other states in the international community.

Places meeting all of these criteria are clearly states, and are normally simply called *countries*, a term referring to the commonly recognized independent political units of the world. Technically, the term "state" is not synonymous with "country," as some countries do not meet all the formal requirements of a state. The United Nations recognizes over 160 independent countries, all of which are technically states (see the Cultural Dimensions box). The United States, the Soviet Union, Japan, Mexico, or Canada are countries that are clearly states. Each has a politically organized body of people and occupies a distinctive territory with a government competent to establish law and maintain itself against intruders. Some countries such as the State of Bahrain, the State of Israel, the Spanish State, or the Independent State of Western Samoa formally recognize their technical status as states in their official long names. Others, such as the Union of Soviet Socialist Republics and the Federal Republic of Germany, do not.

The existence of a politically recognized entity does not always confer upon it the category of state even though they are popularly viewed as a country. Hong Kong is a political unit, but it is a colony of another country and state—the United Kingdom—and will become a province of China in 1997. Other widely recognized political units such as Vatican City, Andorra, San Marino, or other tiny countries are sometimes referred to as *micro states*. Technically they do not qualify as states since they normally rely on other larger powers to provide some of the functions associated with the state, in particular, defense and police, but they are independent countries.

Some areas that have claimed the status of countries or states are not recognized by all other states. The Republic of China (ROC or Taiwan) and the People's Republic of China (PRC) are a case in point. Both claim to be the legitimate representative of the Chinese people, and the PRC even claims Taiwan as part of its national

CULTURAL DIMENSIONS

Countries: A Problem of Definition

Defining the number of independent countries or states in the world is more complex than it appears.* Different organizations define "independent" in different ways, and the number of "countries" varies from 247 to 159 depending on whom you ask. The United Nations includes 159 "countries," but indicates that "not all countries are members of the U.N., but the U.N. has no standard policy on the number of independent states in the world." The U.S. State Department lists 167 countries in its *States of the World*, but since it was published, the State Department has recognized two more countries, St. Kitts and Nevis, and Brunei.

The National Geographic Society recognizes 174 "de facto" independent countries (which include the four "independent" tribal homelands in South Africa and Northern Cyprus) on its world map. The society does not claim that these are all actually officially independent. The U.S. Defense Mapping Agency recognizes 164 independent states, but the Central Intelligence Agency lists 247 "entities" in its factbook. How many sovereign independent states are there in the world? It depends on your definition.

---

*Adapted from Larry Thompson, "It's a Small World—Isn't It," *Washington Post*, November 1988.

---

territory. The United States recognized Taiwan as the official Chinese state and country and ignored the existence of the PRC from 1949 to 1971, maintaining that the most populous country in the world simply did not exist except as an appendage of the tiny Republic of China. The United States recognized both Chinas after 1971, but only in the last decade has it accepted the PRC as the legitimate representative of the great majority of Chinese.

Nearly all countries, (whether or not they possess all of the qualities of a state), are subdivided into smaller subdivisions for administrative or other functions. These subdivisions of countries are referred to by a variety of names, including states (as in the United States, Brazil, India, and Mexico). These territorial subdivisions of countries are found in most independent countries and are generally hierarchical. The hierarchy in the United States extends downward from states to counties, from counties to cities and towns. Other countries have similar hierarchies; for example, the Soviet Union has fifteen republics as the next level of organization. These republics in turn are subdivided into autonomous *oblasts, okrugs, krays,* and *rayons.*

The designation of a place as a country or state is an example of a functional region in geography.

The country exists because of the ties and interconnections that link its various subdivisions together into a common whole (see the Cultural Dimensions box).

Origin and Development of States or Countries

Historical examples of states or countries include the Roman Empire, the Greek Empire, the Chinese dynasties, the Incan and Aztec empires of the Americas, and a host of others. Each country or state has a cradle area where it began. This cradle, often called the *nuclear area*, is the location (hearth) from which the state's power diffused. In Italy, the seven hills of Rome were the center of the Roman world empire and are still the political capital of the state. In France, an island in the Seine River—the Ile de la Cité—was the cradle of the state concept and is still the location of much that is central to the French nation (Figure 11–4).

The differential distribution of the people of a state or country leads to one area becoming the core or *ecumene.* Normally found in the nuclear area, the ecumene contains the most dense population cluster, major transportation network, and major industrial, financial, and political areas

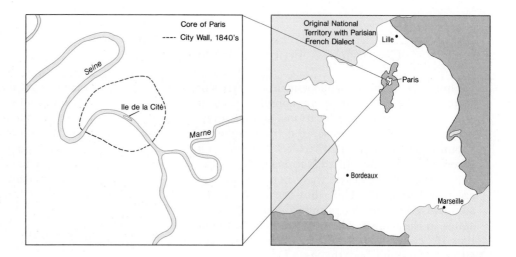

FIGURE 11–4 Ile de la Cité This map of the Ile de la Cité shows its relationship to Paris.

of the country. Normally the political capital is found in the ecumene, as is the case in Paris. Expansion outward from the nuclear area through the ecumene is associated with the development of the formal territory of the country and its boundaries.

Nation

Although the terms are often used synonymously, a *nation* is distinct from a state or country. The term "nation" is applied to a group of people, as the French nation. Although political geographers and political scientists use the term to mean a reasonably large group of people who share im-

The center of Paris is the Ile de la Cité, the island in the Seine River (left center of picture).

portant cultural bonds, its precise definition varies with the individual defining the term. Typically, nation means a reasonably large group of people, sharing one or more important cultural traits, such as religion or language, who are distinguishable from others who do not share their culture. The term "nation" is roughly equivalent to the term "people" when used as a singular noun, as the Korean nation or people. Members of a nation tend to identify with one another and distinguish between themselves and other cultures. The operational definition is sometimes fuzzy because of the uncertainty of size required to constitute a nation.

Nations rely on emotional ties to maintain identity based on cultural characteristics such as language or religion. Members of a particular nation share a common ancestry and historical events that are important to them. Boundaries of nations do not always coincide with countries or states; the diffusion of the Germanic language across Europe, for example, was not accompanied by the development of only one German state based on language (Figure 11–5). The different history and events associated with the independent states of Europe led to their recognition as countries separate from one another. A nation can predate the country or state; thus the German nation existed long before a German state or country was created. The Jewish nation was also a widely recognized social and political unit long before the Jewish State of Israel was formed.

FIGURE 11-5 The German Nation and Boundaries of States where German is Dominant

The Nation-State

Another term used in discussing the political organization of space is *nation-state*, which refers to the case where the boundaries of a state or country correspond to those of a nation. Relatively few such clear examples exist today. One example is the nation-state of Japan. The vast majority of Japanese view themselves first as Japanese; they speak the Japanese language, have emotional ties to Japan, and live within the boundaries of Japan. Another example of a nation-state is Denmark. Occupying the Jutland Peninsula, Denmark is a country whose borders correspond almost identically with those of the nation. Sweden, Norway, Uruguay, Egypt, and New Zealand are all nation-states.

Most states are not nation-states. Individuals in the United Kingdom, for example, may view themselves as Irish, Welsh, Scottish, or English. Residents of India may regard themselves first as

Sikhs or as members of some other culture rather than Indian.

Problems of defining the term "nation-state" become even more complex when a country such as the United States is considered. The great majority of the residents of the country clearly view themselves as distinct from other peoples and regard themselves first and foremost as Americans. Other observers likewise see residents of the United States as Americans, regardless of what state they come from or what language they speak. Some residents of the United States, however, have family or other ties to another country and another nation. Recent Vietnamese migrants or Spanish migrants, for example, are Americans. Or are they? Some observers maintain that the development of the emotive tie to a people and a new land for migrants occurs only in the second-generation population. Because nations tend to be informal regions, and because there are questions over which countries are truly nation-states, states are the common region used to examine the impact of culture on international organization of space (see the Cultural Dimensions box on page 424).

Unity in States and Nations

A number of forces combine to affect the unity and permanence of both nations and states. Ideally, a state or nation-state is characterized by cultural homogeneity. The greater the homogeneity within the unit, the greater the internal strength of the state. Homogeneity comes from *uniformity* and *coherence.*

Uniformity refers to the general degree to which the beliefs, customs, rights, and obligations of all citizens of a state are similar. Although these do not have to be identical, they should be generally uniform in the ideal state. The laws in the United States, for example, provide for equal treatment of all people, free speech, the right to representation for taxation, and the freedom of worship. A second aspect of homogeneity is coherence. Coherence refers to the degree to which the various parts of a state are interrelated. The presence of different languages, different religions, racial or ethnic prejudice, or physi-

CULTURAL DIMENSIONS
The Name of a State

Pardon the people of Burma if they seem somewhat unsure of the name of their country. Until September 1988 it was officially the "Socialist Republic of the Union of Burma." A military coup at that time prompted a name change to simply the "Union of Burma." In May 1989 the new government determined that some residents of the country might be offended by being labeled "Burmese" since the Burmese are actually only the largest cultural group in the country. The country is now the "Union of Myanma," a name the government hopes will quiet some of the approximately twelve ethnic groups whose guerrilla forces have demanded ethnic autonomy since the country gained independence from the British in 1948.

Other countries have faced problems with their names, some for centuries. Holland, for example, is one of several countries that is commonly known by more than one name, only one of which is technically correct. In fact, the correct name of Holland is the "Kingdom of the Nether-

lands." It has eleven provinces, including North Holland and South Holland. The Holland provinces are the most important of the eleven provinces, and they constitute the historic hearth of the country. Currently, the Holland provinces contain much of the country's industry and the three largest cities. Many residents of the two Holland provinces simply refer to the country as Holland, a practice unpopular with residents of some other provinces, especially the north. For them the country is "the Netherlands," regardless of how many people call it "Holland."

Other European countries face similar problems. Many English people call their country "England," but England is only one part of the "United Kingdom" even though the English are numerically dominant in relation to the Scottish, Welsh, and Ulster (North Ireland) residents. Nevertheless, although England dominates the political, economic, and cultural affairs of the United Kingdom, the state is not England.

cal barriers such as mountains, seas, or deserts may form obstacles to coherence. The presence of such barriers sometimes threatens the existence of the country.

Forces that unite a state or country and provide coherence are sometimes referred to as *centripetal forces.* Common language, common religion, common economy or economic well-being, or common history (nationhood) serve as centripetal forces. By contrast, *centrifugal forces* refer to those forces that tend to upset the coherence of a state. Multiplicity of languages, religious differences, racial or ethnic prejudice or conflict, unequal distribution of wealth, or lack of uniform laws and quality of life are all examples of centrifugal forces. Physical barriers such as the fragmentation of a country among several islands may also be a centrifugal force.

Countries employ a variety of means to try to develop coherence. These include a common ed-

ucational system, a common language, programs to redistribute wealth or to improve the economy of a specific area, and mass media. Examples of such programs are legion. The Philippines, for example, adopted Tagalog as an official language to replace the colonial languages of Spanish and English and myriad native dialects. Only recently has Tagalog gained acceptance as the Filipino language, and education still relies heavily on English or Spanish. The lack of a unifying language remains a centrifugal force in the Philippines, a problem intensified by the many far-flung islands that make up the country.

Lack of coherence and uniformity in a country can lead to regionalism or sectionalism. *Regionalism* refers to the tendency for members of a specific subsection of a country to identify with that region. In the United States it is possible to recognize the West, the South, and other such regions. Nevertheless, recognition that one is a member of this region is secondary to being a member of the American nation as a whole (Figure 11–6).

Sectionalism is an intense form of regionalism, in which members of one section of the country disagree with fundamental values or laws that combine to make the country distinct. Members of a section may feel that they have been mistreated economically, may have a different language, or may have a different view of the history surrounding the evolution of the country. An acute form of sectionalism is associated with demands for autonomy or complete independence. Examples of sectionalism leading to separatist demands include the Sikhs of the Punjab in India, the French-speaking Canadians in Quebec, the members of the pre–Civil War South in the United States, and the Basques and Catalans of Spain. Sectionalism is difficult to overcome, and is one of the most important centrifugal forces hindering the homogeneity in a State.

Another centrifugal force is *polyethnicism*. A polyethnic state is one that has several ethnic groups within its borders, usually with each occupying a distinct territory. Polyethnic states may have groups from two or more nations within

Kurdish refugees in Turkey. The Kurds view themselves as a nation, but they occupy portions of several different countries including Turkey, Iran and Iraq.

their borders, or they may be composed of a number of tribes. The relative support of these groups for the state varies. In many African countries, the presence of a number of tribes in specific territories serves as a major centrifugal force. The presence of a strong central government in other

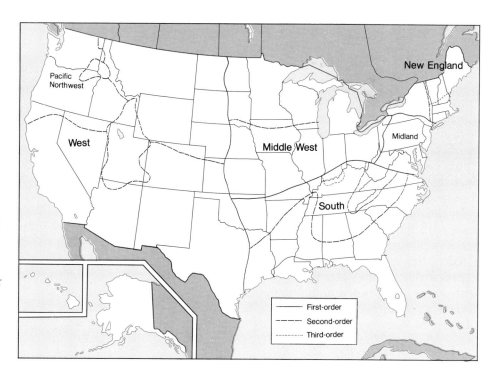

FIGURE 11–6
Regionalism in the United States The United States can be regionalized in many ways. This scheme recognizes broad first order divisions such as the south, and second and third order subdivisions such as the Appalachian region within the south.

countries, such as the Soviet Union, makes the multiplicity of ethnic groups characterized by the individual republics less of a threat to the country's continued existence. The presence of groups in the United States, such as the Indian tribes, Blacks, and Hispanics, may or may not create a polyethnic country. To the extent that distinct ethnic groups occupy their own territory and pay allegiance first to their own group, they are a centrifugal force.

Plural Societies

The terms *plural state* or *plural society* are used by political geographers to refer to areas in which members of various ethnic groups are intermingled and do not have their own territories. The United States has a plural society associated with a fairly high degree of coherence and uniformity because of the wide acceptance of one language and lack of official (*de jure*) restrictions that prevent minorities from enjoying all the benefits of the society.

In Southeast Asia plural societies composed of native peoples and foreign traders (Arab, Indian, or Chinese) became more complex because of population movement directed by the European colonial powers who brought Chinese to Vietnam, Thailand, Singapore, and Indonesia, and Indians to Malaysia and Burma. With the end of colonialism and the emergence of the countries of Southeast Asia as independent states, plural societies have presented numerous challenges. Historically, Europeans had held the highest paid positions in commerce and government in each country, and imported laborers from India or China often filled the mid-level positions, while the native peoples were largely excluded from the commercial and political life of the country. After independence, the native majority population dealt with the problem of cultural pluralism in a variety of ways. Malaysia and Burma each created a federation in which each group was granted rights. Although they attempted to create states with some uniformity and homogeneity, both Malaysia and Burma discriminated against ethnic Chinese by denying them the right to vote, even those who were born in the country. Vietnam

forced hundreds of thousands of Chinese to leave after the end of the Vietnam War.

Exclaves and Enclaves

Where cultural groups occupy distinct territories, such as the Chinese in Malaysia, the centrifugal force and the corresponding challenge for the country are even greater. Many of the countries created in the twentieth century are polyethnic societies containing exclaves or enclaves with associated centrifugal forces.

An *exclave* exists when a territory outside the boundaries of a state is occupied by peoples of its culture. An *enclave* is a group located within the boundaries of a state but with a different culture and ties to a different state. Examples of exclaves and enclaves dot the political map of the world. West Berlin is an exclave of the Federal Republic of Germany (West Germany) and an enclave of the German Democratic Republic (East Germany). Enclaves and exclaves are often associated with *irredentism*, the desire to gain or regain territory. Irredentism has been used to justify conflict to regain lost territory, as in Africa where Somalia has waged war with Ethiopia, Kenya, and Djibouti to unify territory occupied by the Somalis.

Nationalism

The coherence and uniformity leading to a homogeneous society are often associated with strong support on the part of the people for the existence and growth of their state or country. This attitude, which is known as *nationalism*, characterizes the unifying belief of a group of individuals loyal to a country. Nationalism may or may not be associated with a country, as a nation does not need a territory in order to exist. Technically the country of Poland, for example, ceased to exist in the eighteenth century when its territory was divided among its neighbors. The Polish state was recreated in 1919, but the Polish nation had existed throughout this time due to the intense nationalism of the Polish people. The Jewish nation did not exist as a state for nearly two millennia, and even today Jewish nationalism is not uniform throughout the Jewish state or na-

The Royal Canadian Mounted Police is an important icon to Canada. For non-Canadians, the "Mounties" are a readily recognizable symbol of the country.

tion. The Palestine Liberation Organization (PLO) has as its goal the establishment of a Palestinian state, but much of the territory historically associated with the Palestinian people is now part of the state of Israel.

Factors that lead to nationalism are associated with the forces that bring homogeneity, such as common language, religion, beliefs, and history. The symbols of nationalism are sometimes called *icons*; they help unify a people behind the creation or existence of the state or country. Icons include such things as a flag, a national anthem, or images of specific places such as Washington, D.C., and its government buildings. The entire assemblage of symbols in which a people believe is known as their *iconography*; the presence of these symbols facilitates the continued existence of the country. The importance of iconography is illustrated in Canada, where the centrifugal effect of two official languages is partially offset by a common iconography in the flag, hockey, and the history and traditions of the country.

Religion is a dominant force in the lives of many of the world's people. Many children learn to read from holy texts, and common proverbs or sayings are derived from religious writings. Many countries of Europe have Catholic political parties and most support the dominant church with taxes. Re-

ligious iconography is incorporated into the nationalistic iconography in these countries. Where religion is the dominant cultural factor, as in countries where Islam is preeminent, religious iconography may prevail over nationalistic iconography.

The icons of some countries serve as symbols by which other people identify them. The flag of the United States, the monarchy of England, and the hammer and sickle of the Soviet Union are examples of icons that are well known far beyond the borders of the country.

The Boundaries of Countries

Each country is separated from its neighbors by boundaries, the generally invisible lines that indicate the extent of a state's territory. Boundaries are important because they indicate the area of theoretical sovereignty of an individual state or country. Since boundaries divide countries, they must be agreed upon or they become the focal point for conflict.

Boundaries often originate as zones of transition from one cultural core to another rather than as precise, measured lines. Such zones are known as *frontiers*. On a frontier that is either uninhabited or has only a few isolated pioneers, no individual state or country exercises complete sovereignty. Growth of the state concept and associated

Until it was suddenly opened in November of 1989 the Berlin Wall was one of the most visible boundaries in the world.

nationalism have led to the formalization of most border areas as specific boundaries rather than frontiers. Frontiers still exist in areas unsuited for settlement, as in the countries of the Arabian peninsula, where the actual boundary lines between Saudi Arabia, Yemen, the People's Democratic Republic of Yemen, Oman, the United Arab Emirates, and Kuwait are still not settled (Figure 11–7). These boundaries have not needed to be defined more clearly because the frontier zone has been used by nomadic people rather than permanent inhabitants.

Another place where boundaries remain vague is Antarctica, which is claimed by a number of countries. The first formal claim to Antarctic territory was made by Britain in 1908. Since then six other countries have made formal claims: New Zealand in 1923; France in 1924; Australia in 1933; Norway in 1939; Chile in 1947; and Argentina in 1943. Other countries have established research bases in Antarctica, and in 1961 a treaty was adopted by twelve countries, most of whom occupy a sector of the territory (Figure 11–8). Six other countries subsequently joined these twelve in agreeing that Antarctica would be used for peaceful purposes, that military activities would not be permitted, and that freedom of

scientific investigation and cooperation would be maintained. Nevertheless, the original claimants of territory still insist that their rights to Antarctica's resources should be recognized.

Australia, France, and the United Kingdom regard their territories in Antarctica as colonial holdings, complete with postage stamps. Chile and Argentina have incorporated their Antarctic claims into their national territories. Norway refuses to recognize the idea of sectors and will not place a northern or southern boundary on its claims. Other countries—notably Japan, the United States, and the Soviet Union—make no formal territorial claims and refuse to recognize any existing claims. Such frontiers and conflicts over actual demarcation of territory have been resolved in many other areas by actual occupation and settlement of the land. It is possible that Antarctica may continue as an international property, but it is probable that conflict over demarcation of the boundaries will ultimately develop.

Boundaries may be invisible lines, or they may be visible in the landscape, such as those coincident to mountains or coastlines. Landscape features associated with boundaries include markers, fences, guardhouses, police stations, or other obstacles to free passage. Some boundaries, such as that around East Berlin (until its opening in 1989), serve as reflective barriers to diffusion. Other boundaries are permeable, but even so there is usually a recognizable change in the landscape when a boundary is crossed. Signs, toponyms, or businesses may reflect language changes, quality of buildings and life-styles may illustrate differences in economic standards, or crop-livestock combinations reflecting different government support for agriculture may change.

The development of boundaries involves four processes, which ideally occur sequentially: definition, delimitation, demarcation, and administration. The first stage (*definition*) involves the description of the location of the boundary being established in as much detail as possible. *Delimitation* of the boundary occurs only when cartographers, using large-scale maps and air photographs plot the boundary as accurately as possible. *Demarcation* is the actual on-site survey of the precise boundary line. Delimiting or

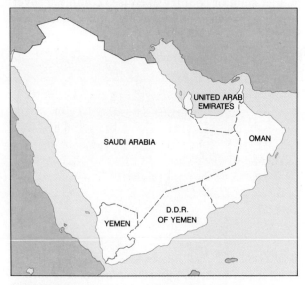

FIGURE 11–7 Frontier Zones and Boundaries in the Arabian Peninsula Frontier zones are shown by dotted lines.

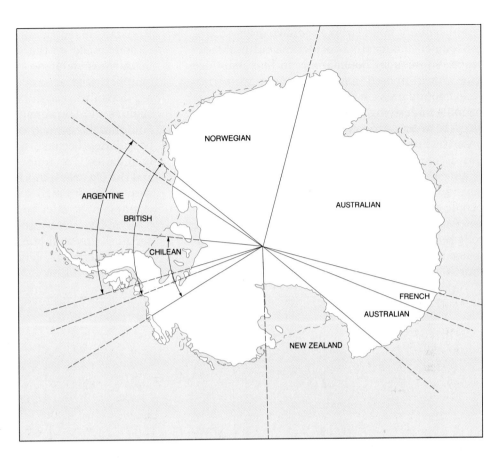

FIGURE 11–8
Territorial Claims in Antarctica The territory claimed by individual countries is not necessarily accepted by all other countries.

demarcating a boundary involves some of the most complex geographical, engineering, and political processes involved in the relationships between countries. Boundary demarcation has not been completed for many of the boundaries of individual countries, since it is a very expensive process; consequently, where there are not boundary problems, states may delay demarcation indefinitely. *Administration* is the process of establishing a regular procedure for maintaining boundary markers and settling minor local disputes over the boundary and its effects.

Ideally a boundary should include within it all of the people who share the same culture. Cultures included in the boundary should be compatible and complete with no large numbers of the group excluded from the national area. A boundary should also facilitate trade and interchange within a country, provide protection, and minimize the amount of energy the state has to devote to defending its boundaries. The types of bound-

aries that have been used in efforts to approach this ideal are very common and are classified in a number of ways.

Types of Boundaries

One classification of boundaries divides them on the basis of their time of establishment. *Antecedent* boundaries were drawn before an area was populated. Many antecedent boundaries are geometrical and follow a line of latitude, a meridian, or another straight line between two known points. The boundary between the United States and Canada west of the Lake of the Woods in Minnesota is a straight line following the forty-ninth parallel. The boundaries of many of the U.S. states are also geometric lines. The boundaries in Africa were originally drawn by Europeans, and when the African countries became independent, they seldom did or could alter the boundaries.

Subsequent boundaries are drawn after an area is occupied and ideally consider the cultural landscape. Subsequent boundaries can be subdivided into superimposed and consequent. *Superimposed* boundaries are drawn without regard to the region's underlying cultural patterns. The boundaries between African countries were superimposed by European colonial powers and as often as not divide tribal groups rather than unite them (Figure 11–9). *Consequent* boundaries attempt to consider the underlying cultural variety of the people involved. The division between Pakistan and India in 1947, for example, was based on the religion of the people. The Treaty of Independence declared that areas where 50 percent or more of the inhabitants were Islamic would be in Pakistan, independent from areas in India where Hindus made up the majority. Other consequent boundaries rely on language differences; the boundaries of the republics of the Soviet Union are drawn to recognize specific linguistic groups such as the Ukrainians, Georgians, and so forth. Sometimes boundaries are referred to as *relict* boundaries because they no longer function as boundaries. The relict boundary between Mexico and the United States before the American Southwest was obtained by the United States is an example of a relict boundary.

Boundaries can also be classified as *physical* or *cultural*. Physical boundaries follow some aspect of the physical landscape, whereas cultural boundaries follow a cultural division such as

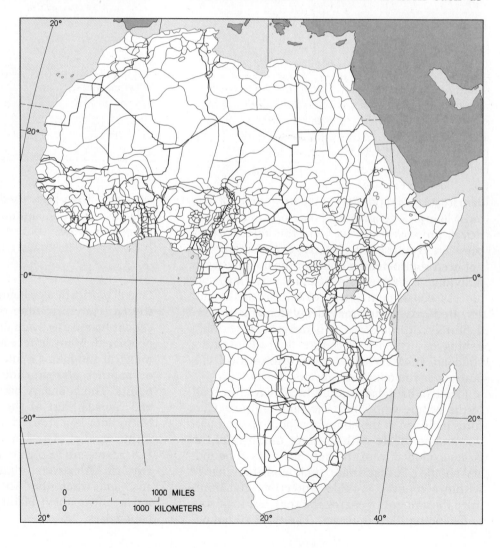

FIGURE 11–9 Tribal and National Boundaries in Africa The boundaries of most African countries do not correspond with the territories of tribal groups or language families.

The Black Sea is an example of a physical barrier, marking part of the boundaries of the Soviet Union, Romania, Bulgaria and Turkey.

language or religion. Boundary lines are often drawn along the axis of certain types of physical features such as rivers, mountains, oceans, lakes, deserts, or swamps.

Mountain barriers are often boundaries because they are permanent, have low population densities, and are difficult to cross. For example, the Pyrenees Mountains between Spain and France have formed a boundary for generations, the oldest continuous boundary in Europe. The high peaks of the Pyrenees have prevented the Spanish culture to the south from expanding beyond the mountain crest into the French area, and the French from expanding to the south. The line along a mountain range is rarely as uniform and precise as it appears on a map and generally represents almost a frontier zone. In the case of the Pyrenees, isolation and difficulty of transportation made the mountains a good boundary, but also enabled the Basque people to use them as a refuge from the expanding French and Spanish cultures.

Other examples of physical boundaries include deserts and swamps. Both are barriers to movement and serve to isolate one area from another, simply because few people live in swamps or deserts. Boundaries between Saudi Arabia and its neighbors, like those in many North African countries, are drawn through desert areas. These serve as suitable boundaries unless valuable resources are discovered; then the problem of de-

marcation becomes difficult and may result in conflict. Thus as long as there is nothing of perceived value in the desert, it makes an excellent boundary.

Water Boundaries

Water provides one of the most common forms of physical boundaries. Oceans, lakes, and rivers are used as boundaries for several reasons. First, they are obvious features on the earth's surface. Second, they may serve as a dividing line between cultures. Third, they may serve defensive purposes. The most effective water boundary is the ocean itself, which by its size offers all these advantages. The United States and Europe, and Latin American and African countries, do not engage in competition over their Atlantic boundaries. Smaller bodies of water are less desirable as boundaries. The boundary between the United States and Canada was initially based on the head waters of the Mississippi River. The original attempts to delimit this boundary assumed that the Lake of the Woods in Canada was the source of the Mississippi, and U.S. representatives insisted on including the Lake of the Woods as part of the Mississippi drainage. A part of the Lake of the Woods shoreline has remained as an antecedent boundary, creating a U.S. exclave where part of Minnesota is accessible by land only from Canada.

Rivers are even less suitable than lakes as boundaries. The Rio Grande between Mexico and the United States is neither a permanent nor a reflecting barrier. The river has changed its course frequently, leaving Spanish-speaking exclaves in the United States and English-speaking exclaves in Mexico. Variation in the flow of the Rio Grande from year to year and season to season, reflecting its origin in an arid and semiarid zone, makes it of little use as a barrier to illegal movement of population. Some observers maintain that if one of the definitions of a state or country is that it can control its territory, then the United States is no longer a state because it cannot control the border between the United States and Mexico and prevent illegal immigration.

A final problem with rivers as boundaries is that a river tends to unify a region rather than divide it. A river and its tributaries often served as a route of transportation and communication in past times. Dividing the river between two adjacent states often led to conflict over use of the river (for example, the Amur River between the People's Republic of China and the Soviet Union) or cut off one country's access to the ocean. The latter problem occurred in the Netherlands and Belgium. The unwillingness of the government of Belgium to allow the Dutch to use the Scheldt River forced the Dutch to construct a new channel for the Rhine River to make it more navigable. The resulting change in transportation and communication is in large part responsible for the emergence of Rotterdam as the largest port in Europe. *International rivers* used for navigation by several countries, such as the Danube River also may create problems. The rights of upstream states to divert water from an international river or of downstream states to prevent navigation by states farther upstream are difficult issues to resolve.

A final challenge presented by the use of water bodies as boundaries is the extent of the territorial claim of a state. At what point does the land of the United States end and become the Atlantic or Pacific oceans? Is it at the high tide mark? The low tide mark? The median between the high and the low tide? How far into the sea does the United States have the right to protect itself from foreign vessels? These issues have led to conflicts and an ongoing United Nations Commission on the Law of the Sea.

Traditionally, several zones have been recognized in regard to territorial control over the oceans. Originally, a 3-mile limit was accepted, representing the greatest range of land-based cannon, but modern weapons have rendered this limit meaningless. Most states now claim a 12-mile limit for absolute territorial sovereignty, but some states claim up to 200 miles. Indeed, some Latin American states first claimed a 200-mile limit in the 1960s and 1970s and began to harass or capture American fishing boats that came closer. Ultimately, international negotiation established a 200-mile economic zone in which the adjacent state has partial control, in that other countries cannot fish or develop other resources in this zone without the permission of the adjacent state.

The seaward limit of a state's territory is important because it affects the *high seas* and free shipping upon them. The high seas are those that are open to free movement by ships from all nations. The Black Sea illustrates the ongoing debate over what constitutes high seas and territorial control. The Soviet Union claims that the Black Sea is a body of inland water to which only the adjacent states should have free access. The basis for this claim is the narrowness of the opening at Istanbul, Turkey. The United States refuses to accept this interpretation and dispatches one or two naval vessels each year to make a circuit of the Black Sea just outside the 3-mile limit. Incidents of Soviet and U.S. ships "bumping" during this annual excursion illustrate the importance associated with recognizing 3 versus 12 miles as the limit of absolute sovereignty.

Related to the problem of territorial control of the oceans is the problem of territorial control of the air above a state. The issue of vertical territorial control was unimportant until the development of balloons and aircraft. Today, most states prohibit aircraft from foreign countries from flying over their land without specific permission. Failure to obtain such permission can have tragic results, as in the destruction of a Korean Airlines plane by the Soviet Union in 1986. A related issue deals with the overflight of a country by satellites in space. If one country has the ability to defend itself from such overflights through destruction of the satellites, does it then have the right or duty to warn other countries that it will do so? Such issues illustrate the continuing difficulty of delimiting, demarcating, and defending boundaries on the basis of physical features.

Cultural Boundaries

A variety of cultural features have been used to demarcate the territory of one country from another. Common cultural boundaries include those based on language, religion, tradition, or economic or political system. The boundary between the Federal Republic of Germany and the German Democratic Republic, for example, reflects contrasting philosophies of government. The boundary between India and Pakistan reflects two different religions—Hinduism and Islam. The boundary between German and Polish territory is a reflection of language differences. The boundaries of some African states, such as Rwanda and Burundi, represent traditional tribal loyalties.

Cultural features provide useful boundaries because they are often more meaningful to the peoples occupying the land than any other form of boundary. The division of central Europe after World War I attempted to use the cultural characteristics of the peoples (i.e., nationality groups) to delimit states in the region. President Woodrow Wilson's vision of the League of Nations was that it would allow each national group to have its own self-identity and recognition.

With the help of the cultural geographer Isaiah Bowman, Wilson and the League of Nations created the new countries of Poland, Czechoslovakia, Yugoslavia, and Albania, even though the U.S. Senate refused to join the League. These divisions did not completely fit the ideal criteria of boundaries since each individual country did not include all the members of its constituent nation. Some countries combined several groups that have a similar language. Czechoslovakia, for example, is composed of a western Czech portion and an eastern Slovak portion. Both Czech and Slovak are Slavic languages, and although they are quite similar, the resulting country did not possess a single national identity. Similar problems existed in Yugoslavia, where several slavic language groups were combined into one country that also included important minority groups (such as the Albanians); this created an amalgamation of cultures that have since maintained an uneasy existence (Figure 11–10).

Another example of the problems associated with creating boundaries based on cultural variables is illustrated by the Indian subcontinent. It was divided along religious lines at the time of partition, creating a West Pakistan Islamic state, an East Pakistan Islamic state, the Hindu state of India, and the disputed territory of Kashmir. Demarcation of the boundaries caused tremendous human suffering. The boundaries cut across railroad lines, irrigation systems, and separated groups from their sacred sites. Many Muslims and Hindus found themselves on the wrong side of

FIGURE 11–10 National Boundaries Compared to Cultural Groups in Eastern Europe

the new borders. Conflict between the ruling maharajah and the people broke out in Kashmir where the majority of the people were Muslims and wanted to be part of Pakistan, but the maharajah was Hindu and wanted to be part of India. Dispute over Kashmir continues to the present time.

The partition of India and Pakistan resulted in the movement of an estimated fifteen million refugees, as well as associated carnage as adherents of one or the other religion were attacked by members of the other. Three to five hundred thousand individuals lost their lives in the conflict over the boundaries. Furthermore, Pakistan consisted of two parts separated by India, creating a centrifugal force that could not be overcome by the centripetal effect of a common religion. After twenty-five years of increasing disagreement between the two halves, East Pakistan declared itself independent and became Bangladesh in 1972.

Another example of the problems associated with boundaries based on cultural phenomena can be found in the division between Germany and Poland at the end of World War I. While President Wilson and the League of Nations' plan

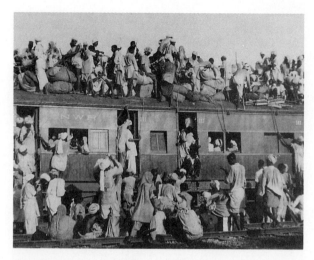

The division of the Indian sub-continent into Muslim Pakistan and Hindu India caused millions to flee their ancestral homes. In this 1947 photo Muslim refugees flee India for new homes in Pakistan.

to provide national recognition was worthwhile, it did not serve other goals of political groups, such as gaining access to the sea. Germany's East Prussia was separated from the balance of Germany by the Polish corridor established to provide Poland a port on the Baltic Sea. Isolated from its economic base in Germany, East Prussia suffered economic upheaval and a depression. The Germans used this unfortunate arrangement as a justification for the invasion of Poland to reunite the German-speaking peoples in the late 1930s.

Recognition that the boundaries were superimposed and did not fit the actual cultural landscape led to a revision of the boundaries of Poland and Germany at the end of World War II. The primary change involved moving Poland's eastern border west in response to the desire of the Soviet Union to regain territory occupied by the Ukrainian people that had been incorporated into Poland in 1919. To eliminate the possibility that future rulers of Germany would invade the Soviet Union to reunite German-speaking peoples, the Soviet Union engaged in a mass movement of population to make the population distribution fit the boundaries. Poles were moved west from the area occupied by the Soviet Union, Germans were moved west from the Polish territory of former Germany, and East Prussia was divided between the Soviet Union and Poland. The Soviet Union orchestrated

these mass movements to ensure that the cultural boundaries and the population distribution coincided with the political boundaries.

Other states have dealt with the problem of multiple linguistic and religious groups in an entirely different manner. For example, Switzerland includes four language groups: German (73 percent), French (20 percent), Italian (5 percent), and Romansch (2 percent). Moreover, the Swiss people are both Catholics and Protestants. Switzerland has enjoyed eleven centuries without conflict among its various ethnic groups in spite of this, because it was organized on the basis of economic activity rather than cultural identity. The Swiss occupied the main passes between southern and northern Europe. The state's evolution was based on efforts to improve roads and to formalize the collection of tolls from travelers. The economic advantage of being part of Switzerland has been the major centripetal force, outweighing the centrifugal force of multiple languages and religions.

The Shape of Countries

The *shapes* of countries can bring problems or advantages of their own. Political scientists, political geographers, and others are interested in the size and shape of countries. *Size* was a particular concern in the early part of the twentieth century as the large countries, especially the United States and the Soviet Union, expanded their influence and control. Some observers believed that the larger countries would simply absorb the smaller states of the world. Since World War II, however, more small countries have been created, and the large countries have not changed their territorial extent. Nevertheless, size is of interest to cultural geographers for a variety of reasons. A larger country may contain more resources, but may also incorporate diverse cultures with associated potential for conflict. Small countries or states have the potential problems of a limited resource base and a small population, making the state difficult to defend.

Characteristics such as standard of living, relative level of peace, and world importance seem to have little correlation with size alone. Some

TABLE 11–1
Size, Population, and Gross Domestic Product of Selected Countries

COUNTRY	SIZE (SQ. MILES)	POPULATION (MILLIONS)	GROSS DOMESTIC PRODUCT (BILLIONS)
Canada	3,831,033	26.1	412.8
France	211,208	55.9	724.1
Japan	145,834	122.7	2,664.0
Netherlands	16,042	14.7	175.3
United States	3,679,245	246.1	4,486.2
USSR	8,600,383	286.0	2,356.7

small countries such as the Netherlands have an extremely high standard of living and are of world importance in spite of their size. Japan, which has less than one-tenth the area of the United States, is the third largest industrial and economic power in the world after only the United States and the Soviet Union (Table 11–1). At the other extreme are large, poor states such as Chad or the Sudan.

Shape

Five basic shapes of states are compact, prorupted, elongated, fragmented, and perforated. The *compact* shape is associated with a state that has a circular or nearly circular shape. A circular shape offers the shortest boundary area to defend, minimizes communication and transportation links from the center to any point, and provides for greater unity. France and Poland are examples of compact states. Each has a relatively short border and a roughly circular shape (Figure 11–11).

Prorupted states exist where a compact state has a large projecting extension. Many states have a proruption that serves as a buffer or provides access to some unique or desirable geographical feature. The proruptions may offer the advantage of access to a specific location, but they also provide a longer border, are subject to annexation by adjacent countries, and increase the transportation and communication costs of a country. Poland between 1919 and 1939 had a long proruption giving it an outlet to the sea. Afghanistan has a proruption included when it was created as a buffer state between czarist Russia and the British colonies of India. Colombia has a proruption giving it a port on the Amazon River, which is

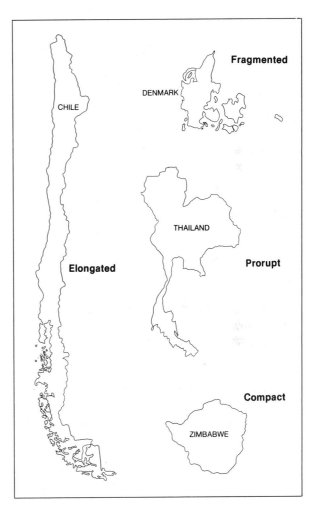

FIGURE 11–11 Theoretical Shapes of Countries

open to international shipping (Figure 11–12). Bolivia has a proruption giving it access to the Paraguay River and thus access to the Atlantic.

FIGURE 11–12 Colombia's Access to the Amazon
Access to the Amazon River is provided for Colombia by a proruption where the community of Leticia is located.

Elongated states have a narrow, extended shape. Examples include Chile, Vietnam, Italy, Norway, and Sweden. The difficulties inherent in an elongated shape are apparent. An elongated shape maximizes the border areas, maximizes the transportation and communication costs (unless they can be done by sea), and makes the country subject to division by invading forces. Problems of maintaining unity among people scattered over such long distances are also common.

Fragmented states are physically splintered into a number of separate pieces. Indonesia, the Philippines, Japan, and the British Isles are fragmented states. The difficulties presented by physical fragmentation include greater boundary areas, transportation and communication problems, and the likelihood of great heterogeneity among the people of the state causing sectionalism and lack of coherence.

Perforated states have numerous enclaves within them. When the residents of these enclaves are hostile to the government or align themselves with outside forces, they present problems of unity. The governments of perforated states may not have effective control of highly nationalistic enclaves, leading to conflict in some cases. Perforated states include South Africa and the United Arab Emirates (see Figure 8–23 on page 315).

Multinational States

The presence of more than one distinct cultural group in a country, with each demanding recognition of its distinct nationality, is termed *multinationalism*. The northern part of Belgium, for example, is dominated by people speaking Flemish, a language closely related to Dutch. The people in the south speak French and are very ethnocentric about their language. As discussed earlier, the continued demand for regional autonomy in Belgium reflects the multinationalism of the country.

Some countries have so many individual cultural groups that recognition of each nation is difficult. India, for example, recognizes twenty-

Cultural divisions in Brussels, Belgium have not prevented its development as an important urban center in Europe.

three separate groups who have been granted political representation as indiviudal states based on language.

Yugoslavia is a multinational state with poly-ethnic divisions. The contemporary boundaries of Yugoslavia stem from its establishment in 1919 by the League of Nations as a homeland for people speaking southern Slavic dialects, and even today approximately 85 percent of Yugoslavia's population speak a Slavic dialect. The country, which was part of the Roman and Byzantine Empires, has long been a crossroads between Asia and central Europe. The cultural pattern is further confused by the dominance of the Orthodox Catholic church in the east and south and the Roman Catholic church in the west and north. Today there are six states in Yugoslavia: Slovenia, Croatia, Bosna-Hercegovina, Serbia, Montenegro

and Macedonia. The Slovenians and the Croats are Roman Catholic, while the Serbs, Montene-grins, and Macedonians are mainly Greek Ortho-dox (Figure 11–13).

The cultural picture is further complicated by the Albanians who occupy portions of Yugoslavia. An estimated 1.5 million Albanians live in the Kosmet region of the Southwest bordering Alba-nia. They are primarily Muslim and speak a language (Illyrian) distinct from the Slavic tongue. The linguistic complexity of Yugoslavia is indi-cated by the Slovene and Croat use of the Latin alphabet, while the Serbs (who are Orthodox in religion) use a Cyrillic script identical to that used in the Soviet Union except for a few letters; this difference persists even though Serbian and Croatian are identical in spoken form. The Re-public of Bosnia-Hercegovina is occupied by both

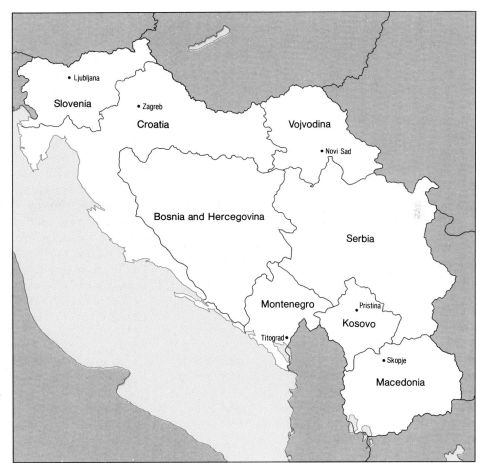

FIGURE 11–13 States of Yugoslavia The individual states of Yugoslavia represent cultural differences.

The cultural contrasts between the various groups in Yugoslavia is intensified by economic conditions. The south is the poorest, most rural area of the country.

Serbs and Croats, and Montenegro is made up of peoples closely related to the Serbs but recognized as a distinct nationality. Yugoslavia also has Hungarians (Magyars), German-speaking peoples, Romanians, Bulgarians, Turks, and gypsies (Table 11–2).

As a consequence of these various groups, the Yugoslavian government faces a continuing problem of multinationalism, compounded by the differential level of economic development among the republics. The northern republics of Serbia and Croatia are the wealthiest, and the

TABLE 11–2
Nationalities of Yugoslavia

NATIONALITY	PERCENTAGE OF POPULATION
Serb	36.3
Croat	19.7
Muslim	8.9
Slovene	7.8
Albanian	7.7
Macedonian	5.9
Yugoslav	5.4
Montenegrin	2.5
Hungarian	1.9
Other (Turks, Romanians, Bulgarians, Gypsies, etc)	3.9

Source: CIA *World Factbook 1988.*

southern republic of Macedonia is the poorest. Even the wisest leaders would have a difficult time maintaining the independence and national integrity of a country with such a high level of heterogeneity and lack of internal coherence (see the Cultural Dimensions box).

Stateless Nations

The presence of multiple nationalities in one country is often a result of another political problem, that of *stateless nations.* Lack of congruity between the boundaries of nations and states may result in groups of people (nations) who have no territory of their own. Such peoples are found in a number of countries of the world, and in each they present distinct challenges. Stateless nations include the Kurds, the South Moluccans of Indonesia, and the Palestinians of the Middle East.

The Kurds live in Iran, Iraq, Syria, and Turkey, though they have demanded an independent state for decades. There are approximately fifteen million Kurdish people, representing between 10 and 20 percent of the populations of the countries they occupy. The Kurds are non-Arabic, Sunni Muslims, but have their own language, dress, and customs. To date, none of the countries they live in has been willing to grant them independence, and the Kurds have suffered from repeated hostility and persecution. The conflict between Iran and Iraq from 1980 to 1988 led the Iranians to support an independence movement among the Iraqi Kurds, but none of the countries involved has been willing to create an independent Kurdistan (Figure 11–14).

The Palestinians of the Middle East are an example of a stateless nation created in this century. The Palestinians had lived for centuries in what became Palestine, a region that came under the control of the British under the mandate system established by the League of Nations after World War I. The British encouraged the development of the area as a homeland for both Palestinians and Jews. In November 1947 the United Nations proposed the division of Palestine into Arab and Jewish states, with Jerusalem and its suburbs as an international zone. The pro-

CULTURAL DIMENSIONS

Life and Politics in Yugoslavia

The antagonism between Serbians and Albanians, who are the dominant ethnic group in Kosovo Province of Serbia, erupted into violence in March 1989 after Serbian strongman Slobodan Milosevic railroaded a series of constitutional changes restricting the province's autonomy through the Kosovo parliament.*

Serbs sang and danced in Belgrade, celebrating their new hegemony over Kosovo. But thousands of Albanians took to the streets in a dozen towns in Kosovo Province. They were met by helmeted riot police, armored personnel carriers, even army tanks. The resulting violence was Yugoslavia's worst since World War II.

As people begin resuming their lives, this large shadow of repression—along with evidence of continuing widespread human rights abuses—lingers. Serbs and Albanians alike believe that another explosion is only a matter of time.

The violence already has widened the gap between the country's conservative, Serb-dominated eastern half and its liberal western half consisting of Croatia and Slovenia. After the Serb victory in Kosovo, the Croats and Slovenes fear that the Serbs will try to impose their will on them, too.

In response to the criticisms, Serbs have begun boycotting Slovene goods, complaining that they have been misunderstood. The revolt, Serbs say, was an organized armed uprising, aided and abetted by the Albanian government in Tiranë.

The Albanians believe the riots represented an outburst by a population reduced to second-class status with little to lose. Unemployment runs as high as 40 percent in some parts of Kosovo Province.

For more than two years, Serbs led by Milosevic have held massive demonstrations to stop alleged persecution of Kosovo's 200,000 Serbian minority by its 1.8 million Albanians. They demanded control over Kosovo's police and judiciary. When the changes actually came, the tinderbox caught fire.

Both Albanian and Serb residents describe the riots as "a civil war." Against rock-hurling youth and unidentified sharpshooters, army units surrounded towns and MIG fighter jets screamed low overhead. In scenes reminiscent of the Palestinian *intifadah*, defiant Albanian youths, their faces covered with bandannas, raced through narrow allies filled with clouds of tear gas. Automatic weapons clattered.

When it all was over, the official count numbered 24 dead, including two police officers. Witnesses, both Serbian and Albanian, put the total death toll at well over 100. Hundreds more were injured.

Hopes for conciliation seem remote. Some independent Albanian voices call for compromise, but Serbs, stung by the recent violence, reject any dialogue.

*Adapted by permission from the *Christian Science Monitor,* April 19, 1989, p. 3.

---

posed division left the proposed Arab state only 1.5 percent Jewish, but the proposed new Jewish state of Israel was 45 percent Arab (See Figure 5–19 on page 192).

On May 14, 1948, Israel declared independence. Egypt, Syria, Lebanon, Jordan, Iraq and Sudan declared war on Israel, and the fighting continued until 1949. A strong unity of purpose among the Jews enabled them to defeat the numerically superior Arab forces. The armistice of 1949 created the new state of Israel, but 750,000 Palestinian Arabs fled to neighboring Arab countries.

Continued hostility between the Israeli Jews and their Arab neighbors led to the Six-Day War of June 1967 when Israel destroyed the air forces of Egypt, Jordan, Syria, and Iraq. Israel expanded its control to the Suez Canal and seized the West Bank of the Jordan River from Jordan. Hundreds of thousands of Palestinian refugees fled into Jordan, Lebanon, and Syria, adding to those who had been living in refugee camps since 1949.

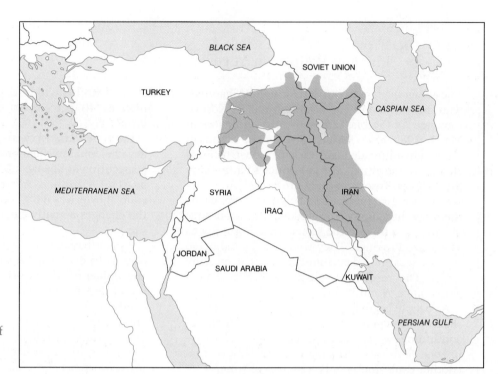

FIGURE 11–14
Kurdistan This map shows the area proposed for an independent state of Kurdistan by Kurdish nationalists.

Subsequent conflicts caused additional Palestinian refugees, and today several million Palestinians constitute a homeless nation in the Middle East. Scattered among the various Arab states and Israel, they share a growing feeling of irredentism, which threatens the Israeli state. Palestinian irredentism is manifest in repeated conflicts with Israel and in support by revolutionary Islamic movements.

The Insurgent State

The conflict within Israel and between Israel and its neighbors over the Palestinian issue exemplifies the conflicts found in states where the boundaries of the nation are not the same as those of the political unit. Another type of condition exists when a group within a state is devoted to overthrowing the existing leadership. This is known as the *insurgent state.*

The turmoil associated with the end of the colonial era and the development of independent states in Africa, Latin America, and Asia has been accompanied by conflict over control of a particular state. These conflicts occur when there is a plural society or when one group of people does not accept the existing leadership of the country. Countries inhabited by several nationalities may experience armed conflict accompanying efforts by individual cultural groups to gain recognition as separate states. Nationalism is often a justification for such conflict.

Attempts by groups to separate from the recognized state and gain control of their own territory range from polite negotiations to armed conflict. An insurgent state arises when a separatist group seizes control of part of the territory of the state. Within this territory, residents are governed by laws made by the members of the insurgent group.

One example of an insurgent state is in Sri Lanka, called Ceylon when it was a colony of the United Kingdom. With an area about equal to that of West Virginia, Sri Lanka has a population of some seventeen million. Culturally Sri Lanka is unlike India or any other adjacent countries. The population is predominantly Buddhist (75 percent) and is polyethnic. Tamils and Sinhalese are the largest ethnic groups, with the Sinhalese majority (70 percent) historically dominant. The Sinhalese are descendants of migrants from northern India who occupied the island over 2,000 years ago (Figure 11–15).

Historically concentrated in the northern half of the island, the Sinhalese were forced south as subsequent Tamil migrants from southern India came to dominate the north of Sri Lanka. Tamil numbers increased through British-sponsored migration, which brought Tamil workers to tea plantations in the south, bringing the Tamils into direct contact with the Sinhalese. In 1978, India agreed to relocate the "plantation" Tamils from southern Sri Lanka back to India. In 1983 Tamil extremists supporting an independent Tamil state in North Sri Lanka rioted, triggering Sinhalese retribution and destruction of life and property among the Tamils. Since 1983, continued conflict has marked the demands of the insurgent Tamils for an independent state. At the present time, an Indian peacekeeping force attempts to maintain

FIGURE 11–15 Distribution of Ethnic Groups in Sri Lanka The complex ethnic composition of Sri Lanka is the basis for the proposed creation of a Tamil state.

some harmony between the two groups, but there are effectively two states rather than one in the tiny island country of Sri Lanka.

Insurgent states may be temporary phenomena whose creators are ultimately brought into the government of the existing state, or the supporters of the insurgent state may become the dominant force in the country. Communist guerrillas in South Vietnam created an insurgent state that was eventually reunited with North Vietnam, creating one country. Insurgent states will no doubt continue to exist because of the variety of nationalistic and political views that characterize many of the nearly ninety countries that have gained independence since the end of World War II.

Trusteeships

Another type of territorial organization is associated with areas administered by international agencies. The best known of these were the mandates of the League of Nations and the trusteeships of the United Nations. After World War I, the League of Nations established mandates in territories formerly controlled by Germany and Turkey. Under the League of Nations, Southwest Africa (Namibia), for example, was a mandate of the Union of South Africa. Countries granted mandates were to help the mandate territory develop economically and politically with the end goal of political independence. The mandates from the League of Nations were transformed into United Nations trusteeships after World War II. The Southwest African mandate became the Southwest African Trusteeship, which finally gained independence in 1989. All but one of the other U.N. trusteeships have gained independence in the last two decades, including such island states as Kiribati, the Solomon Islands, and Vanuatu.

Politics and Conflict

Conflicts are often associated with cultural heterogeneity in a state. Conflict may be in the form of revolution in which governments are overthrown or simply demonstrations by those agitating for change. European political conflicts include those in Northern Ireland and Yugoslavia and those fomented by the Basque peoples of France and Spain and groups such as the Scottish Nationalist Party and the Plaid Cymru of Wales, who are demanding local autonomy. Periodic violence has occurred in both Scotland and Wales, most recently in the latter where the homes of non-Welsh-speaking vacationers have been burned.

The geographical significance of these regional demands for separatism in Europe is obvious. The conditions that lead to success or failure by separatist movements often include geographical conditions. Isolation from the rest of the country by geographic barriers, simple distance, or culture leads to an increased sense of separation. The combination of such separation with lack of integration into the economic framework of the nation and an oftentimes lower standard of living results in demands for autonomy or complete separation.

The process of gaining complete or partial independence from the rest of the state is known as *devolution*. Devolution is the opposite of the unifying tendency that has characterized the emergence of states to replace the colonial holdings of the world in the post–World War II era. The devolutionary movements of Europe are a continued attempt by isolated groups to attain political boundaries reflecting their cultural identity; such movement illustrate the importance of such cultural phenomena as language, religion, and ethnicity in the political affairs of a country.

Conflicts in Asia

Compared to the political conflicts in Asia, those of Europe seem minor. The wars associated with efforts to reshape the map of Asia have dominated the post–World War II era in scale, intensity, and destructiveness. In each case the conflicts can be traced to problems associated with the European colonial era and efforts to create boundaries that reflect the underlying cultural characteristics of the region Three examples of cultural differences in Asia illustrate the nature and import of conflicts that have or are affecting the people and boundaries of the region.

China is the largest (in area and population) country in Asia. It was never occupied by colonial powers in the way that India and Southeast Asia were, but it was controlled by foreign powers. From 1644 to 1911, China was nominally controlled by the Qing (Manchu) Dynasty, the last of the great Chinese dynasties that had begun nearly four thousand years before (Table 11–3).

The last Qing emperor abdicated his throne in 1911, and from that time until 1949 parts of China were ruled by a variety of warlords. Communists under Mao Zedong and nationalists under Chiang Kaishek were the two dominant forces at conflict before World War II, but the Communists were much weaker.

After World War II ended, the Communists gained control in 1949 and declared the independent People's Republic of China. Chiang Kaishek and his army fled to the island of Taiwan where they continued the Republic of China. Many Western nations (especially the United States) continued to recognize the Republic of China as the official representative of the hundreds of millions of Chinese residing on the mainland

TABLE 11–3
The Chinese Dynasties

Xia (HSIA)	2205–1766B.C.
SHANG	1766–1122B.C.
Zhou (CHOU)	1122– 770B.C.
Chunqui (SPRING & AUTUMN ANNALS)	770– 476B.C.
Zhanguo (WARRING STATES)	476– 221B.C.
Qin (CHIN)	221– 206B.C.
HAN	206B.C.–A.D.220
Sanguo (THREE KINGDOMS)	A.D.220– 265
Jin (TSIN)	A.D.265– 420
SOUTHERN & NORTHERN	A.D.420– 589
SUI	A.D.589– 618
TANG	A.D.618– 907
Wutai and Shiguo (FIVE DYNASTIES & TEN KINGDOMS)	A.D.907– 960
Song (SUNG)	A.D. 960–1280
YUAN	A.D.1280–1368
MING	A.D.1368–1644
Qing (CHING)	A.D.1644–1911
The Republic of China	A.D.1912–1949

until the 1970s when the United Nations formally recognized the People's Republic of China as the official Chinese state.

Tension between the People's Republic of China and the Republic of China persisted through the 1970s, but in the 1980s the People's Republic of China has been more open in its relationship to the West and more importantly to its own people and has encouraged contact between the two countries. The People's Republic of China has always maintained that Taiwan is an integral part of the People's Republic, but it is too early to tell how the revolutionary changes since Mao's death will affect these two countries.

On the southern coast of China is Hong Kong, a relic of colonialism in Asia. Hong Kong, a crown colony of the United Kingdom located on the northeast side of the Xi River, is one of the most crowded areas in the world, with nearly six million people in its 400 square miles (1,036 square kilometers) of territory. Most of the population is crowded into the 32 square miles (82 square kilometers) of the island of Hong Kong itself. The population is 98 percent Chinese and relies on China for most of its food and water. China ceded Hong Kong island to the British in 1841 as a treaty port; Britain also acquired the mainland portions of Hong Kong on a ninety-nine year lease in 1898. The changing government and attitudes in the People's Republic of China led to an agreement in the fall of 1984 under which Hong Kong is to become part of China under the concept of "one country, two systems." The Chinese government maintains that Hong Kong will retain its present economic rights, freedoms, and life-styles for fifty years after rejoining China in 1997.

Southeast Asia has been the location of several conflicts related to the cultural history of the region. The most important of these occurred in Vietnam, a former French colony. French colonial rule supported the development of an elite local class who spoke French and who had representation in Paris; eventually revolutionary and separatist movements emerged from among these elite natives. The emergence of a strong Communist party under the Vietnamese nationalistic leader Ho Chi Minh led to an insurgent state. Although France tried to maintain control of its colony, in 1954 the Geneva Peace Accord divided the country into North and South Vietnam. The French withdrew their forces from South Vietnam in 1954, and a decade later American fears of Communist expansion led to U.S. involvement in a conflict that lasted until 1975. From the end of the Vietnam War until 1989, the Vietnamese occupied the neighboring state of Kampuchea (Cambodia), and conflict continues in that country.

Afghanistan represents one of the most recent and most destructive examples of conflict in Asia. Its most recent importance dates from the early twentieth century when it emerged as a buffer state between the expanding Russian empire of the north and the expanding British empire from India (Table 11–4). Following independence in World War II, a revolution overthrew the monarchy and in 1979, Soviet troops invaded Kabul, the capital. From that time until 1988 a period of guerrilla movements and conflict ensued. Some five million Afghans fled across the border, and another two million Afghans were displaced within the country. In April 1988, the Soviet Union signed an accord agreeing to withdraw its 115,000 troops. It is unclear at this point what will become of the country. The conflicting parties of guerrillas ensure that fighting will continue.

Conflict in Africa

Political conflict is endemic in Africa as well. Most African countries gained political independence in the 1960s or later. European control of Africa after 1884–1885—when Africa was divided into spheres of influence by the West European powers at the Congress of Berlin—was reflected in names such as French West Africa, British East Africa, the Belgian Congo, and German East Africa. The European division of Africa into colonies established territorial boundaries that were superimposed across ethnic patterns. The rigid boundaries around each European colonial holding superimposed the European concept of territoriality upon the African continent. The boundaries that were created have little relationship to cultural or physiographic features, but are arcs of circles, lines of latitude, or meridians. Such boundaries were simple for the European diplomats to draw, but they cut through the traditional territories of tribal and cultural groups

TABLE 11–4
Historical Events in Afghanistan

17 July	1973	Lieutenant General Sardar Muhammad Droud Khan in a coup became founder, president, and prime Minister of the Republic of Afghanistan.
	1977	One-party state with republican constitution established.
27 April	1978	"Saur Revolution" established the Democratic Republic of Afghanistan with Nur Muhammad Taraki as secretary-general.
December	1978	Friendship treaty with USSR signed.
September	1979	Taraki ousted by Hafizullah Amin.
25–26 December	1979	5,000 Soviet troops airlifted to Kabul. Amin killed.
27 December	1979	Amin replaced by Karmal.
January	1980	85,000 Soviet troops in Afghanistan.
Spring	1980	First clashes between Soviet troops and Mujahidin.
May	1986	Karmal replaced by Najibullah.
	1987	U.N. estimates 5 million Afghan refugees. 1 million Afghans killed. 12,000–30,000 Soviet deaths.
15 April	1988	Soviets agree to withdraw from Afghanistan.
March	1989	Withdrawal of Soviet troops complete.

of the African continent. Nearly two hundred tribal groups are bisected by present political boundaries based on these colonial boundaries. The potential for conflict resulting from Africa's superimposed boundaries is obvious. The newly emergent states face strong centrifugal forces and enjoy few centripetal forces, and there has been repeated conflict in Africa (Figure 11–16). The continued conflicts are representative of the types of problems facing the African countries as they attempt to develop a strong sense of national identity to overcome cultural fragmentation associated with tribalism in Africa.

Political Conflicts in the Middle East

Although the conflict between Israel and the Palestinians has dominated the political scene in the Middle East since 1947, it is not the only conflict in the area. Other conflicts include the Kurds' quest for an independent state, the war between Iran and Iraq, and the continued ethnic fratricide of Lebanon. Lebanon was once a peaceful region known as the Switzerland of the Middle East. The constitution of the country, which is based on a 1932 census that showed a Christian majority, stipulates that the president will be Christian and the prime minister Muslim. Rapid population growth among the Muslims, augmented by an influx of Palestinian refugees from Israel and the exodus of Christians, has changed the religious composition of Lebanon, and today Muslims constitute a majority. The territory of Lebanon is divided among a host of individual groups, in essence creating a landscape of terror and a political map con-

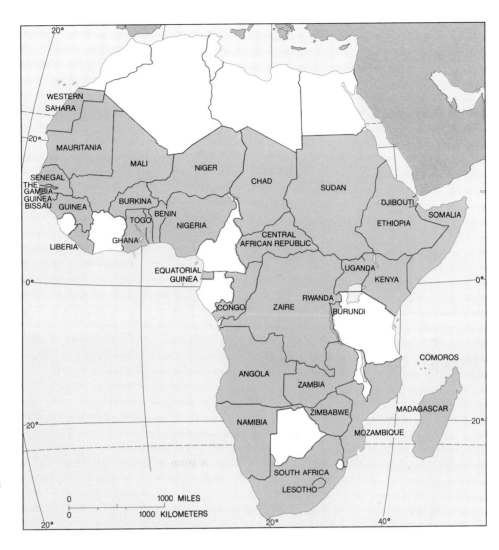

FIGURE 11–16 African Countries Where Armed Conflicts Have Occurred, 1960–1989

Beirut, Lebanon was once a beautiful city, but political conflict has reduced much of it to a wasteland.

trolled by individual warlords (See Figure 11–17 on page 446).

Other Political Conflicts

The political difficulties that involve entire countries are but a few of the conflicts in the world. They normally make national and international news because of their destruction of life and property. Other conflicts that are less publicized also reflect the importance of politics in the cultural geography of the world. As one example, the waters of the Colorado River and its tributaries make contemporary life-styles possible in the states of the American Southwest. Some states—notably California, Nevada, and Arizona—rely

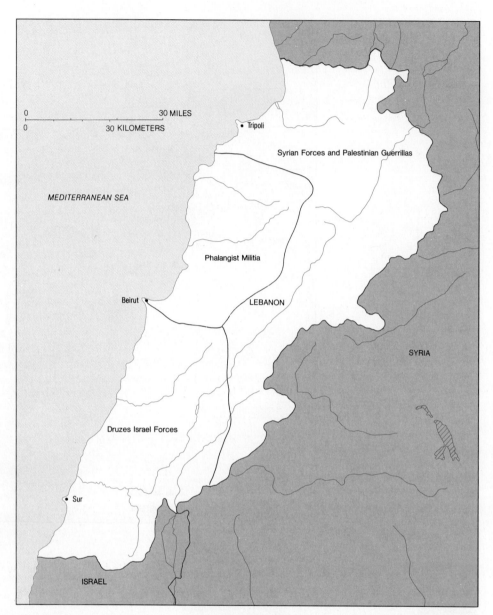

0   30 MILES

0   30 KILOMETERS

• Tripoli

Syrian Forces and Palestinian Guerrillas

MEDITERRANEAN SEA

Phalangist Militia

Beirut

LEBANON

SYRIA

Druzes Israel Forces

• Sur

ISRAEL

FIGURE 11–17 Political Map of Lebanon This map shows the territory controlled by different factions in Lebanon in 1988.

heavily on the Colorado's water for agriculture, industry, and other activities. California's population growth has made it especially reliant on Colorado River water, leading to conflict with other, upstream states over water rights. Conflict with the states at the Colorado's headwaters ultimately led to a legal agreement known as the Colorado River Compact, signed in 1922.

The compact divides the waters of the Colorado River between the upper basin states (Wyo-

ming, New Mexico, Utah, and Colorado) and the lower basin states (California, Arizona, and Nevada). The upper basin states, with smaller populations, did not initially use all of the water the Colorado Compact allocated them, but in the last two decades, large reclamation projects in these states have led to demands for more water than the Colorado River carries. Problems arising from the division of the waters of the Colorado among U.S. states are compounded by the rights of Mexico to a certain portion of its flow and the concentration of minerals in the Colorado River water that Mexico receives that harms Mexican crops and lands.

At an even more local level, each individual town or city is subdivided into a variety of polit-

ical territories that are often at odds with one another. The map of Salt Lake Valley (Figure 11–18) illustrates one facet of such a division. Each individual community in the Salt Lake Valley has its own organization to provide water. Some communities, such as Salt Lake City, have

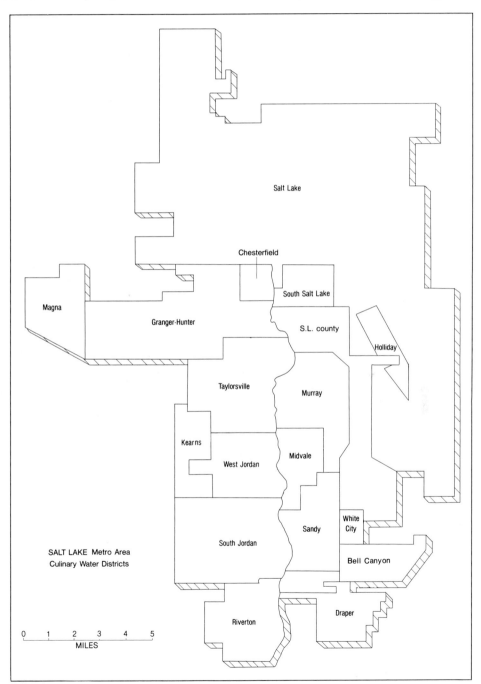

FIGURE 11–18 The Salt Lake Valley Territories of water districts in the Salt Lake Valley illustrate the impact of local political sub-divisions.

several. Each has different water rights, and during drought periods some portions of cities or individual towns lack water while others have a surplus. Such administrative divisions are reluctant to give up any of their rights for the benefit of the entire group. Divisions into school districts, zoning areas, or other political regions affect the lives of individuals of an area in similar fashion.

Overcoming Political Fragmentation

Conflicts related to political issues and control of territory have prompted efforts to minimize confrontation. Several kinds of organizations attempt to promote harmony among states. A common categorization recognizes international organizations, regional organizations, and subsets of each, including economic or military organizations or alliances.

The United Nations is the most widely known international organization that attempts to minimize political conflict in the world. Founded in 1945, its membership now includes nearly all of the independent countries of the world. In addition to playing a major role in the processes leading to the decline of colonialism, the United Nations assists new states; mediates between countries through the International Court of Justice and the International Law Commission; establishes programs to facilitate trade and provide access to the sea for trade and economic development; and promotes peace in areas at war with one another.

Another type of international organization attempting to promote international cooperation is the *Commonwealth*, originally the British Commonwealth of Nations. The Commonwealth is a free association of nearly forty

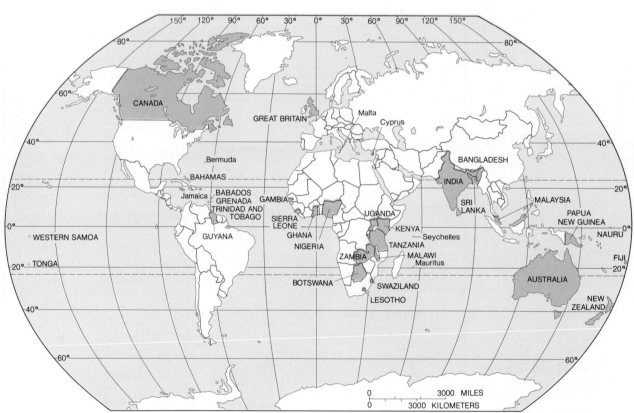

FIGURE 11–19 The Commonwealth Formerly known as the British Commonwealth, the Commonwealth is an example of a voluntary international political organization.

countries from every continent (Figure 11–19). Their only common characteristic is recognition of the British king or queen as head of the Commonwealth, their former status as British colonies, use of the English language and legal system, and their agreement that membership in the Commonwealth brings important benefits. Historically, membership in the Commonwealth or its predecessor allowed free migration to the United Kingdom (essentially halted since 1963), preference in trade with the United Kingdom, and military protection. Today the Commonwealth provides a forum for interaction among the member states and for cooperation in trade and other activities.

The French community is another international body based on former colonial relationships. It consists of countries that chose to be independent from France but retained member-ship in the French community. It has largely ceased to function as originally envisioned, and today serves primarily as a reminder of the former glories of France, and the importance of the French language in some areas of the world (Figure 11–20).

Another well-known international organization is the Organization of Petroleum Exporting Countries (OPEC). Originally organized in 1960 by Iran, Iraq, Kuwait, Saudi Arabia, and Venezuela, it has since been joined by other oil-producing states (Algeria, Ecuador, Gabon, Indonesia, Libya, Nigeria, Qatar, and the United Arab Emirates). Headquarters for the OPEC organization is in Vienna, Austria, a neutral site readily accessible by all members. OPEC has attempted to stabilize oil prices and to ensure that oil-producing countries receive a fair price for their product (Figure 11–21).

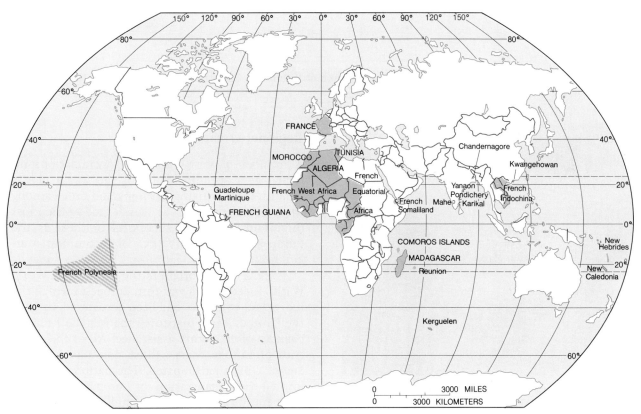

FIGURE 11–20 The French Community The French community provides an international organization for former French colonies.

FIGURE 11–21 Members of the Organization of Petroleum Exporting Countries (OPEC)

The continued use of French in Algeria is a visible evidence of its membership in the French community.

Regional Organizations

A number of very important organizations directed at a specific region of the world play a major role in bringing harmony and unity among the political units of the world. One of the most important and most successful of these is the European Community (EC) or "Common Market" founded in 1955.

The EC had its origin in the period after World War II. Faced with a war-torn Europe in which the individual countries had been devastated, in 1947 the United States instituted the Marshall Plan, a package of economic assistance. All of the countries of Europe except those dominated by the Soviet Union participated. The nations involved formed the Organization for European Economic Cooperation (OEEC) as a vehicle for dispersing American aid. In 1950, Robert Schumann, the foreign minister of France, noted that Germany had coal needed for the manufacture of iron and steel, while France had iron ore, yet neither had a sufficient supply of both. He suggested that the nations of Europe cooperate in manufacturing iron and steel and remove barriers that would prevent this cooperation (Figure 11–22). Schumann's plan was formalized in the European Coal and Steel Community (ECSC). At first only West Germany, Italy, France, Belgium, the Netherlands, and Luxembourg agreed to join. The ECSC was tremendously successful, and in 1955 the six countries involved suggested that Europe consider a common economic community. Although all nations of Europe were once again invited to join, only the original six nations became mem-

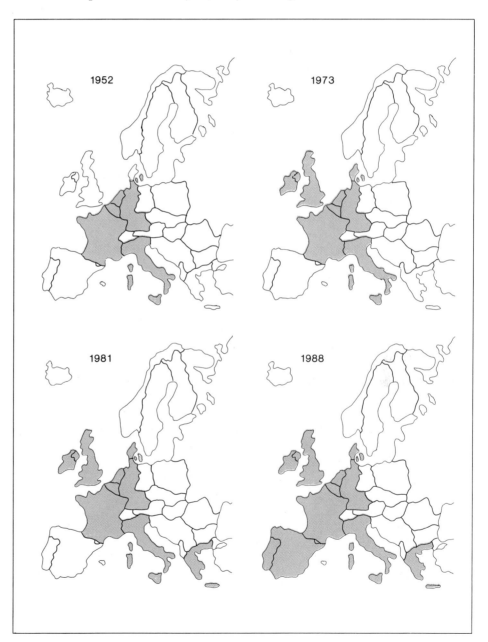

FIGURE 11–22 The Development of the European Community, 1952–1988

bers of the European Economic Community (EEC) in 1955.

The economic success of the EEC led to its expansion to include the United Kingdom, Ireland, and Denmark in 1972. Greece became a full member in 1985, and Spain and Portugal were allowed to become members in 1986. Austria, Sweden, Switzerland, and Finland have special trading relationships with the Common Market, while Norway, which was offered membership in 1972, has chosen to remain outside because the Norwegians fear other nations will fish in their territorial waters and challenge their role as a surplus oil producer. The name of the organization was changed to the European Community in 1988.

Regional organizations such as the EC are primarily concerned with economic activities. The EC has removed many tariff barriers between individual member nations, imposing a standard tariff on all imported goods. In 1992 the EC countries are to eliminate all tariff barriers within the region, but continued differences in language, religion, history, and national identities make it unlikely that the EC will overcome the political fragmentation of Europe to create a United States of Europe.

Nearly every major region in the world boasts some form of regional organization. For example, the Organization of American States (OAS), headquartered in Washington, provides an umbrella organization for all of the independent states of the Western hemisphere except Canada and Guyana (who have chosen not to join). The OAS has a much more limited role than the EC in the economies and political life of the member nations. The primary role of the OAS has been to facilitate activities of interest to members of the organization such as agriculture, health, and industry.

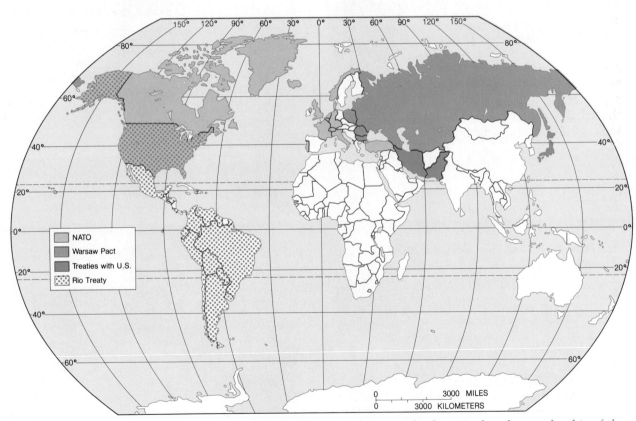

FIGURE 11–23 World Military Alliances Military alliances are constantly changing, but the membership of the North Atlantic Treaty Organization and the Warsaw Pact have changed only slightly in the past three decades.

Military Alliances

Military alliances are another form of supranational organization. They involve agreements between states to provide protection and cooperation in defense-related issues. The number and variety of military alliances is large and constantly changing. The largest and strongest is the North American Treaty Organization (NATO), created in 1949 in Washington by the United States, Canada, and ten Western European countries. Greece and Turkey joined in 1952, and West Germany in 1955. Initially a group of states organized in response to the perceived Communist threat from the Soviet Union, it has persisted to the present. France withdrew from the military activities of NATO in 1966, leading to the shift of the headquarters from Paris to Brussels, Belgium (Figure 11–23). NATO not only unites the members in defensive military agreements, but has a joint military force consisting of units from all member countries except Iceland (which has no military force).

The Warsaw Treaty Organization (Warsaw Pact) is in many ways the Communist counterpart of NATO. Created in 1955 by the Soviet Union and its Eastern Europe satellites in response to the rearming of West Germany and its admission to NATO, the Warsaw Pact has its headquarters in Moscow. Functioning much like NATO, it has been involved in military actions including the occupation of Czechoslovakia in 1968 and of Hungary in 1955. Like NATO, the Warsaw Pact holds military maneuvers and cooperates in defensive military activities.

Conclusion

Even if only the independent countries are considered, the complexity of the political map of the world is sometimes overwhelming to the casual observer. Recognition that there are subsets to each of these units, as well as organizations that transcend the boundaries of individual units, further complicates the world map. It is important to remember, however, that these lines, which may seem arbitrary, indicate underlying relationships of the cultural geography of the world. Boundaries around nations, individual political units within nations, and the various regional, national, and international organizations indicate the changing perceptions, goals, and values of the human population. In spite of the fragmentation that this presents, it is possible to recognize broad regions that have similar cultural characteristics. These similarities may be political, linguistic, religious, or represent some other cultural variable, but each grouping creates a distinctive regionalization of the world.

QUESTIONS

1. Compare and contrast formal and informal spatial organizations.
2. Discuss how a state differs from a nation.
3. Identify some examples of nations and states that have unity or lack unity and explain why you place them in that category.
4. Discuss centrifugal forces and give examples of their effect on present-day countries.
5. What are some problems associated with a plural state? Justify your response.
6. Indicate the types of boundaries and the strengths and weakness of each.
7. What are the advantages and disadvantages of multinational states?
8. Compare and contrast a stateless nation and nationalism.
9. Why do insurgent states exist today?
10. Discuss major conflicts in the world today giving an example from each of the major regions of the world.

SUGGESTED READINGS

Agnew, J. A. ''An Excess of National Exceptionalism: Towards a New Political Geography of American Foreign Policy,'' *Political Geography Quarterly*, 2 (1982): 151–66.

Archer, J. Clark and Peter J. Taylor. *Section and Party: A Political Geography of American Presidential Elections, from Andrew Jackson to Ronald Reagan*. New York: John Wiley & Sons, 1981.

Archer, J. Clark and Fred M. Shelley. *American Electoral Mosaics, Resource Publications in Geography*. Washington, D.C.: Association of American Geographers, 1986.

Archer, J. C. (with G. T. Murauskas, F. M. Shelley, P. T. Taylor, and E. R. White). "Counties, States, Sections, and Parties in the 1984 Presidential Elections." *The Professional Geographer* 3 (August 1985): 279–87.

Baram, Michael S. *Environmental Law and the Siting of Facilities: Issues in Land Use and Coastal Zone Management*. Cambridge, Mass.: Ballinger, 1976.

Bateman, Michael, and Raymond Riley, eds. *The Geography of Defense*. Totowa, N.J.: Barnes and Noble, 1987.

Bennett, D. Gordon, ed. *Tension Areas of the World: A Problem-Oriented World Regional Geography*. Champaign, Ill.: Park Press, 1982.

Bensel, R. F. *Sectionalism and American Political Development 1880–1980*. Madison, Wisconsin: University of Wisconsin Press, 1984.

Bergman, Edward F. *Modern Political Geography*. Dubuque, Iowa: Wm. C. Brown, 1975.

Blacksell, Mark, *Post-War Europe: A Political Geography*. Boulder, Colo.: Westview Press, 1977.

Boal, Frederick W., and J. Neville H. Douglas, eds. *Integration and Division: Geographical Perspectives on the Northern Ireland Problem*. London and New York: Academic Press, 1982.

Bodman, Andrew R. "Regional Trends in Electoral Support in Britain, 1950–1983." *The Professional Geographer* 3 (August 1985): 288–95.

"Boundaries and the Cultural Landscape." *Regio Basiliensis*, symposium volume 22 (December 1981).

Brickman, Ronald, Sheila Jasanoff, and Thomas Ilgen. *Controlling Chemicals: The Politics of Regulation in Europe and the United States*. Ithaca, N.Y.: Cornell University Press, 1985.

Brown, Curtis M., Walter G. Robillard, and Donald A. Wilson. *Boundary Control and Legal Principles*. New York: John Wiley & Sons, 1986.

Brunn, Stanley D. "Cities of the Future," in Stanley D. Brunn and Jack F. Williams, *Cities of the World*, pp. 453–89. New York: Harper & Row, Pub., 1983.

———. *Geography and Politics in America*. New York: Harper & Row, 1974.

Burghart, A. F. "The Bases of Territorial Claims." *Geographical Review* 63 (April 1973): 225–45.

Burnett, Alan D., and Peter J. Taylor, eds. *Political Studies from Spatial Perspectives*. New York: John Wiley & Sons, 1981.

Busteed, M. A., ed. *Developments in Political Geography*. London: Academic Press, 1983.

Chay, John, and Thomas E. Ross, eds. *Buffer States in World Politics*. Boulder, Colo.: Westview Press, 1986.

Clark, Gordon L. "The Spatial Division of Labor and Wage and Price Controls of the Nixon Administration." *Economic Geography* 61 (1985).

Clarke, Colin, and Tony Payne, eds. *Politics, Security and Development in Small States*. London and Winchester, Mass.: Allen and Unwin, 1987.

Cohen, Saul B. "A New Map of Global Geo-Political Equilibrium: A Development Approach." *Political Geography Quarterly* (1982): 223–41.

Cox, Kevin R. *Location and Public Problems: A Political Geography of the Contemporary World*. Chicago: Maaroufa Press, 1979.

Cox, Kevin R., David R. Reynolds, and Stein Rokkan, eds. *Locational Approaches to Power and Conflict*. New York: Halsted Press, 1974.

Dale, E. H., "Some Geographical Aspects of African Land-Locked States." *Annals of the Association of American Geographers* 58 (September 1968): 485–505.

de Blij, Harm J., and Martin Ira Glassner. *Systematic Political Geography,* 4th ed. Toronto: Wiley, 1988.

Dickens, Peter, Simon Duncan, Mark Goodwin, and Fred Gray. *Housing, States and Localities*. New York: Methuen, 1985.

Downing, David. *An Atlas of Territorial and Border Disputes*. London: New English Library, 1980.

DuMars, Charles T., Marilyn O'Leary, and Albert E. Utton. *Pueblo Indian Water Rights: Struggle for a Precious Resource*. Tucson: University of Arizona Press, 1984.

Easterly, Ernest S., III. "Global Patterns of Legal Systems: Notes Toward a New Geojurisprudence." *Geographical Review* 67 (1977): 209–20.

Francis, John G., and Richard Ganzel, eds. *Western Public Lands: The Management of Natural Resources in a Time of Declining Federalism*. Totowa, N.J.: Rowan & Allenheld, 1985.

Fryer, Donald D. "The Political Geography of International Lending by Private Banks." *Transactions of the Institute of British Geographers*, n.s. 12 (1987): 413–32.

Gellner, Ernest. *Nations and Nationalism*. Ithaca, N.Y.: Cornell University Press, 1983.

Gottmann, Jean, ed. *Center and Periphery: Spatial Variation in Politics*. Beverly Hills, Calif.: Sage, 1980.

Gottman, Jean. *The Significance of Territory*. Charlottesville: University of Virginia Press, 1973.

Gradus, Yehuda. "The Role of Politics in Regional Inequality: the Israeli Case." *Annals of the Association of American Geographers* 73 (September, 1983): 388–403.

Gray, Colin S. *The Geopolitics of Super Power*. Lexington, Ky.: University of Kentucky Press, 1988.

Harrison, R. T., and D. N. Livingstone. "The Frontier: Metaphor, Myth and Model." *The Professional Geographer* 2 (May, 1980): 127–32.

Henry, Norah F. "Regional Dimensions of Abortion-facility Services." *The Professional Geographer* 1 (February 1982): 65–70.

Jay, Peter. "Regionalism as Geopolitics." *Foreign Affairs* 58 (1979): 485–514.

Johnson, Hildegard Binder. *Order Upon the Land: The U.S. Rectangular Land Survey and the Upper Mississippi Country.* New York: Oxford University Press, 1976.

Johnston, R. J. *Geography and the State.* New York: St. Martin's Press, 1982.

_____. *The Geography of Federal Spending in the United States of America.* New York: John Wiley & Sons, Research Studies Press, 1980.

_____. *Political, Electoral, and Spatial Systems.* Oxford, England: Clarendon Press, 1979.

_____. "Political Geography and Welfare: Observations on Interstate Variations in Aid to Families with Dependent Children Programs." *The Professional Geographer* 4 (November 1977): 347–52.

Kennan, George F. *The Nuclear Delusion: Soviet-American Relations in the Atomic Age.* New York: Pantheon Books, 1982.

Kliot, Nurit, and Stanley Waterman, eds. *Pluralism and Political Geography—People, Territory and State.* New York: St. Martin's Press, 1983.

Knight, David B. "Identity and Territory: Geographical Perspectives on Nationalism and Regionalism." *Annals of the Association of American Geographers* 72 (December 1982): 514–31.

Lacoste, Yves. "An Illustration of Geographical Warfare: Bombing of the Dikes on the Red River, North Vietnam." *Antipode* 5 (May 1973): 1–13.

Logan, Bernard I., Jr. "Evaluating Public Policy Costs in Rural Development Planning: The Example of Health Care in Sierra Leone." *Economic Geography* 61 (1985).

Lustick, Ian. *Arabs in the Jewish State: Israel's Control of a National Minority.* Austin: University of Texas Press, 1980.

Mackinder, Sir Halford J. *Britain and the British Seas.* London: Heineman, 1902.

_____. "The Geographical Pivot of History." *Geographical Journal* 23 (April 1904): 421–37.

McDonald, Albert. "Mines in a Lawless Sea." *Geographical Magazine* 44 (September 1982): 501–3.

Mathieson, R. S. "Nuclear Power in the Soviet Bloc." *Annals of the Association of American Geographers* 70 (June 1980): 271–279.

Modelski, G. *Long Cycles in World Politics.* Seattle: University of Washington Press, 1987.

O'Loughlin, John. "The Identification and Evaluation of Racial Gerrymandering." *Annals of the Association of American Geographers* 72 (March 1982): 165–84.

Ose-Kwame, Peter, and Peter J. Taylor. "A Politics of Failure:

The Political Geography of Ghanaian Elections, 1954–1979." *Annals of the Association of American Geographers* 74 (December 1984): 574–89.

O'Sulivan, Patrick. "A Geographical Analysis of Guerilla Warfare." *Political Geography Quarterly* 2 (1982): 139–50.

_____. *Geopolitics.* New York: St. Martin's Press, 1986.

O'Sullivan, Patrick, and Jesse W. Miller. *The Geography of Warfare.* London: Croom Helm, 1983.

Ovendale, Ritchie. *The Origins of the Arab-Israeli Wars.* London and New York: Longman, 1984.

Parker, Geoffrey. *A Political Geography of Community Europe.* London: Butterworth, 1983.

_____. *Western Geopolitical Thought in the Twentieth Century.* New York: St. Martin's Press, 1985.

Parker, W. H. Mackinder. *Geography as an Aid to Statecraft.* Oxford: Clarendon Press, 1982.

Pepper, David, and Alan Jenkins, eds. *The Geography of Peace and War.* Boulder, Colo.: Westview Press, 1985.

Peterson, J. E. *Yemen: The Search for a Modern State.* Baltimore: Johns Hopkins University Press, 1982.

Prescott, J. R. V. *The Political Geography of the Oceans.* Newton Abbot, England: David and Charles, 1975.

Pringle, D. G. *One Island? Two Nations? A Political Geographical Analysis of the National Conflict in Ireland.* Letchworth: Research Studies Press, John Wiley & Sons, 1985.

Richmond, Anthony H. "Ethnic Nationalism: Social Science Paradigm." *International Social Science Journal* 39 (February 1987): 3–18.

Rogge, John R. *Too Many, Too Long: Sudan's Twenty-Year Refugee Dilemma.* Totowa, N.J.: Rowman & Allenheld, 1985.

Rose, Richard. "National Pride in Cross-National Perspective." *International Social Science Journal* 37 (1985): 85–96.

Schoenberger, Erica. "Foreign Manufacturing Investment in the United States: Competitive Strategies and International Location." *Economic Geography* 61 (1985).

Smith, Terence R., and Perry Shapiro. "On the Spatial Equity of the California Sales Tax: Estimates of Imbalances in Tax Dollar Flows." *Economic Geography* 55 (1979): 135–46.

Soffer, Arnon, and Julian V. Minghi. "Israel's Security Landscapes: The Impact of Military Considerations on Land Uses." *The Professional Geographer* 38 (February 1986): 28–41.

Swaney, James A., and Frank A. Ward. "Optimally Locating a National Public Facility: An Empirical Application of Consumer Surplus Theory." *Economic Geography* 61 (1985).

Tata, Robert J. "Poor and Small Too: Caribbean Mini-States." *Focus* 29 (November-December 1978): 1–11.

Taylor, Peter J., and John W. House, eds. *Political Geography: Recent Advances and Future Directions*. London: Croom Helm, 1984.

Thrall, Grant Ian. "Spatial Inequities in Tax Assessment: A Case Study of Hamilton, Ontario." *Economic Geography* 55 (1979): 123–34.

Waterbury, John. *Hydropolitics of the Nile Valley*. Syracuse, N.Y.: Syracuse University Press, 1979.

Zagarri, Rosemarie. *The Politics of Size: Representation in the United States, 1776–1850*. Ithaca, N.Y.: Cornell University Press, 1987.

Cultural Realms: Regionalizing the Mosaic of Culture

. . . The region is much more than a passive laboratory. It has its own specific physical environment, its own history, its own culture, all of which condition the expression of spatial or environmental relations.
E. J. TAAFFE

Introduction

The pattern created by global variations in language, economic life, political organization, religion, and other variables of interest to cultural geographers makes a map of great complexity. Viewed on a worldwide scale, the variety of individual countries and cultural phenomena may appear to have little order. Viewed on a large-scale map, even an individual community or country exhibits similar cultural variations. The diverse political organizations, ethnic backgrounds, and religious preferences in even a single community create a complex, fragmented mosaic when each of the cultural variables is examined separately.

These cultural variables may, however, be generalized into regions characterized by some underlying unity or interrelationship to provide a relatively simple but meaningful division of the world. Such a division allows the complexity of the world's cultural geography to be organized into readily recognizable units. Cultural geographers have used the term *cultural realm* to describe a large area where some fundamental unity is recognizable from the organization and integration of significant cultural traits. Delimitation of cultural realms implies that they are distinct from surrounding areas and, like all regions, share some internal unifying characteristics—that is, each realm displays an assemblage of characteristics concentrated in a particular area of the globe that make it different from all other areas.

Delimiting Cultural Realms

Dividing the world into cultural realms raises the same problems associated with defining regions of any type or size: the location and nature of boundaries, the criteria to be used in defining the boundaries and region, the scale of the investigation, and the date the region represents. Boundaries of a region of any type or at any scale must accurately reflect the distribution of the unifying phenomena or assemblage of factors that have been used to characterize it.

Boundaries

Drawing the boundaries of cultural realms is difficult because the boundary is often located in a zone of transition to another cultural realm. Where a great mixing of cultures has occurred over an extended period of time, defining the precise boundary within the zone of transition is very complex. South of the Sahara the expansion of the Islamic religion over the last millennium has led to a complex mixing of Muslim and non-Muslim peoples, making boundaries in part arbitrary (Figure 12–1).

The problem of establishing boundaries for cultural realms is complicated by the scale at which they are examined. At a global scale many of the smaller cultural variations are ignored by boundaries drawn to indicate the major cultural realms.

A current map of cultural realms is a representation of the earth at the end of the twentieth century. Cultural realms drawn at other times would be much different. A map of the cultural realms of the earth two millennia ago would include the Roman Empire, the great civilizations of the Indian subcontinent and China, the Egyptian civilization, and the Germanic, Slavic, African, and American areas characterized by individual civilizations or tribes (Figure 12–2).

In addition, the boundaries of cultural realms should not simply reflect the present pattern of political control, because political dominance by one group may be at variance with the underlying culture. South Africa, for example, is dominated by a white minority of European ancestry, but it is generally defined as part of the African cultural realm, not the European. Decisions about the location of the boundaries of world cultural realms reflect the criteria used by those defining the realms.

World Cultural Realms

Numerous people have divided the world on the basis of culture: anthropologists on the basis of the level of development of individual groups; historians on the basis of the past; and geographers on the basis of both the past and present. Geographers' cultural realms (particularly those

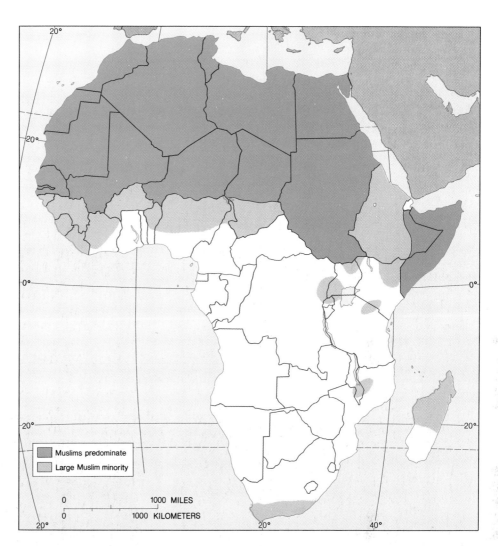

FIGURE 12–1 Muslims in Sub-Saharan Africa
Muslims are a majority or an important minority in these areas of Africa.

related to climatic realms) generally relate to economic development or to specific cultural phenomena such as religions, legal systems, or types of government (Figure 12–3). This text recognizes seven broad cultural realms, which cover areas of various sizes, with great differences in population (Table 12–1). The European (also known as the *Occidental*) has the largest population and covers the largest total land area, reflecting its wide diffusion. The Polynesian-Melanesian (South Pacific) has the smallest population, but covers a large area of the earth's surface, since its population is restricted to the scattered islands of the region.

TABLE 12–1
Population of Cultural Realms

CULTURAL REALM	POPULATION (MILLIONS)
European (Occidental)	1,544
East Asian (Oriental)	1,302
South Asian (Hindu)	963
Islamic	546
Southeast Asia	433
Sub-Saharan Africa	450
Polynesian-Melanesian	6

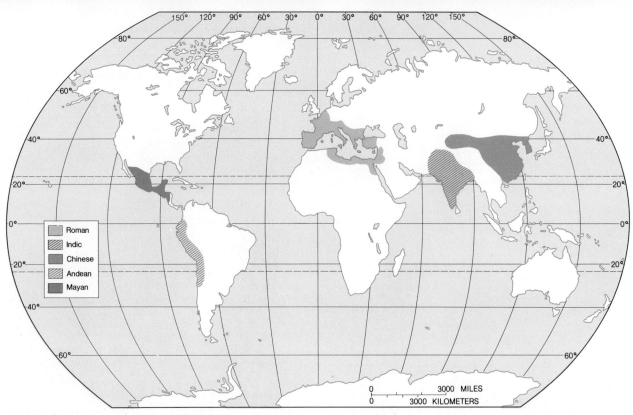

FIGURE 12–2 Cultural Realms in 0 A.D. Five cultural realms could be recognized as dominating specific regions.

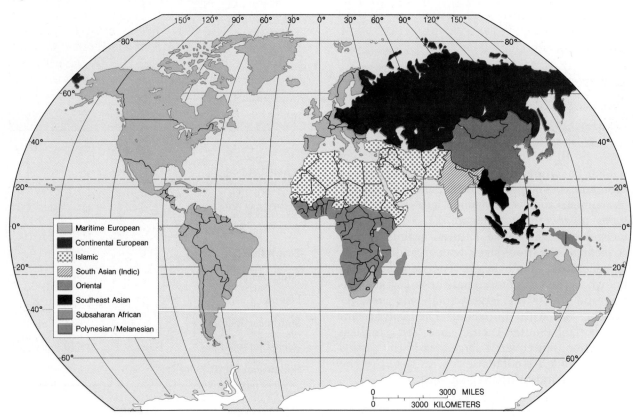

FIGURE 12–3 World Cultural Realms circa A.D. 1990

TABLE 12–2
Subdivisions of the European Cultural Realm

MARITIME EUROPE
Mediterranean Europe Subrealm:
Mediterranean Centre of Europe
Latin America
Western and Northern Europe Subrealm:
Countries of Western Europe
Countries of Northern Europe
Anglo-America
Australia-New Zealand
South Africa
CONTINENTAL EUROPE
Countries of Eastern Europe
Soviet Union

Each of the seven cultural realms can be subdivided into distinct subsets. The European realm, for example, can be subdivided into a continental and a maritime region, with each having subdivisions (Table 12–2). Each of the individual cultural realms will be examined to show its present size, characteristics, and the cultural history and ecology that have created the present global pattern.

The European Cultural Realm

The European cultural realm reflects the origin and diffusion from Europe of systems of government, religion, and life-styles. Prior to A.D. 1500, the European cultural realm was one of the world's smallest. Concentrated in the western portion of present-day Soviet Union and Europe, it diffused widely because of technological advances. The diffusion of European culture was facilitated by the general absence of highly developed and organized civilizations in most of the areas to which it diffused. The aboriginal population in Africa was both small and fragmented into tribal groups, which facilitated European dominance. The European invaders met a larger group of native peoples in North and South America, but the decimation of the native population by European-borne diseases and the lower level of technology and political organization among the

American groups made the European takeover relatively easy. In Asia where the Europeans encountered civilizations that were politically better organized, and the population occupied the arable land more intensively, the Europeans were unable to replace the native culture. Although they gained control of the trading and political systems of much of Asia, Europeans were never able to colonize it as they did the less intensively occupied lands (See Figure 5–17 on page 186).

Only in the Andes and Central America of Latin America did Europeans effectively replace or destroy the core of preexisting and highly developed civilizations. In the other areas, they simply added another layer of administration or economic control to existing civilizations or eliminated the people, as they did in the Caribbean. In tropical Africa, Southeast Asia, and East Asia, Europeans basically gained control of the economy rather than displacing the local population.

The importance of the European cultural realm today is even greater than its size suggests. The diffusion of cultural traits from the Occidental world in the form of technology, fashions, or other Western goods gives the realm an importance beyond its size. The impact of European or Western culture on the rest of the world is pervasive and is still increasing, particularly in urban areas.

The individual subdivisions of the European cultural realm have had varying impacts on the

Tall buildings in downtown Cairo, Egypt are but one example of the impact of Western or European culture on cities of the less industrialized world.

world. Maritime Europe has had a different impact on the world than continental Europe, and the two subdivisions of maritime Europe (Mediterranean and Western and Northern Europe) have also had varying impacts on the areas into which they have diffused. Examination of each of the subsets indicates both the importance of the European cultural realm and the contrasts among its subdivisions.

Mediterranean Europe and Latin America

The Mediterranean division of the European cultural realm has a long and rich history of civilization. The Mediterranean Basin was the core of the Greek (*Hellenistic*) civilization and the subsequent Roman civilization and empire. Later, in the Middle Ages, the Mediterranean was one center of commercial, artistic, and scientific and technological advance within Europe, the other being the Arab world. The development of trade by the merchants of city states such as Genoa and Venice created commercial networks that connected the entire Mediterranean Basin. The expansion of geographical knowledge by the Portuguese and Spanish during the Age of Discovery in the fifteenth century allowed the Mediterranean countries to expand their influence far beyond the Mediterranean.

The first colonization efforts of the European cultural realm occurred after the "discovery" of the New World by Christopher Columbus. The Spanish Conquest of Middle and South America in the early sixteenth century allowed the Spanish to gain control of the Indian civilizations from Mexico to Peru.

The Spanish control of the complex civilizations of the Aztecs of Middle America and the Incas of the Andes brought important changes to the cultural map of Latin America. The Spanish conquest led to tremendous loss of life as disruption of the existing civilizations was compounded by the introduction of European diseases, which reduced the native population by as much as 90 percent. Spanish institutions of land, wealth, religion, language, and political organization modified the existing Indian civilizations. The Spanish government established the Council of Indies to provide guidelines under which individuals were

granted rights to explore, conquer, and colonize. One of the principles developed by the council was the *encomienda system* of land tenure. Whereas the Indians of Mexico and Central and South America had held their lands in common, the *encomienda* system granted Spaniards the right to require labor from Indians on a tract of land in return for the Spanish providing religion, protection, and "civilization." The *encomendero* was required to teach the Indians European culture, the most important aspect of which was the Roman Catholic religion.

Although the *encomienda* system did not initially grant ownership of the land to the *encomendero*, in time the lack of private ownership among the Indians led to ownership of the land by the individual granted the *encomienda*.

The Spanish used the institutions of the Catholic church and the army to conquer and "civilize" Latin America. The army protected the Spanish colonizers and subdued the Indians, and the Catholic church brought Christianity and other aspects of European culture to the Indians. The three elements—landed aristocracy, army, and church—became the basis of power in Latin America. Members of the ruling class were separated both socially and economically from the majority of the inhabitants of Latin America.

The cultural variables brought by the Spanish to Latin America (language, religion, technology, social class) are unevenly distributed. The resulting cultural fragmentation in the Spanish-occupied portions of Latin America differentiates between Europeans, Indians, and *mestizos* (peoples of mixed ancestry).

The lands controlled by the Portuguese were distinctly different from those of the Spanish. The Portuguese did not find advanced civilizations or large numbers of native inhabitants, nor did they discover the precious metals that spurred the Spaniards elsewhere. As a result, the Portuguese focused on the development of crops for export, and because of transportation needs the colonists remained concentrated along the coastal areas of today's Brazil.

Independence movements began in Latin America early in the nineteenth century. The revolution against Spain primarily involved the Spanish colonial elite. Distance from Spain, disagreement over governments, and the complex nature of colonial groups led to independent countries focused on each of the population nuclei in the Spanish colonies (Figure 12–4). Brazil did not become fragmented, in part because it lacked a multiplicity of populated nuclei, and in part because the Portuguese royal family moved to Brazil during the Napoleonic Wars in Europe.

Latin America today presents a contrast between the highlands of the Andes Mountains with their large proportion of Indian populations (Peru, Ecuador, Bolivia), the nations of the mid-latitudes (Chile, Argentina, Brazil) with their large proportion of European descendants, and those of Central America (Mexico, Guatemala) with a large proportion of both *mestizo* (mixed) and Indian populations (Figure 12–5). The southernmost countries of Latin America are most like the core of the European cultural realm. Argentina and Uruguay each have populations that are more than 85 percent of "pure" European ancestry. The most densely populated core area of

FIGURE 12–4 Spanish Viceroyalties and Independent Countries of Latin America The vice royalties established by the Spanish colonizers of Latin America are broadly similar to the independent nations that ultimately developed.

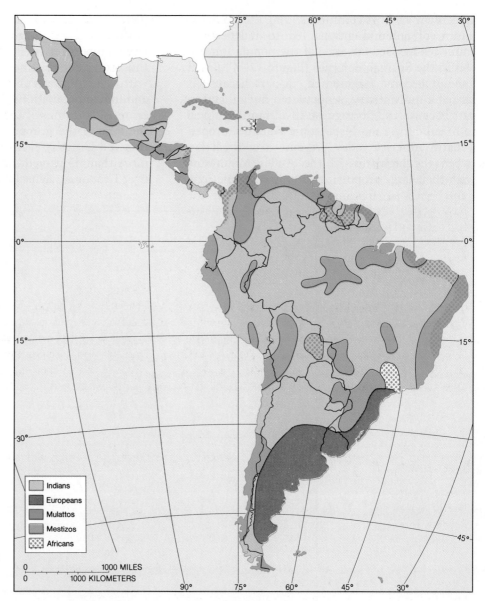

FIGURE 12–5 Cultures of Latin America This generalized map shows the distribution of cultural groups in Latin America.

Legend:
Indians
Europeans
Mulattos
Mestizos
Africans

0   1000 MILES
0   1000 KILOMETERS

Brazil (from Rio de Janeiro south) and central Chile (centered on Santiago) also have high proportions of people of "pure" European ancestry. One quarter of the Chilean people claim pure Spanish ancestry, but the rest of the country has a much higher proportion of *mestizos* than Argentina or Uruguay.

The Indian population is concentrated in the high Andes. In the zone of Indian dominance from southern Colombia to northern Chile/northwestern Argentina, subsistence agriculture, a feudal socioeconomic structure, and a high incidence of landless laborers combine to create standards of life far below that in other areas of the European Realm. High proportions of Indians are also found in the highlands of southern Mexico and in Guatemala.

The *mestizo* population dominates Latin America numerically. Mexico, Venezuela, Colombia, and Brazil are representative of countries with high proportions of *mestizos*. The *mestizo* population is sometimes regarded as transitional between those who are part of the European elite and the Indians. They are transitional in economic terms, being more a part of the commercial economy of the growing urban centers than the Indians, but less so than the Europeans; *mestizos* are also culturally transitional, with the use of a European language rather than an Indian dialect distinguishing them from the Indian culture regardless of their ancestry.

The situation in the hearth of the Mediterranean Europe cultural region is somewhat different. The countries of Portugal, Spain, and Italy have much higher standards of living, much less fragmented cultural makeups, and more stable political settings than does Latin America. In recent years the inclusion of the Mediterranean countries within the European Community (EC) has raised their standard of living, while the quality of life and level of technology in Latin America have risen more slowly. In both Mediterranean Europe and its offshoot, Latin America, however, the cultural traits of the Catholic church, Spanish or Portuguese languages, and late arrival of the democratic revolution are character-

istic. The level of economic development, per capita income, and industrialization in both regions has also lagged behind that of Western and Northern Europe as a result of the late arrival of the Scientific, Industrial, and Democratic Revolutions (Table 12–3). The presence of authoritarian governments in the Mediterranean region until the twentieth century, small, fragmented farms, and economic dominance by a small elite (particularly in Spain and Portugal) effectively separate the Mediterranean countries from the other countries in the core of the European cultural realm.

The Western and Northern European Sub-realm

The countries of Western and Northern Europe have had a profound effect on both the European cultural realm and the rest of the world. The diffusion of peoples, languages, technology, values, and political organizations from this portion of the European realm has affected the entire world. The importance of Western and Northern

TABLE 12–3
Comparisons of Europe and America

COUNTRY	LIFE EXPECTANCY (YEARS)	1988 GNP (PER CAP.) IN U.S. DOLLARS	LITERACY (PERCENTAGE OF POPULATION)	INFANT MORTALITY (PER 1,000)
Canada	76	14,100	99	9
United States	76	17,500	99	9
Austria	73	10,000	98	11
Belgium	75	9,230	98	11
Denmark	75	12,640	99	8
Finland	75	12,180	almost 100	6.2
France	75	10,740	99	9
Germany, West	75	12,080	99	11
Iceland	77	13,370	100	6.1
Ireland	73	5,080	99	11
Luxembourg	74	15,920	100	12
Netherlands	76	10,050	99	8
Norway	77	15,480	100	8
Sweden	77	13,170	99	6
Switzerland	77	17,840	99	8
United Kingdom	74	8,920	99	9
Spain	76	4,840	99	9
Portugal	73	2,230	99	15.8
Italy	75	8,570	99	10.1
Latin America	66	1,720	78	57

Europe to the rest of the world is illustrated by the four major revolutions that took place in this area between the fifteenth and eighteenth centuries: the Democratic Revolution, which occurred between the seventeenth and nineteenth centuries; the later Agricultural Revolution of the eighteenth century; the Industrial-Technological Revolution of the eighteenth to twentieth centuries; and the Scientific Revolution, which began in the seventeenth century.

Even today, the importance of the Western and Northern European subrealm in the world is obvious. Its pioneering efforts to unify economies through the EC, central role in the North Atlantic Treaty Organization, continued importance as a financial, cultural, and artistic center, and the dominance of west European languages (French and English) in international commerce and politics testify to its significance.

The most important region to which the Western and Northern European culture diffused is Anglo America, which in turn has become a secondary hearth for subsequent diffusion to much of the rest of the world. The eastern seaboard of the United States and Canada was colonized by several European nations, including England, Sweden, the Netherlands, France, and Spain (Figure 12–6). Over time, however, despite the multiplicity of colonizing countries, Anglo America emerged as two large states. Both are pluralistic societies made up of immigrants from many parts of the world, but their overwhelming cultural characteristics are those brought from Western and Northern Europe: democracy, capitalism, materialism, Christianity, and the dominance of English. Nevertheless, the term "Anglo America" should not obscure the fact that this is not a monocultural region of English descendants. The ethnic makeup of Anglo America is complex, for it includes descendants of immigrants from Africa, Latin America, Asia, and the islands of the Pacific as well as all the countries of Europe. The United States and Canada are plural societies characterized more by their values and their adherence to the general European culture than by a single distinctive cultural phenomenon that is all-pervasive, such as religion or language.

Anglo America's emergence as two large states seems an unlikely ending to the process of colo-

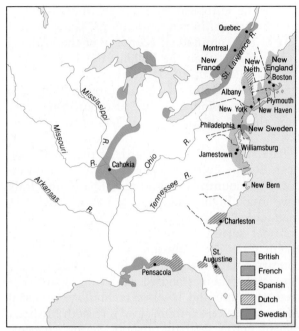

FIGURE 12–6 European Colonization Attempts in Anglo America

nization and settlement that began almost four hundred years ago. The factors that led to only two instead of numerous states include the absence of a readily available source of wealth like the gold and silver of Latin America, the lack of any large, highly organized native civilizations, and the absence of a formalized institution to spread the European culture as the Roman Catholic church did in Latin America. Consequently, Anglo America could grow and expand during the period of settlement and Canada and the United States could subsequently evolve into two large political units and develop their own systems largely independent from their European source regions.

The characteristics of the Western and Northern European cultural subrealm—free enterprise, religious tolerance, democracy, and pluralistic societies—have been emphasized to an even greater degree in America. Moreover, the Industrial Revolution, which had its origin in Europe, was further intensified in Anglo America. Today, the United States is both the largest single industrial producer in the world and the largest and wealthiest market.

The geographic factors that combined with the European culture to allow the United States to emerge as the dominant economic power in the world are its great size, rich resources, central location between the Atlantic and the Pacific for trade, lack of long boundaries with hostile neighbors, and accessibility provided by rivers and ports. The large landmass with relatively few occupants and rich resources combined with the expansion of democratic ideals and capitalism to attract large numbers of migrants.

Anglo America has become the primary source region of a highly modified European culture termed *American*, which has diffused widely. Spreading ideas of mass production and consumption, convenience foods (from fast-food restau-

Some of the traditional elements of English city scapes are found in Australian cities as a result of a conscious attempt by migrants to recreate familiar landscapes.

American products in Nigeria are one example of the importance of the United States as a source of cultural diffusion.

rants to frozen dinners), and American versions of English, music, television, clothing, and other cultural variables, Anglo America has replaced Western and Northern Europe as a hearth for maintaining and diffusing European culture to the rest of the world.

Australia and New Zealand: Outposts of Western and Northern Europe

Far removed from the shores of Europe are two countries that are an integral part of the Western and North European subrealm. Australia and New Zealand are similar in many ways to their original colonizing country, the United Kingdom. Some geographers refer to them as *replicative societies* because they maintain many characteristics of the United Kingdom. The development of towns, governments, and societal values conscientiously followed the model of England. Like other European colonies, both nations have developed economies based on an industrial model with literacy, high incomes, large amounts of leisure time, individualism and materialism, and Christianity.

South Africa: A West European Colonial Relic

South Africa represents a distinctly different outlier of the Western and Northern European cul-

European cultural elements have been superimposed over the African culture in South Africa.

tural subrealm. South Africa is dominated by a minority European population, whose religious, political, economic, and linguistic values have been grafted onto the majority African population. The persistence of the European colonial culture affects the entire country, even though the majority are part of the African cultural realm.

South Africa remains an anomaly in the world. Controlled by its minority European cultural group, it is a potential conflict area of the world. Rising black consciousness, growing black numbers, and increased impatience with the racist policies of the South African government suggest that armed conflict will increase within South Africa. Although the South African government is attempting to develop a Bantu middle class that shares the European culture, the majority of the Bantu population of South Africa belong to the African culture realm and use the European culture only for interaction with Europeans or with other tribes speaking a different African language or dialect.

Continental Europe as a Cultural Subrealm

Inland from the coastal margins of Europe lie the continental countries of Czechoslovakia, Hungary, Romania, Poland, Bulgaria, and the Soviet Union. These countries were not involved in the great colonization of the rest of the world by European nations, but migrants from Continental Europe were an important part of European colonization, especially to the Americas. Today Continental Europe is dominated by the Soviet Union, and the countries of the subregion differ in the degree to which they have adopted the Agricultural, Democratic, Industrial, and Scientific Revolutions of western Europe.

The dominant characteristic of Continental Europe is its adoption of a centrally planned economic model based on the theories of Karl Marx and Vladimir Illyich Ulyanov (Lenin). It is important to note that Continental Europe has always differed from southern, western, and northern Europe. Feudalism persisted much longer, and serfdom was not abolished in czarist Russia until the same decade as slavery was abolished in the United States. The Industrial Revolution came late to Continental Europe. Or-

The Orthodox Churches of Eastern Europe traditionally were a major symbol of the cultural realm.

thodox Christianity, which characterizes the bulk of the countries of Continental Europe, has always emphasized the close relationship of church and state.

Although czarist Russia adopted some of the characteristics of the rest of the European cultural realm after the time of Peter the Great (1689–1725), it has always had an outlook much different than that found elsewhere in Europe. Developing for the most part after the invasion of the Mongols under Ghengis Khan in the thirteenth century, Russian culture places the individual in a subservient role with respect to the government. Consequently, a strong centralized government with little democratic tradition is the norm. Western values of materialism, leisure, capitalism, and democratic and egalitarian societies are at variance with the Orthodox church, central planning of the economy, and Communist governments found in Continental Europe.

The East Asian (Oriental) Cultural Realm

The East Asian or *Oriental* cultural realm includes nearly as many inhabitants as the Euro-

TABLE 12–4
Population and Area of Asian Countries

COUNTRY	POPULATION (MILLIONS)	AREA (SQUARE MILES)
China	1087.0	3,718,783
Hong Kong	5.7	410
Japan	122.7	145,834
North Korea	21.9	46,540
South Korea	42.6	38,025
Macao	0.4	6
Mongolia	2.0	604,250
Taiwan	19.8	13,900

pean realm, but they are concentrated in a much smaller area of the earth's surface (Figure 12–7). The Oriental realm includes China, Korea, Japan, and Taiwan. China alone has more than one-fifth of the world's population, and its area is exceeded only by the Soviet Union and Canada (Table 12–4). The cultural characteristics of the Oriental realm include the ideas of Buddhism, Confucianism, and Taoism, agrarian economies, an early stage of the Industrial Revolution, and the dominance of centrally planned economies and Communist political control. South Korea, Taiwan, and Japan do not share all of these charac-

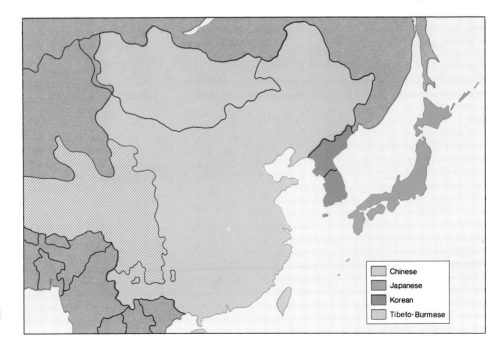

FIGURE 12–7 East Asian (Oriental) Cultural Realm The Oriental cultural realm contains at least four distinct cultural groups.

Chinese
Japanese
Korean
Tibeto-Burmese

teristics, but they contain a minority of the population of the East Asian realm.

The characteristics of the East Asian cultural realm reflect the long history of the Chinese dynasties and the associated Asian religions. The intertwining of religion and the political system has created a culture distinctly different from that found in the European cultural realm. The culture of the East Asian realm reflects the values and traditions of the teachings of Confucius and Taoism combined with elements of Buddhism. The influence of Confucianism with its emphasis on proper relationships led to a practical life-style that emphasized diligence and respect. The result has been what is called the *Confucian meritocracy*. Adopted in China, Korea, and Japan, the Confucian meritocracy is associated with a society in which hard work, proper relationships, and respect for leaders are the norm.

Variations in the East Asian Cultural Realm

East Asia has always been dominated by China. Separated from Europe and South Asia by mountains and deserts, and with a strong ethnocentric belief that the Chinese were superior to other peoples, China has looked inward throughout most of its history. Territorially, China has occupied the same area for the last two thousand years (from the end of the Han dynasty in A.D. 214). The commonality of Chinese ideograms for writing in China, Korea, and Japan has provided some cultural homogeneity not found in the European cultural realm. Confident of their own superior-

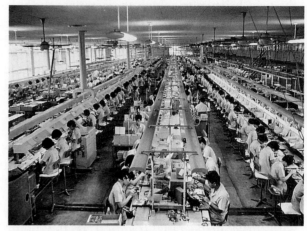

The industrial might of Japan is an anomaly in the East Asian cultural realm.

ity, the Chinese viewed themselves as the center of the earth, the "Middle Kingdom." Not until it was forced to cede control of part of its area to European colonial powers in the eighteenth and nineteenth centuries did China recognize the growing world power of Europe.

Like China, the Japanese refused to interact with Western civilization until the nineteenth century. A U.S. naval expedition under Commodore Perry arrived in Tokyo in 1853, and Perry's display of Western technology caused the Japanese rulers to begin a conscious course of adopting the Western industrial model. The adoption of Western technology ultimately led the tiny Japanese nation into a colonial era in which they occupied Korea and parts of China. After World

TABLE 12–5
Japan's Role in the World (in billions of dollars)

REGION OR COUNTRY	GROSS DOMESTIC PRODUCT	EXPORTS	IMPORTS
Japan	1,418.8	400.0	407.6
Latin America	796.3	129.4	69.9
Canada	336.7	107.0	100.6
Europe	4,105.6	1,172.0	1,183.8
Africa	429.5	130.0	99.2
Persian Gulf	247.0	88.4	63.0
Asia and Middle East	390.0	32.5	80.9
Pacific Basin	476.9	298.9	263.7
United States	3,344.6	362.0	505.2
Australia/New Zealand	212.6	30.0	26.4

Source: U.S. News and World Report, June 20, 1988.

War II, a Western democratic model of government was introduced into Japan by the United States. Since that time Japan has emerged as the third greatest industrial power after the United States and the Soviet Union (Table 12–5).

Changes in the twentieth century have led to distinctive political and economic differences in elements of the East Asian cultural realm, but relatively little change has occurred in the underlying unity of the region. The most important changes have been associated with the transformation of China under the rule of its Communist government, the emergence of Japan as one of the world's most important industrial powers, and the growing economic importance of Hong Kong, Taiwan, and South Korea (Figure 12–8). The latter

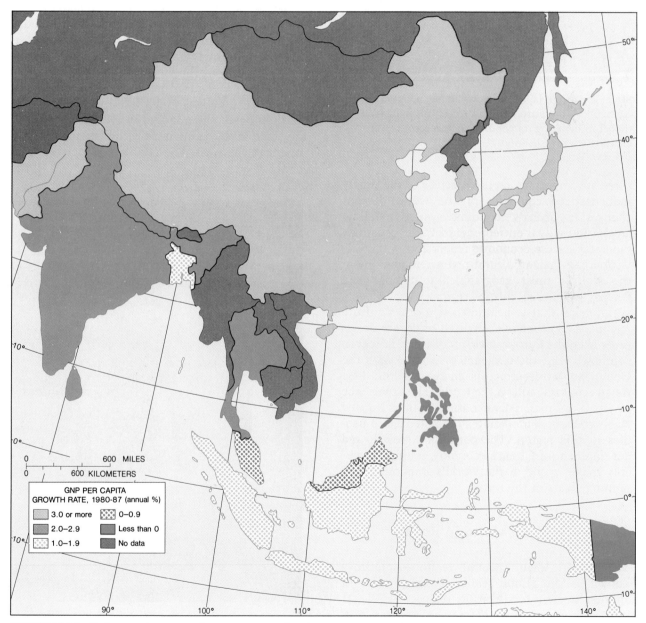

FIGURE 12–8 Per Capita GNP of East Asian and Other Economies The countries of East Asia illustrate how economic improvement varies from country to country.

Street in Tokyo. Japan has consciously adopted elements of the European and American industrial model in utilizing its population as a resource in creating a strong economy.

three are important representatives of the newly industrialized countries of the world. Recent changes in China's economic system to allow greater individual entrepreneurship are causing a modern Chinese economic revolution.

The East Asian cultural realm remains one of the world's most geographically distinct and densely populated culture regions with an economic level that is behind that of other culture regions. The conscious adoption of economic aspects from the European cultural realm, however, combined with the diligence associated with the Confucian work ethic, is increasing the East Asian economic role in the world. The experience of Japan, a small island nation with a limited resource base, demonstrates the potential of peoples in this region. The continued growth and development of China's economy could conceivably make it one of the world's important economic powers in the future.

The South Asian Cultural Realm

The South Asian cultural realm (sometimes known as the Hindu realm) is the product of another ancient civilization (Figure 12–9). At least a thousand years before the emergence of Europe as a dominant world force in exploration and colonization, India was the site of a flourishing civilization that developed social and economic orders that persisted until European colonialism destroyed important elements of its fabric. Central to the Indian civilization was Hinduism, a "religion" that transcends the traditional concept of religion and embodies a way of life that still largely determines the actions of each of its followers.

Hinduism does not extend to all parts of the cultural realm, however. Bangladesh is included in the South Asian cultural realm even though it is overwhelmingly Islamic because it has been a part of the long history of Indian civilization. Although Bangladesh practices the Islamic religion, culturally it is much like India. Village life predominates, the standard of living has remained low, and it is separated from the Islamic cultural realm to the west. The conflicts associated with

Hindu temples are an immediately recognizable element of the landscape in the Hindu cultural region.

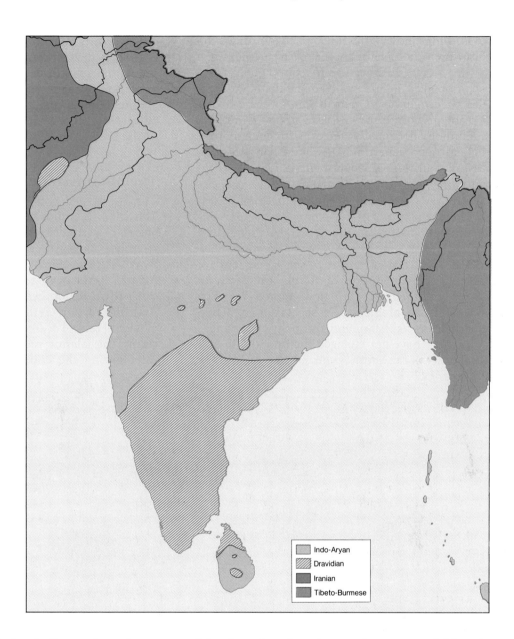

FIGURE 12–9 The South Asian Cultural Realm and Surrounding Cultures

the transformation of the Indian subcontinent into independent countries led to the separation of Bangladesh from India, but it continues to be in India's sphere of influence.

The island nation of Sri Lanka is also part of the South Asian cultural realm. Even though it has attempted to forge a separate political path since independence, it is still culturally, politically, and economically tied to India. Pakistan, however, is not included in the South Asian cultural realm. Pakistan is an extension of the

Islamic cultural realm to the west and historically has been a bridge between the South Asian cultural realm and the Islamic realm. Islamic invaders brought Islam to India and to Bangladesh, but Islam was always a minority religion in India.

Although the South Asian cultural realm has a few outliers in such places as Fiji and the Caribbean (Trinidad), it is a relatively small area that has never engaged in colonization or expansion. Examination of a map of the cultural realms of the world a millennium ago would reveal that the South Asian realm has grown the least among the cultural realms of the world.

India and Bangladesh are among the most complex cultural regions in the world if examined at a local scale. At the time of independence in 1947, the Indian subcontinent was splintered into over 550 princely states, with hundreds of separate dialects and fifteen major languages. Further compounding this complexity was the division between the Islamic and Hindu populations.

The one element that has provided an element of unity in India and the South Asian cultural realm is the long history of civilization. The Indian civilization predates the great age of Greece, the foundation of Rome, the birth of Christ, and the flight of the Jews from Egypt. The independence of India, the separation of Bangladesh from Pakistan, and the political changes since 1947 have not changed the underlying cultural unity of the South Asian cultural realm. The entire area is characterized by the lower level of economic development associated with the less industrialized world. Literacy rates, life expectancy, per capita income, and general quality of life are much lower than those found in the European cultural realm or in areas such as Japan or Taiwan in the East Asian realm.

The Main Islamic Cultural Realm

The main Islamic cultural realm includes one hearth of early domestication and civilization. It stretches across the arid and semiarid regions of North Africa, the Middle East, and Southwest Asia to Pakistan (Figure 12–10). The culture of this entire region reflects the influence of the Arab and Persian civilizations and the dominance of Islam (Table 12–6). Like the European cultural realm, the Islamic realm has expanded dramatically during the course of its existence. Islam

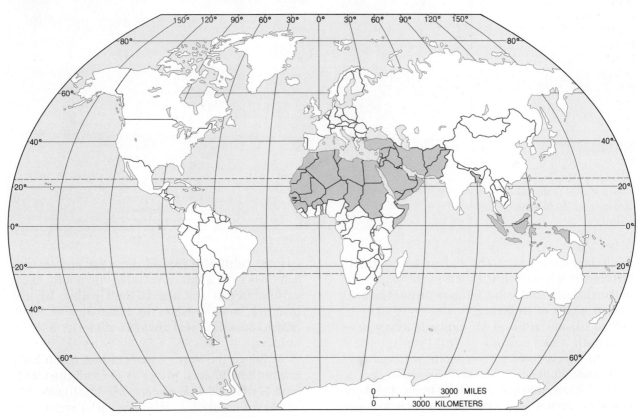

FIGURE 12–10 The Islamic Cultural Realm

TABLE 12–6
Number and Percent of Muslims by Country

COUNTRY	ISLAMIC POPULATION (IN MILLIONS)	% ISLAMIC	COUNTRY	ISLAMIC POPULATION (IN MILLIONS)	% ISLAMIC
Afghanistan	14.4	99	Malaysia	6.5	50
Albania	1.8	70	Maldives	0.1	100
Algeria	17.3	97	Mali	3.5	60
Bahrain	0.3	91	Mauritania	1.4	96
Bangladesh	70.8	85	Mauritius	0.2	17
Benin	0.5	16	Mongolia	0.1	10
Brunei	0.1	60	Morocco	18.3	99
Bulgaria	0.9	11	Mozambique	1.0	10
Burma	1.3	3	Niger	4.2	85
Cameroon	1.0	15	Nigeria	31.3	47
Camoro Is.	0.3	80	Oman	0.8	100
Chad	2.1	50	P. D. R. of Yemen	1.6	90
China	17.9	1.6	Pakistan	72.3	97
Cyprus	0.1	18	Qatar	0.1	100
Djibouti	0.1	94	Samali	3.4	99
Gambia	0.5	90	Saudi Arabi	7.2	95
Gaza	0.4	98	Senegal	4.3	82
Ghana	2.0	19	Sierra Leone	1.0	30
Guinea	3.1	65	Singapore	0.4	15
Guinea-Bissau	0.2	30	Sri Lanka	1.0	6
Egypt	35.4	91	Sudan	11.7	72
Ethiopia	11.8	40	Syria	6.8	87
India	80.0	13	Tanzania	3.8	24
Indonesia	123.2	90	Thailand	2.0	4
Iran	34.1	98	Tunisia	6.1	99
Iraq	11.2	95	Turkey	41.4	98
Ivory Coast	1.8	25	U. A. E.	0.2	92
Jordan	2.7	93	Upper Volta	1.4	22
Kuwait	1.0	93	U. S. S. R.	50.0	19
Lebanon	1.4	51	West Bank	0.6	80
Liberia	0.3	15	Western Sahara	0.1	80
Libya	2.6	98	Yemen	5.5	99
Malawi	0.8	15	Yugoslavia	4.1	19

expanded from its hearth in Arabia to become one of the most important cultures in the world. Expulsion of the Islamic Moors from Spain in 1492, the expansion of the Russian empire to Central Asia, and colonization associated with European expansion have slightly diminished its size, but it remains one of the spatially largest cultural realms in the world. The dominance of Islam across this region provides a degree of uniformity to an area that is otherwise fragmented into a wide variety of states and cultures.

Three types of life-styles are found across this region: city, village, and nomad. The principal focus of Islamic life is the city. Religious life in the Islamic city focuses on the mosque and activities relating to the daily prayers. In the golden era of Islam (A.D. 750–1400), the Islamic cities were the sites of major institutions of learning throughout North Africa, the Middle East, and Southwest Asia. Baghdad, Istanbul, Cairo, and Tehran were major centers of scholarly activity where scientific advances rivaled those of any other area of the world. The major Islamic cities such as Cairo and Istanbul remain important centers of commerce, politics, finance, industry, and education.

Nevertheless, the majority of the residents of the Islamic realm live in villages where they produce basic crops of dates, vegetables, and small grains for local consumption or trade with the cities. Villages in the Islamic world are characterized by the extended family, wide variations between rich and poor, and control of wealth and lands by a privileged few. The location of the Islamic realm in an arid to semiarid region makes water the critical resource across the region. The control of water and its allocation and use are of paramount importance to village life.

Several millennia before Islam, development of social organizations to ensure adequate water for the villages of the arid and semiarid lands now belonging to the Islamic realm fostered the rise of political organizations and civilizations. The Tigris and Euphrates Rivers, the Nile, and other sources of water led to the hydraulic civilizations of the region. Management of water distribution fostered development and improvement in mathematics, engineering, and soil science. Cooperation to obtain water in the more arid regions resulted in some of the world's greatest engineering feats.

The final major group in the Islamic world are the nomads. For many Western observers nomads

Jerusalem illustrates the importance of culture in explaining the geography of the world. The city is a sacred site in the Islamic cultural realm, yet is controlled by Israel creating tension and potential conflict.

are the familiar stereotype of the Islamic world, but in fact, nomads have never comprised more than 5 to 10 percent of the population of that world. Opportunities for education, medical care, and a better quality of life are combining with government inducements to further diminish the total numbers of nomads across the Islamic realm (see the Cultural Dimensions box).

The impact of European influence on the world of Islam has been varied. In some countries, notably Turkey under Ataturk, the leaders adopted the European model of industrialization and education. A fundamentalist conservative Muslim philosophy in other areas, notably Saudi Arabia and Iran, has resisted westernization, slowing industrialization and economic advance.

Islam itself serves as both a unifying and a divisive force throughout the realm. The subdivisions of the Islamic realm reflect the differences between the Shi'ite and Sunni Muslims as well as the minor Islamic subdivisions. The cultural contrasts among Arabs, Turks, and Persians are intensified by varying interpretations of the Koran and its requirements. In addition, the rejuvenation of Islam, particularly among fundamentalist groups, in the last decade has led to increased demands that Islam not be contaminated by European culture and has resulted in significant tensions in nearly every country. Individual countries within

Cairo, Egypt is one of the most important centers of commerce, education, culture, and politics in the Islamic cultural realm.

From Camels to Land Rovers

The Sultanate of Oman lies along the Arabian Sea.* It is part of the Islamic cultural realm, and religion is a central part of its citizens' lives. The life-styles of some of the people are being transformed, however.

The land area of Oman is dominated by dry climates. Less than 1 percent of the total area, around oases, can be cultivated intensively; the remaining deserts, wadis, and mountains are the home of a mobile, animal-breeding population of Bedouins and mountain nomads. The country was more or less isolated from the outside world until 1970 when the young Sultan Qaboos came to power and introduced policies aimed at modernizing the country. Omani oil was first exported in 1967, and oil revenue has grown steadily since that time.

A tendency toward permanent settlement, toward the reduction of migratory activity, and thus to the acceptance of nonagricultural employment can be noted among the nomads and Bedouins of Oman. Despite the freedom of choice and opportunity offered to the individual, the condition of numerous settlements, the abandonment of oasis gardens and wells, the failure of many transportation enterprises, and the emigration of a large portion of the male working force to neighboring countries testify that, on the whole, development has not had positive results.

The Bedouins of inner Oman, for example, used to stay in the southern wadi region and at the edge of the Rub-Al-Khali and the Wahibah sands during the winter, where their camps were located near small wells. They spent the summer months near the oases at the foot of the Oman Mountains in the north. In years of extreme need, it was possible to follow a path through the mountains to the coastal plain in the northern region.

These Bedouins now spend the whole year near the oases or by the modern, well-built roads that connect them, and they rarely move their camps. They prefer to settle within an hour's drive by Land Rover from the newly established schools and hospitals. The shelters that are built in these new settlements are constructed of cement block, a far cry from the felt tents that served the Bedouin for centuries.

Some of these Bedouins have settled in the middle of the broad dry streambeds near the edge of the mountains and have planted date groves. These are primarily members of the Duru tribe in whose territory the oil deposits have been found. This tribe profits from the oil economy and is able to finance the costly layout of gardens (US$6,900–8,900 per acre) and the maintenance of wells. Agriculture on this basis is not yet profitable. In some areas heavy irrigation, increasing salinity, and the poor quality of the soil have made cultivation impossible, and new groves and wells must be established. In some places groves have been abandoned and small, newly laid-out gardens established.

The Duru tribe exemplifies recent changes. The usual migration to distant grazing areas now occurs only occasionally and is done by truck or Land Rover. Small animals and the family are transported in modern vehicles, and the few camels that are still raised are turned out to free pasture. In the last ten to fifteen years, however, most of the tribe has settled near the oasis of Tanan, in Wadi Aswad, and near Awaifi. The style of housing has been adapted to this new mode of life; the old huts of palm branch mats have given way to block-style huts of clay, plywood, and cement blocks. In some settlements all the new housing is constructed of these modern building materials.

One reason for the reduced migration is that the land surrounding the oases is claimed by the oasis inhabitants and the Bedouins are denied any rights of ownership. In order to preserve their right to the huts in which they have lived for generations, the Bedouins must now remain year-round to prevent the oasis inhabitants from tearing down the huts during their absence. Since there are few permanent job opportunities in the oases, the economic situation of some Bedouin families is extremely precarious; in fact, many women have sold their silver jewelry and have turned to weaving simple, primitive carpets for a living.

_____

*Adapted from Rainer Cordes and Fred Scholz, *Bedouins, Wealth, and Change: A Study in Rural Development in the United Arab Emirates and the Sultanate of Oman* (New York: United Nations Natural Resources Program Technical Series, NRTS-7/UNUP-143). Used by Permission.

the Islamic realm also vary somewhat in terms of the degree to which democracy, industrialization, and personal freedoms are allowed. Nevertheless, the unifying element of Islam prevails across the realm. The importance of the religion combines with the concentration of petroleum in the Persian Gulf to make this one of the most important cultural realms in the world.

The diffusion of Islam has created major exclaves in Bangladesh and Indonesia, which differ from the main Islamic realm. Although both are Islamic, they combine elements of an older civilization with their culture. The Islamic cultural realm merges to the south with the African region with its animist and Christian groups. The

boundary of the Islamic realm divides many countries of the Saharan region. The trade and interaction between North African and sub-Saharan groups over time has spread Islam to the Sudan, south along the Indian coast, and across the Sahara.

The Southeast Asian Cultural Realm

South of China and east of the South Asian cultural realm lies an area of cultural fragmentation. The region of Southeast Asia consists of a variety of ethnic and racial subgroups with little unity (Figure 12–11). Peripheral to the great civ-

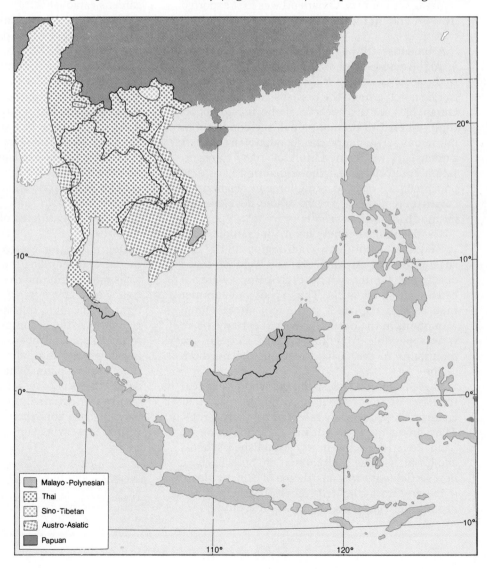

FIGURE 12–11
Cultures of Southeast Asia
The Southeast Asian cultural realm contains several major sub-groups

Legend:
- Malayo-Polynesian
- Thai
- Sino-Tibetan
- Austro-Asiatic
- Papuan

ilizations of India and China, Southeast Asia has been a recipient of cultural elements from each. Diffusion of Buddhism from India and migration of Chinese nationals have contributed to the particular cultural mix that characterizes Southeast Asia. Buddhism dominates the mainland; Indonesia and Malaysia are Muslim, and the Philippines are Roman Catholic. In addition, there are smaller areas of Hinduism, Christianity, and various tribal ethnic religions.

Precolonial Southeast Asia relied on an agricultural economic system supplemented by handicraft activities and metal working using local tin and copper ores. The generally uniform life-style was based on agriculture in the humid tropical climate that prevailed over much of the region. Sedentary villages cultivating rice as a staple crop supplemented by a variety of vegetables and fruits or shifting (slash and burn) agriculture in less favorable areas remain dominant throughout Southeast Asia. Rice is so important that it is estimated that as many as three thousand varieties, with growing seasons ranging from sixty days to one year, are produced in Southeast Asia (see the Cultural Dimensions box). The combination of reliance upon rice, the dominance of agricul-

---

CULTURAL DIMENSIONS
Rice—The Staff of Life

With nearly 2 billion people depending on it as a staple in their diet, rice may be the single most important agricultural crop in the world. Despite its international significance, rice remains something of a mystery crop to the people in the United States who yearly consume only ten pounds per capita, about half as much as the yearly consumption of ice cream. Two out of every three tons of U.S. rice are sold overseas, making the United States the second leading exporter with about 20 percent of world trade. While the United States is a major rice exporter, its export of rice accounts for less than 2 percent of the world's production.

Harvested on about 346 million acres (140 million hectares) throughout the world, world rice production in recent years exceeded 400 million tons of rough rice or 275 million tons of milled rice. The vast majority of this rice is consumed by farm families. In less industrial countries, where over half of the world's rice is produced, probably only one-fourth enters commercial channels. About 12 million tons or less than 5 percent of the rice produced is traded internationally. Approximately 90 percent of the world's rice crop is produced in Asia. Almost half of this crop is unirrigated, so the delicate balance between world rice supply and demand depends crucially on the Asian monsoon.

There are distinct markets based on different rice types, quantities and methods of processing. Preferences for different rices are so strong in some countries that consumers refuse to eat all but their familiar variety. Glutinous rice (also known as waxy or sweet rice) when cooked, forms a gelatin-like mass without distinct grain separation. Most rice-consuming areas in Asia produce small amounts of glutinous rice for use in desserts, ceremonial foods, and sweet dishes. In northeast Thailand and Laos, however, it is the staple food.

Aromatic or scented rice is grown mostly in the Punjab area of central Pakistan and northern India and is called basmati rice. Small quantities of it are also grown in Thailand and sold principally to Hong Kong and Singapore. The volume of aromatic rice traded is limited to about 300,000–400,000 tons annually, and it is sold at prices roughly double that of high-quality, long-grain rice. When cooked, basmati rice grains expand to about twice their original size.

Japonica rice is semi-sticky and moist when cooked. This round-shaped grain is grown in Japan, Korea, Taiwan, China, Australia, the Mediterranean area, Brazil, and California. Indica rice is a long-grained rice grown principally in China, south and southeast Asia, and the southern United States. Indica rice when cooked becomes fluffy, and shows high volume expansion and grain separation. Indica and Japonica rice make up the vast majority of rice produced in the world.

Source: Foreign Agriculture. December 1983, p. 14–17.

ture, and its transitional location make the region distinctly different from other cultural realms in the world.

The Sub-Saharan African Cultural Realm

Because of its particular assemblage of cultural characteristics, Africa south of the Sahara is commonly recognized as a distinct cultural realm. Sub-Saharan Africa differs from North Africa primarily in terms of language, religion, and race. Sub-Saharan Africa is dominated by Negroid peoples whereas Caucasoid characteristics are domi-

nant in North Africa. North Africa is overwhelmingly Islamic, while sub-Saharan Africa is not. Although Islam is widespread in the countries that border the southern margins of the Sahara, further south the influence of Islam fades rapidly.

While Arabic dominates North Africa, sub-Saharan Africa is a mosaic of languages and language families (Figure 12–12). Racially, three broad groups are found in sub-Saharan Africa, the Negroid, Pygmies, and Bushmen. The religious geography of sub-Saharan Africa is equally complex. Adjacent to the southern margins of the Sahara, Islam dominates, but the non-Islamic peoples of sub-Saharan Africa have traditionally been animists. The expansion of European influ-

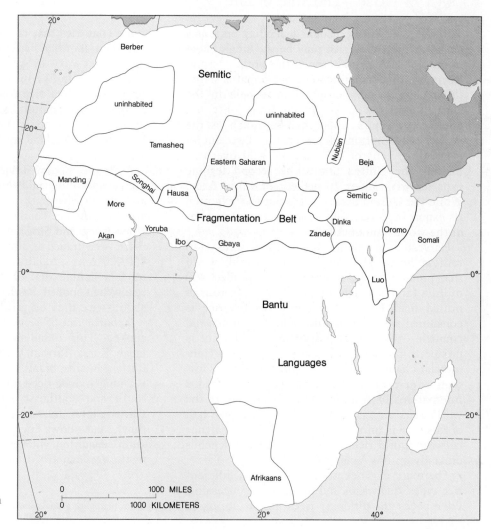

FIGURE 12–12
Languages of Sub-Saharan Africa This map of the major languages of Africa indicates the zone of linguistic fragmentation in central Africa.

ence into Africa in the nineteenth century was accompanied by the diffusion of Christianity, both Protestant and Roman Catholic. Today, both Christianity and Islam are rapidly increasing in sub-Saharan Africa, but they are still minorities.

The political map of Africa reflects the continent's cultural fragmentation. Tribalism remains the norm, with more than five hundred distinct tribal groups. There are fifty-three independent or quasi-independent political states in Africa, most of them in sub-Saharan Africa. The countries of sub-Saharan Africa range in size from the 967,500 square miles (2,505,800 square kilometers) of Sudan and the 905,567 square miles of Zaire (2,345,410 square kilometers) to tiny countries

such as Swaziland with only 6,704 square miles (17,363 square kilometers). Regardless of size, political fragmentation within these countries is an ever-present reality. Revolution, separatist movements, and insurgent states all typify the cultures of sub-Saharan Africa.

Basic economic practices provide the only broad regional cultural patterns of sub-Saharan Africa (Figure 12–13). The majority of Africa's population south of the Sahara is engaged in agriculture. On a percentile basis, sub-Saharan Africa is the most agrarian region in the world. Its agriculture can be divided into four broad groups: nomadic herding, shifting cultivation with intensive crop production, sedentary culti-

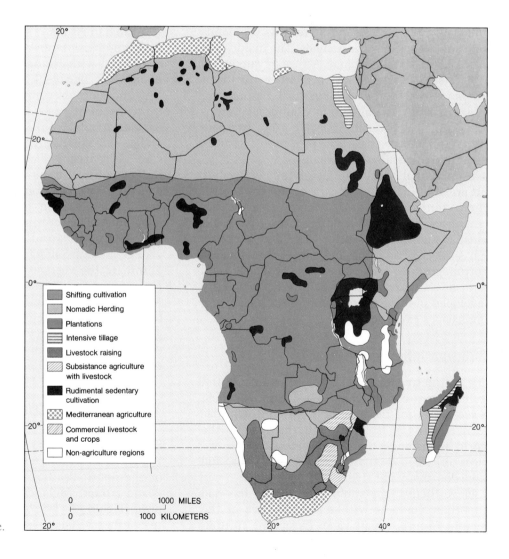

FIGURE 12–13
Predominant Economic Activity in Africa
Nomadic herding and rudimentary sedentary agriculture pre-dominate.

Legend:
- Shifting cultivation
- Nomadic Herding
- Plantations
- Intensive tillage
- Livestock raising
- Subsistance agriculture with livestock
- Rudimental sedentary cultivation
- Mediterranean agriculture
- Commercial livestock and crops
- Non-agriculture regions

0   1000 MILES
0   1000 KILOMETERS

vation, and commercial production including plantations.

In recent years there has been a shift from nomadic herding to more sedentary forms of agriculture. This move to sedentary life is in part the result of governmental attempts to deal with problems of overgrazing or the loss of livestock and economic support associated with the long series of drought in the southern margins of the Sahara and the Sahel. Other reasons include the desire for access to modern medicine, education, and other amenities provided by village or town residents. Moreover, rapidly growing rural populations of sedentary farmers increasingly compete with the pastoralists for lands. Expanding agriculture among the sedentary farmers cuts the traditional migration routes and occupies lands adjacent to water sources or better pasture lands traditionally used by the nomads. The majority of the residents of sub-Saharan Africa are supported by subsistence farming systems, either shifting or sedentary. Basic crops include millet, sorghum, cassava, yams, rice, and corn. Commercial crops such as oil palm, cacao (cocoa), cotton, and peanuts supplement subsistence production in areas with favorable soils and climate.

Commercial agriculture involves fewer people but provides major export earnings for the region. There are three main areas of commercial export agriculture: coastal West Africa (cacao, palm oil, bananas, rubber, and peanuts), East Africa (coffee, tea, sisal [a fiber crop], and cattle), and Nigeria (cotton, peanuts, cacao, copra [dried coconut], and palm oil). Commercial agriculture has generally developed in response to European influence and emphasizes export crops.

European Influence on Africa

The distinctiveness of the sub-Saharan cultural realm in part results from its isolation. For most of recorded time the interior of the continent has been effectively isolated from the rest of the world by distance, the dense vegetation and diseases of tropical environments, and falls and rapids on the major rivers.

Nevertheless, the sub-Saharan realm has had important contacts with representatives of the Islamic, South Asian, and European cultural realms. Pastoral nomads and traders from North Africa brought Islam to Africa south of the Sahara, while traders from the Arab world, Persia, and India sailed down the eastern coast. European contact with sub-Saharan Africa began with Portuguese attempts to reach Asia in the fifteenth century. For the next four hundred years contact was largely restricted to coastal locations as the slave trade developed in West Africa, and Cape Colony was established at the southern tip of Africa as a way station for Dutch ships making the months-long voyage to Asia.

European exploration of the interior of sub-Saharan Africa in the mid-nineteenth century and the scramble by European countries to obtain colonies superimposed the rigid boundaries of the colonial holdings over the existing cultures of sub-Saharan Africa. The independence of the African countries after World War II was generally associated with formal adoption of the colonial boundaries. The resulting potential for tribal conflict and national disunity remains as a heritage from the European occupation of much of sub-Saharan Africa.

The sub-Saharan African cultural realm is in a state of transition. The dominance of the societies and political units by Western-educated elites and the continued conflict between groups over political orientation or control make the sub-Saharan African cultural realm one of the most turbulent in the world. One of the major forces for change in sub-Saharan Africa is the desire to eliminate the vestiges of colonialism. A second force is the conflict between Marxist and non-Marxist groups vying for political power. A final force is that associated with the continued growth of the majority black population in South Africa and the South African government's attempts to maintain its policy of apartheid.

The Oceania Cultural Realm

East of Southeast Asia are the island groups of the Pacific. This region, which is known as Oceania, or the South Pacific, can be divided into three island groups: Melanesia, Micronesia, and Polynesia (Figure 12–14). *Melanesia* is located closest to Southeast Asia. The islands of Melanesia are

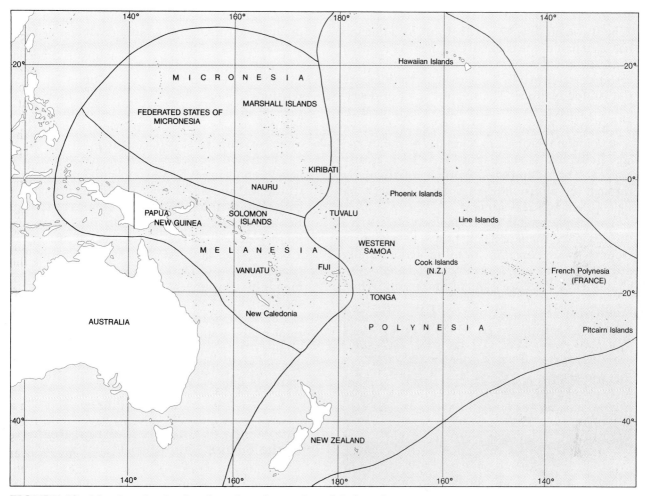

FIGURE 12—14 The Islands of Melanesia, Micronesia and Polynesia

large, with a tropical climate similar to that of the mainland rim of Southeast Asia and the Indonesian Archipelago. Melanesia extends from Southeast Asia to Australia and consists of a number of large islands, including New Guinea. *Micronesia* consists of a few volcanic and complex high islands and many tiny atolls (coral islands with a reef surrounding the lagoon). Guam, a volcanic island, is the largest with an area of only 210 square miles (550 square kilo-meters), while many of the atolls rise only a few feet above the tide level and cover only a few square miles. *Polynesia* covers the largest area of the South Pacific, but it is comprised of a very small amount of land. Polynesia includes both low coral atolls and volcanic islands. Today the residents of this

cultural realm can be subdivided into a Polynesian-Micronesian subset and a Melanesian subset.

Like Africa, the Oceania (South Pacific) cultural realm has been isolated from the main forces for cultural change in the world. Until the advent of European sailing technology that allowed two-way trade, the area remained unknown to the broader world. The European discovery of the islands and peoples of the Polynesian, Melanesian and Micronesian area led to changes in their culture.

The Polynesian-Micronesian islands are more economically advanced, have more contact with the outside world, and have a standard of living that is lower than that found in the European cultural realm, but above that of Melanesia (see Table 12–6 on page 475). All of this broad region has been part of the colonial holdings of European countries and the subsequent mandates and trusteeships established by the League of Nations and the United Nations. The relative level of development reflects the resource base of an individual island or country, as well as the type of colonial or post-colonial relationship it has had with Europeans. Individual island nations have been greatly affected by European culture, particularly through the diffusion of European languages, religions, and political systems.

The displacement of original cultures gives this region a transitional nature. The region of Oceania is in transition from an agriculturally oriented society with generally low technology to an outlier of the European cultural realm. While

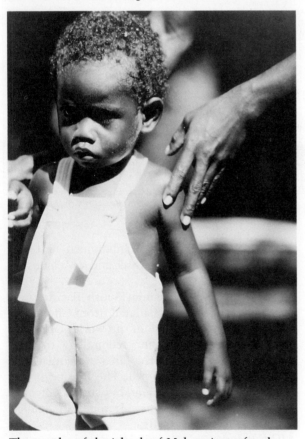

The peoples of the islands of Melanesia are faced with the loss of their cultural distinctiveness as the European cultural realm extends its influence in the region.

most islands rely on production of agricultural products (particularly tropical products) or mineral resources for their economic well-being, they also rely heavily on relations with the countries to which they were tied as colonies or trust territories. Some islands, for example, have experienced an absolute decline in population in the last decade as a result of migration to the United States or other trust parent nations.

The region is one in which the native cultures have been changed by Western influence (see the Cultural Dimensions box). This influence comes both from their former role as colonies or territories and from the view that tourism is a panacea for their poverty and economic woes. Tourism has a darker side, however, for the societies of the island peoples of Oceania. Changing values, demands for consumer items, new attitudes toward the individual and society, and diffusion of Western cultural variables are transforming this cultural realm dramatically. It is too early to determine the future of the region, but it is clear that as a unique culture, the group is threatened by the spread of European values and norms.

The Winds of Change: Cultural Realms in the Future

Some cultural realms seem to be relatively permanent and impervious to change, whereas others have changed their boundaries dramatically in a few centuries. Nevertheless, each of the world's cultural realms is currently being affected by forces that will change its distribution and boundaries in the future.

The major trends that will help shape the cultural realms of the world in the next decades include the following:

1. The diffusion of elements of the European cultural realm, including its values and attitudes.
2. The defensive reaction of non-European cultural realms as they attempt to maintain their cultural integrity.
3. Changes in demographic characteristics affecting growth rates, migration patterns, and economic developments that are asso-

CULTURAL DIMENSIONS
On The Jungle Road with Colgate Toothpaste

Cultural convergence is occurring even in isolated areas as ideas diffuse from the industrial world.* In the jungles of Papua, New Guinea, a group of actors travel the island selling everything from Colgate toothpaste to Dunlop rubber boots.

Hawking household wares to over a million people living in rural villages scattered over the rugged highlands region poses a challenge for companies here. Until they came into contact with Europeans in the 1930s, highlanders lived without metal or the wheel.

Today, the area hosts plenty of stores with modern products. But how to advertise them? Televisions are scarce. Radio and television reception is poor. And the consumers are seldom literate. Enter Wokabout Marketing—a one-of-a-kind amalgamation of theater, commercials, and sales demonstrations.

After some trial and error, Wokabout has become a hit. It weaves product demonstrations into a slapstick-comedy sketch on modern village life. For two years, it has drawn large (200 to 1,500 people), appreciative audiences. Today, the powder-blue Wokabout Isuzu truck halts on the outskirts of Lae, a small urban center known as the gateway to the highlands. Small children and curious adults settle in the grass to watch the five-member troupe set up. A plywood stage fifteen feet wide unfolds from the truck. Wireless microphones are hooked up. A tape deck blares music while props are positioned. Even in the oppressive noonday sun, a crowd of more than 300 has quickly gathered.

The free performance is in pidgin English—the most widely spoken language in the country. The audience is soon howling at Junior's antics as he refuses to brush his teeth. But many are also learning for the first time just what a toothbrush is and how to use it. Junior, Mom, Dad, and Uncle eventually cluster around a bucket of water to brush and sing the Colgate theme song.

For the next hour, peals of laughter accompany the actors' fumbled attempts at using Morteen insect repellent, Cold Power laundry soap, and Gillette shaving cream. Rapt attention is given to Mama explaining the joys of cooking Nambawan canned mackerel and Trukai rice. The show ends with a scramble for free samples tossed from the stage.

Wokabout doesn't advertise products it considers detrimental to health such as tobacco and alcohol. However, since companies may add three products to the skit, sometimes the second and third choices (such as puffed cheese snacks and deodorant) are of dubious value to rural villagers. Religious and political advertising is refused as being too risky; tribal disagreements here are often settled with machetes and bows and arrows.

They've never been threatened, but life for the Wokabout cast is not easy. Last year, they staged nearly three hundred performances for a quarter of a million people. Six days a week, three weeks a month, the troupe jounces through the mountains on dirt roads.

_____

*Adapted from David C. Scott, "On the Jungle Road with Colgate Toothpaste," (by permission from *The Christian Science Monitor*, January 9, 1989.)

ciated with demands for socio-cultural change.

4. Changing patterns of political influence on a world basis.

Cultural realms differ in their geographic extent and importance in each time period. At the present time the world's cultural realms reflect the great changes since the European Age of Discovery. The European conquest of many re-

gions of the world and subsequent diffusion led to the economic and technological domination of the European cultural realm. This domination in turn increased both the influence of the European realm and its rate of diffusion.

The continued dominance of the European cultural realm may be neither certain nor desirable. The diffusion of elements of European culture—fast foods, fashion, and the nuclear family, for example—may or may not be beneficial to other cultures. Some cultures react defensively as they attempt to protect those cultural elements that they feel are integral parts of their life-style. One of the best examples of this is the Islamic cultural realm.

Fundamentalist movements in the Islamic world are rejecting the influence of the European West with its materialistic and destructive (in their perception) tendencies. The continued hostility between individual states in the Islamic world should not mask the growing power of the Islamic cultural realm as a whole. If the centrifugal forces of nationalism and sectionalism within Islam can be overcome, the Islamic realm could conceivably expand dramatically in the next century. Logical areas of expansion might include southward into the sub-Saharan realm, eastward into Melanesia from Indonesia, and east into Southeast Asia from Bangladesh.

Population changes are also affecting the cultural realms of the world. Rapid population growth in Asia, Africa, and Latin America are associated with both changes in the economy and the destruction of life-styles as overpopulation transforms the traditional rural landscape and life of much of these regions. Moreover, the population growth that results in migration to the cities often removes people from their traditional support systems, making them vulnerable to the attractions of other cultures. The need to provide jobs, improve the standard of living, and provide education, medical care, and the other amenities that create a better life-style for their growing populations is causing many of the countries of Latin America, Africa, and Asia to adopt elements of European culture.

Growing populations also result in migration to other countries, thus diffusing the elements of the native culture. Migration of Asians and Latin Americans, for example, has created another element in the complex mix of cultures found in the United States and Canada. Although these migrants adopt some of the elements of the American variant of the European culture, they still maintain many of their old values and traditions. Migration of Indians from India to the United

States, for example, has been accompanied by the establishment of Hindu temples, schools to teach traditional values, and societies to promote and retain values from the home culture.

Political factors are also affecting the geopolitical map of the world. For example, the Islamic realm is torn by the continued conflict between Shi'ite and Sunnite, Christian and Muslim, and Persian and Arab. Such conflicts have signifance far beyond the boundaries of the Islamic realm because it is the site of more than half the world's known petroleum reserves.

Some cultural realms, such as the Sub-Saharan, may be threatened by continued diffusion from adjacent cultural realms. It is unlikely that the Sub-Saharan realm itself will change dramatically in its geographic extent because it lacks a formal mechanism for diffusion such as the Islamic (religious conversion) and European (colonialism and economic expansion) cultural realms have.

The Southeast Asian and Oceania cultural realms likewise will probably not increase dramatically. The cultural characteristics of the Oceania realm could possibly decrease until it becomes a part of the European cultural realm or its outlier in Australia and New Zealand. The Southeast Asian cultural realm faces the same difficulty as the sub-Saharan, the lack of an organized means for diffusion. Population growth in the East Asian and South Asian cultural realms and their expansion may continue to decrease the geographic extent of Southeast Asia.

Conclusion

Regardless of what other changes occur, the world's cultural realms continue to modify their size and boundaries through diffusion. Changes in the cultural realms are but one manifestation of the changes currently affecting the earth as the home of human beings. These changes in the cultural and physical geography of the world will make the cultural realms of subsequent generations as different from today's realms as today's are from those of the past.

QUESTIONS

1. What is a cultural realm?
2. Describe the development of the European (Occidental) cultural realm.
3. Compare and contrast the impact of Mediterranean Europe and Western Europe on the creation of cultural realms in the world.
4. Compare and contrast Australia and New Zealand with South Africa.
5. Describe the differences between Western and Continental Europe.
6. Discuss the primary characteristics of the East Asian realm.
7. What are the characteristics of the South Asian cultural realm?
8. Discuss the differences between the main Islamic cultural realm and its major exclaves.
9. Describe the characteristics of the sub-Saharan African cultural realm.
10. Which cultural realm is the most diverse? Support your response.

SUGGESTED READINGS

Chisholm, Michael. *Modern World Development: A Geographical Perspective.* Totowa, NJ: Barnes and Noble, 1982.

Cole, John P. *The Development Gap: A Spatial Analysis of World Poverty and Inequality.* New York: John Wiley & Sons, 1980.

Crow, Ben and Alan Thomas. *Third World Atlas.* Philadelphia: Open University Press, 1985.

DeSouza, Anthony R., and Phillip Porter. *The Underdevelopment and Modernization of the Third World.* Washington, D.C.: Association of American Geographers, 1974.

Dickenson, J. P., C. G. Clarke, W. T. S. Gould, R. M. Prothero, D. J. Siddle, C. T. Smith, E. M. Thomas-Hope, and A. G. Hodgkiss. *A Geography of the Third World.* New York: Methuen, 1983.

Forbes, D. K. *The Geography of Underdevelopment: A Critical Survey.* Baltimore: Johns Hopkins University Press, 1984.

Ginsberg, Norton S. *Atlas of Economic Development.* Chicago: University of Chicago Press, 1961.

Grossman, Larry. "The Cultural Ecology of Economic Development." *Annals of the Association of American Geographers* 71 (June 1981): 220–36.

Hoffman, George W., ed. *A Geography of Europe: Problems and Prospects,* 5th ed. New York: John Wiley & Sons, 1983.

Jumper, Sidney R., Thomas L. Bell, and Bruce A. Ralston. *Economic Growth and Disparities: A World View.* Englewood Cliffs, N.J.: Prentice-Hall, 1980.

Markusen, Ann R. *The Politics of Regions.* Totowa, N.J.: Rowman and Littlefield, 1987.

Myrdal, Gunner. *Rich Lands and Poor.* New York: Harper and Bros., 1957.

Smith, David M. *Where the Grass is Greener: Living in an Unequal World.* London: Hutchinson University Library, 1981.

Szentes, Tamas. *The Political Economy of Underdevelopment,* 4th ed. Budapest, Hungary: Akadémiai Kiadó, 1983.

The Changing Human World

The world has come a long way from the mid-1970s, when environmental concerns were considered something that only the rich could afford to worry about. Today, they are concerns no one can afford to ignore.   L. E. BROWN

THEMES

The changing world.
The impact of world urbanization on the changing cultural geography.
Industrialization and technological change.
Diffusion of cultural elements and cultural convergence.
Environmental changes.

CONCEPTS

Duration
Direction
Magnitude
Cultural convergence
Aging world
Overpopulation
Erosion
Urban flight
Giant cities
Conflict
Modernization theory
Suburbanization
Desertification
Environmental degradation
Greenhouse effect
Green Revolution

Introduction

The cultural geography of the world is constantly changing. The processes and forces associated with the human use of the earth continue to transform each place. Some of the forces that affect the cultural geography of the world have been discussed in previous chapters: the increase in population, the increasing proportion of urban residents, the rapidly growing cities of the less industrialized world, the changing political map of the world, the changing boundaries of the cultural realms, and the diffusion of languages and religions. These forces for change have the potential to affect the entire cultural fabric of the world.

Change can be examined in terms of *duration*, *direction*, *magnitude*, and *rate*. *Duration* refers to whether a change is temporary or long-lasting. The *direction* of change is the movement toward increased organization or disintegration of a culture. The *magnitude* of change is the degree to which changes affect a segment or the entirety of a group or space. The *rate* of change refers to whether it is fast, slow, continuous, orderly, or uneven. Duration, direction, magnitude, and rate in combination determine how change will affect the cultural geography of the world.

The Rate of Change

It is important to note that the rate of change appears to be constantly accelerating. The futurist writer, Alvin Toffler, in his book *Future Shock* (1970) pointed out that the last 50,000 years of human existence represent only 800 average human lifetimes (62 years). Our ancestors emerged from caves sometime during the 650th lifetime, while printing began only 6 lifetimes ago. The electric motor has been used in only the last 2 lifetimes, and most of the material goods we use today (computers, televisions, and microwaves, for example) were developed only in the present lifetime. Nearly as much change has occurred since your birth as occurred in all human history before it. Ideas, inventions, and ways of doing things are being transformed so rapidly that

changes in your lifetime will be rapid and far-reaching. Thus the rate of change and its continuing acceleration are among the most important factors that will affect the future cultural geography of the world.

The accelerating rate of change is particularly apparent in the areas of transportation and communication. Changes in these areas since the invention of the printing press, such as the development of sailing ships with the ability to tack into the wind and the development of the railroad, have transformed the cultural geography of the world by allowing greater interaction and diffusion of people and ideas. Mass movement of people by air is a phenomenon of only the last three to four decades. The ever greater interaction between peoples and places associated with the electronic age is of even more recent origin. As computers, satellites, and other electronic means of communication become more sophisticated, the potential for interaction and diffusion will continue to increase.

Forces for Change: Cultural Variables

Cultural elements that may change the geography of the world in future decades include population growth and change, urbanization, industrialization and technological innovation, diffusion and change in individual phenomena such as religion or language, and continued *cultural convergence*. Examination of these elements indicates the nature and rate of change that might be associated with them.

Population Change

The rate and magnitude of population growth in the last century are sufficient to call it a revolution. The addition of one billion people to the world's population in only ten years (1978–1988) illustrates both the rate and magnitude of that change. Although birth rates are generally dropping—except in parts of Africa—the global population is so large that it continues to increase at a growing rate despite the lower birth rates. Examination of the duration of the population change suggests that it will not end soon. Al-

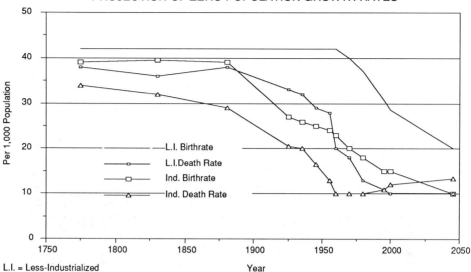

PROJECTION OF ZERO POPULATION GROWTH RATES

FIGURE 13–1 Date of Reaching Zero Population Growth at 1980 Growth Rates Projected trends in world birth rates and dates of reaching zero population growth.

L.I. = Less-Industrialized
Ind. = Industrialized

though birth rates are decreasing in most of Africa, Asia, and Latin America, they are still far above the level needed to maintain a stable or nearly stable global population number (Figure 13–1).

The impact of the population growth is worldwide. At the level of an individual country or region, the doubling of population in economies such as those in Africa is associated with cultural and physical geographic changes that revolutionize the individual societies by placing greater

pressure on resources, causing people to migrate and adopt new techniques or to lower their standard of living. The highest population growth rates are found in Africa, and there have been few indications that the rapid rate of growth in Africa will slow. Population growth rates in Asia and Latin America, by comparison, have decreased, yet the higher base population found in Asia means that Asia will be even more heavily populated in coming decades (Figure 13–2). Even the much discussed Chinese effort to stabilize popu-

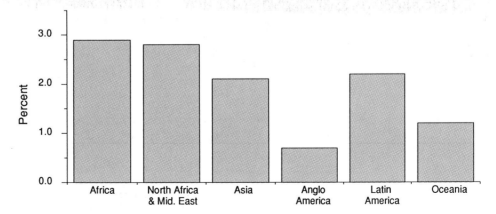

POPULATION GROWTH RATES BY REGION, 1988

FIGURE 13–2 Population Growth Rates in Major World Regions, 1988

lation at 1.2 billion is now unlikely to succeed since changes in China's centrally planned economy and the emphasis on individual entrepreneurship in the contract farming system have made more children desirable. After a decade of amazing success in lowering population growth rates, the Chinese are once more facing increased growth.

Latin America may well see a reduced growth rate over the course of the next decades. Nearly every country in Latin America has adopted a formal population policy; even the Catholic church (which has traditionally opposed such programs) openly supports some programs (Figure 13–3).

Another major change, which may affect the earth is the continued movement of the countries of the industrialized European cultural realm into stage four of the demographic transition. Increasing numbers of elderly in the United States and Canada mirror more marked trends in the Neth-

erlands, Belgium, United Kingdom, Japan, and other industrial countries. The world's elderly population (aged 65 and over) totaled 290 million in 1987 and is growing at a rate of 2.4 percent per year, faster than the global population growth rate and equal to the growth rate in many African countries. Current projections indicate a population of 410 million elderly by the end of this century (Figure 13–4 a and b).

Fifty-four percent of the world's elderly live in the less industrialized countries. By the year 2025, these countries will be home to nearly 70 percent of the world's elderly, a group whose average income, education, and literacy are lower than those of younger groups. The distribution of the elderly in the industrialized world is different from that in the less industrialized world. The elderly are urban residents (70 percent) in the industrial countries and rural (70 percent also) in less industrialized countries. Equally important

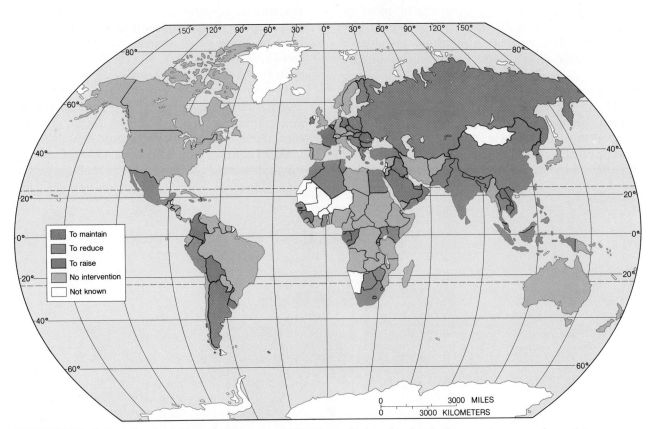

FIGURE 13–3 Population Policies by Country, 1989 The policies of individual countries range from doing nothing to increasing or decreasing the birth rate.

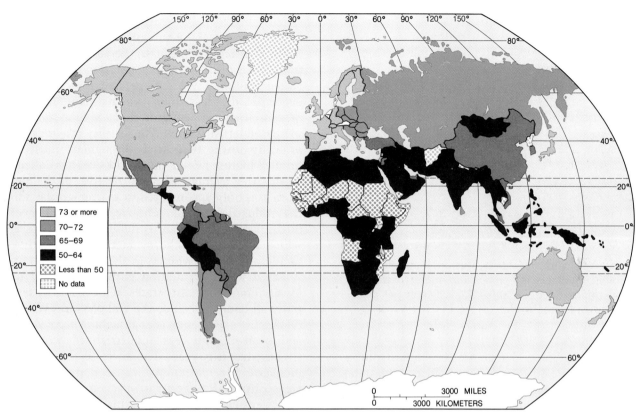

FIGURE 13–4a Life Expectancy of the World Life expectancy in the world (13–4a) is closely related to the growth in the number of elderly (13–4b).

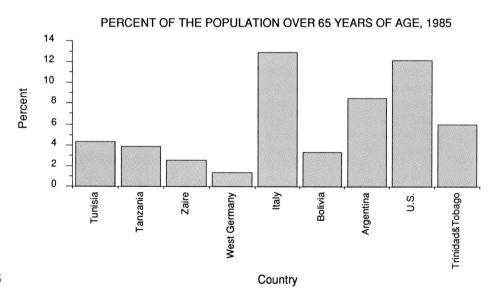

FIGURE 13–4b
Percent of the Population
over 65 Years of Age, 1985

New housing for the elderly, Weston, Massachusetts. Growing numbers and proportions of elderly in both industrialized and less industrialized countries pose challenges for the world.

as a demographic change are the "oldest old"—those who are 80 years of age or older—who today constitute 14 percent of the world's elderly. Nineteen percent of the elderly population in the industrial world are oldest old, but only 11 percent in the less industrialized world. The highest totals of oldest old are in the United States and China, each with about six million. By the year 2005, 31 percent of elderly Americans are projected to be age 80 or older, the highest proportion in the world. As many as 70 percent of the oldest old are female.

The significance of this *aging world* to cultural geography should be apparent. Growing numbers of elderly, and particularly the oldest old, affect the labor force, the medical industry, the economy in general through differing demands, the housing and built environments as their need for accessible living space increases, and the social role of governments in providing for the elderly.

The magnitude of the population issue facing the world has been discussed earlier, but it is helpful to place the issue in context to illustrate how it will affect the world's cultural geography. Less than a generation ago (1950) there were only 2.5 billion people, but global numbers more than doubled by 1988. If present growth rates continue, another billion will be added by the end of the

century, making the earth home to a projected 6.16 billion people. The world population is projected to reach 8.2 billion in the year 2025. Of the projected 3 billion world population increase between 1988 and 2025, 97 percent is expected to occur in the less industrialized regions. The present high rate of growth is *not* because of increased fertility rates. The fertility rate today is probably lower than that throughout most of human history. Even in the less industrialized world, the average number of live births per woman is only 5, compared to an average of 2.3 in the industrial world. The average number of births per adult woman worldwide is less than 5, which is less than the average number of live children borne by women in the past.

Rapid world population growth is caused by lower infant mortality rates achieved in essentially all countries since the end of World War II. Global life expectancy is approaching sixty years, and even in the less industrialized world is slightly over fifty years. The lower death rates and increased life expectancy cause population increase. Ironically, the greatest accomplishment of the twentieth century—assuring that most children born upon the earth can expect a long and full life—is viewed as a problem (*overpopulation*). Lowering the death rate over the past half century reduced unnecessary, tragic, and wasteful infant deaths, an accomplishment that should not be

Amazon Basin, Brazil. Deforestation of tropical areas remains a problem in spite of growing concern over the impact of loss of tropical forests.

overshadowed by the challenges of the resulting population growth.

The dilemma for the world is the momentum that population growth has attained and the impact of increased numbers on the general human condition. Each succeeding age group is larger than the previous in rapidly growing populations, since rapid population growth has a built-in momentum that requires several generations to return to zero. In the less industrialized world, the number of women in the premarriageable age groups is usually twice that of their mother's generation. Even when the fertility rate is halved, the growth rate remains constant if the number of mothers has doubled. Even if the average family size were reduced to two surviving children per woman in the less industrialized world, world population would still increase an average of 66 percent before a stable population was achieved.[*]

Social inertia also helps to ensure continued population growth. Limiting the number of children in the family profoundly affects society in general. Kinship duties and relationships, religious beliefs, and personal perception of well-being may all be changed by smaller family size. It is difficult to demonstrate to an individual farmer that a smaller family is in fact an economic advantage. The farmer with three children may compare unfavorably to one with six who is able to cultivate more land, rely on his or her children for care in old age, and acquire prestige from having a large family. The fear of being left without surviving children is so great in many less industrialized societies that the desire for many children persists even when there is no economic necessity for large families.

Population growth affects the cultural and physical resources of the world as increasing demand for resources leads to environmental deterioration. Environmental deterioration ranges from destruction of tropical rain forests to the degradation of air, water, land, and landforms in the industrial world. Additional problems include depletion or exhaustion of mineral and fuel resources, decline in fish harvest from the oceans, and the destruction of the world's cropland by soil

Pollution of the earth's air, water and land base is an issue that transcends political boundaries.

erosion or strip mining. Increased population will affect the quality of the environment and the quality of life for all who call earth home.

The present rapid population growth rate in the world varies by geographic areas. Growth of the elderly population in the industrial world and of the youthful population in the less industrialized world present different problems, but both change the world's cultural geography.

The continued growth of the world's population will cause the cultural landscape to become more crowded, both in cities and rural regions. Subsequent changes will affect the distribution of land and population, have environmental and social impacts, and change the nature, rate, and extent of urbanization. Growing populations in less industrialized areas cause ever more people to leave the rural setting for cities. The changes in housing requirements and infrastructure in industrial nations with ever older populations will affect housing, urban services, and other aspects of the cultural geography.

World Urbanization: Changing the Cultural Geography

The growth of urbanization presents a dual phenomenon: the emergence of giant cities in the less industrialized world and *urban flight* in much of the industrial world as the electronic age and transportation free companies and individu-

---

[*]Nathan Keyfitz, "On the Momentum of Population Growth," *Demography* 8 (February 1971).

TABLE 13–1
Population of World's Largest Metropolitan Areas (in millions)

	1975		1985		2000
New York	19.8	Mexico City	18.1	Mexico City	26.3
Tokyo-Yokohama	17.7	Tokyo-Yokohama	17.2	São Paulo	24.0
Mexico City	11.9	São Paulo	15.9	Tokyo-Yokohama	17.1
Shanghai	11.6	New York	15.3	Calcutta	16.6
Los Angeles–Long Beach	10.8	Shanghai	11.8	Bombay	16.0
São Paulo	10.7	Calcutta	11.0	New York	15.5
London	10.4	Buenos Aires	10.9	Seoul	13.5
Buenos Aires	9.3	Rio de Janeiro	10.4	Shanghai	13.5
Rhine-Ruhr	9.3	Seoul	10.2	Rio de Janeiro	13.3
Paris	9.2	Bombay	10.1	Delhi	13.3
Rio de Janeiro	8.9	Los Angeles–Long Beach	10.0	Buenos Aires	13.2
Peking	8.7	London	9.8	Cairo/Giza/Imbaba	13.2

als from a central location. The greatest change associated with urbanization is the change in location of the world's largest urban agglomerations between 1950 and the year 2000. In 1950, twenty-three of the world's thirty-five largest urban agglomerations (66 percent) were in the industrial world. By 1985, this situation had completely reversed with twenty-three of the thirty-five largest being located in the less industrial world. Projections by the United Nations indicate that by the year 2000, seventeen of the world's twenty largest metropoli will be in less industrialized nations. (Table 13–1). Only 78 metropolitan areas exceeded one million people in 1950, but 258 were that large by 1985, and United Nations projects there will be 511 exceeding one million inhabitants by 2010. An estimated 46

percent of the world's population was classified as urban at the end of the 1980s, but by the year 2000, that figure will exceed 50 percent. Half of these people (25 percent of the world's total) will be found in cities with over one million inhabitants (Figure 13–5).

The distribution and trends for urban spatial organization differ between the industrial and less industrialized world. The core areas of many metropolitan regions in the industrial world are actually losing population, and the number of metropolitan areas is projected to increase only slightly. Metropolitan growth in the less industrialized countries is projected to lead to 486 cities of over one million by the year 2025. The metropolitan growth in the less industrialized world is likely to be greatest in Latin America and Africa.

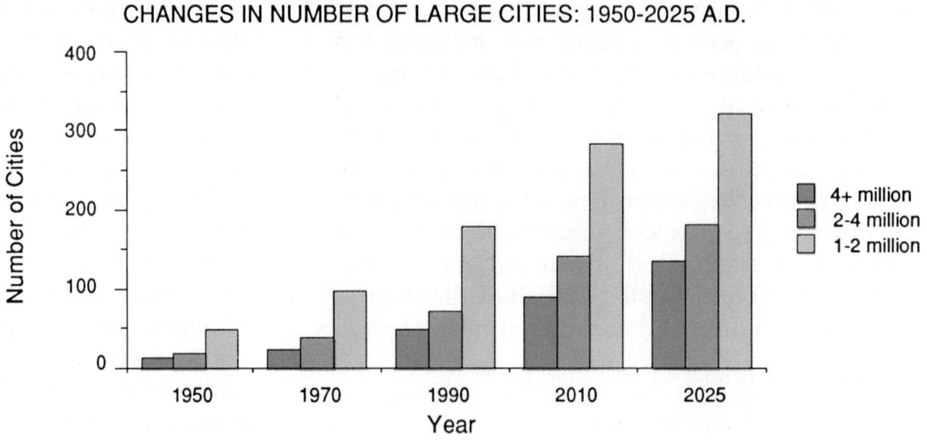

CHANGES IN NUMBER OF LARGE CITIES: 1950-2025 A.D.

FIGURE 13–5 Growth of Large Cities Growth in the number of large cities in the world varies from region to region, but the less industrialized world will experience the greatest growth.

Both will have nearly half of their urban populations concentrated in urban agglomerations exceeding one million by the year 2025.

The urban growth projected for the next generation will present major problems:*

1. Insufficient housing
2. High rates of unemployment and underemployment
3. Decline in nutrition and housing quality
4. Inadequate sanitation and water supplies
5. Overloaded and congested transportation systems
6. Air, water, and noise pollution
7. Inadequate budgets in municipalities
8. Rising crime and other social problems
9. A general deterioration of the perceived quality of urban life

Grim projections about the urban geography of the world in the next generation have not slowed city growth. The reasons for the continued flood of migrants to the cities are obvious: the growth of population in rural areas of the less industrialized world; the limited level of rural development; and the perceived job opportunities, social and cultural amenities, and opportunities for individual betterment of cities.

Slowing the rate of urban growth and related problems in the giant cities of the less industrialized world will require major changes in rural economies to allow more people to be supported by agriculture. For rural areas to absorb greater numbers of people, they must either industrialize, expand agricultural land, or increase production on presently used land.

Industrialization, the model that most less industrialized countries hope for, proceeds slowly in rural areas. Extension of agricultural land is possible in some of the less industrialized countries, but in many, the present level of population has already pushed agricultural expansion to its limits. Intensification of agricultural practices is equally difficult in the less industrialized world. Use of irrigation and labor-intensive farming, including double cropping, are already the norm

*The above list is from Mattei Dogan and John D. Kasarda, *The Metropolis Era*, vol. 1; *The World of Giant Cities* (London: Sage Publications, 1988), p. 19).

Transplanting rice seedlings, Chiang Mai, Thailand. The majority of the world's population is still engaged in traditional agriculture reliant on animate power sources.

and rural population densities are exceedingly high in Asia. The average acreage cultivated by the peasant farmer of Asia (excluding Japan) averages only one acre (0.4 hectares).

Consequently, increasing the intensity of cultivation or expanding acreage is difficult. More land could be converted to agriculture in both Indonesia and the Philippines, but many farmers are reluctant to participate in programs to resettle rural migrants because of social or family ties. Lack of capital or doubts about the potential success of resettlement efforts also make potential participants reluctant. Expansion of agriculture in Africa is limited by harsh environmental conditions. Much of the continent is either too dry or too wet, with leached, infertile soils. These problems are exacerbated by the presence of nagana, which handicaps raising domestic livestock, limits the availability of meat, and causes trypanosomiasis in humans. The United Nations estimates that over 3.8 million square miles (10 million square kilometers) of Africa, an area greater than that of the United States, has only limited utility for agriculture due to nagana.

Faced with continued population growth, leaders of less industrialized nations turn to industrialization as a panacea for their country's problems. New industries are attracted to cities because of government incentives, raw material supplies, and labor. The combination of stagnant rural economies and increased opportunities in the urban settings make the continued growth of the world's largest cities very likely.

Industrialization and Technological Change

The forces that affect growth and change in cities reflect changes in technology, especially those associated with industrialization. The diffusion of industrialization affects the populations and economies of both the industrial and less industrialized realms. Growing populations in the less industrialized world serve as a magnet for labor-intensive industries from the industrial world. The resulting flood of low-priced manufactured goods from industrializing countries further enhances the quality of life in the European realm and its outposts.

The loss of jobs in labor-intensive industries in the traditional industrial world is associated with dramatic changes in urban and economic landscapes. Old centers for manufacturing textiles, boots and shoes, iron and steel, and other products that are increasingly concentrated in less industrialized countries are faced with unemployment, population loss, and obsolescent landscapes.

The region from the Great Lakes to New York has been described by some as a "Rust Belt" because of the decline of the long-established metal industries. Jobs are lost in the auto industry as more cars are imported. Some regions have lost entire industries (such as the textile and boot and shoe industries in New England), as large companies from the industrial world establish subsidiaries in the less industrialized world (Figure 13–6).

The threat of low-price imports to industrial countries often leads to demands for tariff barriers to protect domestic industries. Such barriers discriminate against the poorer countries that are only beginning to industrialize. If all of the industrial countries adopt such tariff barriers, the less industrialized countries will find it difficult to industrialize.

At the same time, the lowering of tariff barriers and the concomitant increase in industrialization in the less industrialized world disrupt cultural geography. The development of an industrial complex to assemble televisions or cameras, manufacture textiles or clothing, or refine iron and steel has an uneven effect on a country. Such industries are normally concentrated in the cities, further separating the rural poor from urban residents. But even in the cities, only a few of the residents enjoy the advantages of such industrialization. In consequence, a small middle class is created in the cities of the less industrialized world, with views and life-styles markedly different from those of the majority urban poor. The resultant dichotomy is often the basis for political unrest and conflict.

The diffusion of Western technology to the less industrialized world may be viewed as a mixed blessing. Many observers believe that the diffusion of the industrial model is simply an attempt to exploit the cheap labor of the third world (see the Cultural Dimensions box on page 500). Nevertheless, countries such as Taiwan and South Korea that have attracted industries with cheap

Ford Lio Ho Motor Company, Chungli, Taiwan. Diffusion of industrialization is leading to greater links betweeen the industrialized countries and the rest of the world.

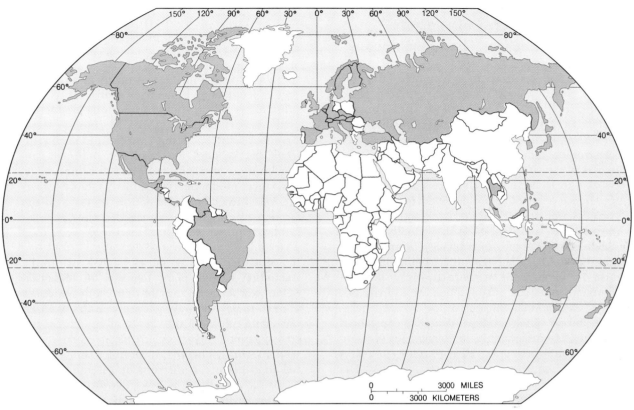

FIGURE 13–6 Supranational Firms: Distribution of MacDonald's, 1989 Supranational firms are located in numerous countries far beyond the one in which corporate headquarters are found.

labor have expanded upon this initial industrialization, raising living standards and well-being. Most of the less industrialized countries, faced with rapidly growing populations and rural population pressure, look to the European cultural realm as a source for technology to improve their quality of life (Table 13–2 on page 501).

Diffusion of Religion

Changes in the religious geography of the world are hard to predict. The continued diffusion of Christianity to Africa, the diffusion of Buddhism, Hinduism, and Islam to the United States and Western Europe by immigrants, and the increased militancy of some Islamic peoples are sure to affect the world in the future. Rising forces of nationalism in African countries tend to offset the adoption of Christianity by the middle and upper classes, but both Islam and the militancy of

Islamic fundamentalists continue to diffuse into sub-Saharan Africa.

The migration of Asian peoples to the United States and to a lesser extent to European countries results in diffusion of individuals believing in Asian religions, but to date has not resulted in any major change in the religious map of the world. The present role of religion seems to be one of sustaining existing cultural realms, rather than bringing about fundamental changes. The great changes in the world's cultural realms associated with the diffusion of the Europeans and the expansion of Islam do not appear to have a modern successor.

Diffusion of Language

Diffusion of language, like the diffusion of religion, tends to be slow. The rapid change in the linguistic map of the world associated with Euro-

CULTURAL DIMENSIONS
Cheaper Than Machines

Kim Hai Kwee, 19, sits tensely hunched over her microscope.* Jaunty music plays in the background and her head aches dizzily from hours of soldering tiny gold wires to almost invisible chips of silicon. Kim is a worker in the export-oriented electronics industry. Some 90 percent of the workers are young women aged 18 to 23. They work 8 to 10 hours a day, six days a week. Many developed severe eye problems during their first year of employment: chronic conjunctivitis, near-sightedness, and astigmatism. Virtually anyone who stays on the job more than 3 years must wear glasses, and is called "grandma" by her friends.

In one investment brochure the Malaysian government touts the "manual dexterity of the oriental female" as an incentive for Western investors. In one factory, manual dexterity consists of dipping assembled units in large open vats of sulfuric and nitric acid. Heavy fumes are everywhere and the floors are wet and slippery. The women wear boots and gloves that sometimes leak, causing burns. The workers in these plants are exposed to some of the most dangerous acids and solvents, such as trichlorethylene, xylene and benzene, which cause nausea and dizziness. They have also been linked to cancer, and to liver, kidney and lung disease.

"I am sick because of acid concentrated," one woman, Maznah, revealed in broken English. She had been dipping components in acid rinse for three years. The company refused to allow her to transfer to other work. When she visits the camp doctor, he tells her she has "flu." Over 40 percent of the women come from rural areas to work for the $2-a-day wages. When asked why they had left their families, they say there are no jobs back in the village. Most had never worked before.

The electronics industry continually recruits young, rural women—which does nothing to help relieve the steady buildup of jobless in the cities. Women workers have an average stay in the Malaysian factories of 1-2 years; the annual turnover rate is as high as 80 percent. The electronics firms glamorize Western female stereotypes, stressing female passivity and emphasizing a pa-

ternalistic discipline. One company arranges cosmetics lectures and organizes sports teams. Another offers free lipsticks for reaching production goals. Beauty contests ("Miss Motorola," "Miss National Semiconductor") are common and reinforce the Western ideals of consumerism and modernity over traditional Muslim values.

To meet production goals, many factories operate round the clock. Production quotas goad the women to ever higher targets. "If they say one hundred and we can do it, then next week they give us a lot more to do," Azizah, 23, told one investigator. These pressures pay off in profits. According to one plant manager in Malaysia, "One worker working for one hour produces enough to pay the wages of ten workers working one shift, plus all the costs of materials and transportation." Factory managers praise the women workers highly. According to one personnel manager at an instrument assembly plant: "This job was done by boys two or three years ago. But we found that girls do the job as well and don't make trouble like the boys. They're obedient and pay attention to orders. So our policy is to hire all girls."

It's no wonder the workers are praised for their docility when strikes are forbidden and the death penalty invoked for inciting labor unrest. Performance bonuses, often as much as one fifth of the monthly wage, are often revoked for one day of tardiness, sickness or even personal leave or vacation. Some companies state they will be automating their assembly plants in the next ten years. But as long as wages are low and workers available and docile, the chances of being replaced by automate bonding machines are slim. As Maefun, a Hong Kong assembly worker put it: "We girls are cheaper than machines. A machine costs over $2,000 and would replace only two of us. And then they would have to hire a machine tender, for $120 a month."

---

*Source: Diana Roose, "Cheaper than Machines" (Copenhagen: World Conference of the United Nations Decade for Women, 1980.)

TABLE 13–2
Changes in GNP of Selected Countries

COUNTRY	1988 PER CAPITA GNP (U.S. DOLLARS)	GROWTH RATE 1982–1987 (PERCENT)
Bangladesh	160	3.7
Bolivia	540	− 2.7
Chile	1,320	− 0.1
Egypt	760	5.3
Japan	12,250	3.8
Korea (South)	2,370	8.8
Mozambique	210	− 7.0
Niger	260	− 2.5
Norway	15,480	3.8
Spain	4,840	2.1
Sweden	13,170	1.9
United States	12,500	3.0

pean colonialism and migration is over. Today, the adoption of European languages because of increased interaction and trade between the industrial and less industrial worlds is the basis for diffusion. Diffusion of English terms to almost every society is common, but the diffusion of other languages is much less obvious. The recognition by the United Nations of several languages as official (including English, Chinese, French, Spanish, and Russian) has not made them equally attractive.

English, due to its association with technological and scientific advances, tends to be chosen as a second or replacement language. The Chinese language is diffusing very slowly, essentially within areas of China traditionally inhabited by non-Chinese, such as Tibet. The Chinese government is encouraging the study of English as a second language today to allow greater access to the industrial world and its ideas.

Diffusion of the other large language groups, Spanish and Portuguese, primarily reflects the movement of migrants from Latin America to North America. Acculturation and assimilation generally lead to the adoption of English by the second generation of Spanish migrants, even when Spanish is still spoken at home, because English is essential to their becoming part of the economy and society of the United States.

The net effect of the diffusion of languages for the foreseeable future will be to strengthen the role of English in the international scene. Although it is unlikely that whole societies will adopt English, it appears that English will be increasingly important in international trade and communications, strengthening the role of English as a worldwide *lingua franca*.

Diffusion of Ideology and Political Systems

The slow evolution of religion and language is unlike the diffusion of changes in the political geography of the world. Population growth and technological change are often associated with an increasing discrepancy between the rich and poor, the "haves" and "have-nots" in individual countries and between major world regions. The emergence of many new countries in the post–World War II era created the potential for sudden and dramatic political change. It can be expected that the future will bring increased examples of insurgent states, increased conflict within individual countries between groups favoring differing political-economic systems, and greater demands by individual culture groups for autonomy or recognition (Figure 13–7).

Economic factors are an underlying cause of many conflicts, although these conflicts are not necessarily between the Marxist centrally planned system and the capitalistic free enterprise system of the United States and Europe. Increasingly, political conflicts deal with issues

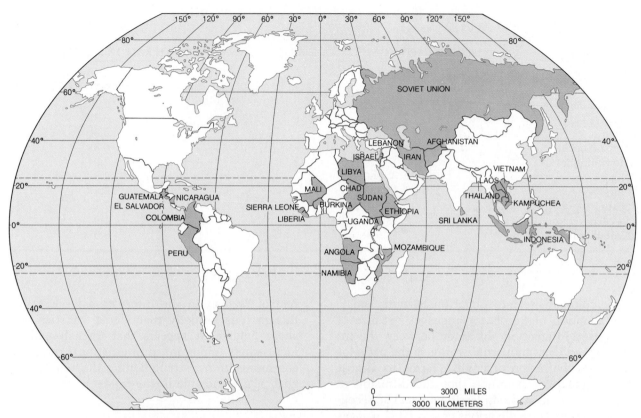

FIGURE 13–7 Countries Experiencing Conflict Eastern Europe's political turmoil of 1989 is not shown.

relating to the economy rather than to the political process itself. Demands by the poor of Latin American and African countries often lead them to groups that promise reallocation of wealth. Historically, these groups have been Marxist oriented, since the basic theories of Marx postulated that capitalism led to ever greater wealth for a few and increased suffering for the masses.

The diffusion of Communism and centrally planned economies modeled in some fashion on the experience of the Soviet Union altered the political map of the world between 1917 and 1987 (Figure 13–8). The emergence of the Soviet Union and People's Republic of China and conflicts in Korea, Vietnam, Afghanistan, and elsewhere played a major role in shaping the present cultural geography of the world, especially with regard to military alliances and military spending (Table 13–3). The reflective barriers placed around their respective countries by the Communist governments of the Soviet Union and China prevented the diffusion of some ideas from outside for decades; at the same time they actively promoted diffusion of their own ideology within their borders and to other countries. At the beginning of the last decade of the twentieth century, it appears these barriers may be weakening. New openness in the Soviet Union under the policies of "Glasnost" (openness) and economic reform (referred to as "Perestroika") may indicate a fundamental change in Soviet willingness to exchange ideas with the outside. A similar process has been occurring in China in the last few years as elements of capitalism are reintroduced into the Chinese economy.

The increased acceptance of elements of free enterprise may be one of the major factors changing the cultural geography of the world in the next generation. It is important to note that free enterprise is not synonymous with democracy and multiparty political representation. Elements of the free enterprise system are associated with a

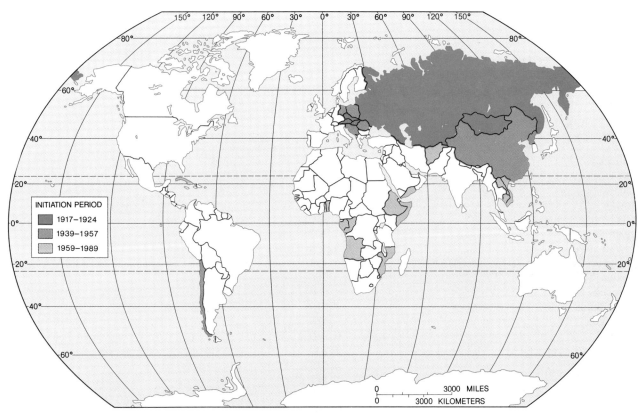

FIGURE 13–8 The Diffusion of the Communist Model of Government Chile's socialist government was quickly overthrown, but Eastern Europe is only now experiencing reform.

TABLE 13–3
Military Spending by Country

COUNTRY	VALUE (MILLIONS OF DOLLARS)	PERCENTAGE OF GNP
United States	237,100	6.3
Algeria	1,403	2.7
Argentina	2,327	3.7
Australia	4,657	2.9
Austria	921	1.3
Belgium	2,669	3.1
Brazil	1,778	0.8
Bulgaria	4,532	7.8
Canada	7,604	2.2
Chile	817	4.2
China	24,040	7.5
Taiwan	3,867	6.6
Cuba	1,600	5.9
Czechoslovakia	7,642	5.8
Denmark	1,395	2.4
East Germany	10,680	6.3
Egypt	5,122	13.5

TABLE 13–3
continued

COUNTRY	VALUE (MILLIONS OF DOLLARS)	PERCENTAGE OF GNP
El Salvador	251	6.1
Finland	790	1.5
France	22,350	4.1
Greece	2,664	7.2
Hungary	3,286	4.1
India	7,141	3.6
Indonesia	2,211	2.6
Iran	11,360	7.2
Iraq	14,640	42.5
Israel	7,206	27.1
Italy	10,110	2.7
Japan	12,700	1.0
Kuwait	1,438	5.3
Malaysia	1,193	3.8
Mexico	966	0.7
Morocco	666	5.0
Netherlands	4,510	3.2
Nicaragua	473	13.4
Nigeria	1,210	1.7
North Korea	5,200	22.6
Norway	1,679	2.9
Oman	2,111	27.7
Pakistan	1,990	5.9
Peru	1,450	7.1
Philippines	396	1.2
Poland	13,440	5.7
Portugal	680	3.5
Romania	5,350	4.4
Saudi Arabia	22,220	21.3
Singapore	1,015	5.3
South Africa	3,540	4.2
South Korea	4,590	5.4
Soviet Union	260,000	12.6
Spain	3,633	2.2
Sweden	2,928	3.1
Switzerland	2,203	2.0
Syria	4,255	22.4
Thailand	1,682	3.9
Turkey	2,467	4.5
United Arab Emirates	1,932	7.4
United Kingdom	25,410	5.3
Venezuela	1,067	1.6
West Germany	22,780	3.3
Yugoslavia	1,791	3.6

Source: U.S. Statistical Abstract (Washington, D.C.: U.S. Government Printing Office, 1988).

variety of governmental types, from authoritarian, dictatorial governments to democratic federalistic governments.

Cultural Convergence: Reality or Illusion

The continued diffusion of cultural traits from one region to another has led some to hypothesize that cultural convergence is an evolutionary process by which traditional societies adopt scientific methods and technology, become industrialized, and increase their standard of living to that of the industrial nations. The *modernization theory* assumes that the process of industrialization is associated with simultaneous changes in technology, industry, urbanization, family and society, education, gender roles, and the role of government.

The theory of modernization uses Western experience as a model for less industrialized nations. It assumes that the industrial experience of the European cultural realm can be replicated in other areas through industrialization, acculturation, or induced change. The theory of cultural convergence argues that adoption of modernization, particularly industrialization, will cause places throughout the world to become more alike, as may be observed in the spread of Western technology, the English language, McDonald's drive-ins, and other traits of the industrialized Western world. In this view the diffusion of European cultural traits to the large cities of the less industrialized world through mass media, technology, and political-economic systems represents hierarchical diffusion. This thesis maintains that just as large cities throughout the world now share many similarities in their downtown regions (high-rise office buildings, interaction with the rest of the world, proliferation of chain fast-food establishments) that symbolize diffusion and adoption of Western culture, these traits will also diffuse to rural areas.

Nevertheless, in spite of the diffusion of elements of Western culture to the less industrialized world, today's world is actually becoming culturally more *divergent*. The contrasts between industrial and less industrialized countries are becoming greater in terms of per capita income or quality of life (Table 13–4). Although elements of

TABLE 13–4
Contrast in Per Capita GNP, 1989

COUNTRY	PER CAPITA GNP (U.S. $)
Algeria	2,570
Argentina	2,350
Australia	11,910
Brazil	1,810
Burkina Faso	150
France	10,740
Greece	3,680
Haiti	330
Honduras	740
India	270
Israel	6,210
Japan	12,850
Jordon	1,540
Philippines	570
Poland	2,070
Saudi Arabia	6,930
Solomon Islands	530
South Africa	1,800
Sudan	320
Switzerland	17,840
Soviet Union	7,400
United States	17,500
Zaire	160
Zimbabwe	620

the industrialized world have spread to many areas, for most people in the world the differences between life in a country in Africa and one in Europe are greater today than at anytime in the past. Prior to the Industrial Revolution, for example, life in the agrarian societies of Africa, Asia, or Latin America was remarkably similar to that in Europe. Increasing industrialization, increasing accumulation of wealth, and the continued dominance of the world economic scene by the industrial nations have made them increasingly different from the poorer, more rural, less industrialized nations. This suggests that *cultural divergence* is related to the process and rate of diffusion. Diffusion of technology or other aspects of the industrial world first affects urban locations, while rural areas of the less industrialized world change more slowly. Since the majority of the population of the less developed world is rural, the contrast between the industrialized and less industrialized countries widens.

Some observers maintain that in spite of the differences among the economies and societies of the world, it is possible to view the world as one single economic system with a division of labor in which some nations produce raw materials and some use the raw materials for manufacturing. The proponents of this view argue that the present difference between the industrial and less industrialized countries represents the exploitation of the less industrialized world and its resources by the more industrialized world with its technology. The basis for this thesis is the theory of Karl Marx, which basically argues that in a capitalistic, free enterprise society or economy, a few will always benefit at the expense of the masses.

The proponents of the theory of a one world economic system recognize two classes of coun-tries: core countries that are rich, powerful and free of outside control; and peripheral societies that are weak, poor, and rely on one or two items for their export earnings (Figure 13–9). The one world economic system view implies that the less industrialized countries will continue to be dominated by the industrial countries. Countries that produce a single item, or a few items, for export are rarely able to control the prices they receive. Countries of Latin America that rely on production of coffee, sugar, bananas, tin, or petroleum exemplify this problem (Table 13–5). When market prices change as a result of either oversupply or decreasing demand, the economies of these less industrialized countries are devastated.

Marxist cultural geographers argue that it is the responsibility of the industrial core countries to ensure that the world's wealth is allocated in a way that allows residents in the less industrialized peripheral economies to enjoy a decent stan-

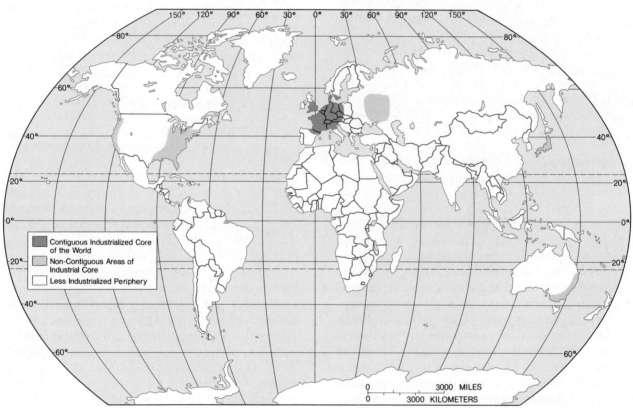

FIGURE 13–9 World Core and Periphery Europe and North America are the core from which many ideas diffuse.

TABLE 13–5
Exports of Latin America, 1989

COUNTRY	MAJOR EXPORT	PERCENTAGE OF COUNTRY'S TOTAL EXPORTS
Bolivia	Natural gas	52
Brazil	Tobacco	21
Chile	Copper	46
Colombia	Coffee	62
Costa Rica	Coffee	34
Cuba	Sugar	77
Dominican Republic	Sugar	18
El Salvador	Coffee	67
Ecuador	Oil	42
Guatemala	Coffee	44
Haiti	Coffee	32
Honduras	Coffee	38
Peru	Coffee	17
Trinidad/Tobago	Oil	79

dard of living. Non-Marxist observers of the same phenomenon argue that supply and demand will ensure that those peoples whose livelihood has been threatened will adjust by producing a crop or raw material that is in demand. Whichever view you adopt, it is clear that diffusion and acculturation are making the world more interdependent. Decisions made in New York affect the peasant farmer in Southeast Asia. Decisions made in the corporate boardrooms of Japan affect the American consumer; environmental hazards (such as frost or hurricanes) affect the price paid by Europeans and Americans for tropical products from Latin America, Asia, and Africa.

Forces for Change: Physical Variables

Future changes will affect not only the cultural geography of the world but also the physical environment through the process of cultural ecology. The changes affecting the physical environment today are among the most important in the history of the earth. Environmental damage caused by acid rain (increased acidity in precipitation caused by industrial emissions), landscape destruction associated with mining or resource extraction, loss of forests, and deterioration of the air, water, and soil base of the world are well documented.

Soil erosion, for example, is a major problem throughout the world. Soil losses in the heartland of the corn belt of the United States average more than ten tons per acre per year, equivalent to a tax of two bushels of soil for each bushel of corn raised in Iowa. Adoption of Western mechanized cultivation techniques in the less industrialized realm may well result in even more erosion in those areas.

Tropical rain forests continue to decrease at a rate that by some estimates equals the area of Massachusetts each month. Increasing acreage lost to *suburbanization*, *desertification*, and ero-

Erosion remains a serious environmental problem throughout the world.

sion lead some to question the ability of the world to continue to feed its ever-growing population. Examination of some of these environmental problems illustrates how they will affect the world's cultural geography in the future.

Issues of Environmental Degradation

Environmental degradation continues to increase at a dramatic rate, causing pollution of air, water, and soils in the industrial world. The population of the United States, for example, increased by some 50 percent in the last quarter century, but the nitrogen oxide emitted by automobiles (the main creators of smog) increased by 63 percent in spite of attempts to limit such emissions. The production and emission of phosphates that stimulate the growth of algae in surface waters and reduce the oxygen available to sustain fish (in a process known as eutrophication) increased 700 percent in the same time period.

The average consumption of material goods did not increase dramatically during this time period, and in fact some of the industrialization trends in the United States changed. Increased environmental degradation seems to be the result of a change in the types of items used in daily life. Increased emission of nitrogen oxides is associated with high performance engines that run at high compressions and temperatures. The increase in phosphate emissions and destruction of water quality can be traced to the change from soap to phosphate-based detergents in the American home and the increased use of manufactured fertilizers on farmlands.

A vital change has been the replacement of leather, glass, wood, cotton, and manure by synthetic petrochemical products including detergents, plastics, synthetic fibers, chemical fertilizers, and pesticides. The increased use of petrochemical products and their subsequent incineration or movement into surface water bodies by leaching have greatly increased the problem of environmental degradation.

Increased petroleum refining, adoption of the industrial model, and the proliferation of the automobile are spreading environmental degradation from the industrial countries to the rest of the world. Continued reliance upon the European

Pollution of the world's water resources continues in spite of widespread publicity about its significance.

industrial model in the newly industrializing countries may create environmental problems even more severe than those currently found in the industrial world. Whereas the industrial world is characterized by at least token efforts to minimize air, water, or soil pollution, many of the newly industrializing countries are unable to pay the costs of maintaining or restoring environmental quality. Thus a vicious cycle is promoted in which the drive to raise the quality of life to Western standards creates ever greater environmental problems, thus reducing the quality of life in newly industrializing regions.

Loss of Agricultural Land

Closely related to the environmental degradation of the air, water, and soil is the loss of agricultural land. Erosion of productive soils by wind and water extends from the subsistence agricultural lands of Africa to the corn belt of the United States. Some geologists estimate that 24 billion tons of topsoil are being lost to erosion each year in the last decades of the twentieth century, three times the rate that would occur without human action. The loss of topsoil through accelerated erosion is partially masked by the nearly tenfold increase in fertilizer use since 1950 and the tripling of irrigated cropland in the same time period. Still, in time loss of topsoil affects the total productivity of the soil. Topsoil is generally a thin veneer over subsoil and rock substrata. Most farmers cultivate a topsoil mantle less than ten to twelve inches thick. The continued erosion of this thin but essential mantle leads to increased use of fertilizer to offset its loss and eventually to declining crop yields.

Soil erosion increases as population pressure increases the demand for food in the less industrialized world. Croplands are pushed onto steeper slopes and other land that is more erodible. Forests are cleared in the tropics, with resultant erosion. In industrial nations, farmers turn to repeated cultivation of single crops to maximize their returns and expand production into less suitable lands with associated increases in erosion.

Examination of the major agricultural countries—the United States, the Soviet Union, China, and India—reveals the severity of the loss of topsoil through erosion. The national resources inventory of 1982 in the United States concluded that 44 percent of U.S. cropland was losing topsoil more rapidly than new soil was being formed. A study for the Chinese government reported that 5 billion tons of soil are lost through its rivers each year. The Indian government estimated that decline in soil fertility through erosion affects 370 million acres (equivalent to the total cropped land in the United States) and that 6 billion tons of soil are lost through erosion each year. The Soviet Union has not released figures on erosion, but other observers suspect that its loss of soil may be greater than that of the United States because many Soviet farmers use land that is less naturally suited to agriculture. (Figure 13–10).

At the same time that topsoil is being lost to erosion, and marginal lands are becoming unsuitable for cultivation, some of the best agricultural lands are being lost to cities and suburbs. Particularly in the industrial world, suburbanization and ownership of single-family homes has removed millions of acres of land from production. In itself the loss of the agricultural land does not threaten the world's ability to feed its peoples,

FIGURE 13–10 Soil Loss in Selected Countries, 1970–1986 Erosion creates a major challenge to the continued successful human use of the earth.

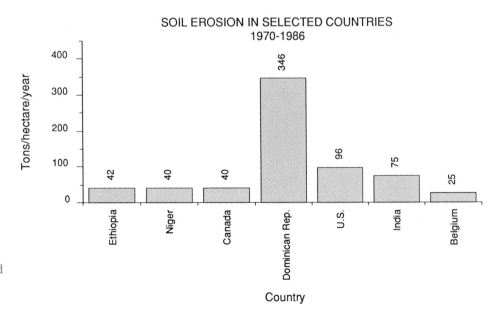

SOIL EROSION IN SELECTED COUNTRIES
1970-1986

Tons/hectare/year

Ethiopia 42
Niger 40
Canada 40
Dominican Rep. 346
U.S. 96
India 75
Belgium 25

Country

Destruction of cropland by erosion affects both agricultural production and contributes sediments to streams and lakes.

but added to the loss of topsoil through erosion, it is a significant contributor to agricultural concerns facing the world of tomorrow. Many suburban homes are built on prime or irreplaceable agricultural lands; such lands are generally level and are located in Florida, California, or other areas with climates suitable for crops that will not grow in other climates.

Climatic Change

At the same time that agricultural land is affected by erosion and suburbanization, some observers are suggesting that climatic change may further modify the surface of the earth and its human geography. One hypothesis regarding climate change involves the so-called greenhouse effect. The greenhouse effect occurs because short-wave solar radiation can penetrate the earth's atmo-

sphere. Reflected from clouds, the earth's surface, or other opaque surfaces, short-wave radiation becomes long-wave radiation that is absorbed by carbon dioxide and water vapor in the atmosphere, making the earth habitable. The process is called the *greenhouse effect* because it is similar to what happens in a greenhouse when light penetrates the glass and heats the interior. Some observers use the term "greenhouse effect model" to refer to a hypothesized global warming since the warming is related to the greenhouse effect of the atmosphere (Figure 13–11). In this model the greenhouse effect refers only to the cumulative warming of the earth's atmosphere due to increased carbon dioxide in the atmosphere caused by fossil fuel combustion.

For those concerned about climatic change, the amount of carbon dioxide in the atmosphere is of critical importance. Climatologists have discovered that the history of the earth for the last few million years has been one of regular cycles of ice ages: roughly 100,000 years of the ice age are followed by an interglacial period of about 10,000 years which in turn is followed by another ice age.

Proponents of the greenhouse effect model point out that although no direct measurements of the carbon dioxide in the atmosphere were made before the twentieth century, drilling in glaciers in the Antarctic suggests that the carbon dioxide level was about 280 parts per million before 1850. In 1986, the carbon dioxide content measured in Hawaii (where measurements have been made continuously for twenty years) was 350 parts per million. The earliest record in Hawaii (1958) showed only 316 parts per million. Based on evidence from glaciers, climatologists suggest that during the glacial periods, the carbon dioxide content was about 250 parts per million, and that present levels may be higher than at anytime since the beginning of the most recent ice age some 100,000 years ago.

These researchers are concerned about the increased combustion of carbon-based fuels (coal, petroleum, natural gas, and wood), which adds carbon dioxide to the atmosphere. They hypothesize that about 58 percent of all carbon dioxide produced by burning since monitoring began in 1958 has stayed in the air. It was assumed in the past that much of the carbon dioxide was trans-

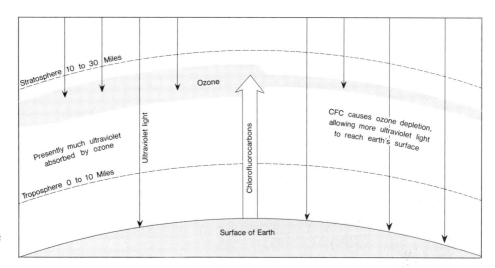

FIGURE 13–11 The Greenhouse Effect and the Ozone Layer

formed into oxygen by vegetation in the tropics, but continued destruction of the tropical forests has limited this role (Figure 13–12).

The implications of climatic change for the geography of the world are significant. Average global temperatures may rise between 1.5 and 4.5 degrees Celsius (2.7 to 8.1 degrees Fahrenheit) if the level of carbon dioxide increases to approximately 500 parts per million. Already, records suggest that there has been an increase of 0.5

degrees Celsius (0.9 degrees Fahrenheit) in the last century.

The impact of global warming on human occupation of the earth varies from place to place. Climatologists suggest that high-latitude regions will warm much more than low-latitude regions, and that the resultant climate changes will differ in terms of both temperature and precipitation (Figure 13–13). The industrialized countries of the higher northern latitudes will suffer most

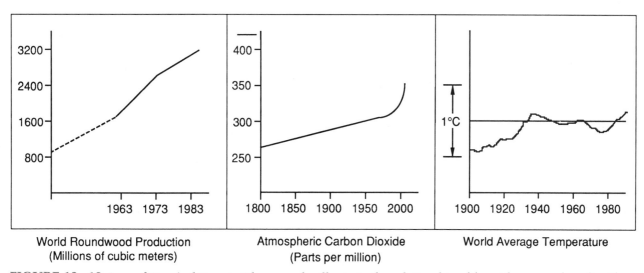

World Roundwood Production
(Millions of cubic meters)

Atmospheric Carbon Dioxide
(Parts per million)

World Average Temperature

FIGURE 13–12 Loss of Tropical Forests These graphs illustrate the relationship of forest loss to carbon dioxide and temperature change.

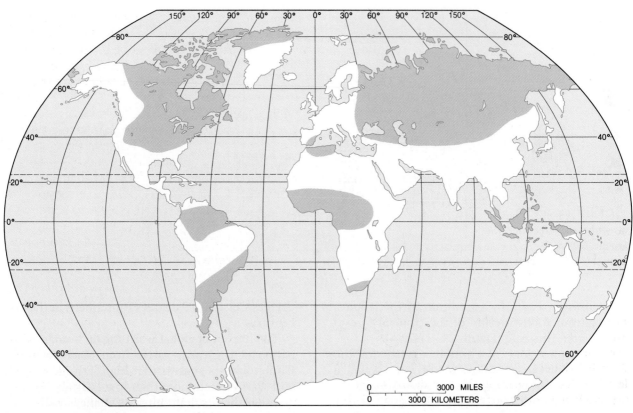

FIGURE 13–13 Hypothesized Changes in Temperature and Precipitation in the Coming Decades Mid- to high-latitude grain producing regions are predicted to be dryer, while tropical areas become wetter.

from the predicted climatic changes. The grain belts of the United States will be hard hit by drought, while in the lower tropical and subtropical latitudes increased precipitation will result in more stable crop production. Increasing temperatures would also cause a rise in the sea level, flooding low-lying coastal areas, increasing the rate of coastal erosion, and requiring construction of dikes to protect coastal cities from wave action during storms.

Another hypothesis of environmental change involves the ozone layer that shields inhabitants of the earth from damaging ultraviolet radiation. Changes in the ozone layer of the atmosphere may be caused by the use of chlorofluorocarbons (CFCs). CFCs are used in refrigerants and as propellants in aerosol sprays, such as hairspray. In the upper atmosphere ultraviolet light breaks off a chlorine atom from a CFC molecule. The chlorine molecule attacks an ozone molecule, breaking it apart. Through the chemical process, the chlorine molecule is freed again, which means

that each CFC molecule can destroy thousands of ozone molecules.

In the late 1980s, scientists discovered a "hole" in the ozone layer over Antarctica, even though calls for decreased reliance on fluorocarbons had begun in the late 1960s. If damage to the ozone layer actually occurs, it could lead to increased incidences of skin cancer as more ultraviolet rays strike the earth. Recently, scientists have found that other gases in addition to the traditional greenhouse gases of carbon dioxide, oxygen, and water vapor play a much more important role than expected. The use of synthetic chlorine compounds, including the fluorocarbons, affect the concentration of these gases and could also contribute to the greenhouse effect model.

Desertification

Another climate change is the human-induced increase in the extent of the world's deserts. Through *desertification* deserts are expanding into

Overgrazing is associated with the process of desertification, yet growing populations in arid and semi-arid areas seems to require expansion of annual populations to feed the additional mouths.

more humid regions along their margins as plant cover and soils are destroyed by overgrazing and erosion. Each year an estimated 59 million acres (27 million hectares) of productive land are rendered useless for farming through desertification. Although it is most severe on the fringes of the Sahara, desertification occurs on every continent (Figure 13–14. See also Figure 6–8 on page 219.)

The impact of desertification upon the cultural geography is obvious. The nomadic peoples in regions north and south of the Sahara are trapped in a vicious cycle. Population growth encourages nomads to keep larger herds, larger herds increase grazing pressure, and increased grazing pressure leads to erosion and desertification, lowering the productivity of the herds. This process is intensified when periodic droughts strike the desert margins. The overgrazed land is incapable of providing enough forage for the animals, and the vegetation is totally destroyed as animals fight starvation. Ultimately, the human population is forced to change its economy and culture.

Estimates by the United Nations Environmental Program suggest that more than a third of the earth's land surface (about 45 million square kilometers or 17.35 million square miles) is threatened by desertification. Nearly 20 percent of the planet's total population are directly affected. The complex interplay of population growth, overgrazing, erosion, climatic change, and desertification affect the inhabitants of all of

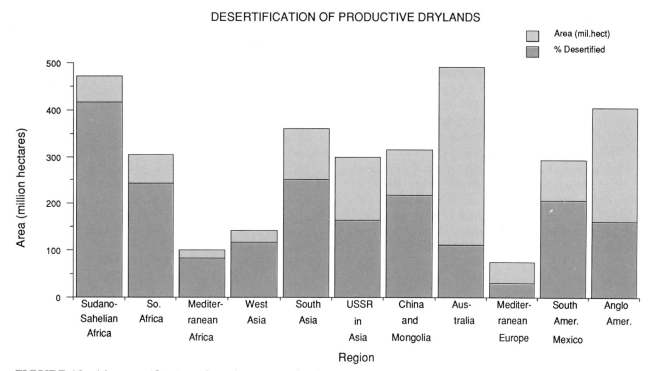

FIGURE 13–14 Desertification of Productive Drylands

the arid and semiarid regions of the world. Desertification is intensified by population growth that exceeds the world growth rate (56 percent growth worldwide in the quarter century from 1960 to 1985, but 81 percent growth in the arid lands). Within this sector of the earth's population, moreover, nearly 62 percent of the total population relies on agriculture for its livelihood.

Loss of the Tropical Rain Forests

Another major environmental change is the destruction of the world's tropical rain forests. The growing population of both sedentary farmers and slash and burn agriculturalists who place greater pressure on the forests is partially responsible. Increased population in shifting agricultural systems results in greater demand for land to rotate or a shorter fallow period. Growing populations of peasant farmers also encroach upon the forest land used by shifting agriculture farmers, pushing them into mature tropical rain forests.

The U.S. demand for cheap, lean meat for hamburgers and pet food causes the clearing of roughly 8,000 square miles of forest each year for cattle grazing. The demand for tropical hardwoods by international companies also destroys the tropical rain forests.

The loss of rain forest is particularly severe in selected countries. West African countries such as Nigeria and Ghana, where population pressure is high, have lost from 80 to 90 percent of their tropical rain forest. Essentially all of the tropical forest in Haiti has been removed as a result of population pressure and demands for cropland and lumber for fuel and building.

The loss of the tropical rain forests has a severe impact on the cultural geography of the world. In particular, it increases the amount of carbon dioxide in the atmosphere contributing to global warming. Yet tropical rain forest destruction continues because the populations of the less industrialized countries continue to grow, and industrial countries continue to exploit their cheap raw materials.

The loss of the tropical rain forests is also alarming because these forests are a repository for many of the plant and animal species of the world. The disappearance of any of the world's species is a crisis that cannot be ignored. More than 25 percent of prescription medicines come from chemical compounds discovered in nature, so the loss of the tremendous gene pool of the tropical rain forests may also seriously limit advances in medicine and science (see the Cultural Dimensions box).

Expanding the Green Revolution

The concern over the loss of the tropical rain forests, desertification, and soil erosion raises the critical issue of feeding the world's population. The grim scenario of war, disease, and famine hypothesized as population checks by Malthus have been avoided in part because of chemical fertilizers and hybrid grains developed in the 1950s that were the basis for the so-called *Green Revolution*. These miracle grains produce two to three times the yields of traditional varieties. They have become the mainstay of agricultural production in the industrialized world and in China and India, but have not yet spread to Africa, which has the most rapidly increasing population in the world. The increased population growth in rural areas of Africa causes migration to the cities, primarily of the male population. Women and children are left to farm, resulting in even lower rates of increase of production or actual decreases in yields.

To generate a Green Revolution for Africa, it is essential that basic research into the major subsistence food crops of Africa (millet, sorghum, and manioc) be undertaken. African farmers need to be better trained and to accept the need for conservation techniques. The role of governments in monopolizing the production of tropical crops and dictating prices received by farmers must change if the rural farmers are to increase their productivity. Reforms to provide land for farmers will also be essential if African agriculture is to be transformed.

Irrigation to water the arid and semiarid regions of Africa is an immediate problem, as is the need for simple farm equipment and basic capital to allow adoption of better techniques. Unless such changes are made, the grim forecast of Malthus may be realized in the Sub-Saharan African cultural realm.

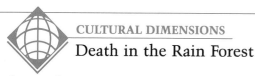

Death in the Rain Forest

The Brazilian Francisco Mendes Filho was shot to death at his home in the little jungle town of Xapuri in the state of Acre on December 22 by a hired gunman.* His death is an immeasurable loss to those who knew and loved him and, because of the scope and impact of his work, to millions who have never even heard of him. For ten years, he had been the leader of the National Council of Rubber Tappers (CNS), representing some 150,000 people who live in, and off, the rain forests that cover 87 percent of Acre.

In 1987, he received the Global 500 Award from the United Nations Environmental Program for his nonviolent, yet remarkably effective, efforts to halt the destruction of the rain forest. Twelve percent of all Brazil's rain forests have already been destroyed, mostly in the last decade, by cattle ranchers, small-scale farmers, and speculators.

Mendes's greatest practical achievement was the creation, through negotiations with the Brazilian government, multilateral development banks, members of the U.S. Congress, and environmental agencies, of four "extractive reserves" in Acre and eight in other Amazon states. These reserves, including more than 5 million acres, are protected for the extraction of rubber, nuts, resins, and other forest products.

Mendes hoped that 40 percent of Acre will eventually be classified as reserves. But he was also concerned that the extractive reserve movement might fail in areas without a strong rubber tappers' union.

The forest where Mendes walked is in a reserve called Cachoeira. Sixty-seven families live there, including Mendes's aunt and uncle. A cattle rancher named Darly Alves claims part of Cachoeira as his own land. In recent months, Mr. Alves had repeatedly threatened Mendes's life, to the point where the governor of the state assigned him two bodyguards. These guards were in his house when Mendes walked out his back door and was shot. Alves's son Darcy admitted guilt a few days after the murder, but police continued to search for his father. He was found hiding in the forest last week and is now in police custody, pending the investigation of his alleged responsibility for this and other killings.

The entire world needs the rain forests Mendes died trying to protect. Yet it should be understood that in Brazil, which contains 30 percent of the earth's remaining tropical forests, inequities of land distribution are a major cause of the increasing destruction of these forests. Small farms, half of all rural properties in Brazil, cover only 3 percent of occupied rural land, while large estates, owned by less than 1 percent of the landholders, occupy 43 percent of the land. And an estimated 335 million hectares are being held, unused, for speculative purposes.

A key role in many violent land disputes in Brazil is played by a landowners' association, the Rural Democratic Union (UDR) whose leaders admit to having stockpiled 70,000 weapons. UDR defends large landholdings and cattle ranches against peasants' attempts to implement existing land reform laws. According to Amnesty International, the organization is responsible for the deaths, disappearance, and torture of hundreds of Brazilian peasants, priests, and union leaders. The day after Mendes's murder, the UDR leader in Acre, Joao Branco, left the country and is now reported to be in Paris.

The world will always owe a huge debt to Mendes, for his work will continue. Those who knew him can be grateful they did and treasure him in their memories.

---

*Adapted from the *Christian Science Monitor*, January 12, 1989, p. 19.

The Destruction of Traditional Societies

Growing populations, desertification, and increasing nationalism are causing important changes in traditional societies. Nomadic peoples are forced to resettle in permanent locations or are prevented from migrating because of enforcement of political boundaries. Population growth forces people from the land into cities, breaking down their traditional extended families. Increased economic interaction with industrial areas creates demands for specialized crops or products, further changing traditional societies. The changes that are occurring in the less industrialized world are profound and pervasive and will remake the cultural geography of the majority of the earth's inhabitants in your lifetime.

Conclusion

The changes associated with the world's physical and cultural geography are remaking the world. Remember that colonialism, which created much of the present cultural map, only ended recently. Europeans controlled 9 percent of the world's land surface in 1500. By 1800 they ruled about 35 percent and by 1878, 67 percent. At the beginning of World War I (1914) Europeans controlled 85 percent of the world's land. European colonialism effectively ended shortly after World War II. The change from a colonial to a postcolonial world should not mask the continued dominance of the world by the European cultural realm. Today's domination is economic and political: economic because the former colonies have trade relationships with their former colonial masters and political because former colonies rely on former European colonial powers for defense.

The United States and the Soviet Union dominated the world in the post–World War II era. Since the creation of the Common Market, the EC and Japan have begun to rival the two superpowers. The increased importance of countries other than the United States and Soviet Union may signal the beginning of an important change in world interrelationships. Creation of trade agreements between individual countries such as France and its former colonies; the increasing importance of countries like China, Brazil, India, and Indonesia because of their size, population, and productivity; and the role of countries like Saudi Arabia that possess a particular resource are changing the traditional industrial versus less industrialized regionalization of the world.

These changes seem to be leading to a world in which the cultures of other groups will expand at the expense of the Europeans. The resurgence of Islam, the spread of Sikh discontent in India, and the growth of revolutionary movements in Africa and Latin America suggest that the cultural geography of the world at the beginning of the twenty-first century will be much different than it is today. The forces for change now operating ensure that you will see dramatic and important changes in the world's cultural geography in your lifetime.

QUESTIONS

1. Discuss the elements of change in cultural geography.
2. What impact will urbanization have on cultural geography?
3. What is the possibility of expanding agricultural resources?
4. What are the advantages and disadvantages of diffusion of Western technology?
5. Which cultural element's diffusion will have the greatest impact upon the world? Why?
6. Will economic conflict change the character of the world? Why or why not?
7. What is the responsibility of the wealthy nations to the less industrialized? Can they assist without cultural convergence occurring?
8. List and describe three environmental problems of the world. What impact will they have on culture?
9. What might we do to overcome the greenhouse effect?
10. Which environmental change will have the greatest impact on the world? Why?

SUGGESTED READINGS

Barr, Brenton M., and Kathleen E. Braden. *The Disappearing Russian Forest: A Dilemma in Soviet Resource Management.* Totowa, N.J.: Rowman and Littlefield, 1988.

Bernard, Frank E. "Planning and Environmental Risk in Kenyan Drylands." *Geographical Review* 75 (January 1985): 58–70.

Boulding, Kenneth E. "The Economics of the Coming Spaceship Earth," in H. Jarret, ed., *Environmental Quality in a Growing Economy,* pp. 3–14. Baltimore: Johns Hopkins University Press, 1966.

Bouvier, Leon F. "Planet Earth 1984–2034: A Demographic Vision." *Population Bulletin* 39 (February 1984).

Brown, Lester R. "Food Resources and Population: The Narrowing Margin." *Population Bulletin* 36 (September 1981).

Brown, Lester R., et al. *State of the World.* New York and London: W. W. Norton, annually since 1984.

Brown, Lester R., and Edward C. Wolf. *Reversing Africa's Decline.* Worldwatch Paper 65. Washington, D.C.: Worldwatch Institute, 1985.

Brown, Lester R., and Pamela Shaw. *Six Steps to a Sustainable Society.* Worldwatch paper 48. Washington, D.C.: Worldwatch Institute, March 1982.

Bryson, Reid, and Thomas Murray. *Climates of Hunger: Mankind and the World's Changing Weather.* Madison: University of Wisconsin Press, 1977.

Bunge, William W. "The Geography of Human Survival." *Annals of the Association of American Geographers* 63 (September 1973): 275–95.

Butzer, Karl W. "Adaptation to Global Environmental Change." *The Professional Geographer* 2 (August 1980): 269–78.

Calzonetti, Frank, and Robert Hanham. "Regional and Temporal Trends in Power Plant Unit Siting, 1912–1978." *The Professional Geographer* 4 (November 1983): 416–26.

Carbon Dioxide Assessment Committee, U.S. National Research Council. *Changing Climate.* Washington, D.C.: National Academic Press, 1983.

Carr, James H., and Edward E. Duensing, eds. *Land Use Issues of the 1980s.* New Brunswick, N.J.: Center for Urban Policy Research.

Commoner, Barry. *The Closing Circle; Nature, Man and Technology.* New York: Knopf, 1971.

A Continent in Crisis: Building a Future for Africa in the 21st Century. Washington, D.C.: The Population Institute, 1988.

Crosby, Alfred W. *Ecological Imperialism: The Biological Expansion of Europe, 900–1900.* Cambridge, England: Cambridge University Press, 1986.

Cutter, Susan L., and William D. Solecki. "The National Pattern of Airborne Toxic Releases." *The Professional Geographer* 41 (1989).

Fearnside, Philip M. *Human Carrying Capacity of the Brazilian Rainforest.* New York: Columbia University Press, 1986.

Gardner, Lytt I., Jr., John F. Brundage, Donald S. Burke, John G. McNeil, Robert Visintin, and Richard N. Miller. "Spatial Diffusion of the Human Immunodeficiency Virus Infection Epidemic in the Unted States, 1985–87." *Annals of the Association of American Geographers* 79 (March 1989): 25–43.

Gibson, Mary, ed. *To Breathe Freely: Risk, Consent, and Air.* Totowa, N.J.: Rowman and Allanheld, 1985.

Greenberg, Michael R., and Richard F. Anderson. *Hazardous Waste Sites: The Credibility Gap.* New Brunswick, N.J.: Rutgers University, Center for Urban Policy Research, 1984.

Gupta, Harsh K. *Geothermal Resources: An Energy Alternative.* Amsterdam and New York: Elsevier Scientific Publishing Co., 1980.

Hewitt, Kenneth. "Place Annihilation: Area Bombing and the Fate of Urban Places." *Annals of the Association of American Geographers* 73 (June 1983): 257–84.

Jackson, Barbara (Ward), and Réné Dubos. *Only One Earth: The Care and Maintenance of a Small Planet.* New York: W. W. Norton, 1972.

Jackson, James P. "Life & Death in Tropical Forests: The Edge of Extinction." *American Forests* 94 (November/December 1988): 40–49.

Karan, P. P., and Cotton Mather. "Tourism and Environment in the Mount Everest Region." *Geographical Review* 75 (January 1985): 71–92.

Karan, P. P., Wilford A. Bladen, and James R. Wilson. "Technological Hazards in the Third World." *Geographical Review* 76 (April 1986): 195.

Myers, Norman, ed. *Gaia: An Atlas of Planet Management.* Garden City, N.Y.: Doubleday, 1984.

O'Connor, A.M. *The Geography of Tropical African Development.* 2d ed. Oxford: Pergamon Press, 1978.

Odell, Peter R., and Kenneth E. Rosing. *The Future of Oil.* London: Kogan Page, 1980.

Pasca, T. M. "The Politics of Tropical Deforestation." *American Forests* 94 (November/December 1988): 21–23.

Pearce-Batten, Anthony. "The Future of Renewable Energy." *Focus* 29 (January/February 1979): 1–16.

Porritt, Jonathan. *Seeing Green: The Politics of Ecology Explained.* Oxford: Glackwell, 1985.

Postel, Sandra. "Global View of a Tropical Disaster." *American Forests* 94 (November/December 1988): 25–29.

Redclift, Michael. *Development and the Environmental Crisis: Red or Green Alternatives?* London: Methuen and Co., 1984.

Sampson, R. Neil. "ReLeaf For Global Warming." *American Forests* 94 (November/December 1988): 9–11.

Seidel, S., and D. Keyes. *Can We Delay a Greenhouse Warming?* Washington, D.C.: U.S. Government Printing Office, 1987.

Shannon, Gary W., and Gerald F. Pyle. "The Origin and Diffusion of AIDS: A View from Medical Geography." *Annals of the Association of American Geographers* 79 (March 1989): 1–24.

United Nations. *World Map of Desertification at a Scale of 1:25,000,000.* U.N. Conference on Desertification, 1977.

Vietmeyer, Noel. "Animal Farming Saves Forests." *American Forests* 94 (November/December 1988): 40–49.

Weir, David, and Mark Schapiro, *Circle of Poison.* San Francisco: Institute for Food and Development, 1981.

White, Lynn, Jr. "The Historic Roots of Our Ecologic Crisis." *Science* 155 (1967): 1203–7.

Whitmore, T. C. *Tropical Rain Forests of the Far East*, 2d ed. New York: Oxford University Press, 1984.

Wirth, Timothy E. "Addressing the Challenge of Climate Change." *American Forests* 94 (November/December 1988): 12–14.

Worster, Donald. *Nature's Economy: A History of Ecological Ideas*, 2d ed. New York: Cambridge University Press, 1985.

Ziegler, Donald J., James L. Johnson Jr., and Stanley D. Brunn. *Technological Hazards.* Washington, D.C. Association of American Geographers, 1983.

Glossary

absolute location The geographic location defined by latitude and longitude and specific internal geographic characteristics of a place. Used interchangeably with site.

accessibility The ease with which interchange can occur between two places or people.

acculturation Changes occurring in a culture through intercultural borrowing. Changes may include such things as technology, language, and values.

acid rain Acidified rainwater that severely damages plant and animal life. Caused by the oxides of sulfur and nitrogen that are released into the atmosphere when coal, oil, and natural gas are burned.

Afrikaans The white population of South Africa who are descendants of the Boers (Dutch farmers).

Afro-asiatic A family of languages of Southwestern Asia and Northern Africa.

alternative foods Nondomesticated species of wild plants and animals that have potential value as food crops.

aging world Term used to refer to the increasing average age of the populations of the industrialized world and the related growth in numbers of the elderly.

Agricultural Revolution Two periods in which major changes occurred in agriculture resulting in increased yields and new types of farming. In the first humans shifted from gathering wild plants and hunting wild animals to plant and animal domestication. In the second revolution equipment and seeds were improved to increase yields and productivity.

animists People who worship naturally occurring phenomena such as mountains, trees, or animals.

antecedent boundary A political border drawn prior to the settlement of an area.

apartheid Literally, "apartness." The Afrikaans' term given to South Africa's policies of racial separation and the highly segregated socio-geographic patterns they have produced.

Arabic A principal language of the Semites, usually associated with Arabs.

arable Land suited for farming.

arithmetic growth Refers to simple addition as a means of increase.

assimilation The process of integrating one culture into another so that the former loses its distinctiveness.

asymmetrical relationship Relationship in which one participant benefits more than, or at the expense of, the others involved.

Balkan Peninsula A peninsula in southeastern Europe including the countries of Greece, Albania, Yugoslavia, and Bulgaria.

Balkanization The fragmentation of a region into smaller, often hostile, political units.

barriers A boundary or limit; anything that acts to obstruct or prevent passage.

barrios Spanish term meaning neighborhood; commonly used to refer to slums or shanty towns in Spanish-speaking Latin America.

basin irrigation A system of irrigation in which floodwaters of a river are trapped in basins created by building small dikes around fields to retain the silt-rich floodwater.

Basque An ethnic group in northern Spain and southern France with a unique language, dress, diet, religion, and economic standards.

bilingualism Knowledge of more than one language.

biological resistance Limits on population growth caused by human characteristics such as length of gestation.

birth rate The annual number of births per 1,000 individuals within a given population or country.

Blue Nile The Blue Nile's source is in Ethiopia. It joins the White Nile in Sudan to create the Nile.

Boer Farmers of Dutch ancestry in South Africa.

boundary An invisible line that marks the extent of a state's territory.

Brahman The highest group in the Hindu caste system.

break in bulk Location where large cargoes are broken down into smaller units.

Buddhism A religion derived from the teachings of Buddha, who attempted to reform the Hindu belief system.

buffer zone A country or number of countries separating two powerful countries who are adversaries.

cadastral pattern The shapes formed by property borders; the pattern of land ownership.

calorie deficit The total amount of food measured in calories that would need to be added to provide minimum nutritional needs for the malnourished of the world.

Cape Coloured An ethnic group in South Africa composed of descendants from a mixture of the African population and the early European visitors and immigrant Malays.

capitalism An economic system based on the concept of private ownership of property and the freedom of the individual to engage in economic activity and to receive the benefits of that economic activity (profit).

carrying capacity The number of animals, crops, or people an area can support on a continual basis without degrading the environment. The carrying capacity varies with technology, land-use techniques and geographic characteristics.

caste system The strict social segregation of people in Hindu society on the basis of ancestry and occupation.

cathedral cities A European city sufficiently important to be the site of a cathedral in the Middle Ages.

central place Any place providing goods and services to a surrounding nonurban area.

centrally planned economies Political economic systems in which the state controls the total economy.

centrifugal forces Forces that tend to weaken or destroy a country's unity.

centripetal forces Forces that tend to unify a country or area.

centuriation An ancient Roman rectangular survey system, traces of which can still be seen in the cadastral pattern.

chain migration Process by which people migrate to a place because a relative or friend from their community emigrated to that place.

China proper The name applied to humid eastern China.

Christaller model Model that illustrates how cities develop a hierarchical pattern based on their role as central places.

circulation Free movement or passage.

city morphology The internal organization of the roads, factories, homes, and other features of the built environment of a city.

city state A state consisting of a city and its immediately surrounding territory such as Singapore.

clan Cultural groups based on extended family relationships.

collectivization The reorganization of a country's agriculture through expropriation of private holdings and creation of relatively large-scale units, which are farmed and administered cooperatively.

colonialism Political and economic control of a country by a foreign power.

Common Market A group of twelve European countries (as of 1988) that belong to a supranational association to promote their economic interests. Official name is European Economic Community (EEC), commonly shortened to European Community (EC).

communes Settlements based on collective ownership and use of property, goods, and the means of production.

Communism Economic system where all factors of production are owned by the state in the name of the workers.

compact holdings Property ownership characterized by a single parcel of land.

compact state A term used to describe a state that possesses a roughly circular, oval, or rectangular territory.

complementarity Production of goods or services by two or more places in a mutually beneficial fashion.

complementary relations Two or more regions exchanging goods and services that one lacks and the other has produced or developed.

complementary resources Resources whose usefulness is enhanced by the presence of the other.

completed family size The actual number of living children one woman bears in her lifetime.

concentric zone model A social model that depicts a city as five areas bounded by concentric rings.

conflict A controversy, disagreement, or battle between two or more parties.

Confucian meritocracy The traditional bureaucracy of China based on the Confucian ideal that the most competent should be the leaders.

Congo Kordofanian A name given to languages of sub-Saharahan Africa.

conspicuous consumption High levels of consumption of resources and goods associated with individuals or countries.

contract system Recent Chinese system of allowing individuals or members of a production brigade to keep the production or output required over the contract quota.

conurbation Extensive urban area formed by expansion of cities to form one continuous urbanized area.

cooperatives In cooperatives farmers pool their resources to purchase equipment and market their products, but individuals farm their own land.

core area   An area of dense population that forms the urban-industrial heart of a nation; also the cultural, economic, and political center of a nation or group of people.

Corn Belt   A region in the central United States where corn is the dominant crop.

cottage industries   Small-scale production using high labor inputs.

creolization   Combining elements of two languages to form a third distinctive language, such as French Creole.

crop rotation   Changing the crops planted in a regular cycle to maintain or increase the soil fertility and yield.

cultural boundaries   Boundaries that are created as a result of cultural factors or variables.

cultural changes   Changes that occur in the culture of an area.

cultural convergence   The tendency of world cultures to become more alike as the culture and/or technology of the industrialized world is dispersed, particularly to urban areas of less industrialized countries.

cultural determinism   The theory that a person's range of action—for example, food preferences, desirable occupation, rules of behavior—is limited largely by the society within which he or she lives.

cultural ecology   The study of the complex, intricate relationships between the physical environment and people as culture-bearing animals.

cultural fragmentation   The presence of a host of cultural traits in one region or country as opposed to dominance of one trait over a large area.

cultural landscape   The human-made landscape; the visible human imprint on the land.

cultural pluralism   Variety of cultures living in a country, but not necessarily in distinct territories.

cultural realm   A group of countries sharing related cultural traits, such as the Islamic realm.

cultural resistance   Attitides, values, habits, laws, and so forth of a specific culture that prohibit certain items or behaviors.

cultural revolution   A period in China when universities were closed and many youth tried to restore the Communist revolution to a condition of altruism they believed had existed earlier.

culture   A total way of life held in common by a group of people, including such learned features as speech, ideology, behavior livelihood, technology, and government.

culture hearth   Heartland, source area, place of origin of a major culture.

Cyrillic alphabet   The old Slavic alphabet used today in Russia, Bulgaria, and Serbia.

death rate   The ratio between deaths and the number of people in a given population for a given time period.

de facto   A Latin word meaning from the fact; in reality, fact, or actuality.

de jure   A Latin word meaning according to law or by right.

demarcation   In political geography, the actual marking of a political boundary on the landscape by means of barriers, fences, walls, or other markers.

Democratic Revolution   Establishment of government systems characterized by free and open elections, freedom from taxation without representation, freedom from arbitrary acts of those in power, freedom of knowledge, freedom of public assembly and speech, and freedom to vote your conscience through use of a secret ballot.

demographic transition model   Model of population change that suggests countries move from a slow population growth stage with both high birth and high death rates, to a stage of rapid population growth when death rates drop, and then to a stage of slow or negative population growth as birth rates also fall.

dependency ratio   The proportion of a population that is either too young or too old to be economically self-sustaining or produce a surplus.

desertification   The expansion of the desert to moister areas along its margins as plant cover and soils are destroyed through overgrazing and erosion or climatic change.

devolution   In political geography, the disintegration of a nation-state, especially as the result of emerging or reviving regionalism.

dharma   The Hindu concept of duty within society.

dialect   A regional variation of a more widely spoken language.

dialect regions   Areas where a dialect is spoken.

diffusion   The spread of an idea or material object over space.

direct consumption   Direct consumption of cereal grains and the like rather than meat or alcohol products.

direction   The line or course along which a person or thing moves.

domain   A territory of rule or control.

domestication   The transformation of a wild animal or wild plant into a domesticated animal or a cultivated crop to gain control over food production. A necessary evolutionary step in the invention of agriculture.

double cropping   Planting two crops in succession in a field in the same year.

Dravidians   Members of an ancient Australoid race in southern India; also a family of languages spoken in that region.

dry farming   A method of cultivation in which crops are alternated with a fallow period in which water accumulates in the soil.

dual economy   An economy with both a small, intensive modern sector and a large traditional sector. The modern system is tied to the broader world economy while the traditional economy supplies local needs.

duration   The period of time during which something exists or persists.

dynasty   A ruler or ruling party that persists for an extended period of time.

Eastern Orthodox   A division of Catholicism with its headquarters in Constantinople.

economic development   The nature of a country's economy in terms of level of industrialization and modernization.

economic distance   The distance between places measured as the cost of moving goods or people.

economic takeoff   The point at which the economy of a country begins to develop or improve rapidly.

economies of scale   The savings that accrue from large-scale production whereby the unit cost of manufacturing decreases as the level of operation enlarges.

EC   European Community, formerly the EEC.

ecumene   The inhabited portion of a region.

EEC   European Economic Community.

effective national territory   That part of a country that is controlled by the central government, contributes to the economic base of the country, and feels itself part of the country.

ejidos   Cooperative farms in Mexico.

elongated state   A state that is long and narrow such as Vietnam.

emigration   Migration from a location.

encomendero   The Spaniard to whom an *encomienda* was granted.

encomienda   A large grant of land including its Indian occupants. It was designed to "civilize" the Indians and protect them from European exploitation by placing them under the protection of a Spaniard who was to teach, feed, and clothe them in return for their labor.

environmental degradation   Qualitative loss in the environment caused by human actions.

environmental determinism   The theory that the physical environment controls certain aspects of human action.

environmental modification   Any changes to the environment made by humans.

environmental perception   The mental image of the physical environment held by an individual or a group.

environmental possibilism   An approach that says the physical environment may set limits on human actions, but people have the ability to adjust to the physical environment and choose a course of action from many alternatives.

epicanthic   The epicanthic eyefold is an extra fold of skin in the inner corner of the eye, resulting in the distinctive eye shape of the Mongoloid people.

erosion   The wearing down or away of the earth's surface by wind or water.

ethnic group   People sharing a common and distinctive culture.

ethnic religions   Local religions of groups that share beliefs and practices; membership comes through birth into the group.

ethnocentricity   The idea that one's own ethnic group, race, or other group is in some way inherently superior to others.

ethnocentrism   Belief in the superiority of one's own ethnic group.

exclave   A piece of a state separated from the main body of the state by the intervening territory of another state.

exclusivist religions   Religions that have selective membership.

exotic streams   A stream that flows across a region of dry climate and derives its discharge from adjacent upland where a water surplus exists.

exponential growth   The increase associated with increasing any number by some exponent or power. Malthus maintained that populations tended to double, thus the exponent is 2.

expropriation   The process by which a country seizes the property of private citizens, businesses, or corporations. Ownership is then maintained by the government either directly or indirectly.

extended families   Family members beyond the nucleated family of husband, wife, and children.

Far East   An area commonly including the Koreas, Japan, China, and the islands belonging to them. Sometimes used to refer to all of Asia east of Afghanistan.

federal government   A government in which groups or specific areas elect representatives to govern the country.

Fertile Crescent   Crescent-shaped zone of productive lands extending from near the southeastern Mediterranean coast through Lebanon and Syria to the alluvial lowlands of Mesopotamia. One of the world's great source areas for agricultural innovations.

fertility rate   The number of children born to women in the reproductive age group (15–44) as a percentage of women in that group.

filariasis   Infestation of parasites of the blood or tissues. Usually transmitted by mosquitoes.

fjord   Narrow, steep-sided, elongated, and inundated coastal valley deepened by glacier ice that has since melted away, leaving the sea to penetrate.

fragmentation   The continued division of land into smaller and smaller pieces with each generation.

fragmented landholding   Where the farmer's property is divided into two or more nonadjacent pieces of land.

fragmented state   A state whose territory consists of several separated parts, not a contiguous whole.

freehold   Private ownership (of land) with full rights to rent, encumber, sell or bequeath the property.

frontiers   An area not yet fully integrated into a national state.

genetic drift   A chance modification of gene composition that occurs in an isolated population and is accentuated through inbreeding.

genetic isolation   Inbreeding among a population group that occurs when there is no interbreeding with nonmembers of the group.

genotype   A genetic constitution of an organism.

gentrification   Replacement of lower income groups by higher-income people as buildings are restored.

geopolitics   The use of geographical relationships to justify a policy of national expansionism.

gerrymander   To divide an area into voting districts in such a way as to give one political party an unfair advantage in elections, to fragment voting blocks, or to achieve other nondemocratic objectives.

ghetto   A distinct section of a city characterized by a particular ethnic composition.

gravity model   A model that states that the potential use of a good or service at a particular location is directly related to the number of people at a location and inversely related to the distance people must travel to reach the good or service.

greenhouse effect   Heating of the earth's surface as short-wave solar energy passes through an atmosphere transparent to it but opaque to reradiated long-wave terrestrial energy. Also refers to the increased opacity of the atmosphere caused by the addition of increased amounts of carbon dioxide from the burning of fossil fuels.

Green Revolution   The development of higher yield, fast-growing varieties of rice and wheat that has led to increased production per unit area and a temporary narrowing of the gap between population growth and food needs.

Gross National Product (GNP)   The total value of all goods and services produced annually in a country.

growth rate   The net rate of population increase calculated by subtracting deaths and emigrants from births and immigrants.

guru   A spiritual leader of Hinduism.

hacienda   Large estates in Latin America, especially Mexico.

hajj   The pilgrimage by Muslins to Mecca.

harijans   The untouchables in India who are at the bottom of the caste system.

head link   A city that serves as a link to the rest of the world for a country; imports and exports flow through it.

hearth   The region from which innovative ideas originate.

hejira   The flight of Muhammad, prophet of Islam, from Mecca to Medina in A.D. 622; the date on which the lunar-based calendar of Islam begins.

hierarchical diffusion   A type of expansion diffusion; innovations spread from one important person to another or from one urban center to another, temporarily bypassing persons of lesser importance and rural areas.

Hinduism   A formalized set of religious beliefs with social and political ramifications; the dominant religion of Indian society.

hinterland   The area a city serves and draws people from to purchase goods and services.

holistic   The idea that the entirety of actions affects a given place.

hollow frontier   A frontier that moves on without leaving permanent towns, farms, and industry, as in Brazil where the exploitive economy traditionally collapsed creating hollow frontiers rather than permanent colonization.

homelands   Ten areas within the country of South Africa set aside by the white government for settlement by the black population; an outgrowth of apartheid policy. In theory such areas are to function someday as economically independent units.

horizontal nomadism   The periodic movement of herds over long distances in search of pasture.

human ecology   The study of the relationship between biological humans and their physical environment, with particular emphasis on people as agents of environmental change.

hunting and gathering   Peoples of the world who live in a subsistence economy and live off the land by hunting and gathering.

hydraulic civilizations   Social organizations that developed in the Middle East and North Africa to ensure adequate water for the villages through a highly structured distribution system.

hydraulic model   A model of development of civilization that concludes that the need to regulate irrigation water led to laws and governments.

ideograms   A character or symbol representing an idea without expressing a particular word or phrase for it, as in Chinese.

ideological differences   Belief systems and variations between people.

ideology   An individual's or group's system of beliefs.

immigration   The legal or illegal movement of people into a country of which they are not native residents.

industrial cities   Cities whose economy is based on the industry within that city.

Industrial Revolution   A development characterized by the substitution of inanimate (machine) power for animate power sources, increasing productivity and demand for resources.

industrialization   The development and increase in manufacturing and industrial activity.

infant mortality rate   The ratio between deaths in the first year of life and the number of people in a given population. Often expressed as the number of deaths per 1,000 per year.

innovation model   A model arguing that the first cities were closely tied to a breakthrough innovation in technology, food production, or social organization.

institutional framework   The customary or legal processes that set the parameters within which the economy of a country functions.

insular   Referring to an island or locations having islandlike characteristics.

insurgent state   Political and territorial control over part of a country by insurgent (guerrilla) movements that oppose the existing government.

interconnections   The connections or links between countries of the world in communication, transportation, business, cultural values, and political arrangements.

intercropping   The raising of two or more crops in the same field at the same time.

international rivers   Rivers that flow through more than one country.

intervening opportunities   The presence of a nearer opportunity that diminishes the attractiveness of sites farther away.

Irish curve   Rapid population growth followed by a precipitous drop in numbers and stabilization at a lower level.

irredentism   Claims on lands culturally or historically related to a nation that are now occupied by a foreign government.

irrigation   The distribution of water by building channels from source areas of water or storage areas such as lakes.

Islam   Religion founded by the prophet Muhammad in Saudi Arabia around A.D. 624. Islam is the name of the religion and means submission to the will of one God (Allah). Muslim (Moslem) refers to a member of Islam, one who submits himself or herself to the will of Allah.

isolation   The condition of being geographically cut off or far removed from mainstreams of thought and action.

Jainism   A religion in India (an offshoot of Hinduism) that emphasizes the sanctity of life of any creature.

junta   A group of military officers controlling the government; generally they obtain control of the government by force and often exercise power dictatorially.

karma   The Hindu concept of an accounting of good and evil.

kibbutzim   A collective farm or settlement in Israel.

kolkhoz   A collective farm in the Soviet Union in which farming activities are performed by members who share in the profits after meeting quotas required by the central government.

Koran   Writings of the prophet Muhammad that Muslims accept as divine revelations from Allah.

labor specialization   The emergence of specific occupational categories that replaced the general labor requirements of the early Neolithic period.

landforms   The configuration of the land surface into distinctive forms such as hills, valleys, plateaus.

landscape regions   The uniqueness of a region that is expressed in the type or character of the area or region.

land tenure   The form of land ownership and control; includes absentee landlordism, sharecropping, and tenancy.

land use   The way the land is used.

language   Any method of communicating ideas, feelings, or intent through signs, symbols, or gestures.

language family   A collection of individual languages related to each other by virtue of having a common ancestor before recorded history.

latifundia   Large estates in Latin America owned by individuals of European descent, who typically control most of the wealth in the individual countries.

Latin vulgate   The vernacular Latin used by the masses in the Roman empire; it is the source for the Romance languages.

lifeboat earth   The concept that there is only a finite quantity of resources; thus industrialized countries should concentrate only on their own well-being and not assist the less industrialized nations.

life expectancy   Statistically determined number of years that an individual is expected to live.

lingua franca Use of a second language that can be spoken and understood by many peoples in regions with a diversity of languages.

linguistic region An area in which a specific language or dialect dominates.

long lot A unit-block farm consisting of a long, narrow ribbon of land, stretching back from a road, river, or canal.

magnitude The degree of intensity of a given item or subject.

Mahayana One of the major schools of Buddhism; active in Japan, Korea, Nepal, Tibet, Mongolia, and China.

Malayo-Polynesian A family of languages spoken in Indonesia, Melanesia, Micronesia, and Polynesia.

Malthusian equation The early nineteenth-century hypothesis of Thomas Malthus who argued that population growth would always outrun the earth's capacity to produce sufficient food.

Mandarin Dominant Chinese language, spoken mostly in northern China.

map scale The ratio of the actual distance to the distance indicated on a map.

maritime A climate characterized by moderate temperature, medium to high rainfall, and high humidity generally found along coasts.

Marxism The economic doctrine that wealth and control of factors of production should be held by the workers.

Mecca The birthplace of Muhammad in Saudi Arabia; the most holy city of Islam.

megalopolis A group of large cities that have effectively become one large urban area through their growth.

Melanesia A group of islands in the southwestern Pacific Ocean, extending southeastward from the Admiralty Islands to the Fiji Islands.

mental map A map, often including positive or negative images of different areas, as perceived in the mind of an individual person.

meritocracy The idea developed from Confucianism in China based on the concept that only the most qualified should have positions of authority.

Mestizo A person in Latin America of mixed European and Indian blood.

metes and bounds (surveying) A survey system producing irregularly shaped parcels and often oriented to physical environmental features.

metropolis A large urban center.

Micronesia An area of coral and volcanic islands in the Pacific Ocean.

migration A change in residence intended to be permanent.

minifundia Tiny farms in Latin America that provide a subsistence existence upon which a significant portion of the Indian population depends.

mining landscapes Landscapes characterized by the buildings, pits, wasteheaps, and refineries of mining.

mir Lands in Russia that, prior to 1905, were owned and cultivated communally by peasant villagers.

mixed farming A combination of crop cultivation and animal husbandry.

mobility The ability to move from one location to another.

modernization The process of making something modern in appearance, style, or character.

monoculture Large fields of the same crops such as the great wheat fields or corn fields of the Midwestern United States.

monotheistic religion A religion whose adherents believe there is only one God.

mortality rate The number of deaths in a given period of time or place, normally given as number per 1,000 population.

moshav A cooperative farm or settlement in Israel.

multiple nuclei model A model that depicts a city growing from several separate focal points.

mutations An alteration or change.

nagana A destructive disease of the African tropics spread by the tsetse fly.

nation A group of people with a distinct culture. The nation may or may not coincide with political boundaries.

nation-state A country whose population possesses a substantial degree of cultural homogeneity and unity and where the political unit coincides with the area settled by a nation.

nationalism The feeling of pride and/or ethnocentrism focused on an individual's home territory.

nationalization The process by which ownership of private property is assumed by the government.

natural boundaries Boundaries that are due to either physical or cultural characteristics.

natural hazards A process or event in the physical environment that has consequences harmful to humans.

neocolonialism Control of former colonies by colonial powers, especially by economic means.

Neolithic Revolution The change from dependence on gathering wild plants and hunting wild animals to plant and animal domestication and sedentary farming. Also called the First Agricultural Revolution.

Nirvana The concept of unity with the universal spirit in Buddhism.

nomad A person who moves from place to place in a search for food and water for herd animals.

nomadism   Cyclic movement among a definite set of places. Nomadic peoples are mostly pastoralists.

North Atlantic Drift   The relatively warm currents of the Atlantic resulting from the Gulf Stream.

nuclear families   Families that include only the immediate family of father, mother, and children.

occidental   Western; originates from the Latin for "fall," or the setting of the sun in the west.

official language   The legally recognized language of a state.

oil shale   A kind of rock containing oil in a dispersed form.

OPEC   Organization of Petroleum Exporting Countries, a cartel of oil-producing and exporting nations formed in 1960.

Orient   The countries of the Asian continent excluding the Soviet Union.

Oriental   Eastern; originates from the Latin word for "rise," or the direction in which the sun rises.

overpopulation   When population numbers exceed the carrying capacity of an area.

ozone depletion   The reduction of the ozone layer that shields and protects the earth's surface from the effects of ultraviolet radiation found in the sun's rays.

paddy   Field of wet rice; also refers to unhusked rice.

pass laws   South African legislation requiring blacks to carry a work pass with them at all times.

peasants   Rural farmers in a stratified society. Peasants may or may not own land; the term refers to a way of life.

peasant society   A society in which the roles are highly specific and usually established at birth or by living in a specific village.

peasant systems   Individuals are born or participate in specific roles such as carpentry, farming, or production of a specific item.

people's communes   Massive farms with great labor potential, formed by grouping cooperatives together after the Chinese revolution.

periodic markets   Markets that operate at some regular interval (every Sunday, for example) rather than daily.

Persian   A language group in the Middle East, spoken by Iranians.

phenotypes   Environmental and genetical appearance of a species.

physical boundaries   Boundaries that result from physical characteristics such as lakes, rivers, or mountains.

physical quality of life   Degree to which basic human needs of food, shelter, education, and medical care are available in a country.

physical remoteness   Locations that are separated from other locations by great distances or rugged landforms.

piedmont   Hilly, rolling land lying at the foot of a mountain range and forming a transition between mountain and plain.

pilgrimage   A journey to a place of great religious significance by an individual or by a group of people (such as a pilgrimage to Mecca for Muslims).

plane of the ecliptic   An imaginary plane in which the earth's orbit lies.

plantation   A large estate normally owned by a corporation concentrating on large-scale production of a cash crop for export from the tropics.

plantation economy   An economy of a region or country that is based upon plantation agriculture. Normally indicates a separation in occupation between local residents who provide labor and foreign administrators.

plebiscite   An election by the voters of a country or region to determine what course of action they will follow.

PLO   Palestine Liberation Organization, founded in 1962 by displaced Arabs from Israel.

plural society   Society in which various ethnic groups live intermingled throughout a country, although often occupying distinct residential sections of cities. Each ethnic group recognizes its own distinctive cultural traits, icons, values, and life-styles.

point settlements   Settlements located in harsh environmental settings to capitalize on unique site features such as resources. They are generally extractive in nature with little or no hinterland.

polyculture   A farming system in which numerous types of crops are grown together; also known as intercropping.

polyethnicism   Several ethnic groups within a country's borders, with each group occupying a distinct territory.

Polynesia   A scattered group of islands of the eastern and southeastern Pacific Ocean, extending from New Zealand north to Hawaii and east to Easter Island.

polytheistic   Refers to the practice of worshiping more than one god.

population density   The number of people in a given area, usually indicated per square mile or kilometer.

population doubling time   The number of years it takes a given population such as that of a country to double.

population explosion   The rapid growth of the world's human population during the past century, attended by ever shorter doubling times and accelerating rates of increase.

population pyramids A bar graph representing the distribution of population by age and sex.

ports A place where goods are brought into and out of a country.

positive checks Controls on population hypothesized by Malthus, including late marriage or other conscious decisions that limit birth rates.

postindustrial society The characteristics of a nation that has moved from an industrial society to one in which most of its employment is in the service sector.

preventive checks Also known as negative checks. Malthus predicted that war, famine, and disease would ultimately limit the size of a population.

primary activities Work or occupations involved with producing raw materials from farming, forestry, fishing, or mining.

primate city A city that dominates the urban network of a country; by definition it is at least twice as large as the next largest city.

primogeniture The first-born or eldest child of the same parents.

pristine state The natural environment unmodified by people.

producers' cooperatives Chinese village cooperatives of thirty to fifty households in which the villagers combine their resources and their land in an effort to increase productivity.

profane space The portion of the world that is not viewed as sacred.

prorupt state A state with a narrow, elongated land extension leading away from the main body of territory.

push-pull migration A theory used to explain the movement of people from rural areas to urban centers. The migrants are forced out of one area by limited opportunity and attracted to cities by perceived advantages.

quality of life index A measure based on the standard of living indicated by key indices such as literacy, infant mortality, and access to medical and other social services.

race Any group of people more or less distinct by genetically transmitted physical characteristics.

racism The idea that one's own ethnic stock is superior.

rain forest destruction The loss of tropical rain forest through burning, road and dam building, settlement, and air pollution.

raubwirtschaft Literally robber economy; used to refer to mining or other environmentally destructive economic activity.

refugees People who are forced to migrate from a country for political reasons.

region An area on the earth's surface marked by certain properties; a commonly used term and a geographic concept of central importance.

regionalism A strong identity of feeling for a region.

reincarnation The belief that the souls of the dead return to life in new bodies or new forms.

relative location Location of a place or region with respect to other places or regions. Used interchangeably with situation.

relic languages Languages that once occupied a large area but have been overwhelmed by another; for example, the Celtic languages of western Europe.

relocation diffusion The spread of an innovation or other element of culture that occurs with the bodily relocation (migration) of an individual or group that has the idea.

repatriation The process of returning people to the country of their birth or where they hold citizenship.

replicative cities Cities that are similar or modeled after those in another culture.

replicative societies Societies that result from conscious or unconscious efforts to model themselves on another society.

resources The physical elements of the world that are perceived as useful by the population of a region or world.

return migration The permanent return of people to their home community after migrating elsewhere for a time.

revolutionary changes Changes that occur as a result of revolution; dramatic changes.

ritual purity The purifying of the soul by ritual procedures or behavior, especially in Hinduism.

riverine population concentrations The growth of populations along the river basins of the world.

Romance Languages that are based on Latin.

Romansch A romance language found in Europe.

rural-urban migration The movement of people from rural areas to urban areas.

sacred place A place of worship, or any space endowed with characteristics that evoke a spiritual-mystical response from believers.

Sahel Semiarid zone across most of Africa between the southern margins of the arid Sahara and the moister savanna and the forest zone to the south.

salinization The accumulation of salts in the upper levels of soil as salts carried by irrigation water are left at or near the surface as water evaporates.

savanna The wet-dry tropical regions of the world. Also, a grassland with scattered trees and bushes located on the edge of equatorial rain forests.

sawah   Indonesian term for wet rice cultivation.

Scientific Revolution   The tremendous increase in scientific development and discoveries in health and technology of the nineteenth and twentieth centuries.

secondary activities   Occupations or types of work that process raw materials and transform them into products. Also referred to as manufacturing.

secondary domesticates   Plant or animal domesticates that occurred when the original domesticate proved unsuitable for another area.

sectionalism   Excessive devotion to local interests and customs.

sector model   An economic model that depicts a city as a series of pie-shaped wedges.

sedentary   Permanently located in a particular area.

segregation   A measure of the degree to which members of a minority group are not uniformly distributed among the total population.

seigneur   A feudal lord in French Canada who held land by feudal tenure.

Semitic   A cultural group in the Middle East and North Africa. The two major Semitic groups are Arabs and Jews.

sharecropping   Relationship between a large landowner and farmers on the land in which the farmers pay rent for the land they farm by giving the landlord a share of the annual harvest.

shatter belt   Region located between stronger countries (or cultural-political forces) that is recurrently invaded and/or fragmented by aggressive neighbors.

shifting agriculture   Cultivation of crops in forest clearings that are abandoned in favor of newly cleared nearby forest land after a few years.

Shi'ite   An Islamic minority created after the death of Muhammad who believe that the leader of Islam should be a direct descendant of Muhammad. Dominant in Iran and Syria.

sickle-cell anemia   A hereditary anemia of some Africans characterized by the presence of oxygen-deficient sickle cells.

Sikh   An individual who follows the Sikh religion. Sikhs are characterized by adherence to the common surname of Singh, and males wear long hair, a full beard, a turban, and a dagger.

Sikhism   A religious group that combines elements of Islam with Hindu beliefs.

Sino-Tibetan   A linguistic group that includes the Sinitic and Tibeto-Burman language families.

site   The internal locational attributes of a place, including its local spatial organization and physical setting.

situation   The external locational attributes of a place; its relative location with reference to other nonlocal places.

slash and burn agriculture   Clearing and burning the vegetation, planting crops for one or two years until new vegetation begins to overrun the fields, and then moving on to a virgin area to repeat the process. Also known as **shifting agriculture.**

Social Darwinism   The idea that those who are most fit will survive; the richest will surpass.

social distance   A measure of the perceived degree of social separation between individuals, ethnic groups, neighborhoods, or other groupings; a recognition of the difference between social classes.

social inertia   Social change is slow or is difficult to begin.

social revolution   A change that takes place in a society so dramatically that it creates a new social order, as in Japan after the Meiji restoration.

social selection   The conscious or unconscious selection of specific phenotypes by a group. Manifest in laws against racial intermarriage and the like.

socialism   A variety of political and economic theories and systems of social organization based on collective or governmental ownership and administration of the means of production and the distribution of goods.

sovkhoz   A farm in the Soviet Union owned by the state; farm members are paid a set wage regardless of the profitability of the farm.

spatial interaction   The relative amount of interaction between a place or region and other places or regions.

state   The formal name for the political units we commonly call countries.

stateless nation   A cultural group without formal possession of a territory in which they control the government; for example, the Palestinians.

stimulus diffusion   When a specific trait fails to diffuse, but the underlying idea does.

street people   People who live on the street.

structural changes   Changes that are made in the formal structures of a country such as land tenure, political organization, or economic relationships with other areas.

subsistence farmers   Farmers who produce their own food.

subsistence systems   Agricultural systems in which most of the produce is consumed directly by the farm family rather than being traded or marketed.

suburb   A functionally specialized segment of a larger urban complex; external to a central city.

suburbanization   The movement of urban population to the suburbs.

suffrage   The universal and free right to vote.

Sun Belt   The states of the southern and southwestern United States from Florida to California.

Sunnite The major religious group of Islam.

superimposed boundaries Boundaries placed over the existing political and cultural patterns, fragmenting cultural groups and joining peoples of diverse backgrounds.

taboos The prohibition of an object, word, or social custom.

tariff A tax placed on imports or exports into a country. Often imposed on imports to protect an economy from lower cost imported goods.

tariff barriers Taxes, quotas, or health standards placed on exports or imports to retard the flow of goods into a country.

terra incognita Literally unknown land in Latin.

territoriality Persistent attachment of most animals to a specific area; the behavior associated with the defense of the home territory.

territorial shape The form or shape of a country.

tertiary activities Occupations and work that provide services such as banking, tourism, transportation, retailing, finance, or education.

theocracy State whose government is under the control of a ruler who is deemed to be divinely guided or under the control of a group of religious leaders, as in Khomeini's Iran.

Theravada A branch of Buddhism.

threshold The population required to make provision of services economically feasible.

topography The configuration of a land surface including its elevations and the position of its natural and human-made features.

toponyms Any name derived from a place or region.

traditional landscapes Landscapes that are dominated by villages and small farms, typical of less industrialized countries.

transhumance Seasonal movement of people and their livestock in search of pastures. This movement may be vertical (into highlands during the summer and back to lowlands in winter) or horizontal (in pursuit of seasonal rainfall).

transportation landscapes Landscapes dominated by transportation features, such as the freeway system of Los Angeles.

treaty ports Selected ports where Western powers established commercial enclaves from which they traded with China.

triage The world population concept that suggests that the industrialized nations should help those who only need slight aid and let the extremely overpopulated nations fend for themselves.

tribal groups Groups of people organized on the basis of tribes.

tribal religion A religion associated with a specific tribe or group and a specific place.

tribalism The practice of tribal religion.

tripod of culture The use of fire, symbolic speech, and tool making.

trust territories Former colonial holdings assigned by the United Nations to one of the industrialized nations for development assistance.

tsetse fly Insect pest in Africa that infects people with sleeping sickness and animals with nagana.

tundra A zone between the northern limit of trees and the polar region in North America, Europe, and Asia.

underemployed Individuals who have only part-time or seasonal work (as in agriculture) or who are engaged in labor that is only marginally productive.

universalizing religion Religion that has broken its ties to a specific space and maintains that its beliefs are appropriate for all people.

untouchables The lowest level of the caste system in India.

urban flight The movement of the middle and upper classes from the central city to the suburbs.

urbanization The proportion of a country's population living in urban places is its level of urbanization. The process of urbanization involves the movement to and the clustering of people in towns and cities, a major force in every geographic realm today. Urbanization occurs when a sprawling city absorbs rural countryside and transforms it into suburbs.

urban morphology The pattern and relationship of land use, economic activity, and population distribution in urban areas.

urban village Suburban nuclei focused on a transport function that serves as a central place for a suburban area.

vernacular Idioms of a certain region or area.

vertical nomadism The periodic movement of herds to higher or lower areas in search of better pasture.

Von Thünen model Explains the location of agricultural activities in a commercial, profit-making economy. A process of spatial competition allocates various farming activities into concentric rings around a central marketplace, with profit-earning capability the force that determines how far a crop locates from the market.

westerlies Prevailing surface winds in the mid latitudes that generally blow from the west to the east.

wheat belts Areas of the world where the major crop is wheat, such as the Midwestern United States.

zero population growth (ZPG) A stable population that basically maintains itself without increasing or decreasing.

ziggurat In ancient Mesopotamia, a temple shaped as a tall pyramidal tower.

Zionism The desire for a Jewish homeland and its expansion.

Index